Pathogenic *Escherichia coli*

Evolution, Omics, Detection and Control

https://doi.org/10.21775/9781910190777

Edited by

Pina M. Fratamico, Yanhong Liu and Christopher H. Sommers

Eastern Regional Research Center
Agricultural Research Service
United States Department of Agriculture
Wyndmoor, PA
USA

Caister Academic Press

Caister Academic Press
Norfolk, UK

www.caister.com

British Library Cataloguing-in-Publication Data
A catalogue record for this book is available from the British Library

ISBN: 978-1-910190-77-7 (paperback)
ISBN: 978-1-910190-78-4 (ebook)

Description or mention of instrumentation, software, or other products in this book does not imply endorsement by the author or publisher. The author and publisher do not assume responsibility for the validity of any products or procedures mentioned or described in this book or for the consequences of their use.

Cover design adapted from images provided by Pina M. Fratamico.

Ebooks

Contents

Preface

Escherichia coli (*E. coli*) was discovered by the German bacteriologist, Theodor Escherich, in 1885 and was initially called *Bacterium coli commune*. *E. coli* is part of the intestinal microflora of humans and other mammals and is a beneficial organism in the gut, producing vitamins K and B12 and preventing pathogenic bacteria from colonizing in the intestine. *E. coli* is one of the most thoroughly studied prokaryotic species and has served as a model microorganism in molecular genetics investigations, yielding considerable information on bacterial genome evolution. *E. coli* has also evolved pathogenic mechanisms through acquisition of virulence genes, and the diseases caused by different *E. coli* pathotypes depend on possession of specific virulence factors, including adhesins, invasins, and toxins. The goal of this book is to describe the *E. coli* pathotypes and how whole-genome and other DNA sequencing approaches, as well as genetic analyses are allowing fascinating new insights into the evolution of the bacterium, paving the way for the development of novel detection and therapeutic strategies.

The virulence characteristics of the different *E. coli* pathotypes are described in Chapter 1, and Chapter 4 focuses on the virulence mechanisms and types of diseases caused by extraintestinal pathogenic *E. coli*. The importance of horizontal transfer of virulence genes among different pathotypes, leading to the emergence of hybrid hypervirulent *E. coli* strains that may have resistance to multiple antibiotics is highlighted. Analysis of genome sequence data of diverse collections of *E. coli* has indeed revealed the great genetic diversity of *E. coli*, as well as evolution and emergence of *E. coli* lineages due in large part to the acquisition and deletion of DNA elements (Chapter 2 and Chapter 7). Whole-genome sequencing of *E. coli* and other pathogens is transforming public-health surveillance and outbreak investigations, providing an unparalleled level of strain discrimination and the ability to more accurately identify and characterize the organism (Chapter 8). Various sequencing technologies and available bioinformatics tools for assembly, annotation, and analysis of whole-genome sequence data are presented in Chapter 9. Determination of the presence of specific phenotypic traits, such as toxin production, has been employed for detection of *E. coli* pathotypes; however, this can be laborious, time consuming, and costly. Thus, genetic-based methods such as the polymerase chain reaction targeting pathotype-specific markers, which are often virulence-related genes, are increasingly replacing phenotypic-based testing (Chapter 3). Major advances have been made in recent years in the development of rapid methods that allow faster, more sensitive, specific, and expedient detection and identification of pathogens compared to traditional culture-based methods. In Chapter 6, a comprehensive overview of rapid technologies that can potentially be used as portable real-time detection systems is provided. Major advances in nucleic-acid sequencing technologies have allowed the development of metagenomics approaches for analysis of microbial communities, including those in food products. Metagenomics technologies are being utilized as culture-independent methods for studying the impact of perturbations in microbial communites on the growth and survival of pathogens, including *E. coli*, and for detection and characterization of pathogens in food and other types of samples (Chapter 10). 'Omics' technologies, including genomics, proteomics, transcriptomics, and metabolomics are revolutionizing approaches for pathogen detection and control and for treatment of infections. In Chapter 11, the authors describe the use of omics approaches to understand the responses of *E. coli,* particularly Shiga-toxin-producing *E. coli,* to stress conditions and how the information can be used to develop control strategies, such as the identification of novel inhibitors, to decrease survival in food. Omics data can be integrated in systems-level investigations to understand more completely the physiological changes that occur in *E. coli* under stress conditions and to identify potential new drug targets (Chapter 12). Resistance of *E. coli* and other pathogens to multiple antibiotics is occurring at an alarming rate, and thus, alternative strategies to treat infections are needed. In Chapter 5, the authors describe the use of bacteriophages to treat *E. coli* infections as one alternative strategy. The editors anticipate that this volume will be a valuable resource for individuals interested in *E. coli* pathogenesis and control, as well as in the evolutionary dynamics of this micro-organism.

Pina M. Fratamico, Yanhong Liu and Christopher H. Sommers
Eastern Regional Research Center
Agricultural Research Service
United States Department of Agriculture
Wyndmoor, PA, USA

Escherichia coli Pathotypes

1

James L. Smith and Pina M. Fratamico*

Eastern Regional Research Center, Agricultural Research Service, United States Department of Agriculture, Wyndmoor, PA, USA.

*Correspondence: pina.fratamico@ars.usda.gov

https://doi.org/10.21775/9781910190777.01

Abstract

Escherichia coli strains are important commensals of the intestinal tract of humans and animals; however, pathogenic strains, including diarrhoea-inducing *E. coli* and extraintestinal pathogenic *E. coli*, exist. Intestinal *E. coli* pathotypes may cause a dehydrating watery diarrhoea, or more severe diseases such as haemorrhagic colitis and haemolytic uraemic syndrome. The different *E. coli* pathotypes can be transmitted to humans via contaminated food and water, and transmission can also occur through animal and person-to-person contact. The extraintestinal pathogenic *E. coli* reside in the intestinal tract but may escape to cause disease in bodily sites outside of the gut and can cause urinary tract infection, neonatal meningitis, sepsis, and other types of infections. Avian pathogenic *E. coli* are a cause of poultry diseases and are closely related to extraintestinal pathogenic *E. coli* strains that cause human infections. Poultry contaminated with extraintestinal pathogenic *E. coli* and avian pathogenic *E. coli* may be an important source of *E. coli* strains that cause illness in humans at non-intestinal sites; however, additional research is needed to confirm this. Furthermore, hybrid *E. coli* strains have emerged in recent years that harbour virulence genes from more than one pathotype, and these have caused serious outbreaks and infections.

Introduction

Escherichia coli is a Gram-negative, facultatively anaerobic, chemoorganotrophic, non-spore-forming, rod-shaped bacterium that is a member of the family Enterobacteriaceae. *E. coli* is a commensal organism making up part of the intestinal microflora of humans and other vertebrates, birds, and reptiles. *E. coli* strains have been traditionally classified based on serotyping using different antisera to determine the O-(somatic) polysaccharide and H-(flagellar) surface antigens. The O-group defines the *E. coli* serogroup and the O+H combination defines the serotype. There are currently over 180 O- and 53 H-antigens, and thus a large number of possible O+H combinations results in numerous *E. coli* serotypes. A large number of serotypes of *E. coli* are non-pathogenic, but some serotypes have acquired different virulence genes, and thus, they may cause various types of diseases, including diarrhoea and other intestinal diseases, as well as diseases at non-intestinal sites (Croxen *et al.*, 2013). Pathogenic strains of *E. coli* are often referred to as pathotype, referring to strains that causes a specific type of illness and are distinguished from other members of the species by their molecular markers and virulence factors (Scheutz *et al.*, 2014). Some of the characteristics of the pathotypes of *E. coli* are discussed in the following sections of this chapter.

Enteroinvasive *Escherichia coli* (EIEC)

EIEC are intracellular pathogens that are physiologically and taxonomically more related to *Shigella* spp. than to typical *E. coli* strains and cause an invasive inflammatory colitis with profuse watery diarrhoea. The site of infection of both EIEC and *Shigella* is the colonic mucosa. EIEC strains may carry the *sen* gene, which encodes for *Shigella* enterotoxin-2 (ShET-2), which induces inflammation of intestinal epithelial cells (Croxen *et al.*, 2013; Kaper *et al.*, 2004). Genes required for invasion (virulence effector genes, *sen* gene, and T3SS genes) are located on a large (210–230 kb) plasmid, pINV, in EIEC and *Shigella*. EIEC subclones possessing virulence genes and pathogenicity islands of uropathogenic *E. coli*, indicating the possible emergence of hypervirulent EIEC have been described (da Silva *et al.*, 2017) (see section on Hybrid pathogenic *E. coli*).

The infectious dose of *Shigella* is estimated to be 10–200 cells, whereas that for EIEC is estimated to be *ca.* 1000-fold higher (Croxen *et al.*, 2013; Parsot, 2005). The reason for the difference remains unknown. EIEC outbreaks are rarely detected in people in Nottingham, UK, during 2014 (Newitt *et al.*, 2016). Humans are the only known reservoir of EIEC; therefore, most food-borne infections are caused by consuming foods contaminated by individuals infected with the pathogens. Treatments recommended for EIEC infection include rehydration and antibiotics (Croxen *et al.*, 2013). Real-time PCR assays targeting the invasion-associated plasmid (pINV) genes can be used to identify and detect *Shigella* and EIEC. All *E. coli* strains harbour the lactose permease gene (*lacY*) which is absent in *Shigella*. However, both *E. coli* and *Shigella* carry the β-glucuronidase gene (*uidA*). Duplex real-time PCR assays targeting the *uidA* and *lacY* genes or *lacY* and the *ipaH* gene (invasion plasmid antigen H; found in *Shigella* and EIEC) could be used to differentiate *Shigella* spp. from EIEC (Pavlovic *et al.*, 2011; Løbersli *et al.*, 2016).

Diffusely adherent *E. coli* (DAEC)

Diffuse adherence is a pattern of cell attachment in which bacteria cover the entire surface of HeLa or HEp-2 cells whereas, with localized adherence, bacteria form microcolonies in a few sites on the cell culture surface (e.g. a scattered pattern) (Croxen *et al.*, 2013). The diffuse attachment to cell surfaces by DAEC is mediated by both fimbrial (Dr and F1845) and afimbrial (Afa) adhesins (Le Bouguénec and Servin, 2006). The DAEC are a poorly characterized heterogeneous population of *E. coli* that colonize the small bowel and cause watery diarrhoea in children ≤ 5 years of age (Patzi-Vargas *et al.*, 2013). The Afr-Dr adhesins may be the predominant virulence factor in DAEC; however, the mechanism of induction of diarrhoea in DAEC strains remains uncertain. Volunteer studies have not been performed, and thus, the status of DAEC as a cause of diarrhoea is controversial. Demonstration of diffuse adherence to HEp-2 cells is considered to be the gold standard for detecting DAEC. A DNA targeting *daaC*, a gene involved in the production of F1845 fimbriae reacted with DAEC strains; however, the probe also cross-reacted with enteroaggregative *E. coli* strains (Snelling *et al.*, 2009).

Enteropathogenic *E. coli* (EPEC)

EPEC infections are a major cause of diarrhoea, vomiting, and low-grade fever in infants and young children in developing countries. The diarrhoea may be unrelenting and potentially fatal (Ochoa and Contreras, 2011). EPEC possess the LEE (locus of enterocyte effacement) pathogenicity island that is important for virulence and the formation of attaching and effacing (A/E) lesions and intimate attachment of the bacteria to the gut mucosa. The LEE pathogenicity island encodes structural components of type III secretion system (TTSS; injects effector proteins into host cells), regulators, chaperones, and effectors secreted by the TTSS. A region known as Per (plasmid-encoded regulator), found on the EPEC adherence factor plasmid (pEAF), is involved in regulating LEE genes. EPEC also possesses non-LEE-encoded (Nle) effector genes that decrease host immune responses, inhibit phagocytosis, and alter host protein secretion, tight-junction integrity, and vesicle trafficking (Chen and Frankel, 2005; Croxen *et al.*, 2013; Dean and Kenny, 2009).

The full mechanism of how EPEC produce a diarrhoeic response is not clear. The initial attachment of EPEC to enterocytes involves pEAF carrying the *bfp* operon, which encodes the bundle-forming pilus. Bundle-forming pili allow EPEC to form adherent microcolonies, which bind to receptors on host cell surfaces. Strains that have both the *eae* (located on LEE and encodes intimin necessary for induction of the A/E lesions) and *bfpA* genes are termed typical EPEC; those with *eae* but lacking *bfpA* are termed atypical. Typical EPEC strains are a main cause of infantile diarrhoea in developing countries but rarely in the developed world; atypical EPEC strains induce a less severe disease (Croxen *et al.*, 2013; Jafari *et al.*, 2012). A severe watery stool with a large amount of mucus (blood is rare) is characteristic of EPEC-associated disease. Providing fluids is the treatment for the diarrhoea-induced dehydration (Croxen *et al.*, 2013). Humans are the reservoir for typical EPEC whereas both animals and humans are the reservoirs for atypical EPEC; however, the serotypes found in animals generally are not those that cause human disease (Croxen *et al.*, 2013). Foods contaminated by an infected food handler have been associated with EPEC diarrhoea. Most infections are sporadic, and the particular food vehicles are usually unidentified. However, any food or water exposed to the faeces of an EPEC-infected individual may induce diarrhoea (Croxen *et al.*, 2013). Phenotypic tests, including identification of the type of adherence to tissue culture cells and A/E histopathology can be performed to distinguish EPEC from other diarrhoeagenic *E. coli*; however, more preferred methods for detection and identification of EPEC involve PCR-based screens targeting the *eae* and *bfp* genes. Gene sequence data are needed to correctly distinguish between typical and atypical EPEC and to differentiate EPEC strains that cause acute diarrhoea from those that do not (Croxen *et al.*, 2013; Jafari *et al.*, 2012).

Enteroaggregative *E. coli* (EAEC)

EAEC include a heterogeneous population of diarrhoeic *E. coli* strains that show a stacked-brick pattern of bacterial cells (termed aggregative adherence) attached to HEp-2 cells and to each other. The stacked-brick pattern is encoded by genes present on the pAA plasmid. The pAA plasmid carries genes encoding the production of the aggregative adherence fimbriae (AAFs), which mediate adherence of EAEC to the intestinal mucosa (Croxen *et al.*, 2013; Cennimo *et al.*, 2007). The *aggR* gene on the plasmid encodes AggR, a transcription regulator of AAF biogenesis. EAEC strains containing the *aggR* gene are termed typical EAEC, whereas strains lacking *aggR* are termed atypical; however, both atypical and typical strains induce diarrhoea (Kaper *et al.*, 2004). A thick mucus biofilm on the intestinal mucosa and necrotic lesions in the ileal mucosa are outstanding features of EAEC diarrhoea. There are three stages in EAEC pathogenesis: initial adherence of bacteria to the intestinal mucosa and mucus layer; increased mucus production leading to a thick bacterial biofilm on the mucosal surface; and production of toxins that damage the mucosa and induce intestinal secretion (Kaur *et al.*, 2010). Thus, the major feature of EAEC virulence is colonization of the intestinal mucosa with mucoid biofilm formation, elaboration of enterotoxins, and mucosal inflammation. The virulence plasmid, pAA, encodes the AAF fimbriae necessary for adhesion, which is followed by secretion of heat-labile Pet and heat-stable EAST-1 toxins. The *pet* gene located on pAA encodes the plasmid-encoded toxin (Pet) that disrupts the actin cytoskeleton and induces epithelial cell rounding. The plasmid-mediated EAST-1 enterotoxin behaves *in vitro* like the heat-stable toxin of enterotoxigenic *E. coli*. In addition, some EAEC strains produce the *Shigella* enterotoxin ShET1 (Croxen *et al.*, 2013; Kaper *et al.*, 2004). The role that these enterotoxins may play in the induction of diarrhoea by EAEC is not clear. Individuals susceptible to EAEC diarrhoea include children and immunocompromised individuals living in both industrialized and developing countries, and in addition, EAEC strains are a major cause of traveller's diarrhoea. Since the diarrhoea leads to dehydration, oral hydration is the preferred treatment (Croxen *et al.*, 2013; Kaur *et al.*, 2010). Ingestion of foods contaminated by infected food handlers are probably the source of EAEC diarrhoea; however, the reservoir of EAEC is unknown (Croxen *et al.*, 2013). EAEC strains can be identified by demonstrating aggregative adherence in a stacked-brick pattern on HEp-2 culture cells. EAEC can be detected by multiplex PCR targeting the plasmid (pAA)-associated gene *aatA* (encodes AatA, an *E. coli* efflux pump TolC homologue) or the chromosomally located *aaiC* gene (encodes secreted protein AaiC; AaiC secretion is regulated by AggR). However, amplification of the *aatA* or *aaiC* genes does not always differentiate diarrhoeagenic EAEC from non-diarrhoeagenic strains (Coxen *et al.*, 2013; Jafari *et al.*, 2012). Therefore, detection of EAEC will remain challenging because of the heterogeneity of strains within this pathotype.

Enterotoxigenic *E. coli* (ETEC)

ETEC induces a mild self-limiting watery diarrhoea or a more severe purging cholera-like watery diarrhoea. Children in developing countries or in areas with poor hygiene and unclean water are particularly susceptible to ETEC infections, which are associated with notable morbidity and mortality. ETEC is a major cause of traveller's diarrhoea for visitors to developing countries. Toxins produced by ETEC induce a non-inflammatory watery diarrhoea lacking blood, mucus and leucocytes. If diarrhoea is prolonged, it is necessary to administer rehydration therapy (Croxen *et al.*, 2013; Okoh and Osode, 2008). Humans are the reservoir of ETEC, and ingestion of food or water contaminated by an infected person is often the cause of illness. The infective dose is in the range of 10^6 to 10^8 organisms. Anti-motility agents or antibiotics are not recommended for children infected by ETEC. Infected children living in Bangladesh were found to be more malnourished and have stunted growth compared to those with non-ETEC diarrhoea (Croxen *et al.*, 2013; Okoh and Osode, 2008). In addition to ETEC infection of human children, newborn and young animals including calves, lambs, and piglets (but not adult animals) are susceptible to ETEC diarrhoea leading to economic losses to the livestock industry (Croxen *et al.*, 2013). ETEC colonizes the small bowel epithelial surface utilizing various adhesins and colonization factors leading to diarrhoea induced by enterotoxins (Croxen *et al.* 2013; Fleckenstein *et al.*, 2010). ETEC elaborate plasmid-encoded diarrhoea-inducing enterotoxins, including heat-labile enterotoxins (LT-1, LT-2) and heat-stable enterotoxins (STa, STb). LT-1 (designated as LT) and STa (designated as ST) are associated with human disease. LT is an AB_5 toxin closely related to cholera toxin. The B subunits bind to GM1 gangliosides and deliver the A subunit into the host cell where the A1 portion interacts with an intracellular guanine nucleotide protein (Gsα). Inhibition of Gsα GTPase activity induces increased levels of intracellular cyclic AMP. The cystic fibrosis transmembrane regulator (CFTR) chloride channel is activated by the cyclic AMP, which results in electrolyte and fluid loss into the intestinal lumen (Croxen *et al.*, 2013; Fleckenstein *et al.*, 2010; Okoh and Osode, 2008). ST is a small, single peptide toxin that binds to the host cell guanylate cyclase C, resulting in increased intracellular cyclic GMP levels. Cyclic GMP-dependent protein kinase II phosphorylation of the CFTR induces increased chloride secretion and inhibition of NaCl absorption with resultant loss of water due to osmotic diarrhoea. ETEC strains may produce LT or ST or both (Croxen *et al.*, 2013; Fleckenstein *et al.*, 2010; Okoh and Osode, 2008). The *ast* gene of ETEC encodes EAST-1, which as described above, is a heat-stable toxin structurally similar to ST1. It induces an increase in cyclic GMP, but it is not clear if EAST-1 has a role in ETEC-induced diarrhoea (Croxen *et al.*, 2013; Fleckenstein *et al.* 2010). Interestingly, ETEC strains expressing ST were more likely to infect children with A or AB blood types than those with blood type O (Croxen *et al.*, 2013; Fleckensteein *et al.*, 2010). Enzyme-linked immunosorbent assays tests are used to detect the LT or ST toxins, and PCR-based methods are used to target the genes (*lt* and *st*) encoding the toxins (Croxen *et al.*, 2013; Okoh and Osode, 2008).

Shiga toxin-producing *E. coli* (STEC)

STEC strains have also been referred to as enterohaemorrhagic *E. coli* (EHEC) when they cause severe disease, such as haemorrhagic colitis (HC) or haemolytic uraemic syndrome (HUS). They have also been referred to as verotoxigenic *E. coli* (VTEC). STEC serotype O157:H7 and many non-O157 STEC serogroups are important food-borne pathogens that cause diarrhoea, HC, and HUS; death may result from severe disease. At least 90% of STEC infections result in HC; 5–15% of the HC cases progress to HUS and 5–15% of HUS cases result in death. Patients with HUS may develop encephalopathy and suffer long-term consequences, but most HUS patients recover completely. The infectious dose of STEC O157:H7 and also STEC O111 is in the range of 10–100 cells; however, that for non-O157 STEC is not known (Gould *et al.*, 2009). Individuals most susceptible to complications (and death) due to HUS are the very young and the elderly. Treatment with antibiotics for STEC infections is not recommended because they may stimulate the lytic cycle of the bacteriophage that carries the Shiga toxin gene, resulting in release of higher levels of Shiga toxin. Recommended treatments include hydration for diarrhoea, blood transfusion for severe bloody diarrhoea, and dialysis for HUS (Croxen *et al.* 2013; Smith *et al.*, 2014).

The *eae* (*E. coli* attaching and effacing) gene located on the LEE locus (also found in EPEC) encodes the production of intimin found on the bacterial outer membrane. Tir (transmembrane intimin receptor), encoded by *tir*, is translocated by T3SS (bacterial type 3 secretion system) to the host cell membrane. Tir permits attachment of the bacterial intimin protein to the host cell membrane, and this intimate attachment is characterized by the A/E intestinal lesions with localized loss of the microvilli (effacement) and an accumulation of polymerized actin pedestals beneath and surrounding the adherent bacterial cell. The bacteria remain extracellular during infection. The A/E lesion is a prominent feature of STEC O157 and non-O157 STEC intestinal pathology. T3SS, encoded on LEE, forms a pore complex that facilitates the injection of bacterial effector proteins directly into the eukaryotic host cells following attachment. The effector proteins disrupt eukaryotic cell processes and permit STEC to modify the host environment with eventual induction of disease symptoms (Franzin and Sircili, 2015).

The most crucial virulence factor in O157 and non-O157 STEC is a bacteriophage-encoded cytotoxin similar to the *Shigella dysenteriae* Shiga toxin. STEC produce two closely related Shiga toxins, Stx1 and Stx2; Stx2 has *ca.* 60% sequence homology to Stx1. Both toxins have the AB_5 structure. The genetic variants for *stx1* are *stx1a*, *stx1c*, and *stx1d* and for *stx2*, there are seven known variants that include *stx1a–stx1g*. The LD_{50} (in mice) of Stx1 is > 1000 ng, whereas that of Stx2a is 6.5 ng. Thus, strains that carry *stx2a* are associated with more severe disease. Stx-producing strains lacking *eae* can lead to illness; however, the combination of *eae* and *stx* genes is strongly associated with severe illness, including HUS (Brooks *et al.*, 2006). STEC adhere to the intestinal mucosa and secrete Stx, which enters the bloodstream. The toxin is transported to the renal microvascular endothelial cell surface, binds to globotriaosylceramide (Gb_3) followed by endocytosis. The internalized toxin inhibits protein synthesis and induces apoptosis of the renal endothelial cells. Injury to the endothelial cells activates platelets and leucocytes to form fibrin thrombi in the kidney capillaries. Narrowing of the capillaries leads to a reduced blood supply to glomeruli with loss of kidney function and HUS. HUS is characterized by haemolytic anaemia (destruction of red blood cells), acute kidney failure (uraemia), and low platelet count (thrombocytopenia) (Bergan *et al.*, 2012; Croxen *et al.*, 2013). Fermentation of dietary fibre in the colon results in the formation of short-chain fatty acids including butyrate, which increases the expression of Gb3 with receptors for Stx. Mice fed a high-fibre diet showed an increased intestinal colonization by *E. coli* O157:H7 strain 86-24, leading to a 25% increase in mortality compared with mice fed a low-fibre diet. Binding of Stx to colonocytes and renal tissue of mice fed a high-fibre diet was greater than that of mice fed a low-fibre diet. The increased binding of Stx as a result of a high-fibre diet may result in a greater incidence of severe disease induced by O157 and non-O157 STEC (Zumbrun *et al.*, 2013).

The most important reservoir for STEC is the intestinal tract of cattle; therefore, contact with cattle, the cattle environment, and food products originating from cattle are major risk factors for STEC infection to humans. It is probable that all ruminants can act as reservoirs for STEC. Non-ruminants, including cats, dogs, pigs, horses, rabbits, and chickens also can carry STEC. Infected individuals can transmit STEC person to person, and STEC-containing food may result from cross-contamination or by contact from an infected food preparer. Strict hygiene during animal slaughtering, vegetable and fruit growth and harvesting, product handling, and food preparation are the most important factors to consider in the control and prevention of STEC-induced disease (Croxen *et al.*, 2013; Smith *et al.*, 2014).

A large number of food products including meats, vegetables, dairy products, fruit, fruit juices, produce, and nuts have been associated with both O157 and non-O157 STEC infections. Interestingly, food-borne outbreaks are more commonly associated with STEC O157:H7, whereas non-O157 STEC are more often associated with sporadic cases (Kasper *et al.*, 2010). The Centers for Disease Control and Prevention's FoodNet Program in the United States reported that for the years 2000–2010, there were 7694 STEC infections. STEC O157 was the cause of 5688 infections (74.0%) and non-O157 STEC caused 2006 infections (26.1%). STEC serogroups O26, O45, O103, O111, O121, and O145 (the major non-O157 STEC in the United States) were responsible for 1416/7694 (18.4%) of STEC infections (Gould *et al.*, 2013). The global annual incidence of STEC-related illnesses has been estimated to be 2,801,000 cases with 3890 cases of HUS, and 230 deaths (Majowicz *et al.*, 2014).

STEC possess mechanisms to resist the conditions encountered during bacterial passage through the acid stomach and the mild acid conditions of the gut. The three major acid-resistance (AR) mechanisms that protect the cells against exposure to pH 2.0 to 2.5 are the AR-1 oxidative system, the AR-2 glutamate decarboxylase system, and the AR-3 arginine decarboxylase system. The most efficient AR system is the AR-2 glutamate decarboxylase system, which gives STEC the ability to survive and grow in acidic environments such as acidified or fermented foods and to survive in the host during transit through the acidic gastrointestinal tract. In *E. coli* O157:H7, AR-2 and AR-3 protect the cells against oxidative stress during extreme acid challenge (pH 2.5). There are multiple systems in *E. coli* that protect against oxidative stress, and they are induced under different conditions and acid environments. An acid fitness island (AFI) consisting of 12 genes (*slpA* through *gadA*) was identified in *E. coli*; however, the role of all of the AFI genes in acid resistance and the effect of exposure to acid on expression of acid-resistance-related genes are not completely clear (Smith *et al.*, 2014).

Multiplex PCR assays targeting *stx1* (encodes Stx1), *stx2* (encodes Stx2), *eae* (encodes intimin), *ehxA* (encodes enterohaemolysin), and *saa* (encodes an adhesion present in STEC lacking the *eae* gene) and other virulence genes and genetic markers can be used to detect STEC. Thus, PCR can be used to differentiate LEE-containing STEC from those lacking the LEE locus. A real-time PCR diagnostic assay targeting the *stx1*, *stx2*, *eae*, and $rfbE_{O157}$ (encodes *rfbE* O-antigen transporter specific for STEC O157) genes was used to detect STEC strains and to differentiate STEC O157 from non-O157 STEC (Noll *et al.*, 2015). Commercially available real-time PCR-based assays for detection of O157:H7 and other important STEC serogroups have been developed, and the performance of the assays was compared to that of the methods used by the USDA Food Safety and Inspection Service (FSIS) for regulatory testing for the pathogens (Baranzoni *et al.*, 2017; Fratamico *et al.*, 2017). *E. coli* O157:H7 was declared as an adulterant in beef in 1994 by the FSIS, and the top six non-O157 STEC serogroups (O26, O45, O103, O111, O121, and O145) were declared as adulterants in 2012. A multiplex PCR assay was developed to simultaneously detect eight STEC O-serogroups targeting the *wzx* (O-antigen flippase) genes of STEC serogroups O26, O45, O103, O111, O113, O121, O145 and O157. (Croxen *et al.*, 2013; Parsons *et al.*, 2016; Smith *et al.*, 2014).

Enteroaggregative haemorrhagic *E. coli* (EAHEC)

During the summer of 2011, a large outbreak of Shiga toxin-induced diarrhoea, bloody diarrhoea and HUS occurred in Germany and spread to several European countries and North America. More than 4000 cases were reported, including more than 900 cases of HUS, and approximately 50 deaths. The outbreak was due to ingestion of uncooked sprouted fenugreek seeds (CDC, 2013). The strain of *E. coli* responsible for the outbreak was EAEC serotype O104:H4 that had acquired a plasmid for antibiotic resistance and the *stx2a* gene of STEC (but not the *eae* gene) and possessed the *aggR* gene present on plasmid AA of EAEC. The strain was designated EAHEC serotype O104:H4 and showed the stacked-brick phenotype of EAEC and produced Shiga toxin. The EAHEC strain is a pathogen that has combined the virulence properties of the EAEC pathotype with the production of Shiga toxin of the STEC pathotype, and thus can be referred to as a hybrid pathotype of *E. coli* (Navarro-Garcia, 2014).

The EAHEC O104:H4 outbreak was associated with a high severity of illness; approximately 80% of the infected individuals had bloody diarrhoea, 25% with bloody diarrhoea presented with HUS, and severe neurological symptoms developed in about 50% of the HUS cases (Jansen and Kielstein, 2011). Augmented virulence of EAHEC O104:H4 was not related to increased Stx production but rather to more efficient colonization by aggregative adherence fimbriae. The more effective colonization allowed sustained delivery of toxin into the host intestinal cells (Beutin and Martin, 2012). The only known reservoir for EAEC and EAHEC strains is humans. Ruminants, especially cattle, are the primary reservoirs of STEC strains; however, neither EAHEC O104:H4 nor EAEC strains have been detected in the faeces of cattle (Auvray *et al.*, 2012; Wieler *et al.*, 2011). Thus, it appears that EAEC and EAHEC strains are not zoonotic pathogens. EAHEC O104:H4 is transmitted by the faecal–oral route and food, and most of the German outbreak infections were due to eating uncooked sprouts. However, other foods that had been contaminated by infected food handlers were also later implicated. In addition, infections were also transmitted by the person-to-person route. The method used for detection of enteroaggregative STEC O104:H4 involves the amplification of the *stx2*, *aggR*, and wzx_{104} genes to determine the presence of the gene encoding Shiga toxin 2, the presence of a plasmid encoding the AAF fimbriae, and the serogroup O104-specific *wzy* gene (O104 specific O-antigen flippase) (Baranzoni *et al.*, 2014).

Adherent invasive *E. coli* (AIEC)

E. coli strains recovered from Crohn's disease (CD) lesions are able to adhere to and invade cultured intestinal epithelial cells. Therefore, these strains are labelled as adherent invasive *E. coli* (AIEC). AIEC have been implicated in inflammatory bowel diseases (IBD) such as Crohn's disease (CD) and ulcerative colitis (UC). However, a single causative agent has not been positively identified for IBD, and the disease may be caused by a combination of human genetics, intestinal microbiota, environmental factors, and various enteric pathogens such as *Mycobacterium paratuberculosis*, *Campylobacter* species, and AIEC. AIEC are distinct from other *E. coli* pathotypes since they do not have the genes that are associated with ETEC, STEC, EIEC, EAEC, or EPEC. A distinguishing phenotype is that they do not cause diarrhoea; however, AIEC induce inflammation and the production of proinflammatory cytokines such as tumour necrosis factor-alpha and others. AIEC are not only found in humans but appear to be very similar to the same types of *E. coli* in animals. The AIEC pathotype is characterized by the ability to adhere, invade, and replicate in host epithelial cells and macrophages. Strains of AIEC have been isolated from *ca.* 35% of CD cases. Small *et al.* (2013) established an animal model for chronic AIEC infection in streptomycin-treated mice. Their results in mice showed that the AIEC-induced intestinal inflammation was similar to what is observed in humans with Crohn's disease. However, currently AIEC are poorly defined as disease agents and may merely represent those *E. coli* found in the intestinal tracts of patients with IBD (Agus *et al.*, 2014; Coxen *et al.*, 2013; Martinez-Medina and Garcia-Gil, 2014).

Extraintestinal pathogenic *E. coli* (ExPEC)

ExPEC are found in the intestinal tract; however, they do not induce diarrhoea but, rather, cause diseases outside the intestine. Intestinal pathogenic *E. coli* are obligate pathogens, whereas ExPEC are facultative pathogens that behave as harmless commensal organisms until they escape the intestinal tract and invade other areas of the body. ExPEC related to sepsis causes over 40,000 deaths per year in the United States, compared with *c.* 500 deaths caused by *Salmonella*, *Campylobacter*, and *E. coli* O157:H7 combined. In general, the intestinal pathogenic *E. coli* strains have well-defined virulence traits, whereas ExPEC strains are very diverse and have few common virulence factors (Köhler and Dobrindt, 2011). To define ExPEC by virulence properties is difficult. There is a limited understanding of these organisms, since ExPEC and commensal non-pathogenic *E. coli* cannot be differentiated by molecular epidemiological techniques. ExPEC strains have been defined as *E. coli* isolates containing two or more of the following virulence markers as determined by multiplex PCR: *papA* (P fimbriae structural subunit) and/or *papC* (P fimbriae assembly); *sfa/foc* (S and F1C fimbriae subunits); *afa/dra* (Dr-antigen-binding adhesins); *kpsMT* II (group 2 capsular polysaccharide units); and *iutA* (aerobactin receptor) (Smith *et al.*, 2007). *E. coli* strains form six phylogenetic groups labelled A, B_1, B_2, C, D, E, and F based on electrophoretic polymorphism of esterases and other enzymes. Most ExPEC belong to the B_2 and D phylogenetic groups. B_2 and D groups possess virulence factors that allow ExPEC to invade, colonize, and induce disease in non-intestinal bodily sites (Clermont *et al.*, 2013). Human diseases caused by ExPEC include neonatal meningitis, urinary tract infections, necrotizing fasciitis, sepsis, pneumonia, and surgical site infections. ExPEC strains also cause extraintestinal infections in poultry, other domestic animals, and pets. ExPEC appear to be potential zoonotic pathogens since many virulence factors are common between human and animal strains of ExPEC (Markland *et al.*, 2015; Smith *et al.*, 2007). ExPEC strains that cause disease in poultry are known as avian pathogenic *E. coli* (APEC). It is probable that chickens and other animals act as reservoirs for human ExPEC strains (Smith *et al.*, 2007). ExPEC infections are expensive in terms of medical costs and productivity losses for both humans and farm animals. ExPEC strains have been isolated from retail raw meats (beef and pork) and poultry suggesting that these *E. coli* strains may act as food-borne pathogens (Bergeron *et al.* 2012; Köhler and Dobrindt, 2011; Markland *et al.*, 2015; Smith *et al.*, 2007).

Uropathogenic *E. coli* (UPEC)

UPEC is the major cause of urinary tract infections (UTIs) in humans and are a sub-pathotype of ExPEC. UTIs are common bacterial infections in women and are associated with very high health costs in terms of morbidity, treatment, and loss of productivity. It is estimated that the cost of UTIs is $1.6 billion; however, other estimates indicate health care costs closer to $3.5 billion per year (Smith *et al.*, 2007; Ulett *et al.*, 2013). ExPEC have acquired resistance to trimethoprim-sulfomethoxazole, fluoroquinolones, and other antibiotics, thus making ExPEC illnesses such as urinary tract infection difficult to treat, leading to added costs of these infections. UPEC cause *ca.* 80% of community-acquired and *ca.* 30% of nosocomial-acquired UTIs (Smith *et al.*, 2007). UPEC reside in the gastrointestinal tract with commensal strains, and if they gain access to the urinary tract can cause infection. Urethritis is an infection of the urethra, cystitis is an infection of the bladder, and a kidney infection is termed pyelonephritis. Individuals such as the very young, the elderly, females, pregnant women, and the immunocompromised are at increased risk for UTIs. In addition, urinary catheterization is associated with an increased risk of UTI (Smith *et al.*, 2007). UPEC strains possess virulence factors that make them efficient pathogens of the urinary tract. These virulence factors include specific adhesins, several toxins including haemolysin, cytotoxic necrotizing factor type 1 (CNF1), a protease toxin (secreted autotransported toxin; SAT), and iron acquisition systems (Bien *et al.*, 2012; Ulett *et al.*, 2013). It is becoming apparent that an avian reservoir is the probable source of UPEC. Investigations of community-acquired UTIs and outbreaks of UTI suggest common point sources, such as contaminated food products such as chicken. There is a high genetic similarity, including antibiotic resistance and virulence gene patterns, between APEC and ExPEC strains suggesting that strains from poultry play an important role in causation of UTIs (Bergeron *et al.*, 2012; Nordstrom *et al.*, 2013). Treatment of UTIs is becoming more complicated due to increased resistance of UPEC to antimicrobial compounds (Markland *et al.*, 2015; Nordstrom *et al.*, 2013).

Neonatal meningitis *E. coli* (NMEC)

The most frequent cause of Gram-negative-associated meningitis in newborns is NMEC, a sub-pathotype of ExPEC. NMEC account for 20–40% of cases of neonatal meningitis in the USA with a death rate of *ca.* 8%; 30–50% of survivors show severe neurological sequelae. NMEC is acquired by neonates (the neonatal period is first 28 days after birth) during birth or from the environment. The organisms invade and multiply in the bloodstream and then cross the blood–brain barrier followed by NMEC invasion of the meninges and the CNS. The meningitis-causing *E. coli* induce the release of pro-inflammatory compounds resulting in brain oedema and increased intracranial pressure with meningitis and neuronal injury. Most of the neonate survivors have sequelae, which include hydrocephalus, seizures, mental retardation, cerebral palsy, and hearing loss. Neonatal meningitis is treated with appropriate antibiotics (Bonacorsi and Bingen, 2005; Smith *et al.*, 2007; Wijetunge *et al.*, 2015).

Sepsis-associated *E. coli* (SEPEC)

SEPEC is a sub-pathotype of ExPEC that induces sepsis. Sepsis is an overwhelming inflammation of the entire body caused by an immune response elicited by an infection of the bloodstream. Sepsis is life-threatening and may induce tissue damage, organ failure, and death. Sepsis, the body's reaction to septicaemia (infection of the bloodstream), is a major cause of morbidity and mortality with high health care costs in infants, children, and the elderly. SEPEC causes *ca.* 30% of sepsis cases in the United States (Mokady *et al.*, 2005a,b).

Avian pathogenic *E. coli* (APEC)

APEC is also a sub-pathotype of ExPEC, and is responsible for large economic losses in the poultry industry worldwide. APEC strains are heterogeneous populations that are genetically diverse, have a distinctive range of virulence factors, and induce different diseases such as septicaemia, airsacculitis, pericarditis, perihepatitis, salpingitis, cellulitis, and death in various avian species (Dhama *et al.*, 2013). APEC is a zoonotic risk to humans, with some APEC and human ExPEC strains having similar phylogenic backgrounds and sharing a number of virulence genes. Human ExPEC strains have been shown to induce colibacillosis lesions in chicken models, and APEC strains can cause infections in animal models of UTI and neonatal meningitis (Tivendale *et al.*, 2010). Ewers *et al.* (2007) have indicated that APEC potentially can serve as a reservoir of virulence-associated genes for UPEC and NMEC. APEC strains have been isolated from retail poultry meats; therefore, consumption of poultry is a probable source of human disease-inducing APEC/ExPEC (Bergeron *et al.*, 2012).

Hybrid pathogenic *E. coli*

The massive 2011 outbreak caused by EAHEC O104:H4 increased awareness that hybrid *E. coli* pathotypes exist, and the clinical significance of this pathogen indicated that hybrid pathotypes can represent food-borne pathogens with higher virulence potential due to the presence of virulence-related genes from different *E. coli* pathotypes. A number of other EAEC serotypes, including O111:H2, O111:H21, O86:HNM, and O59:H- that acquired *stx2* have been shown to cause disease in humans. EAHEC serotype O111:H2 was associated with an outbreak of HUS that occurred in France involving ten children (Boudailliez *et al.*, 1997; Morabito *et al.*, 1998, 1999). EAHEC serotype O86.HNM infected a Japanese boy who developed HUS and encephalopathy and later died (Iyoda *et al.*, 2000). EAHEC O111:H21 was isolated in 2012 from a 3-year-old Northern Irish girl with HUS and severe cerebral involvement; the strain was also isolated from the child's mother and 4-year old brother, both of whom had diarrhoea (Dallman *et al.*, 2012). EAHEC serotype O59:H- was isolated from a patient with bloody diarrhoea in Germany (Prager *et al.*, 2014).

Serotype O80:H2 harbouring *stx2c* and *stx2d*, as well as *eae*, and a conserved virulence plasmid region carrying *iroN* (salmochelin) and *iucC* (aerobactin) characteristic of ColV plasmids of ExPEC was isolated from a patient presenting with HUS and bacteraemia (Mariani-Kurkdjian *et al.*, 2014). Soysal *et al.* (2016) studied the strains of STEC O80:H2 isolated from 54 French patients during 2005–2014. The STEC O80:H2 isolates contained the S88 plasmid associated with avian pathogenic *E. coli* and neonatal meningitis *E. coli* (both are ExPEC pathogenic *E. coli*), which may explain the bacteraemia reported in some of the HUS patients. Therefore, these O80:H2 strains represent STEC/ExPEC hybrids. Bielaszewska *et al.* (2014) isolated STEC serotype O2:H4 strains from the stools of diarrhoeic patients who also induced experimental UTI in mice. Therefore, these strains were STEC/UPEC hybrids. It is not clear if these STEC/UPEC hybrids caused UTI in human hosts from which the hybrids were isolated.

E. coli serotypes O2:H27 and O101:H- STEC/ETEC hybrids were isolated from two Finnish children with diarrhoea and HUS, respectively. The hybrids produced Stx2a and STa (heat-stable toxin from enterotoxigenic *E. coli*). Several O2:H27 STEC/ETEC hybrids (as well as other serotypes) were isolated from bovine faeces (Nyholm *et al.*, 2015a, 2015b). Leonard *et al.* (2016), also reported animals from which STEC/ETEC strains have been isolated. Thus, the presence of STEC/ETEC hybrids in animals suggests that transfer of these hybrids to humans via food or by animal contact is possible. Monday *et al.* (2006) described the isolation of STEC/ETEC hybrids of serotype Ont:H52 from produce and clinical cases. These isolates are interesting in that they carried the genes for Stx1 and ST; however, immunological testing for Stx1and ST showed that only Stx1 was expressed under laboratory conditions.

An *E. coli* strain isolated from a child with severe diarrhoea harboured the *eae* and *elt* (encodes the heat-labile toxin) genes but not *stx* genes. Therefore, the strain was designated as an EPEC/ETEC hybrid. The strain had an atypical EPEC (aEPEC) genetic background (serotype not given) (Dutta *et al.*, 2015). In a survey of 51 strains of enteroinvasive *E. coli* (EIEC) isolated between 1964 and 1987 from cases of human diarrhoea, da Silva *et al.* (2017) detected the presence of virulence genes usually associated with ExPEC. The EIEC strains, belonging to the most common EIEC serogroups, harboured virulence genes and one to four pathogenicity islands (PAIs) typically found in UPEC. Thus, EIEC strains can form hybrids by incorporating virulence genes and PAIs from ExPEC strains, especially from UPEC strains, and represent the evolution of EIEC into a potentially more virulent pathotype. It is likely that other hybrids of the *E. coli* pathotypes exist and have caused disease but have not been identified and that *E. coli* will continue to evolve.

Evolution of pathogenic *E. coli*

While the study of the evolution of *E. coli* pathogenicity is still in its infancy, there has been some speculation concerning the evolution of STEC. The evolution of *E. coli* O157:H7 may have had its origin in an O55:H7 strain with a predisposition to acquire foreign DNA. A model for the evolution of STEC was proposed (Donnenberg and Whittam, 2001; Feng *et al.*, 1998; Law, 2000; Wick *et al.*, 2005). An EPEC-like ancestor that expressed β-glucuronidase, fermented sorbitol, and contained the LEE locus inserted at *selC* that produced γ-intimin evolved into *E. coli* O55:H7 that is the ancestor of O157:H7. The next step involved the acquisition of *stx2* (via transduction by a toxin-converting bacteriophage), and then the virulence plasmid known as pO157 was gained followed by acquisition of an *rfb* region that specifies the O157 antigen, replacing the O55 *rfb* region. The ability to ferment sorbitol was later lost, and the phage that carried *stx1* was gained. Further evolution of STEC O157:H7 involved the loss of β-glucuronidase activity through mutations of the *uidA* gene (encodes β-glucuronidase) leading to the *E. coli* O157:H7 pathogen that has spread worldwide.

The evolution of the non-O157 STEC is not as clear as that proposed for O157:H7. STEC serogroups O26, O45, O103, O111, O121, and O145 (the major non-O157 STEC in the USA) are similar to O157 STEC in that they possess LEE and the pO157 plasmid, produce Shiga toxin, and can cause illness as severe as that caused by O157:H7, and, thus, it is probable that they evolved in a similar stepwise process as O157 STEC by uptake of phage, chromosomal, and plasmid DNA into a putative ancestor (Law, 2000). Comparison of the insertion sites of LEE into non-O157 STEC serogroups indicated that their evolutionary origins are different from that of STEC O157:H7. The insertion site for LEE in STEC O157:H7 is the *selC* tRNA gene, but Wieler *et al.* (1997) found that the LEE insertion site for the non-O157 STEC serogroups O26 and O111 was not *selC*. The LEE insertion site for O26:H-/H11 is *pheU* tRNA gene (Müller *et al.* 2009) whereas that for serogroup O111 is the *pheV* tRNA gene (Ogura *et al.*, 2009). Müller *et al.* (2009) have shown that the insertion site for LEE in STEC O103:H2 is *pheV*. A number of studies have shown that *E. coli* O26:H11/H- is the second most common serotype after O157:H7 linked to cases of illness due to STEC. To gain a better understanding of the phylogenetic relationships of O26:H11 strains found in France, Delannoy *et al.* (2017) performed whole-genome sequencing and comparative genomics of eight O26:H11 strains carrying *stx2* and compared the sequences to those of O26:H11 strains found in GenBank. Phylogenetic analysis identified four clonal complexes separated in distinct lineages, with one representing a new French O26:H11 clone. They found numerous mobile genetic elements (MGE), including prophages and plasmids that represented up to 20% of the genomes, and they proposed that the strains evolved through separate MGE acquisition events. Other pathogenic *E. coli* probably arose through the uptake of pathogenicity islands, plasmids, transposons, and/or phage and by subtraction of genes to modify the ancestral strain in such a way that a human or animal pathogen was generated (Kaper *et al.*, 2004).

Control of pathogenic *E. coli*

At the present time, there are no vaccines or medications available that can protect the individual from pathogenic *E. coli*-induced illness. Since a large part of illnesses caused by *E. coli* is linked to the ingestion of contaminated foods, *E. coli* contamination must be limited at every step of the food processing chain – from farm to fork. Training and education are essential; individuals involved in the food chain must learn and understand the basics of food safety and hygiene. Meticulous hygiene (i.e. good food-handling practices), as well as good personal hygiene by working personnel is the most effective means of restricting the introduction of pathogenic

E. coli into foods at each step of the food chain from farm to fork.

Modes of pathogenic *E. coli* transmission include ingestion of contaminated food or water, contact with environments containing faeces from infected animals, animal to human and person to person. Scallan *et al.* (2011) estimated that 68% of STEC O157 illness is food-borne, compared with 82% of non-O157 STEC illness, 100% of ETEC ilness and 30% of illness caused by diarrhoeagenic *E. coli* other than STEC and ETEC. Outbreaks may arise in institutions serving food, such as hospitals, nursing homes, daycare centres and restaurants, as well as in homes, if foods are contaminated with pathogenic *E. coli*.

Field crops can be contaminated with pathogens by tainted irrigation water, poorly composted animal manure used as fertilizer, migratory or feral animals, and by unhygienic practices of field personnel. During harvesting of field crops, equipment utilized should be cleaned and sanitized, but the major factor that leads to contaminated vegetables and fruit during harvest is failure of good personal hygiene by harvest workers (Alum *et al.*, 2016). In 2011, the Food and Drug Administration Food Safety Modernization Act was signed into law. It focuses more on preventing food safety problems rather than on actions to be taken when problems occur, and one key objective is the establishment of science-based standards for safe growing, harvesting, packing and holding of produce on farms. Animals grown for food should be raised in clean and stress-free environments, receive pathogen-free feed and water, and be cared for by workers with good personal hygiene. Before slaughter, the hides of animals should be cleaned and careful evisceration is necessary to prevent contamination of the carcass with faecal contents. Proper temperature control and clean wash water is necessary and the workers must have good personal hygiene (Alum *et al.*, 2016).

During processing and preparation of both vegetables and carcasses for food shipment to retail outlets, equipment must be cleaned and sanitized, clean water should be used, good personal hygiene by workers is necessary, and temperature control must be maintained. The processed products must be shipped under sanitary conditions at the correct temperature necessary for the product. At the retail outlet, products must be stored at the correct temperature in a clean environment. At the consumer level (homes, restaurants, institutions), it is necessary to prevent cross-contamination (keep raw foods and cooked foods separate), cook foods properly, keep cooked foods at a proper holding temperature and, most importantly, the preparation environment must be clean and the preparer must practice good personal hygiene (Alum *et al.*, 2016). Clearly, control of *E. coli* infections requires a multidisciplinary approach in animal and plant production and risk-based preventative controls along the entire food supply chain.

References

Agus, A., Massier, S., Darfeuille-Michaud, A., Billard, E., and Barnich, N. (2014). Understanding host-adherent-invasive *Escherichia coli* interaction in Crohn's disease: opening up new therapeutic strategies. Biomed. Res. Int. 2014:567929. doi:org/10.1155/2014/567929.

Alum, E.A., Urom, S.M.O.C., and Ben, C.M.A. (2016). Microbiological contamination of food: the mechanisms, impacts, and prevention. Int. J. Scient. Technol. Res. 5, 65–78.

Auvray, F., Dilasser, F., Bibbal, D., Kérourédan, M., Oswald, E., and Brugère, H. (2012). French cattle is not a reservoir of the highly virulent enteraggregative Shiga toxin-producing *Escherichia coli* of serotype O104:H4. Vet. Microbiol. *158*, 443–445.

Baranzoni, G.M., Fratamico, P.M., Rubio, F., Glaze, T., Bagi, L.K., and Albonetti, S. (2014). Detection and isolation of Shiga toxin-producing *Escherichia coli* (STEC) O104 from sprouts. Int. J. Food Microbiol. *173*, 99–104. https://doi.org/10.1016/j.ijfoodmicro.2013.12.020

Baranzoni, G.M., Fratamico, P.M., Boccia, F., Bagi, L.K., Kim, G.-H., Anastasio, A., and Pepe, T. (2017). Evaluation and performance of the IQ-Check kits and the USDA Microbiology Laboratory Guidebook methods for detection of Shiga toxin-producing *Escherichia coli* (STEC) and STEC and *Salmonella* simultaneously in ground beef. J. Appl. Microbiol. *122*, 809–816.

Bergan, J., Dyve Lingelem, A.B., Simm, R., Skotland, T., and Sandvig, K. (2012). Shiga toxins. Toxicon *60*, 1085–1107. https://doi.org/10.1016/j.toxicon.2012.07.016

Bergeron, C.R., Prussing, C., Boerlin, P., Daignault, D., Dutil, L., Reid-Smith, R.J., Zhanel, G.G., and Manges, A.R. (2012). Chicken as reservoir for extraintestinal pathogenic *Escherichia coli* in humans, Canada. Emerging Infect. Dis. *18*, 415–421. https://doi.org/10.3201/eid1803.111099

Beutin, L., and Martin, A. (2012). Outbreak of Shiga toxin-producing *Escherichia coli* (STEC) O104:H4 infection in Germany causes a paradigm shift with regard to human pathogenicity of STEC strains. J. Food Prot. 75, 408–418. https://doi.org/10.4315/0362-028X.JFP-11-452

Bielaszewska, M., Schiller, R., Lammers, L., Bauwens, A., Fruth, A., Middendorf, B., Schmidt, M.A., Tarr, P.I., Dobrindt, U., Karch, H., and Mellmann, A. (2014). Heteropathogenic virulence and phylogeny reveal phased pathogenic metamorphosis in *Escherichia coli* O2:H6. EMBO Mol. Med. *6*, 347–357.

Bien, J., Sokolova, O., and Bozko, P. (2012). Role of uropathogenic *Escherichia coli* virulence factors in development of urinary tract infection and kidney damage. Int J Nephrol *2012*, 681473. https://doi.org/10.1155/2012/681473

Bonacorsi, S., and Bingen, E. (2005). Molecular epidemiology of *Escherichia coli* causing neonatal meningitis. Int. J. Med. Microbiol. *295*, 373–381.

Boudailliez, B., Berquin, P., Mariani-Kurkdjian, P., Ilef, D., Cuvelier, F.B., Capek, I., Tribout, B., Bingen, E., and Piussan, C. (1997). Possible person-to-person transmission of *Escherichia coli* O111- associated hemolytic uremic syndrome. Pediatr. Nephrol. *11*, 36--39.

Brooks, J.T., Sowers, E.G., Wells, J.G., Greene, K.D., Griffin, P.M., Hoekstra, R.M., and Strockbine, N.A. (2005). Non-O157 Shiga toxin-producing *Escherichia coli* infections in the United States, 1983-2002. J. Infect. Dis. *192*, 1422–1429.

CDC. (2013). Outbreak of *Escherichia coli* O104:H4 infections associated with sprout consumption – Europe and North America, May–July 2011. Morb. Mort. Weekly Report. *62*, 1029–1031.

Cennimo, D.J., Koo, H., Mohamed, J.A., Huang, D.B., and Chiang, T. (2007). Enteroaggregative *Escherichia coli*: a review of trends, diagnosis, and treatment. Infect. Med. *24*, 100-110.

Chen, H.D., and Frankel, G. (2005). Enteropathogenic *Escherichia coli*: unravelling pathogenesis. FEMS Microbiol. Rev. *29*, 83–98.

Clermont, O., Christenson, J.K., Denamur, E., and Gordon, D.M. (2013). The Clermont *Escherichia coli* phylo-typing method revisited: improvement of specificity and detection of new phylo-groups. Environ. Microbiol. Rep. *5*, 58–65.

Croxen, M.A., Law, R.J., Scholz, R., Keeney, K.M., Wlodarska, M., and Finlay, B.B. (2013). Recent advances in understanding enteric pathogenic *Escherichia coli*. Clin. Microbiol. Rev. *26*, 822–880. https://doi.org/10.1128/CMR.00022-13

Dallman, T., Smith, G.P., O'Brien, B., Chattaway, M.A., Finlay, D., Grant, K.A., and Jenkins, C. (2012). Characterization of a verocytotoxin-producing enteroaggregative *Escherichia coli* serogroup O111:H21 strain associated with a household outbreak in Northern Ireland. J. Clin. Microbiol. *50*, 4116-4119.

da Silva, L.C., de Mello Santos, A.C., and Silva, R.M. (2017). Uropathogenic *Escherichia coli* pathogenicity islands and other ExPEC virulence genes may contribute to the genome variability of enteroinvasive *E. coli*. BMC Microbiol. 17:68. doi: 10.1186/s12866-017-0979-5.

Dean, P., and Kenny, B. (2009). The effector repertoire of enteropathogenic *E. coli*: ganging up on the host cell. Curr. Opin. Microbiol. *12*, 101–109. https://doi.org/10.1016/j.mib.2008.11.006

Delannoy, S., Mariani-Kurkdjian, P., Webb, H.E., Bonacorsi, S., and Fach, P. (2017). The mobilome; a major contributor to *Escherichia coli* stx2-Positive O26:H11 strains intra-serotype diversity. Front. Microbiol. *8*, 1625. https://doi.org/10.3389/fmicb.2017.01625

Dhama, K., Chakraborty, S., Barathidasan, R., Tiwari, R., Rajagunalan, S., and Singh, S.D. (2013). *Escherichia coli*, an economically important avian pathogen, its disease manifestations, diagnosis and control, and public health significance: a review. Res. Opin. Anim. Vet. Sci. *3*, 179-194.

Donnenberg, M.S., and Whittam, T.S. (2001). Pathogenesis and evolution of virulence in enteropathogenic and enterohemorrhagic *Escherichia coli*. J. Clin. Invest. *107*, 539–548. https://doi.org/10.1172/JCI12404

Dutta, S., Pazhani, G.P., Nataro, J.P., and Ramamurthy, T. (2015). Heterogenic virulence in a diarrheagenic *Escherichia coli*: evidence for an EPEC expressing heat-labile toxin of ETEC. Int. J. Med. Microbiol. *305*, 47–54. https://doi.org/10.1016/j.ijmm.2014.10.006

Ewers, C., Li, G., Wilking, H., Kießling, S., Alt, K., Antáo, E.-M., Laturnus, C., Diehl, I., Glodde, S., Homeier, T., Böhnke, U., Steinrück, H., Philipp, H.-C., and Wieler, L.H. (2007). Avian pathogenic, uropathogenic, and newborn meningitis-causing *Escherichia coli*: how closely related are they? Int. J. Med. Microbiol. *297*, 163–176.

Feng, P., Lampel, K.A., Karch, H., and Whittam, T.S. (1998). Genotypic and phenotypic changes in the emergence of *Escherichia coli* O157:H7. J. Infect. Dis. *177*, 1750–1753.

Fleckenstein, J.M., Hardwidge, P.R., Munson, G.P., Rasko, D.A., Sommerfelt, H., and Steinsland, H. (2010). Molecular mechanisms of enterotoxigenic *Escherichia coli* infection. Microbes Infect. *12*, 89–98. https://doi.org/10.1016/j.micinf.2009.10.002

Franzin, F.M., and Sircili, M.P. (2015). Locus of enterocyte effacement: a pathogenicity island involved in the virulence of enteropathogenic and enterohemorrhagic *Escherichia coli* subjected to a complex network of gene regulation. Biomed. Res. Int. 2015:534738. doi: 10.1155/2015/534738.

Fratamico, P.M., Bagi, L.K., and Abdul-Wakeel, A. (2017). Detection and isolation of the 'top seven' Shiga toxin-producing *Escherichia coli* in ground beef: comparison of RapidFinder kits to the U.S. Department of Agriculture Microbiology Laboratory Guidebook. J. Food Prot. *80*, 829–836.

Gould, L.H., Bopp, C., Strockbine, N., Atkinson, R., Baselski, V., Body, B., Carey, R., Crandall, C., Hurd, S., Kaplan, R., *et al.* (2009). Recommendations for diagnosis of Shiga toxin-producing *Escherichia coli* infections by clinical laboratories. MMWR Recomm. Rep. *58*, 1–14.

Gould, L.H., Mody, R.K., Ong, K.L., Clogher, P., Cronquist, A.B., Garman, K.N., Lathrop, S., Medus, C., Spina, N.L., Webb, T.L., *et al.* (2012). Increased recognition of non-O157 Shiga toxin-producing *Escherichia coli* infection in the United States during 2000-2010: epidemiologic features and comparison with *E. coli* O157 infections. Foodborne Pathog. Dis. *10*, 453–460.

Iyoda, S., Tamura, K., Itoh, K., Izumiya, H., Ueno, N., Nagata, K., Togo, M., Terajima, J., and Watanabe, H. (2000). Inducible stx2 phages are lysogenized in the enteroaggregative and other phenotypic *Escherichia coli* O86:HNM isolated from patients. FEMS Microbiol. Lett. *191*, 7–10.

Jafari, A., Aslani, M.M., and Bouzari, S. (2012). *Escherichia coli*: a brief review of diarrheagenic pathotypes and their role in diarrheal diseases in Iran. Iran J. Microbiol. *4*, 102–117.

Jansen, A., and Kielstein, J.T. (2011). The new face of enterohaemorrhagic *Escherichia coli* infections. Euro Surveill. *16*, 19898.

Kaper, J.B., Nataro, J.P., and Mobley, H.L. (2004). Pathogenic *Escherichia coli*. Nat. Rev. Microbiol. *2*, 123–140. https://doi.org/10.1038/nrmicro818

Kaspar, C., Doyle, M.E., and Archer, J. (2010). White paper on non-O157:H7 Shiga toxin-producing *E. coli* from meat and non-meat sources. Available from: http://fri.wisc.edu/docs/pdf/FRI_Brief_NonO157STEC_4_10.pdf. Accessed 6 September 2017.

Kaur, P., Chakraborti, A., and Asea, A. (2010). Enteroaggregative *Escherichia coli*: an emerging enteric food borne pathogen. Interdiscip. Perspect. Infect. Dis. 2010: article ID 254159. doi:10.1155/2010/254159.

Köhler, C.D., and Dobrindt, U. (2011). What defines extraintestinal pathogenic *Escherichia coli*? Int. J. Med. Microbiol. *301*, 642–647. https://doi.org/10.1016/j.ijmm.2011.09.006

Law, D. (2000). The history and evolution of other Shiga toxin-producing *E. coli*. World J. Microbiol. Biotechnol. *16*, 701-709.

Le Bouguénec, C., and Servin, A.L. (2006). Diffusely adherent *Escherichia coli* strains expressing Afa/Dr adhesins (Afa/Dr DAEC): hitherto unrecognized pathogens. FEMS Microbiol. Lett. *256*, 185–194.

Leonard, S.R., Mammel, M.K., Rasko, D.A., and Lacher, D.W. (2016). Hybrid Shiga toxin-producing and enterotoxigenic *Escherichia* sp. cryptic lineage 1 strain 7v harbor a hybrid plasmid. Appl. Environ. Microbiol. *82*, 4309–4319.

Løbersli, I., Wester, A.L., Kristiansen, Å., and Brandal, L.T. (2016). Molecular differentiation of *Shigella* spp. from enteroinvasive *E. coli*. Eur. J. Microbiol. Immunol. *6*, 197–205. https://doi.org/10.1556/1886.2016.00004

Majowicz, S.E., Scallan, E., Jones-Bitton, A., Sargeant, J.M., Stapleton, J., Angulo, F.J., Yeung, D.H., and Kirk, M.D. (2014). Global incidence of human Shiga toxin-producing *Escherichia coli* infections and deaths: a systematic review and knowledge synthesis. Foodborne Pathog. Dis. *11*, 447-455.

Mariani-Kurkdjian, P., Lemaître, C., Bidet, P., Perez, D., Boggini, L., Kwon, T., and Bonacorsi, S. (2014). Haemolytic-uraemic syndrome with bacteraemia caused by a new hybrid *Escherichia coli* pathotype. New Microbes New Infect. *2*, 127–131. https://doi.org/10.1002/nmi2.49

Markland, S.M., LeStrange, K.J., Sharma, M., and Kniel, K.E. (2015). Old friends in new places: exploring the role of extraintestinal *E. coli* in intestinal disease and food-borne illness. Zoonoses Public Health *62*, 491–496. https://doi.org/10.1111/zph.12194

Martinez-Medina, M., and Garcia-Gil, L.J. (2014). *Escherichia coli* in chronic inflammatory bowel diseases: an update on adherent invasive *Escherichia coli* pathogenicity. World J. Gastrointest. Pathophysiol. *5*, 213–227.

Mokady, D., Gophna, U., and Ron, E.Z. (2005a). Extensive gene diversity in septicemic *Escherichia coli* strains. J. Clin. Microbiol. *43*, 66–73.

Mokady, D., Gophna, U., and Ron, E.Z. (2005b). Virulence factors of septicemic *Escherichia coli* strains. Int. J. Med. Microbiol. *295*, 455–462.

Monday, S.R., Keys, C., Hanson, P., Shen, Y., Whittam, T.S., and Feng, P. (2006). Produce isolates of the *Escherichia coli* Ont:H52 serotype that carry both Shiga toxin 1 and stable toxin genes. Appl. Environ. Microbiol. *72*, 3062–3065.

Morabito, S., Karch, H., Mariani-Kurkdjian, P., Schmidt, H., Minelli, F., Bingen, E., and Caprioli, A. (1998). Enteroaggregative, Shiga toxin-producing *Escherichia coli* O111:H2 associated with an outbreak of hemolytic-uremic syndrome. J. Clin. Microbiol. *36*, 840–842.

Morabito, S., Karch, H., Schmidt, H., Minelli, F., Mariani-Kurkdjian, P., Allerberger, F., Bettelheim, K.A., and Caprioli, A. (1999). Molecular characterization of verocytotoxin-producing *Escherichia coli* of serogroup O111 from different countries. J. Med. Microbiol. *48*, 891–896.

Müller, D., Benz, I., Liebchen, A., Gallitz, I., Karch, H., and Schmidt, M.A. (2009). Comparative analysis of the locus of enterocyte effacement and its flanking regions. Infect. Immun. *77*, 3501–3513. https://doi.org/10.1128/IAI.00090-09

Navarro-Garcia, F. (2014). *Escherichia coli* O104:H4 Pathogenesis: an enteroaggregative *E. coli*/Shiga toxin-producing *E. coli* explosive cocktail of high virulence. Microbiol. Spectr. *2*(6). https://doi.org/10.1128/microbiolspec.EHEC-0008-2013

Newitt, S., MacGregor, V., Robbins, V., Bayliss, L., Chattaway, M.A., Dallman, T., Ready, D., Aird, H., Puleston, R., and Hawker, J. (2016). Two linked enteroinvasive Escherichia coli outbreaks, Nottingham, UK, June 2014. Emerging Infect. Dis. 22, 1178–1184. https://doi.org/10.3201/eid2207.152080

Noll, L.W., Shridhar, P.B., Shi, X., An, B., Cernicchiaro, N., Renter, D.G., Nagaraja, T.G., and Bai, J. (2015). A four-plex real-time PCR assay, based on rfbE, stx1, stx2, and eae genes, for the detection and quantification of Shiga toxin-producing *Escherichia coli* O157 in cattle feces. Foodborne Pathog. Dis. *12*, 787–794. https://doi.org/10.1089/fpd.2015.1951

Nordstrom, L., Liu, C.M., and Price, L.B. (2013). Food-borne urinary tract infections: a new paradigm for antimicrobial-resistant food-borne illness. Front Microbiol. *4*, 29. doi: 10.3389/fmicb.2013.00029.

Nyholm, O., Halkilahti, J., Wiklund, G., Okeke, U., Paulin, L., Auvinen, P., Haukka, K., and Siitonen, A. (2015a). Comparative genomics and characterization of hybrid shigatoxigenic and enterotoxigenic *Escherichia coli* (STEC/ETEC) Strains. PLOS ONE *10*, e0135936. https://doi.org/10.1371/journal.pone.0135936

Nyholm, O., Heinikainen, S., Pelkonen, S., Hallanvuo, S., Haukka, K., and Siitonen, A. (2015b). Hybrids of shigatoxigenic and enterotoxigenic *Escherichia coli* (STEC/ETEC) among human and animal isolates in Finland. Zoonoses Public Health *62*, 518–524. https://doi.org/10.1111/zph.12177

Ochoa, T.J., and Contreras, C.A. (2011). Enteropathogenic *Escherichia coli* infection in children. Curr. Opin. Infect. Dis. *24*, 478–483. https://doi.org/10.1097/QCO.0b013e32834a8b8b

Ogura, Y., Ooka, T., Iguchi, A., Toh, H., Asadulghani, M., Oshima, K., Kodama, T., Abe, H., Nakayama, K., Kurokawa, K., *et al.* (2009). Comparative genomics reveal the mechanism of the parallel evolution of O157 and non-O157 enterohemorrhagic *Escherichia coli*. Proc. Natl. Acad. Sci. U.S.A. *106*, 17939–17944. https://doi.org/10.1073/pnas.0903585106

Okoh, A.I., and Osode, A.N. (2008). Enterotoxigenic *Escherichia coli* (ETEC): a recurring decimal in infants' and travelers' diarrhea. Rev. Environ. Health. *23*, 135–148.

Parsons, B.D., Zelyas, N., Berenger, B.M., and Chui, L. (2016). Detection, characterization, and typing of Shiga toxin-producing *Escherichia coli*. Front. Microbiol. 7:478. doi: 10.3389/fmicb.2016.00478.

Parsot, C. (2005). *Shigella* spp. and enteroinvasive *Escherichia coli* pathogenicity factors. FEMS Microbiol. Lett. *252*, 11–18.

Patzi-Vargas, S., Zaidi, M., Bernal-Reynaga, R., León-Cen, M., Michel, A., and Estrada-Garcia, T. (2013). Persistent bloody diarrhea without fever associated with diffusely adherent *Escherichia coli* in a young child. J. Med. Microbiol. *62*, 1907-1910.

Pavlovic, M., Luze, A., Konrad, R., Berger, A., Sing, A., Busch, U., and Huber, I. (2011). Development of a duplex real-time PCR for differentiation between *E. coli* and *Shigella* spp. J. Appl. Microbiol. *110*, 1245–1251. https://doi.org/10.1111/j.1365-2672.2011.04973.x

Prager, R., Lang, C., Aurass, P., Fruth, A., Tietze, E., and Flieger, A. (2014). Two novel EHEC/EAEC hybrid strains isolated from human infections. PLOS ONE *9*, e95379. https://doi.org/10.1371/journal.pone.0095379

Scallan, E., Hoekstra, R.M., Angulo, F.J., Tauxe, R.V., Widdowson, M.A., Roy, S.L., Jones, J.L., and Griffin, P.M. (2011). Food-borne illness acquired in the United States – major pathogens. Emerging Infect. Dis. *17*, 7–15. https://doi.org/10.3201/eid1701.P11101

Scheutz, F. (2014). Taxonomy meets public health: The case of Shiga toxin-producing *Escherichia coli*. Microbiol. Spectr. *2*, https://doi.org/10.1128/microbiolspec.EHEC-0019-2013

Small, C.-L.N., Reid-Yu, S.A., McPhee, J.B., and Coombes, B.K. (2013). Persistent infection with Crohn's disease-associated adherent-invasive *Escherichia coli* leads to chronic inflammation and intestinal fibrosis. Nat. Commun. 4:1957. doi:10.1038/ncomms2957.

Smith, J.L., Fratamico, P.M., and Gunther, N.W. (2007). Extraintestinal pathogenic *Escherichia coli*. Food-borne Pathog. Dis. *4*, 134–163. https://doi.org/10.1089/fpd.2007.0087

Smith, J.L., Fratamico, P.M., and Gunther, N.W. (2014). Shiga toxin-producing *Escherichia coli*. Adv. Appl. Microbiol. *86*, 145–197. https://doi.org/10.1016/B978-0-12-800262-9.00003-2

Snelling, A.M., Macfarlane-Smith, L.R., Fletcher, J.N., and Okeke, I.N. (2009). The commonly-used DNA probe for diffusely-adherent *Escherichia coli* cross-reacts with a subset of enteroaggregative *E. coli*. BMC Microbiol. *9*, 269. https://doi.org/10.1186/1471-2180-9-269

Soysal, N., Mariani-Kurkdjian, P., Small, Y., Liguori, S., Gouali, M., Loukiadis, E., Fach, P., Bruyand, M., Blanco, J., Bidet, P., and Bonacorsi, S. (2016). Enterohemorrhagic *Escherichia coli* hybrid pathotype O80:H2 as a new therapeutic challenge. Emerg. Infect. Dis. *22*, 1604–1612.

Tivendale, K.A., Logue, C.M., Kariyawasam, S., Jordan, D., Hussein, A., Li, G., Wannemuehler, Y., and Nolan, L.K. (2010). Avian-pathogenic *Escherichia coli* strains are similar to neonatal meningitis. *E. coli* strains and are able to cause meningitis in the rat model of human disease. Infect. Immun. *78*, 3412–3419.

Ulett, G.C., Totsika, M., Schaale, K., Carey, A.J., Sweet, M.J., and Schembri, M.A. (2013). Uropathogenic *Escherichia coli* virulence and innate immune responses during urinary tract infection. Curr. Opin. Microbiol. *16*, 100–107.

WHO. (2006). Five keys to growing safer fruits and vegetables: promoting health by decreasing microbial contamination. http://apps.who.int//iris/bitstream/10665/75196/1/9789241504003_eng.pdf?ua=1 (accessed 6 September 2017).

Wick, L.M., Qi, W., Lacher, D.W., and Whittam, T.S. (2005). Evolution of genomic content in the stepwise emergence of *Escherichia coli* O157:H7. J. Bacteriol. *187*, 1783–1791.

Wieler, L.H., McDaniel, T.K., Whittam, T.S., and Kaper, J.B. (1997). Insertion site of the locus of enterocyte effacement in enteropathogenic and enterohemorrhagic *Escherichia coli* differs in relation to the clonal phylogeny of the strains. FEMS Microbiol. Lett. *156*, 49–53.

Wieler, L.H., Semmler, T., Eichhorn, I., Antao, E.M., Kinnemann, B., Geue, L., Karch, H., Guenther, S., and Bethe, A. (2011). No evidence of the Shiga toxin-producing *E. coli* O104:H4 outbreak strain or enteroaggregative *E. coli* (EAEC) found in cattle faeces in northern Germany, the hotspot of the 2011 HUS outbreak area. Gut Pathog. 3, 17. https://doi.org/10.1186/1757-4749-3-17

Wijetunge, D.S., Gongati, S., DebRoy, C., Kim, K.S., Couraud, P.O., Romero, I.A., Weksler, B., and Kariyawasam, S. (2015). Characterizing the pathotype of neonatal meningitis causing *Escherichia coli* (NMEC). BMC Microbiol. *15*, 211. https://doi.org/10.1186/s12866-015-0547-9

Zumbrun, S.D., Melton-Celsa, A.R., Smith, M.A., Gilbreath, J.J., Merrell, D.S., and O'Brien, A.D. (2013). Dietary choice affects Shiga toxin-producing *Escherichia coli* (STEC) O157:H7 colonization and disease. Proc. Natl. Acad. Sci. U.S.A. *110*, E2126–2133.

Escherichia coli Genome Plasticity and Evolution

2

David W. Lacher[1], Michael L. Kotewicz[2], Mark K. Mammel[2] and Christopher A. Elkins[2*]

[1]Division of Healthcare Quality Promotion, National Center for Emerging and Zoonotic Infectious Diseases, Centers for Disease Control and Prevention, Atlanta, GA, USA
[2]Division of Molecular Biology, Center for Food Safety and Applied Nutrition (CFSAN), United States Food and Drug Administration, Laurel, MD, USA.

*Correspondence: nme0@cdc.gov

https://doi.org/10.21775/9781910190777.02

Abstract

Escherichia coli is associated with environmental, commensal, and emergent pathogenic niches. Extensive genome sequencing has expanded and refined the landscape for this microbial species beyond traditional taxonomy reliant on fragmented phenotypic and genetic data. Core genome sequences provide a new standard to investigate clonal relationships, diversity, species definition, and strain classification. Specifically, major phylogroups originally established by multilocus enzyme electrophoresis are recapitulated with greater resolution than previously reported and new environmental cryptic lineages sharpen perspectives on species definition. *Shigella* species are interwoven across this high-resolution landscape, again providing further evidence to reject their species distinction. We use *E. coli* O157:H7 as a paradigm for pathogen evolution and offer critical analyses of associated clade and lineage models with respect to core genome phylogenetics. This perspective is further stratified and extended to the strain level to highlight the plasticity of genome architectures in these pathogens. In aggregate, the species landscape is expectedly well populated from a clinical perspective but is under-represented from environmental and commensal niches. Regardless, the robustness in framework and fundamental architecture of core genome analysis should be foundational for future molecular systematic applications.

Introduction

Illnesses associated with pathogenic *Escherichia coli* and *Shigella* species present a significant worldwide public health burden that extends beyond their associated pathotypes and underdeveloped economies. In the developed world, consumption of foods contaminated with Shiga toxin-producing *E. coli* (STEC) is problematic given acquired virulence elements in emerging hybrid pathogens witnessed with the European O104:H4 enteroaggregative STEC outbreak (Denamur, 2011; Jackson *et al.*, 2012; Kupferschmidt, 2011) or recent associations of antibiotic resistance in extraintestinal pathogenic *E. coli* with foods (Manges and Johnson, 2015; Vila *et al.*, 2016). The breadth of such pathogens and their ability to rapidly adapt to novel environmental niches necessitates highly parallel analysis methods to accurately identify and discriminate individual strains, their virulence potential, and history of causing disease. The genetic diversity among the clinical isolates and environmental and food-borne reservoirs associated with disease is important considering a landscape confounded by commensal *E. coli*, unlike other uniformly pathogenic species (Chaudhuri and Henderson, 2012; Tenaillon *et al.*, 2010). Hence, analysis of temporally and geographically diverse collections of isolates suggests that *E. coli* display extensive pan-genomic variability both within and among serogroups (Jackson *et al.*, 2011).

Interestingly, despite these issues, *E. coli* is the most well-characterized species of bacteria from both functional and genetic perspectives, undoubtedly attributable to its status as a commensal, pathogen, and classical molecular genetic mainstay. Since 1997, when the first genome sequence of K-12 MG1655 was released (Blattner *et al.*, 1997) and subsequently compounded by years of whole genome analytical efforts, it became clear that this species displays substantial genomic plasticity. Despite this heterogeneity, molecular-based phylogenies using, for example, multilocus enzyme electrophoresis (MLEE) and multilocus sequence typing (MLST) developed sequentially with continuously sharpened resolution to capture underlying clonality and evolutionary lineages. To their credit, these techniques originally identified broad groupings of strains that have generally remained conserved across the growing *E. coli* genetic landscape for more than 30 years despite significant advances in technology. Indeed, recent developments in next generation sequencing have logarithmically expanded our genomic understanding of this pathogen to the extent that the pan-genome of this species may now have been sampled well enough to provide a unique genomic prospective of the species. Consequently, further sequence information will influence the unique combinatory network, which is the principal component for describing new strains, yet may provide limited changes in the global landscape of the species.

The *E. coli* phylogroups

Early molecular investigations into genetic diversity and substructures in *E. coli* populations involved allozyme electrophoretic profiling (Milkman, 1973, 1975; Selander and Levin, 1980; Whittam *et al.*, 1983). At the time, these investigations assumed a somewhat retrospective role to the intense functional genetic studies occurring with a few relatively well-defined non-pathogenic laboratory strains. The ensuing study of natural isolates revealed an underappreciated genetic diversity that consisted of at least three groups of strains (Whittam *et al.*, 1983). This underlying structure suggested that evolving yet stable clonal lineages exist as shown by others (Ochman and Selander, 1984a; Achtman *et al.*, 1984). Most importantly, the clonal structure supported a neutral theory of genetic fitness as suggested years earlier by serotypes with worldwide distribution and, therefore, stable lineages. Subsequently and conceptually, these observations coalesced into an *E. coli* reference collection (ECOR) of 72 strains stratifying geography, host, pathogenic status (uropathogenic with notable absence of diarrhoeagenic strains), and newly identified genetic diversity (Ochman and Selander, 1984b).

Using an improved MLEE analysis, the ECOR collection resolves phylogenetically into four named groups (A, B1, B2, and D) however an apparent fifth (unnamed) group was also observed (Herzer *et al.*, 1990) that later became group E. This latter group becomes critically important in food and public health safety giving rise to the well-known human enterohaemorrhagic pathogen with serotype O157:H7 addressed later in this chapter. Regardless, these groups demonstrate broad between group distances of 0.31–0.44 when reconstructed from the original ECOR collection MLEE data (Fig. 2.1). Furthermore, the phylogroups appeared to be well conserved in many separate studies employing strains from a variety of sources and pathotypes including the shigellae (Pupo *et al.*, 1997) such that subsequent research interest was generated in molecular genetic methods for their determination (Clermont *et al.*, 2000, 2015). These methods ranged from basic triplex PCR assays with increasing complexity (Clermont et., 2013) to several MLST schemes that employ various housekeeping genes with differing levels of nucleotide diversity (Reid *et al.*, 2000; Wirth *et al.*, 2006; Lacher *et al.*, 2007; Jaureguy *et al.*, 2008). It should be noted that the intent of MLST schemes and extensions thereof involved higher discriminatory power for probing evolutionary pathways, estimates of recombination, and clonal pathogen histories. However, phylogenies generated in this manner, while revealing conflicting signals and networks, generally conserved topologies consistent with the initial ECOR MLEE phylogroupings. Thus, the underlying *E. coli* species substructures were robust and could be reconstructed *a priori* with various levels of discriminatory genetic tools as opposed to *a posteriori* methods designed specifically to recapitulate the MLEE-based findings.

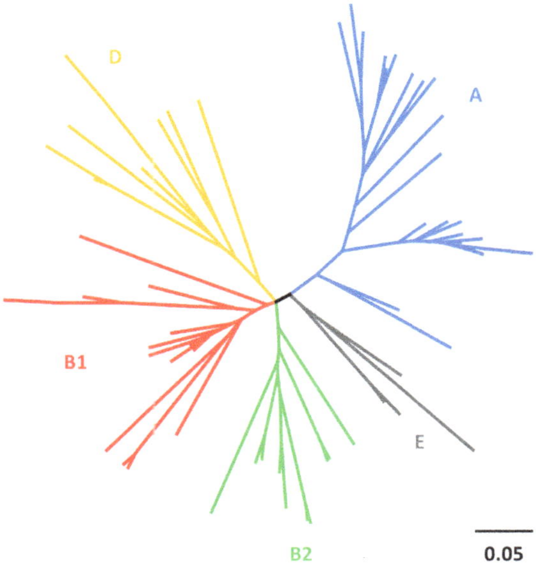

ECOR	A	B1	E	D	B2
A		0.38	0.36	0.44	0.41
B1	0.38		0.31	0.39	0.33
E	0.36	0.31		0.38	0.35
D	0.44	0.39	0.38		0.37
B2	0.41	0.33	0.35	0.37	

Figure 2.1 Reconstruction of ECOR groups as originally reported by Herzer *et al.* (1990). Distance matrix reflects between group distances calculated by averaging those distances generated for respective ECOR strain pairwise comparisons using the same underlying principles involving proportional allele differences. Values within the matrix are colour-coded from low to high on a green to red scale. Neighbour joining tree generated in MEGA.

Development of a genomic framework for the *E. coli* species

Moving to higher analytical resolution beyond MLEE, PCR, and MLST-based techniques required advancements in molecular genetic analytical tools. The genomic era was ushered in with microarray technology as a powerful platform to investigate gene expression in the new field of transcriptomics. However, our laboratories employed microarrays for a different purpose involving whole genome interrogation of enteric pathogens (Jackson *et al*, 2011). We harnessed this technology for genome-scale gene allele profiling (presence/divergence) and observed significant gene heterogeneity with strain collections including ECOR that differed on the order of megabases across the analysis using a commercially available array from Affymetrix. Interestingly, as observed with other methods, the array profiles using maximum likelihood methods were parsed into the five phylogroups interspersed with *Shigella* strains (Jackson *et al.*, 2011), indicating a balance of core phylogenetic signal amidst otherwise horizontally acquired transient or otherwise accessory elements. Prior to high-throughput next-generation sequencing technologies, these tools proved powerful for genomic analysis in discovery-based pursuits. They were further developed with subsequent and increasingly powerful custom designs intended to capture the emerging *E. coli* pan-genome (Elkins *et al.*, 2013).

Currently there are over 5000 complete and partially assembled *E. coli* genomes with nearly a log order higher number of raw next-generation sequence-generated genomic data deposited in the Sequence Read Archive at the National Center for Biotechnology Information public repository (https://www.ncbi.nlm.nih.gov/). Significant efforts have been made to develop a rational approach for analysis of this rapidly expanding whole-genome sequence availability. One high-level distinction involves identifying core conserved genome content across the species apart from pan-genome content as was highlighted with whole-genome microarray profiling. The rationale is fundamentally important for distinguishing evolutionary clonal histories encoded in core elements necessary for cross-comparison of strains apart from emerging pathogenic traits on acquired elements. In parallel, this concept emerged early in the sequencing era suggesting a core genome of approximately 2200 genes and an expanding pan-genome of at least 13,000 genes in a limited sample size of fewer than 17 genomes (Rasko *et al.*, 2008; Tenaillon *et al.*, 2010). Scaling up to over 2000 genomes reveals a relatively stable core and pan-genome of over 90,000 genes (Land *et al.*, 2015). Such measurements depend on critical criteria to determine gene differences, and hence allele parameters, for tallying core and pan-genome (Zhao *et al.*, 2014). For example in our internal analysis, a new gene is considered part of the pan-genome if it displays 95% identity over 90% of the length of the

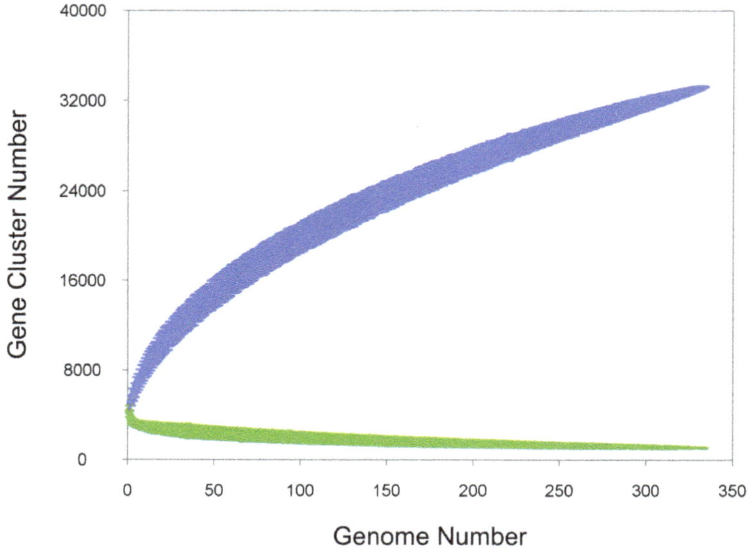

Figure 2.2 The *E. coli* core (green) and pan-genomes (blue). The 344 observed core genome types (represented as 'Genome Number') were identified to computationally condense and simplify whole genome data to a common reference representative with < 0.2% core genetic difference. New genes (represented as 'Gene Cluster Number') are tallied using BLAST criteria of 95% identity over 90% gene length. Generated by PanGP; see Zhao *et al*. 2014.

respective gene. In such a conservative analysis, 344 genome types (representing 3348 genomes) yield a pan-genome of over 32,000 genes (Fig. 2.2).

To develop a genomic phylogenetic framework, we further refined the analysis regime focusing on the core genome with single nucleotide polymorphisms (SNPs) serving as the basis of estimating genetic distances. To determine the core gene set, 4500 open reading frames of *E. coli* K-12 MG1655 were used as reference to query 42 representative closed whole-genome-sequences of *E. coli* and *Shigella* that span the available diversity of the species. Two criteria were employed: (i) a gene was excluded if more than one match occurred in a genome or if the gene was absent in more than one of the representative genomes; and (ii) homology was established from BLAST similarity across a specific query length of respective pairwise comparisons. Preliminary investigations were conducted by varying these parameters to conservatively capture a large enough core capable of displaying the potential diversity across the species while relaxing parameters at the species limits to include the closest known relatives *E. albertii* and *E. fergusonii*. Values were selected to establish a core backbone of 2542 genes covering 2.4 Mbp.

Using the concatenated 2.4 Mbp scaffold, SNPs could then be extracted to develop a phylogenetic landscape. SNPs were identified using 136 closed and 3212 draft genomes with BLAST alignments to each position in the K-12 MG1655 core. In addition to the instance-based criterion, mismatches were ignored if found in an alignment gap respective to K-12 MG1655. In total, 826,737 SNP sites were identified as a basis to rationally analyse and compare genomic data at strain-level resolution. Although presented as just one approach that may take similar forms across laboratories, there is no standardized methodology to compare species-level whole-genome data as it is inherently interdependent of the desired outcome.

Hallmark features of the species landscape

The complexity of the more than 800,000 SNPs can be reduced to 344 representative genotypes to provide a high-level overview of genome diversity across the species. The original five phylogroups (A, B1, B2, D, and E) were again evident and recapitulated within this highly refined analysis (Fig. 2.3). There is also some evidence for a phylogroup F (not shown) as has been described by others (Jaureguy *et al*., 2008). Furthermore, *Shigella* spp. are found

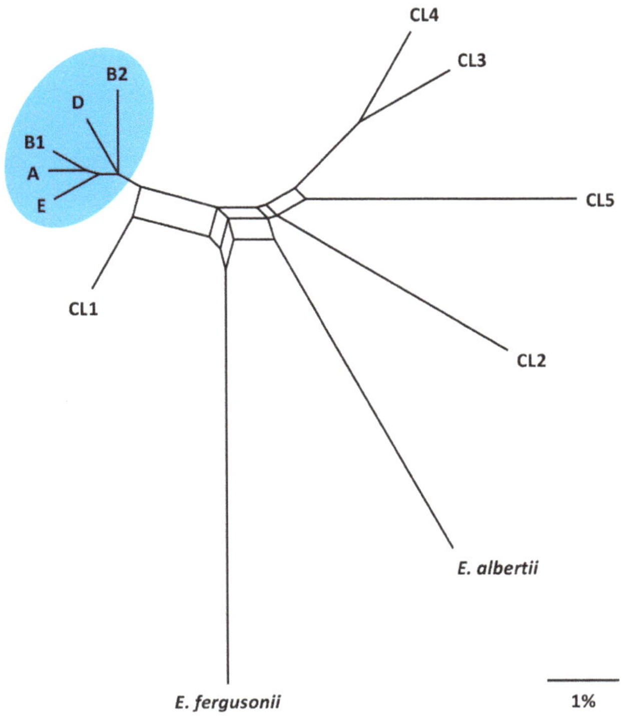

Group	A	B1	E	D	B2	CL1	CL2	CL3	CL4	CL5	Ea	Ef
A		1.1	1.4	2.0	2.3	3.2	7.5	6.7	6.5	8.1	9.0	9.5
B1	1.1		1.5	2.0	2.3	3.2	7.5	6.6	6.5	8.1	9.0	9.6
E	1.4	1.5		2.0	2.3	3.2	7.5	6.7	6.5	8.1	9.0	9.6
D	2.0	2.0	2.0		2.2	3.0	7.5	6.7	6.5	8.1	9.0	9.5
B2	2.3	2.3	2.3	2.2		3.3	7.6	6.8	6.7	8.2	9.0	9.6
CL1	3.2	3.2	3.2	3.0	3.3		7.7	7.0	6.9	8.3	9.2	8.8
CL2	7.5	7.5	7.5	7.5	7.6	7.7		7.4	7.5	8.2	9.4	11.2
CL3	6.7	6.6	6.7	6.7	6.8	7.0	7.4		3.1	7.0	9.2	11.0
CL4	6.5	6.5	6.5	6.5	6.7	6.9	7.5	3.1		7.2	9.2	10.9
CL5	8.1	8.1	8.1	8.1	8.2	8.3	8.2	7.0	7.2		9.6	11.9
Ea	9.0	9.0	9.0	9.0	9.0	9.2	9.4	9.2	9.2	9.6		12.1
Ef	9.5	9.6	9.6	9.5	9.6	8.8	11.2	11.0	10.9	11.9	12.1	

Figure 2.3 Model core SNP genotypes that span the genus *Escherichia*. *E. coli sensu stricto* are highlighted in blue and distance matrix values represent per cent divergence over the 2.4 Mbp core backbone. Networked tree structure generated in SplitsTree using the neighbour net method. Values within the matrix are colour-coded from low to high on a green to red scale.

in multiple phylogroups including B1 and E (not shown). In addition to these phylogroups of *E. coli sensu stricto*, several cryptic lineages have been observed in the environment (Walk *et al.*, 2009). Extended MLST analysis of atypical *Escherichia* isolates identified five novel phylogenetic clades, referred to as cryptic lineages, from environmental, human, and animal sources. Phenotypically, only cryptic lineage 3 was found to possess discriminating biochemical markers involving the loss of sucrose and sorbitol fermentation along with the loss of lysine utilization (Walk *et al.*, 2009).

Juxtaposed to the classical phylogroups with *E. albertii* and *E. fergusonii* as nearest species, the cryptic lineages are informative for providing a perspective on species boundaries. The breadth of the landscape as presented (inclusive of *E. albertii* and *E. fergusonii*) spans approximately 12% sequence difference across the core genome whereas *E. coli sensu stricto* spans < 2.5% (Fig. 2.3). Furthermore, within the five phylogroups there appears to be demarcation with the closest intergroups being A, B1, and E at < 1.5% core difference, while B2 and D are approximately 2% apart. Using these metrics, it would seem reasonable to include cryptic lineage 1 within the *E. coli* species considering this entire group can be related within approximately 3% difference. In support of this perspective, the nearest neighbours to this species group (phylogroups A through E and cryptic lineage 1 inclusive) are cryptic lineages 3 and 4 that more than double the core genome difference at approximately 6.7%.

It is noteworthy that this same conclusion regarding cryptic lineage 1 was obtained earlier in a targeted study of whole-genome phylogeny with the cryptic strains. Cryptic lineage 1 is unique in that it displayed significant horizontal transfer of core genes between the genomes of the *E. coli* phylogroups in contrast to the other cryptic lineages (Luo *et al.*, 2011). On the contrary, Luo *et al.* (2011) also concluded that the cryptic lineages should be classified as *E. coli* species because they share the same *E. coli* core gene set. While this may be correct as determined in their analysis of 'core', the strength of our analysis involves the extensive profiling of thousands of publicly available genomes such that genetic distance becomes a focused quantifiable power. Nonetheless, it is possible with further depth in genome sequence and strain biogeography, the landscape may resemble more of a continuum that lacks notable demarcations to define species *per se*.

Intraphylogroup genomic diversity

Moving to a higher level of resolution using the core genome framework, the phylogroup level reveals significant genetic diversity given the current public availability of genome sequences. For ease of visualization and suggestive of probabilistic recombinant communities/genetic pools, the respective within phylogroup terminal branches were collapsed to approximately < 0.2% sequence diversity in the core genome, including some minimal case-by-case instances above this threshold when skewing resulted from obvious localized recombination. Using this construction, phylogroups B1 and E are key to understanding *E. coli* pathogens of current intense interest but from different perspectives (Fig. 2.4). Phylogroup B1 reveals significantly broader diversity than phylogroup E in both non-pathogen content, as well as notable pathogens. Hence, it contains enterohaemorrhagic *E. coli* (EHEC) serogroups of concern such as O26:H11, O111:H8 and H11, O45:H2, O103:H2 and H25, and O121:H19 as well as enteroaggregative STEC O104:H4. On the contrary, phylogroup E contains O145:H28, *Shigella dysenteriae* type 1, and O157:H7 – the most notable and common pathogen associated with haemorrhagic colitis and haemolytic uraemic syndrome. With the latter, it is significantly rich in representation with a deep-rooted bias given its clinical significance and ease of culture-based identification with non-sorbitol-fermenting and β-glucuronidase-negative phenotypes. In a quantitative representation and comparative sense, the framework contains approximately 2000 and 500 genomes representing the known diversity in B1 and E, respectively, whereas depth in O157:H7 alone is on the order of 400 genomes.

The strength of comparative core genome phylogenies is useful for clonal and evolutionary linkages. The power of such a framework can be augmented with overlays of useful accessory information. Here we demonstrate, as an example, that phylogroup B1 displays a general richness in circulating virulence genes amongst the various lineages not seen to this extent in other phylogroups (Fig. 2.4). However, examples for the various virulence biomarkers and pathotypes can be found in every phylogroup, including the cryptic lineages. Considering the tree is pruned to relative genome types, it should be noted that such serial overlays do not directly indicate the existence of a respective strain containing all such factors but, rather, that the factors exist in a 'circulating pool' of close genetically related strains. Therefore, it can be used as a predictive strategy to relate a history of disease and genome types ripe for pathogen development as has been witnessed recently with certain hybrid pathotypes (Leonard *et al.*, 2016; Rasko *et al.*, 2011). Regardless, this type of analysis can be done for any number of features of interest, including assessing serotype concordance, antibiotic resistance, plasmid content, or various metadata, including geographical and source attributions, etc.

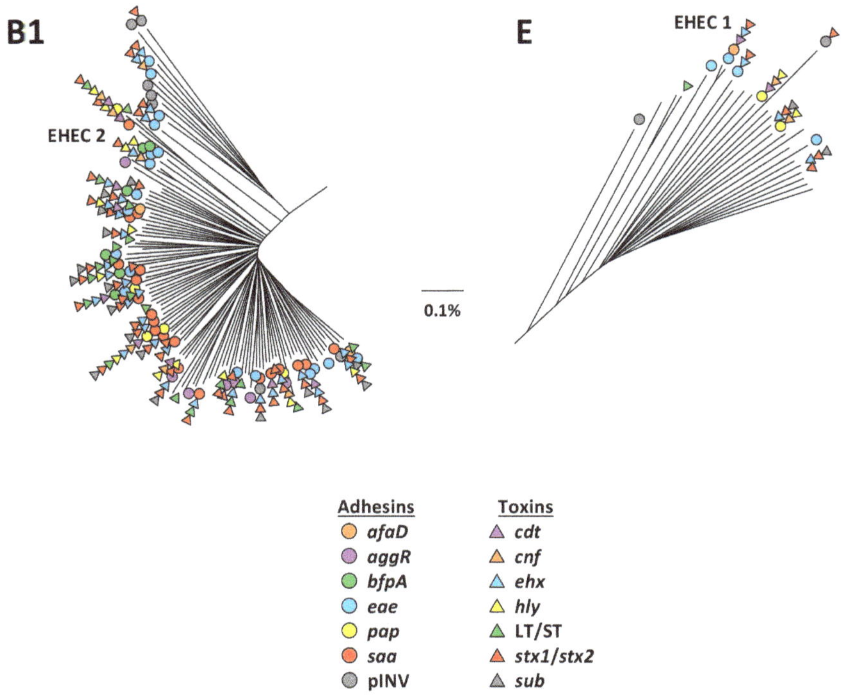

Figure 2.4 Examples of intraphylogroup diversities derived from core gene analysis highlighting phylogroup B1 and phylogroup E with selected overlays of key virulence factors associated with various pathotypes. While technically not an adhesin, the enteroinvasive *E. coli/Shigella*-associated invasion plasmid (pINV) was included to balance the two sets of virulence factors. Presence of pINV was determined by the presence of any combination of the *icsP*, *mkaD*, and *ospB* loci. Distance scale bar relative to 2.4 Mbp of concatenated core backbone genes. A total of 110 phylogroup B1 genotypes representing 1818 genomes (includes O26/O111 EHEC 2 pair) and 27 phylogroup E genotypes representing 503 genomes (includes O55/O157 EHEC 1 pair) are shown. Neighbour joining trees constructed with MEGA.

Refinement of O157:H7 cladistics

Since the first outbreak of O157:H7 in 1982 (Riley *et al.*, 1983), this serotype has remained the predominant EHEC pathogen in North America, as well as other certain geographical locations worldwide. Over the years, there have been intense research efforts to characterize this pathogen and the genetic basis for its emergence (Sadiq *et al.*, 2014). Initially, it was considered a rather rare serotype and, similar to studies discussed earlier with natural populations of *E. coli*, phylogenetic analysis of various O157 strains with MLEE revealed a clonal nature to the pathogen (Whittam and Wilson, 1988; Whittam *et al.*, 1988). Molecular genetic analysis over a number of years along with increasing availability of O157:H7 strains elucidated key molecular characteristics of its evolutionary emergence (Feng *et al.*, 1998). A stepwise model accounted for acquisition and loss of serotyping markers from an O55:H7 ancestral strain, acquisition of Shiga toxin phages and the EHEC virulence plasmid as well as loss of the sorbitol and β-glucuronidase phenotypes. Further genetic underpinnings using enhanced molecular subtyping such as MLST and pulsed-field gel electrophoresis identified key intermediate archetype strains and clonal complexes in support of the model (Feng *et al.*, 2007). More refined analysis using genome sequencing-based microarrays identified key SNPs for resolving the resulting nine clades (Manning *et al.*, 2008; Lacher, 2011). The impetus for such an analysis was related to potential differences in apparent virulence of O157:H7 clade members. A notable example was observed with the 2006 spinach-associated outbreak, which resulted in significant levels of hospitalization with the contaminating strain belonging to clade 8 (Manning *et al.*, 2008; Eppinger *et al.*, 2011).

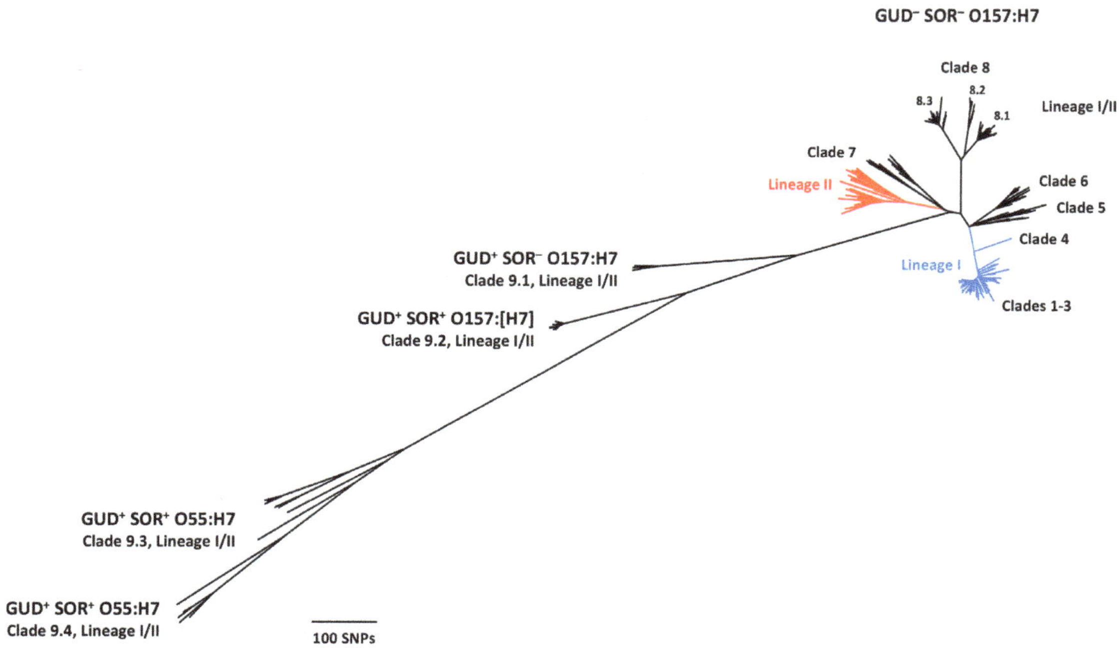

Figure 2.5 Core genome SNP phylogeny of O55:H7 and O157:H7 strains with previously described stepwise clade and lineage models overlaid. Lineage I strains are shown in blue, while lineage II strains are in red. All remaining strains are classified as lineage I/II (black). Diagnostic phenotypes for β-glucuronidase activity (GUD) and sorbitol fermentation ability (SOR) were inferred from *uidA* and *srlE* genotypes, respectively. Neighbour joining tree constructed with MEGA.

Another view of evolutionary clonal substructure was proposed in parallel to the clade structure describing O157:H7 lineages. As first reported, natural strand-based DNA octamer repeats across the genome served as the basis for classifying O157:H7 genomes and revealed a non-random association with strains of human and bovine origin (Kim *et al.*, 1999). Similar in rationale to clade virulence, two lineages (I and II) would parse with some apparent association with virulence potential but with less discriminatory power. Methodological improvements were made to speed analysis and expand testing of this approach with development of the lineage-specific polymorphism genotyping assay based on six genetic markers, known as the LSPA-6 assay (Yang *et al.*, 2004). Subsequently, with higher resolution studies involving comparative genomic hybridizations of strains using microarrays, this concept was expanded to include a lineage I/II group (Zhang *et al.*, 2007). The two classification schemes for O157:H7, clades and lineages, offer different levels of resolution yet were approached from significantly different perspectives and intended aims from their inception. Clades present a perspective derived from an evolutionary point of view; lineages were derived from linkages to a pathogen reservoir, i.e. bovine versus human.

A core genome phylogeny based on hundreds of thousands of SNP markers can be magnified to the very tips of the species landscape for this pathogen with replete specificity for O157:H7 (Fig. 2.5). As with the phylogroup designations, core genome cladistics supported much of the initial efforts to model O157:H7 using lower resolution and fragmented molecular and phenotypic methods. Thus, significant concordance is observed in strain cladistics with clades described as part of the stepwise model, which are monophyletic (Fig. 2.5). Further refinements can be made with this level of genome data revealing subclustering within clades 5, 7 and 8 (Fig. 2.5). Indeed, three clade 8 subgroups, denoted as clades 8.1, 8.2 and 8.3 become clearly distinct. When overlaid with the lineage concept, it is apparent that the lineages represent a higher order classification scheme with lineage I containing strains belonging to clades 1, 2, 3 and 4, while lineage II contains a subset of clade 7 strains (Fig. 2.5). The lineage scheme, however, can be confounding given the evident para- and polyphyletic nature that is due to the apparent (over)reliance on the *folD-sfmA* intergenic marker for strain disposition (Yang *et al.*, 2004; Zhang *et al.*, 2007). As a result, lineage I/II contains strains belonging to clades 5, 6, 7 (non-lineage II) and 8, as well as strains with the ancestral wild-type sorbitol and β-glucuronidase phenotypes classified as clade 9 (Fig. 2.5).

It will remain confounding to associate virulence with the genetic substructures described above for O157:H7. In part, this is due to the interplay of the human host and properties of different O157:H7 strains. Hence, the propensity to develop haemorrhagic colitis versus haemolytic uraemic syndrome cannot be treated independently as a property of O157:H7. In the context of food safety, the relative pathogenicity of an environmental or food-derived strain is an estimate based on a molecular risk assessment. The strength of core genome analysis will enable a better path forward beyond serotyping for determining a history of illness and disease with predictive potential (Lupolova *et al.*, 2016).

Genome architecture and plasticity framework

A whole-genome sequence-based phylogeny of *E. coli* gives a broad evolutionary view of how the species has changed and diverged, one view of its plasticity, but whole-genome sequence is more often than not a misnomer. *E. coli* genomes are represented in a number of different forms in GenBank, with a minority assembled into a complete circular chromosome at a high level of sequence coverage. The majority of *E. coli* genomes are sets of contigs ranging from under a dozen to often several hundred. The contig sets are notable for gaps at repetitive elements, including transposons and prophages, which are problematic in assemblies of the short 100- to 500-bp sequence reads that are produced by most next-generation sequencing platforms. The use of single-molecule real-time sequencing is increasing the number of complete circular *E. coli* genomes (Koren and Phillippy, 2015), but most are still presented as contig sets with numerous gaps. Using bioinformatic tools, these contig collections are effectively studied and arranged into phylogenetic groupings with detailed SNP comparisons and accurate calculations of genome differences. In complete genomes, however, the gaps become resolved into highly polymorphic sites and a different perspective emerges, one in which insertions, deletions, and inversions dominate. The comparative alignments of complete chromosomes along their entire lengths clearly demonstrate a wholly different view of *E. coli* genomic plasticity (Fig. 2.6), one that is separate from the long-term species-wide phylogeny and the high resolution of strain differences obtained by whole-genome sequences and SNP variation.

Genome architecture and plasticity in phylogroup E

At the start of the boom in *E. coli* comparative genomics, as no surprise, the first two genomes sequenced after the release of K-12 MG1655 were the archetype strains from the two widely publicized

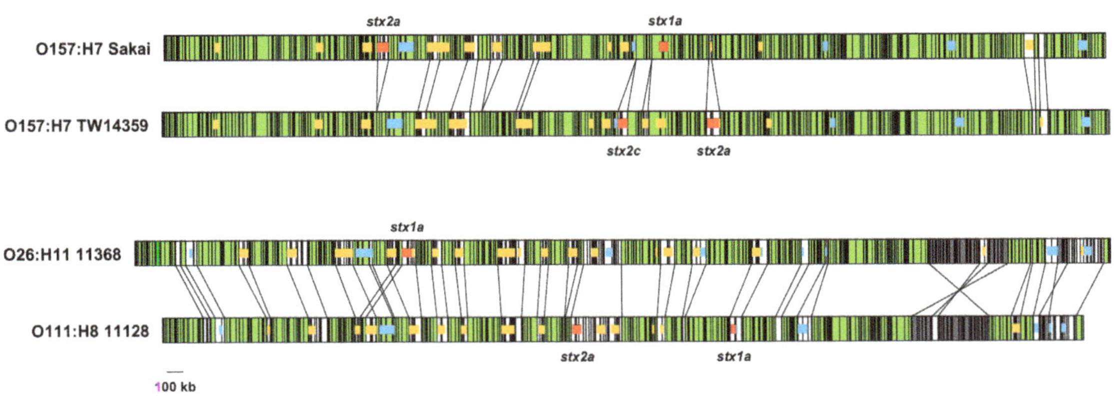

Figure 2.6 Alignment of linearized *in silico Bam*HI restriction maps of EHEC reference strains. Vertical black lines within each map represent *Bam*HI cut sites. Aligned fragments are shown in green, while unaligned ones are white. Contiguous aligned fragments are indicated by the lines connecting the maps. Chromosomal inversions are indicated by the grey fragments. Prophages and integrative elements are shown by the orange and blue boxes, respectively, with the Shiga toxin-encoding prophages in red.

O157:H7 food-borne outbreaks in 1982 and 1996, namely strains EDL933 (Perna *et al.*, 2001) and Sakai (Hayashi *et al.*, 2001). As mentioned previously, O157:H7 strains fall into phylogroup E, and the two most prominent features of these pathogen genomes were (i) the increase in genome size by almost a megabase compared to K-12 MG1665 and (ii) the high number of insertion and deletion polymorphisms (indels) notable in pairwise comparisons. The terms 'mosaic' and 'peppered' have become common to describe the plasticity of these and other pathogenic *E. coli* genomes including the subsequent sequencing of uropathogenic *E. coli* strain CFT073 (Welch *et al.*, 2002). Early pairwise comparisons of genomes have given way to more comprehensive tabulations of variant indel sites. For example, there are more than 50 common prophage insertion sites (Steyert *et al.*, 2012) with interesting deleted or cryptic variant prophages commonly found in a genome lacking generally complete functional prophages. In comparing the first O157:H7 sequence of EDL933 to the K-12 MG1655 sequence, Perna *et al.* (2001) determined 333 O157 specific and four K-12 specific islands, referred to as O and K islands, respectively. With the sequencing of another O157:H7 isolate, Sakai, and subsequent *E. coli* pathotypes, it became clear that most O and K islands were not O157 and K-12 specific, but rather chromosomal sites of variation, i.e. indel hot spots. In tallying indels in the Sakai outbreak strain, differences between the two O157 isolates were equally striking, including a 429 kbp inversion found in EDL933 between cryptic prophages O and P (Kotewicz *et al.*, 2008). In later studies, the classification of such events would depend more on the specific genes in the islands such as tellurite resistance, antibiotic resistances, or Shiga toxin containing prophages than classification as O or K islands.

Within the O157:H7 group, the linearized *in silico* restriction maps of clade 1 strain Sakai and clade 8 strain TW14359 from the 2006 spinach-associated outbreak show nearly a dozen indel differences (Fig. 2.6). In the first third of the aligned O157:H7 chromosomes, Sakai contains a Shiga toxin 2a prophage inserted at *wrbA* whereas TW14359 lacks a prophage at that locus. To the right of the centres of the two aligned chromosomes, TW14359 contains two Shiga toxin-containing prophages where no prophage insertions are found in Sakai. The first prophage, containing the Shiga toxin 2c allele, is inserted into the exonuclease *sbcB*, and the other, carrying the Shiga toxin 2a allele, is inserted into the *argW* tRNA locus. Between these two positions on the chromosome, Sakai contains a prophage inserted at *yehV* carrying the Shiga toxin 1a allele. Interestingly, although TW14359 contains a prophage at this locus, it does not carry Shiga toxin genes.

In addition to this sequence-based comparison of complete genomes, a similar broad view of *E. coli* genome plasticity has been demonstrated using optical maps. Although of lower resolution, the majority of indels seen with whole-genome sequences have been resolved and characterized in pairwise or multiple alignments of the optical maps of different strains (Kotewicz *et al.*, 2007). In addition to indels, the other notable evidence of genome plasticity observed in these optical maps was the high number of chromosomal inversions, mainly at homologous prophages, but also in some cases involving other repetitive elements. Wide variations in prophages and notably the Shiga toxin-containing phages identified in genome sequences were graphically illustrated by the optical map comparisons (Kotewicz *et al.*, 2008). Subsequent whole-genome optical maps demonstrated that nearly one-third of O157:H7 strains analysed showed an inversion, and of those strains, one-third contained a second inversion. Although sequencing has far surpassed optical mapping, the resolution of inversions is not easily derived from short read sequence data. However, with appropriate reference alignments of carefully sequenced strains, chromosome structures including indels can be easily resolved and verified with *in silico* restriction maps. Such visualizations of complete genomes provide an overview of positions of prophages, pathogenicity islands, and their variations, as well as the clear demarcation of chromosomal inversions.

Genome architecture and plasticity in phylogroup B1

In the EHEC pathotype classification scheme, strains with the O157:H7 serotype belong to the EHEC 1 clonal group, which is found in phylogroup E. Another group of strains capable of causing a clinically indistinguishable disease, known as the EHEC 2 clonal group, is found in phylogroup B1. The EHEC 2 clonal group is more diverse serologically than EHEC 1 with the group containing multiple serogroups including O26 and O111 which are part of the 'Big Six' serogroups that have been linked to food-borne illnesses and are regulated by the US Department of Agriculture. The genome plasticity present within phylogroup B1 can be illustrated by the alignment of the *in silico* restriction maps of the EHEC 2 reference genome strains 11368 (O26:H11) and 11128 (O111:H8) (Fig. 2.6). In addition to over 20 variant prophage and pathogenicity island indels, the map alignments revealed two inversions in the terminal region of the chromosomes, one that is approximately 500 kbp in length and a smaller one of about 70 kbp. In a comprehensive examination of EHEC 2 strains, optical maps were generated for ten O26:H11, ten O111:H8, and six O118:H16 strains and compared to the reference *in silico* maps shown in Fig. 2.6. The optical maps of the O26:H11 and O118:H16 strains possessed, on average, larger chromosomes than the O111:H8 strains (5.7 vs. 5.4 Mbp). Comparative analyses of the 28 strains identified 265 markers within regions known to contain prophages and other integrative elements for strain identification and subtyping purposes. Of these, 239 were indels, 22 were inversions, and four were suggestive of double crossover events. As a group, the O26:H11 strains were the most variable, while lower levels of within serotype variation were observed for both the O111:H8 and O118:H16 strains. Cluster analysis of the map polymorphisms resolved each of the serotypes with the O26:H11 and O118:H16 strains more similar to each other than either was to O111:H8. Thus, we conclude that indel and rearrangement events, dominated by prophages, have played a large role in shaping the chromosomal architecture of the EHEC 2 clonal group during its evolution and diversification. Comparative alignments give a clear picture of the variation of certain critical virulence factors including prophages and pathogenicity islands within the chromosome. As mentioned previously, these snapshots of the plasticity of the complete *E. coli* chromosome and its structural variation are quite different from the long-term phylogenetic view of the species and the high resolution view of strain differences obtained by whole-genome sequences and SNP variation.

Gap analysis and landscape bias: future directions

Given the bioinformatic power of public sequence databasing and the current status of *E. coli* genomics outlined above, crowdsourcing of data becomes as important, if not more so, than *de novo* acquisition attributable to individual research efforts. The genomic landscape will need better representation of environmental or commensal strains to balance the clinical/disease bias with currently available genomes. As highlighted in this chapter, the mere existence and genetic relationships of the cryptic lineages calls into question the genomic disposition of species definitions and more importantly may revolutionize bacterial systematics and taxonomy. For example, while not the focus of this chapter, multiple events giving rise to the shigellae are interwoven across the landscape in several phylogroups. Hence, with more representation stratified across phylogroups and increased intraphylogroup depth as has been accomplished with O157:H7, will the landscape represent a continuum, blurring lines between other closely recognized species? Will this continuum assume more of a quasispecies concept that exists for viruses and will quantitative concepts developed with *E. coli* principally hold for other bacterial genera?

Along these lines, whole-genome sequences can also be mined for genetic underpinnings of the various serotypes. Currently, the molecular genetic markers for all 53 known H-types have been determined along with more than 260 named and unnamed O-types (inclusive of the shigellae) which exceed the available traditional antibody-based serotyping panels. Further, bioinformatic resources for rapid determinations of serotype, virulence, antibiotic resistance, etc. from raw short-read or partially assembled sequencing data are being made available (i.e. https://cge.cbs.dtu.dk/services/). In some cases, serotypes fail to align with the evolutionary relationships revealed by the core genome phylogeny and call into question the reliance on an antigenic manifestation for primary strain identification. It is important here to suggest a unified approach to naming and cataloging strains in a genomic era. One such suggestion might involve establishment of a nomenclature strategy (such has been done with the Enzyme Commission) and based on landscape core 'genome types' that would unify a myriad of available molecular diagnostic schemes.

The approaches described can be applied to other pathogens. However it is becoming clear that strain-level epidemiological association based on whole-genome sequence comparisons may need to be empirically and quantitatively determined for each species. Hence, how many SNPs apart would be required at a whole genome scale for a robust determination for outbreak inclusivity/exclusivity for *E. coli*? Regardless, the overall architecture of the SNP-defined phylogenetic landscape for *E. coli* is generally stable with incorporation of new genome accessions. Moreover, unique genomic signatures can be identified that stratify the desired taxonomic resolution, which may be important in metagenomic pursuits. Specifically for STEC, additional challenges still exist to qualitatively mine and identify virulence content indicative for potential disease severity such as haemolytic uraemic syndrome. This aspect remains unclear at this point and is unlikely that any simple universal marker for pathogenicity will be found and exploited for public health purposes; composite analysis remains a current requirement.

References

Achtman, M., Mercer, A., Kusecek, B., Pohl, A., Heuzenroeder, M., Aaronson, W., Sutton, A., and Silver, R.P. (1983). Six widespread bacterial clones among *Escherichia coli* K1 isolates. Infect. Immun. *39*, 315–335.

Blattner, F.R., Plunkett, G. 3rd, Bloch, C.A., Perna, N.T., Burland, V., Riley, M., Collado-Vides, J., Glasner, J.D., Rode, C.K., Mayhew, G.F., *et al.* (1997). The complete genome sequence of *Escherichia coli* K-12. Science *277*, 1453–1462.

Chaudhuri, R.R., and Henderson, I.R. (2012). The evolution of the *Escherichia coli* phylogeny. Infect. Genet. Evol. *12*, 214–226. https://doi.org/10.1016/j.meegid.2012.01.005

Clermont, O., Bonacorsi, S., and Bingen, E. (2000). Rapid and simple determination of the *Escherichia coli* phylogenetic group. Appl. Environ. Microbiol. *66*, 4555-4558.

Clermont, O., Christenson, J.K., Denamur, E., and Gordon, D.M. (2013). The Clermont *Escherichia coli* phylo-typing method revisited: improvement of specificity and detection of new phylo-groups. Environ. Microbiol. Rep. *5*, 58–65.

Clermont, O., Gordon, D., and Denamur, E. (2015). Guide to the various phylogenetic classification schemes for *Escherichia coli* and the correspondence among schemes. Microbiology *161*, 980–988.

Denamur, E. (2011). The 2011 Shiga toxin-producing *Escherichia coli* O104:H4 German outbreak: a lesson in genomic plasticity. Clin. Microbiol. Infect. *17*, 1124–1125. https://doi.org/10.1111/j.1469-0691.2011.03620.x

Elkins, C.A., Kotewicz, M.L., Jackson, S.A., Lacher, D.W., Abu-Ali, G.S., and Patel, I.R. (2013). Genomic paradigms for food-borne enteric pathogen analysis at the USFDA: case studies highlighting method utility, integration and resolution. Food Addit. Contam. Part A Chem. Anal. Control Expo. Risk Assess. *30*, 1422–1436.

Eppinger, M., Mammel, M.K., Leclerc, J.E., Ravel, J., and Cebula, T.A. (2011). Genomic anatomy of *Escherichia coli* O157:H7 outbreaks. Proc. Natl. Acad. Sci. U.S.A. *108*, 20142–20147. https://doi.org/10.1073/pnas.1107176108

Feng, P., Lampel, K.A., Karch, H., and Whittam, T.S. (1998). Genotypic and phenotypic changes in the emergence of *Escherichia coli* O157:H7. J. Infect. Dis. *177*, 1750–1753.

Feng, P.C., Monday, S.R., Lacher, D.W., Allison, L., Siitonen, A., Keys, C., Eklund, M., Nagano, H., Karch, H., Keen, J., and Whittam, T.S. (2007). Genetic diversity among clonal lineages within *Escherichia coli* O157:H7 stepwise evolutionary model. Emerg. Infect. Dis. *13*, 1701–1706.

Hayashi, T., Makino, K., Ohnishi, M., Kurokawa, K., Ishii, K., Yokoyama, K., Han, C.G., Ohtsubo, E., Nakayama, K., Murata, T., Tanaka, M., Tobe, T., Iida, T., Takami, H., Honda, T., Sasakawa, C., Ogasawara, N., Yasunaga, T., Kuhara, S., Shiba, T., Hattori, M., and Shinagawa, H. (2001). Complete genome sequence of enterohemorrhagic *Escherichia coli* O157:H7 and genomic comparison with a laboratory strain K-12. DNA Res. *8*, 11–22.

Herzer, P.J., Inouye, S., Inouye, M., and Whittam. T.S. (1990). Phylogenetic distribution of branched RNA-linked multicopy single-stranded DNA among natural isolates of *Escherichia coli*. J. Bacteriol. *172*, 6175–6181.

Jackson, S.A., Kotewicz, M.L., Patel, I.R., Lacher, D.W., Gangiredla, J., and Elkins, C.A. (2012). Rapid genomic-scale analysis of *Escherichia coli* O104:H4 by using high-resolution alternative methods to next-generation sequencing. Appl. Environ. Microbiol. *78*, 1601–1605.

Jackson, S.A., Patel, I.R., Barnaba, T., LeClerc, J.E., and Cebula, T.A. (2011). Investigating the global genomic diversity of *Escherichia coli* using a multi-genome DNA microarray platform with novel gene prediction strategies. BMC Genomics *12*, 349. https://doi.org/10.1186/1471-2164-12-349

Jaureguy, F., Landraud, L., Passet, V., Diancourt, L., Frapy, E., Guigon, G., Carbonnelle, E., Lortholary, O., Clermont, O., Denamur, E., *et al.* (2008). Phylogenetic and genomic diversity of human bacteremic *Escherichia coli* strains. BMC Genomics *9*, 560. https://doi.org/10.1186/1471-2164-9-560

Kim, J., Nietfeldt, J., and Benson, A.K. (1999). Octamer-based genome scanning distinguishes a unique subpopulation of *Escherichia coli* O157:H7 strains in cattle. Proc. Natl. Acad. Sci. U.S.A. *96*, 13288–13293.

Koren, S., and Phillippy, A.M. (2015). One chromosome, one contig: complete microbial genomes from long-read sequencing and assembly. Curr. Opin. Microbiol. *23*, 110–120. https://doi.org/10.1016/j.mib.2014.11.014

Kotewicz, M.L., Jackson, S.A., LeClerc, J.E., and Cebula, T.A. (2007). Optical maps distinguish individual strains of *Escherichia coli* O157: H7. Microbiology *53*, 1720–1733.

Kotewicz, M.L., Mammel, M.K., LeClerc, J.E., and Cebula, T.A. (2008). Optical mapping and 454 sequencing of *Escherichia coli* O157:H7 isolates linked to the US 2006 spinach-associated outbreak. Microbiology *154*, 3518–3528. doi: 10.1099/mic.0.2008/019026-0.

Kupferschmidt, K. (2011). Germany. Scientists rush to study genome of lethal *E. coli*. Science 332, 1249–1250. https://doi.org/10.1126/science.332.6035.1249

Lacher, D.W., Steinsland, H., Blank, T.E., Donnenberg, M.S., and Whittam, T.S. (2007). Molecular evolution of typical enteropathogenic *Escherichia coli*: clonal analysis by multilocus sequence typing and virulence gene allelic profiling. J. Bacteriol. *189*, 342–350.

Lacher, D.W. (2011). The evolutionary model of *Escherichia coli* O157:H7. In Population Genetics of Bacteria: A Tribute to Thomas S. Whittam, S.T. Walk and P.C.H. Feng, ed. (ASM Press, Washington, DC), pp. 227–239.

Land, M., Hauser, L., Jun, S.R., Nookaew, I., Leuze, M.R., Ahn, T.H., Karpinets, T., Lund, O., Kora, G., Wassenaar, T., *et al.* (2015). Insights from 20 years of bacterial genome sequencing. Funct. Integr. Genomics *15*, 141–161. https://doi.org/10.1007/s10142-015-0433-4

Leonard, S.R., Mammel, M.K., Rasko, D.A., and Lacher, D.W. (2016). Hybrid Shiga toxin-producing and enterotoxigenic *Escherichia* sp. cryptic lineage 1 strain 7v harbors a hybrid plasmid. Appl. Environ. Microbiol. *82*, 4309--4319.

Luo, C., Walk, S.T., Gordon, D.M., Feldgarden, M., Tiedje, J.M., and Konstantinidis, K.T. (2011). Genome sequencing of environmental *Escherichia coli* expands understanding of the ecology and speciation of the model bacterial species. Proc. Natl. Acad. Sci. U.S.A. *108*, 7200-7205.

Lupolova, N., Dallman, T.J., Matthews, L., Bono, J.L., and Gally, D.L. (2016). Support vector machine applied to predict the zoonotic potential of *E. coli* O157 cattle isolates. Proc. Natl. Acad. Sci. U.S.A. *113*, 11312–11317.

Manges, A.R., and Johnson, J.R. (2015). Reservoirs of extraintestinal pathogenic *Escherichia coli*. Microbiol. Spectr. 3(5). https://doi.org/10.1128/microbiolspec.UTI-0006-2012

Manning, S.D., Motiwala, A.S., Springman, A.C., Qi, W., Lacher, D.W., Ouellette, L.M., Mladonicky, J.M., Somsel, P., Rudrik, J.T., Dietrich, S.E., *et al.* (2008). Variation in virulence among clades of *Escherichia coli* O157:H7 associated with disease outbreaks. Proc. Natl. Acad. Sci. U.S.A. *105*, 4868–4873. https://doi.org/10.1073/pnas.0710834105

Milkman, R. (1973). Electrophoretic variation in *Escherichia coli* from natural sources. Science *182*, 1024–1026.

Milkman, R. (1975). Allozyme variation in *E. coli* of diverse natural origins. In Isozymes V4: Genetics and Evolution, C.L. Markert, ed. (Academic Press, New York), pp. 273–285.

Ochman, H., and Selander, R.K. (1984a). Evidence for clonal population structure in *Escherichia coli*. Proc. Natl. Acad. Sci. U.S.A. *81*, 198–201.

Ochman, H., and Selander, R.K. (1984b). Standard reference strains of *Escherichia coli* from natural populations. J. Bacteriol. *157*, 690–693.

Perna, N.T., Plunkett, G. 3rd, Burland, V., Mau, B., Glasner, J.D., Rose, D.J., Mayhew, G.F., Evans, P.S., Gregor, J., Kirkpatrick, H.A., Pósfai, G., Hackett, J., Klink, S., Boutin, A., Shao, Y., Miller, L., Grotbeck, E.J., Davis, N.W., Lim, A., Dimalanta, E.T., Potamousis, K.D., Apodaca, J., Anantharaman, T.S., Lin, J., Yen, G., Schwartz, D.C., Welch, R.A., and Blattner, F.R. (2001). Genome sequence of enterohaemorrhagic *Escherichia coli* O157:H7. Nature. *409*, 529–533.

Pupo, G.M., Karaolis, D.K., Lan, R., and Reeves, P.R. (1997). Evolutionary relationships among pathogenic and nonpathogenic *Escherichia coli* strains inferred from multilocus enzyme electrophoresis and *mdh* sequence studies. Infect. Immun. *65*, 2685–2692.

Rasko, D.A., Rosovitz, M.J., Myers, G.S., Mongodin, E.F., Fricke, W.F., Gajer, P., Crabtree, J., Sebaihia, M., Thomson, N.R., Chaudhuri, R., *et al.* (2008). The pangenome structure of *Escherichia coli*: comparative genomic analysis of *E. coli* commensal and pathogenic isolates. J. Bacteriol. *190*, 6881–6893. https://doi.org/10.1128/JB.00619-08

Rasko, D.A., Webster, D.R., Sahl, J.W., Bashir, A., Boisen, N., Scheutz, F., Paxinos, E.E., Sebra, R., Chin, C.S., Iliopoulos, D., *et al.* (2011). Origins of the *E. coli* strain causing an outbreak of hemolytic-uremic syndrome in Germany. N. Engl. J. Med. *365*, 709–717. https://doi.org/10.1056/NEJMoa1106920

Riley, L.W., Remis, R.S., Helgerson, S.D., McGee, H.B., Wells, J.G., Davis, B.R., Hebert, R.J., Olcott, E.S., Johnson, L.M., Hargrett, N.T., *et al.* (1983). Hemorrhagic colitis associated with a rare *Escherichia coli* serotype. N. Engl. J. Med. *308*, 681–685. https://doi.org/10.1056/NEJM198303243081203

Reid, S.D., Herbelin, C.J., Bumbaugh, A.C., Selander, R.K., and Whittam, T.S. (2000). Parallel evolution of virulence in pathogenic *Escherichia coli*. Nature *406*, 64–67. https://doi.org/10.1038/35017546

Sadiq, S.M., Hazen, T.H., Rasko, D.A., and Eppinger, M. (2014). EHEC Genomics: Past, Present, and Future. Microbiol Spectr 2, EHEC–0020–2013. https://doi.org/10.1128/microbiolspec.EHEC-0020-2013

Selander, R.K., and Levin, B.R. (1980). Genetic diversity and structure in *Escherichia coli* populations. Science *210*, 545–547.

Steyert, S.R., Sahl, J.W., Fraser, C.M., Teel, L.D., Scheutz, F., and Rasko, D.A. (2012). Comparative genomics and *stx* phage characterization of LEE-negative Shiga toxin-producing *Escherichia coli*. Front. Cell Infect. Microbiol. 2, 133. doi: 10.3389/fcimb.2012.00133.

Tenaillon, O., Skurnik, D., Picard, B., and Denamur, E. (2010). The population genetics of commensal *Escherichia coli*. Nat. Rev. Microbiol. *8*, 207–217. https://doi.org/10.1038/nrmicro2298

Vila, J., Sáez-López, E., Johnson, J.R., Römling, U., Dobrindt, U., Cantón, R., Giske, C.G., Naas, T., Carattoli, A., Martínez-Medina, M., *et al.* (2016). *Escherichia coli*: an old friend with new tidings. FEMS Microbiol. Rev. *40*, 437–463. https://doi.org/10.1093/femsre/fuw005

Walk, S.T., Alm, E.W., Gordon, D.M., Ram, J.L., Toranzos, G.A., Tiedje, J.M., and Whittam, T.S. (2009). Cryptic lineages of the genus *Escherichia*. Appl. Environ. Microbiol. *75*, 6534–6544. https://doi.org/10.1128/AEM.01262-09

Welch, R.A., Burland, V., Plunkett, G. 3rd, Redford, P., Roesch, P., Rasko, D., Buckles, E.L., Liou, S.R., Boutin, A., Hackett, J., Stroud, D., Mayhew, G.F., Rose, D.J.,

Zhou, S., Schwartz, D.C., Perna, N.T., Mobley, H.L., Donnenberg, M.S., and Blattner, F.R. (2002). Extensive mosaic structure revealed by the complete genome sequence of uropathogenic *Escherichia coli*. Proc. Natl. Acad. Sci. U.S.A. *24*, 17020-17024.

Whittam, T.S., Ochman, H., and Selander, R.K. (1983). Multilocus genetic structure in natural populations of *Escherichia coli*. Proc. Natl. Acad. Sci. U.S.A. *80*, 1751–1755.

Whittam, T.S., Wachsmuth, I.K., and Wilson, R.A. (1988). Genetic evidence of clonal descent of *Escherichia coli* O157:H7 associated with hemorrhagic colitis and hemolytic uremic syndrome. J. Infect. Dis. *157*, 1124–1133.

Whittam, T.S., and Wilson, R.A. (1988). Genetic relationships among pathogenic *Escherichia coli* of serogroup O157. Infect. Immun. *56*, 2467-2473.

Wirth, T., Falush, D., Lan, R., Colles, F., Mensa, P., Wieler, L.H., Karch, H., Reeves, P.R., Maiden, M.C., Ochman, H., *et al.* (2006). Sex and virulence in *Escherichia coli*: an evolutionary perspective. Mol. Microbiol. *60*, 1136–1151.

Yang, Z., Kovar, J., Kim, J., Nietfeldt, J., Smith, D.R., Moxley, R.A., Olson, M.E., Fey, P.D., and Benson, A.K. (2004). Identification of common subpopulations of non-sorbitol-fermenting, β-glucuronidase-negative *Escherichia coli* O157:H7 from bovine production environments and human clinical samples. Appl. Environ. Microbiol. *70*, 6846-6854.

Zhang, Y., Laing, C., Steele, M., Ziebell, K., Johnson, R., Benson, A.K., Taboada, E., and Gannon, V.P. (2007). Genome evolution in major *Escherichia coli* O157:H7 lineages. BMC Genomics. *8*, 121.

Zhao, Y., Jia, X., Yang, J., Ling, Y., Zhang, Z., Yu, J., Wu, J., and Xiao, J. (2014). PanGP: a tool for quickly analyzing bacterial pan-genome profile. Bioinformatics *30*, 1297–1299. https://doi.org/10.1093/bioinformatics/btu017

Diarrhoeagenic *Escherichia coli*: Virulence Genes and Other Markers for Detection and Typing

3

Stefano Morabito* and Rosangela Tozzoli

European Union Reference Laboratory for *Escherichia coli*, Istituto Superiore di Sanità, Rome, Italy.

*Correspondence: stefano.morabito@iss.it

https://doi.org/10.21775/9781910190777.03

Abstract

Escherichia coli is a usually commensal bacterium that harmlessly colonizes the gastrointestinal tract of mammals, where it exert san advantageous effect on the host. Nonetheless, some strains have evolved the ability to induce disease in humans in many anatomical sites. Among these, diarrhoeagenic *E. coli* (DEC) causes enteric/diarrhoeal illness, sometimes with systemic complications. DEC is subdivided into pathotypes on the basis of the colonization mechanism and the toxins elaborated, which determine the clinical, pathological, and epidemiological features of the disease induced. DEC pathotypes include enterotoxigenic (ETEC), enteroinvasive (EIEC), enteropathogenic (EPEC), enteroaggregative (EAEC), and Shiga-toxin-producing *E. coli* (STEC). In order to differentiate pathogenic *E. coli* strains from commensal *E. coli*, pathotype-specific genetic markers are generally employed for the diagnosis of the infections and strain characterization. This chapter describes the different DEC pathotypes, in order to provide a general picture of their virulence mechanisms and the associated genetic determinants, with the aim of illustrating the molecular targets used for their identification and typing.

Introduction

Diarrhoeagenic *Escherichia coli* (DEC) are pathogenic *E. coli* that have evolved the ability to induce enteric/diarrhoeal disease in humans (Nataro and Kaper, 1998). The groups of *E. coli* that cause gastroenteritis are classified into named pathotypes, including enterotoxigenic (ETEC), enteroinvasive (EIEC), enteropathogenic (EPEC), enteroaggregative (EAEC), and Shiga-toxin-producing *E. coli* (STEC). Individual strains of each pathotype possess a distinct set of virulence-associated characteristics, comprising either colonization factors or toxins, which determine the clinical, pathological, and epidemiological features of the diseases they cause. STEC is the only group with a recognized zoonotic origin, with ruminants being recognized as their main animal reservoir. As far as the other *E. coli* pathotypes are concerned, they are usually transmitted person-to-person through a typical faecal–oral route, but the consumption of food and water secondarily contaminated by infected individuals also plays an important role in spreading the infections. Therefore, the identification of diarrhoeagenic *E. coli* is not only confined to the diagnosis of human infections and the characterization of isolated strains, but also involves analysis aimed at identifying these pathogens in the vehicles of infection.

Since pathogenic *E. coli* are not discernible from commensal ubiquitous *E. coli* based on their phenotype, usually genetic markers are used for both detection and typing of such strains. In principle, the detection of the presence of virulence genes as diagnostic markers characterizing the strains belonging to the different pathotypes is exploited to determine the presence of specific pathogenic

E. coli in the sample or clinical specimen. This chapter will describe the different DEC pathotype characteristics, in order to provide a general picture of their virulence mechanisms and the associated genetic determinants, and it will illustrate the genetic targets used for the identification and typing of DEC. Moreover, particular focus will be dedicated to characterization of STEC strains since this DEC pathotype is of major public health concern and comprise food-borne pathogens often implicated in large community outbreaks of severe disease. Given the importance of STEC, many efforts have been directed towards the development of proper detection methods both in the clinical samples and in the vehicle of infection, as well as for deploying typing methods to tackle these pathogens.

Diagnostic markers for enterotoxigenic *E. coli* (ETEC)

Enterotoxigenic *Escherichia coli* (ETEC) are one of the most common causes of infantile and adult diarrhoea in low-income countries where the infections caused by these strains are endemic. Additionally ETEC are a leading cause of traveller's diarrhoea, affecting people from industrialized regions travelling in such areas (Gonzales-Siles and Sjöling, 2016). It has been estimated that ETEC cause 2.5 million cases of infection and 700,000 deaths in children below five years of age each year, worldwide (Fischer Walker *et al.*, 2010; Kotloff *et al.*, 2013; Lamberti *et al.*, 2014; Liu *et al.*, 2012; Qadri *et al.*, 2005). ETEC-induced diarrhoea is associated with consumption of contaminated water in areas where there is deficiency of systems to treat drinking water and that are characterized by a low-level general sanitation status (Curtis *et al.*, 2000); however, ETEC have also been associated with outbreaks of food-borne gastroenteritis (MacDonald *et al.*, 2015) and are regarded as an emerging cause of outbreaks in developed countries, such as the USA (Dalton *et al.*, 1999), Denmark (Ethelberg *et al.*, 2010), South Korea (Cho *et al.*, 2014), Norway (MacDonald *et al.*, 2015), and Japan (Konishi *et al.*, 2011). ETEC were initially identified as causes of human diarrhoea in the 1950s, and soon after, the quest for the identification of the ETEC virulence-associated traits began, with the aim of understanding the pathophysiology of the infections and to develop methods for their identification. The main characteristic of ETEC is the ability to produce one or two plasmid-borne enterotoxins, the heat-labile toxin (LT), and/or the heat-stable toxin (ST) that both affect the ion channels in the epithelial membrane (Fleckenstein *et al.*, 2010), causing the typical acute watery diarrhoea associated with ETEC infection. Besides the production of the toxins, ETEC isolates typically express numerous colonization factors (CFs) that mediate adhesion to the epithelium (Gaastra and Svennerholm, 1996). The most prevalent CFs in human infections are CF A/I and the '*coli* surface antigens' 1 to 6 (CS1–CS6), although CS7, CS14, and CS17 are also common (Gaastra and Svennerholm, 1996; Qadri *et al.*, 2005; von Mentzer *et al.*, 2014). These factors act as adhesins facilitating the adherence of the bacteria to the intestinal mucosa and play a role in the host specificity of the ETEC strains (Gaastra and Svennerholm, 1996; Qadri *et al.*, 2005).

Initially, ETEC identification was based on the phenotypic detection of the toxins in the supernatants from single *E. coli* colonies. This approach involved laborious procedures such as the rabbit ileal loop test (Evans *et al.*, 1973), the intra-gastric percutaneous injection in suckling mice (Giannella, 1976), or the assessment of the cytopathic effect induced in CHO or Y1 adrenal cell monolayers (Donta *et al.*, 1974; Guerrant *et al.*, 1974). Several immunoassays, which are more practical and easy to use than the mentioned tests have been developed for ST detection, and these showed a good correlation with those conducted on animals and cultured cells (Cryan, 1990; Giannella *et al.*, 1981). The easiest and currently the most common approach for the identification of ETEC relies on the detection of either the LT and/or ST-coding genes as one or both the toxins may be expressed by ETEC strains (Fleckenstein *et al.*, 2010; Reis *et al.*, 1982; Sack, 2011; Spangler, 1992; Torres *et al.*, 2001). Different PCR assays aimed at detecting these virulence genes have been described in the literature, including conventional PCR assays targeting the genes *lt* and *st* (Stacy-Phipps *et al.*, 1995) and a number of multiplex PCR assays for the detection of different diarrhoeagenic *E. coli* types (Aranda *et al.*, 2007; Toma *et al.*, 2003; Vidal *et al.*, 2004). Moreover, with the advent of real-time PCR, multi-parametric tests that include the detection of ETEC, as well

Table 3.1 Genes, with their location and function, commonly found in EAEC (Adapted from EFSA Opinion 2015)

EAEC gene	location	Function/Protein
aggR	pAA	Master regulator of a set of EAEC plasmid virulence genes, including aggregative adherence factors, fimbriae AAF/I-AAF/V.
aaR	pAA	AggR-activated regulator.
aatA-P	pAA	ABC protein responsible for transporting the dispersin protein out of the outer membrane.
aap	pAA	Secreted protein named dispersin.
aggA	pAA	AAF/I
aafA	pAA	AAF/II
agg3A	pAA	AAF/III
agg4A	pAA	AAF/IV
agg5A	pAA	AAF/V
aaiC	Chromosomal	AaiC, secreted protein. Type 6 secretion system. Mode of action unknown
ORF3/4	pAA	Co-regulated two–gene cluster with homology to isoprenoid synthesis genes
capU	Chromosomal	Hexosyltransferase homologue
pet	pAA	108 kDa autotransporter protein that functions as a heat-labile enterotoxin and cytotoxin
sigA	Chromosomal	IgA protease-like homologue, enterotoxin and cytotoxin
pic	Chromosomal	Mucinase, immunomodulation, colonization, lectin-like haemagglutinin
sepA	pAA	Shigella extracellular protein, enterotoxin
sat	pAA	Secreted autotransporter toxin. Enterotoxin and cytotoxin, impairment of tight junctions, autophagy
astA	pAA	Enteroaggregative heat-stable toxin (EAST1), which has similarities to *E. coli* STa enterotoxin.
eilA	Chromosomal	*Salmonella* HilA homologue, activates the bacterial surface protein Air.
air	Chromosomal	Possible aggregation and adherence

as other enteric pathogens have been deployed (Liu *et al.*, 2013; Youmans *et al.*, 2014). It has been proposed that the diagnosis of ETEC strains should include, besides the detection of the ETEC ST and LT-coding genes, the identification of the presence of other virulence genes such as *clyA, eatA, tia, tibC, leoA,* and *astA* (EAST1 toxin) (Fleckenstein *et al.*, 2010).

As the ETEC-associated diarrhoea may be caused by the ingestion of contaminated food and water (Blanco *et al.*, 1995; Huerta *et al.*, 2000; Levine *et al.*, 1993; Olsvik *et al.*, 1991; Rooney *et al.*, 2004), detection of ETEC is also carried out in non-human samples for the identification of the vehicles of infection in the framework of epidemiological investigations. Usually the markers exploited for these analyses are the same as those used for the detection of human infections and include the genes encoding the heat stable and heat labile enterotoxins. An international standard method developed in the framework of the International Organization for Standardization (ISO), typically used to assay food and water samples is not available. In the European Union (EU), the EU Reference Laboratory for *E. coli*, in Rome, Italy, has published on its website a procedure called 'Detection of enterotoxigenic *Escherichia coli* in food by real-time PCR amplification of the *lt, sth,* and *stp* genes, encoding the heat-labile and heat-stable enterotoxins' (EU-RL VTEC_Method_08_Rev 0, 05/07/2013, http://www.iss.it/binary/vtec/cont/EU_RL_VTEC_Method_08_Rev_0.pdf). This procedure aims at identifying the LT-coding gene together with those encoding the human and porcine variants of the ST-coding gene and details the procedure for the enrichment of food samples and the subsequent screening for the presence of the *st* and *lt* virulence genes by real-time PCR. Following a positive screening result, the enrichment culture is streaked onto selective agar plates (i.e.

MacConkey or TBX) for testing single colonies for the presence of the same targets. The primers and probes indicated in the EURL VTEC method are those described by Liu and colleagues (Liu *et al.*, 2013) who deployed an enteric TaqMan Array Card for detecting 19 different enteropathogens, including ETEC.

Diagnostic markers for enteroinvasive *E. coli* (EIEC)

Enteroinvasive *Escherichia coli* (EIEC) strains can cause dysentery and were first recognized as diarrhoeal causative agents in volunteer studies conducted in early the 1970s (DuPont *et al.*, 1971). EIEC strains are biochemically and genetically closely related to *Shigella* spp., and it has been observed that EIEC and *Shigella* strains also share virtually identical pathogenetic features (Nataro and Kaper, 1998). Both organisms have been shown to invade the colonic epithelium, a phenotype mediated by both plasmid and chromosomal loci. In addition, both EIEC and *Shigella* spp. elaborate one or more secretory enterotoxins that may play roles in diarrhoeal pathogenesis. Like *Shigella* spp., EIEC spp. are usually non-motile and unable to decarboxylate lysine and to ferment lactose (Nataro and Kaper, 1998), although exceptions to this biochemical and phenotypic profile have been described (Michelacci *et al.*, 2016).

A limited set of O-antigens are found in EIEC strains, including O28ac, O29, O112ac, O121, O124, O135, O136, O143, O144, O152, O159, O164, O167, and O173 (Cheasty and Rowe, 1983; Gross *et al.*, 1983; Guth *et al.*, 1989; Lan *et al.*, 2004; Matsushita *et al.*, 1993; Orskov *et al.*, 1991; Silva *et al.*, 1980; Voeroes *et al.*, 1964). Three of these EIEC-associated O-antigens, O112ac, O124, and O152, are shared with strains belonging to *Shigella* spp. (Cheasty and Rowe, 1983). Moreover, differentiation between EIEC and *Shigella* organisms with the same serotype often proves very difficult due to their almost identical biochemical and physiological traits. EIEC can be distinguished from other *E. coli* strains by testing their capacity to invade cells by the Serény test or by identification of bacterial invasion-associated proteins or genes via specific tests. However, these methods are not generally used for routine diagnosis, and EIEC strains are mainly provisionally identified by O-serogrouping with commercially available antisera in routine diagnostic laboratories (Beutin *et al.*, 1997).

The main virulence-associated mobile genetic element of EIEC is a large plasmid of about 220 kb, named pINV, which is very similar to the large virulence plasmid found in *Shigella* (Formal *et al.*, 1983; Harris *et al.*, 1982; Sansonetti *et al.*, 1982). *Shigella* strains, which naturally lost the plasmid did not cause keratoconjunctivitis in guinea pigs (Sansonetti *et al.*, 1982), whereas an *E. coli* K-12 derivative containing a 230-kb pINV from an enteroinvasive *E. coli* shared all the invasive characteristics of the parental enteroinvasive strain (Small *et al.*, 1987), underlining the importance of the plasmid in virulence. Several genes on this plasmid are involved in the complex process of invasion, as well as in EIEC survival in the gastrointestinal environment. In the invasion process, however, multiple bacterial genes are involved, some of the latter being present on the chromosome. Thirty-eight genes present on a 31-kb fragment on pINV encode most of the factors involved in the pathogenesis of EIEC infections. These regulate the process of invasion and diffusion by cell spreading, while inhibiting the autophagy of the internalized cells (Gomes *et al.*, 2016).

The diagnosis of EIEC is mainly done through cultural and phenotypic testing. EIEC present optimal growth on culture media used for Enterobacteriaceae such as MacConkey agar, xylose-lysine-deoxycholate (XLD) agar, and Hektoen enteric (HE) agar, but their identification may not take advantage of the use of selective media such as *Salmonella/Shigella* agar (SS) or bismuth sulfite agar, particularly for some serogroups (Silva *et al.*, 1980). Additionally, EIEC does not show a typical pattern of carbon source usage, which can be used to identify the pathotype. The ability to ferment lactose varies in the different strains; some EIEC strains cannot ferment lactose or use this source very slowly (72 h), making it difficult to differentiate them from *Shigella* (Baudry *et al.*, 1990). The complex biochemical patterns of EIEC strains and the difficulties in discriminating them from *Shigella* spp. strains, account for the common use of O-antisera to identify these pathogens (Gomes *et al.*, 2016).

The easiest way to arrive at the diagnosis of EIEC is through the use of molecular assays to identify specific virulence genes. The most commonly used test is the PCR for detection of the presence of the *ipaH* gene, which is present in multiple copies (4–10) on the large virulence plasmid of EIEC and *Shigella* (Ashida and Sasakawa, 2015; Gomes *et al.*, 2016). In addition to the detection of *ipaH*, other assays may be carried out to identify the presence of the invasion-associated locus *ial*, involved in cell penetration by *Shigella* and EIEC (Sousa *et al.*, 2013). It has been proposed that screening for the presence of the *uidA* and *lacY* genes could be useful to differentiate EIEC from *S. flexneri*, as the former are supposed to be positive for both genes, while the latter should lack the *lacY* operon (Pavlovic *et al.*, 2011). This assay, however, may not be effective in identifying EIEC strains that do not ferment lactose. A simple and rapid test used for the diagnosis of EIEC/*Shigella* infections is based on colorimetric detection of periplasmic apyrase (ATP-diphosphohydrolase) enzyme activity in patient stools (Sankaran *et al.*, 2009). This feature is required for the localization of the IcsA protein at one pole of the bacterial cell, which in turn recruits the host actin-regulator N-WASP, governing the process of intracellular motility and spreading via actin polymerization in the form of comet tails (Mauricio *et al.*, 2017).

Although EIEC is typically transmitted by inter-human contact, cases and outbreaks of infections have been reported that were caused by the consumption of contaminated food (Escher *et al.*, 2014). The international organization for standardization (ISO) has published a horizontal cultural method for the detection of *Shigella* spp. in food and animal feeding stuffs (ISO 21567:2004). The method can be used for the detection of EIEC, as well, and it has been proposed that in the revision of the standard, EIEC should be included in the field of application. Finally, a laboratory procedure for the detection of enteroinvasive *E. coli* in food by real-time PCR amplification of the *ipaH* gene has been made available by the EU Reference laboratory for *E. coli* (http://www.iss.it/binary/vtec/cont/EU_RL_VTEC_Method_07_Rev_0.pdf).

Diagnostic markers for enteraggregative *E. coli* (EAEC)

Enteroaggregative *E. coli* (EAEC) atrains are pathogenic *E. coli* affecting children and adults worldwide, responsible for cases of acute and persistent diarrhoea (Navarro-Garcia and Elias, 2011). This group of pathogenic *E. coli* was associated with diarrhoeal disease for the first time in 1987 by Nataro and colleagues (Nataro *et al.*, 1987). Early studies highlighted the association between EAEC and persistent diarrhoea in children, predominantly in low-income countries (Nataro and Kaper, 1998). More recently, EAEC strains have been identified as an important cause of acute diarrhoea in children and adults in a wider range of countries, including Europe, and in travellers returning from developing countries. EAEC infection is usually transmitted by the faecal–oral route but also through the consumption of contaminated food and water (Itoh *et al.*, 1997; Pai *et al.*, 1997; Scavia *et al.*, 2008). EAEC is primarily characterized by the ability to adhere to human intestinal epithelium with a typical pattern known as aggregative adherence (AA). This phenotype is observed by carrying out an adhesion assay with the human epithelial type 2 (HEp-2) cell line where EAEC attach with a typical pattern named 'stacked-brick layer' (Nataro and Kaper, 1998). The cell culture assay represents the gold standard for the identification of these strains; however, given the complexity of the procedure, its execution is generally carried out at reference laboratories only, while other diagnostic tests, mainly based on PCR detection of EAEC-specific virulence genes have been developed for the benefit of the front-line diagnostic laboratories.

Most of our knowledge about EAEC pathogenesis and virulence is based on data accumulated from studies with EAEC prototype strain 042, since its association with human diarrhoea in a study conducted in volunteers (Nataro *et al.*, 1995). Since then, several studies aimed pinpointing the association of virulence genes with EAEC have been carried out, all involving the analysis of panels of isolates from cases and controls, mainly from developing countries (Boisen *et al.*, 2012; Nataro and Kaper, 1998; Regua-Mangia *et al.*, 2004; Sarantuya *et al.*, 2004). The factors and related genes identified as being involved in EAEC virulence include adhesins, toxins, and secreted proteins. However, none of these factors is always present in all EAEC strains, making the diagnostic procedures complicated. Many of these virulence factors, including those mediating AA, are conveyed on high-molecular-weight plasmids termed pAA (Vial *et al.*, 1988).

The large virulence plasmid of EAEC, pAA, harbours a number of determinants that together with factors encoded by chromosomal loci appear to be involved in pathogenesis. The genes and toxins commonly found in EAEC are listed in Table 1. Among these, *aggR* plays a key role encoding a global regulator of EAEC virulence genes (Nataro *et al.*, 1994). The division of EAEC strains into either typical or atypical subgroups was proposed (Huang *et al.*, 2004). This classification is based on the presence or absence of the *aggR* gene encoding the EAEC master regulator, with the strains expressing AggR more likely to cause diarrhoeal disease. A number of AggR-regulated genes have been described in the prototype EAEC strain 042. The genes encoding the aggregative adherence fimbriae family (AAF/I-AAF/V), also involved in biofilm-forming ability, were the first found to be regulated by AggR (Nataro *et al.*, 1994), followed by *aap* (encoding the dispersin surface protein) (Sheikh *et al.*, 2002), and the Aat secretion system, which is required for transport of dispersin to the bacterial surface (Nishi *et al.*, 2003). AggR also activates expression of the Aai type VI secretion system (T6SS) in EAEC strain 042 (Dudley *et al.*, 2006), even if the role of Aai in EAEC virulence remains unknown (Morin *et al.*, 2013). Factors that are not under AggR control include the Air adhesin, a regulator termed EilA, the EAEC heat-stable toxin EAST-1 (encoded by the *astA* gene), and a set of toxins termed serine protease autotransporters of Enterobacteriaceae (SPATEs) (Boisen *et al.*, 2012).

The first molecular tool for detection of EAEC consisted of a DNA fragment of pAA used as probe (CVD432), which proved to be highly specific for detecting EAEC strains. This probe was developed by Baudry *et al.* (1990) who tested fragments derived empirically from the plasmids of EAEC strains 17-2 and 042 and found that a 1.0 kb plasmid-derived Sau3a fragment hybridized with 56 (89%) of 63 EAEC strains (defined by the HEp-2 assay) (Baudry *et al.*, 1990). The correlation of the positivity to the CVD432 probe with the AA phenotype seems to vary. As a matter of fact, some studies reported a correlation of 89% (Baudry *et al.*, 1990; Kang *et al.*, 1995), while in other studies the sensitivity was substantially lower (Fang *et al.*, 1995). A PCR assay using primers complementary to the CVD432 probe was designed (Schmidt *et al.*, 1995). This simple PCR tool showed a similar sensitivity and specificity when compared to the cell culture adhesion and the hybridization assays, returning a 97.7% concordance. An additional PCR primer pair was designed based on the CVD432 probe sequence and was included into a multiplex assay for the detection of diarrhoeagenic *E. coli* (Toma *et al.*, 2003). The assay was 100% accurate in identifying the 30 EAEC strains used in the study (Toma *et al.*, 2003). Similar findings were reported when the same primers were used in different combinations of multiplex assays conducted either on artificially contaminated stool samples (Aranda *et al.*, 2007) or on a reference strain collection (Brandal *et al.*, 2007).

In a study reported by Tsai *et al.* (2003), EAEC isolates displaying the AA pattern of adhesion on cultured HeLa cells were assayed for the presence of a panel of genes present on the large plasmid pAA by PCR. In particular, specific PCR tests were designed to detect the sequences of the *astA*, *aafA*, and *aggA* genes (Tsai *et al.*, 2003), and the results of the tests were compared with those obtained with the primers designed based on the CVD432 probe (Schmidt *et al.*, 1995). The panel of test strains used in the study included six *E. coli* strains showing the AA adhesion and about 400 negative strains. The results showed that the PCR test targeting the *aggA* gene was comparable to that using the primers deployed on the CVD432 probe, with both being successful in identifying all six EAEC, while the PCR tests amplifying the Aaf and the EAST1-encoding (*astA*) genes were not successful with all the strains, being positive with one and four of the six EAEC isolates, respectively (Scheutz *et al.*, 2012). Currently, there is no consensus on a single PCR assay for use in the routine testing of EAEC isolates. Notably, the most recent proposal for standardization of methods for EAEC identification considers the possibility of simultaneously using markers from both the plasmid and the chromosome into PCR assays. The main schemes propose the association between the genes *aaiA*, *aaiG*, *aggR*, and *aatA* (Andrade *et al.*, 2014) or *aaiC* and *aggR* (BIOHAZ, 2013; Scheutz *et al.*, 2011).

Following the identification of EAEC capable of producing Shiga toxins (Stx), and particularly after the Stx-producing *E. coli* O104:H4 outbreak occurred in Germany in 2011 (Werber *et al.*, 2012), much attention has been posed in developing assays to detect such hybrid strains. The obvious strategy was to couple the detection of Stx-coding genes with EAEC-specific targets. In the EU, a microbiological criterion has been issued regarding the Stx-producing *E. coli* (STEC) in sprouts (EU Reg. 209/2013, (EC, 2013), which requires the producers to check the end product for the presence of STEC belonging to the serogroups associated with the most severe forms of the infections, which in the regulation include the O104:H4 serotype. For the latter STEC, characterization as EAEC is also required. For the sake of responding to the European regulatory requirements and in the absence of international standards for the identification of EAEC, the EU RL VTEC has developed molecular methodologies to screen food samples for the presence of EAEC by the detection of targets designed within the *aggR* and *aaiC* genes sequences (http://www.iss.it/binary/vtec/cont/EU_RL_VTEC_Method_05_Rev_1.pdf) and for the identification of the genes associated with the O104:H4 serotype (http://www.iss.it/binary/vtec/cont/EU_RL_VTEC_Method_04_Rev_1.pdf).

Diagnostic markers for enteropathogenic *E. coli* (EPEC)

EPEC are a significant cause of infantile diarrhoea in low income countries, and even if the frequency of EPEC infections has decreased in industrialized countries as a consequence of drinking water treatment and general hygiene standards improvement, they still continue to be an important cause of diarrhoeal disease (Nataro and Kaper, 1998). A hallmark phenotype of EPEC is the characteristic colonization mechanism, which induces a distinctive histopathology known as the attaching and effacing (A/E) lesion, characterized by the effacement of the intestinal microvilli and the intimate attachment of the bacteria to the host epithelial surface (Nataro and Kaper, 1998). The ability to produce A/E lesions has also been detected in certain strains of Stx-producing *E. coli* (STEC) and in strains of other bacterial species (Albert *et al.*, 1991; Albert *et al.*, 1992; Schauer and Falkow, 1993).

The genes responsible for the production of A/E lesions are located on a 35-kb pathogenicity island called the locus of enterocyte effacement (LEE) (McDaniel *et al.*, 1995), which harbours the genes encoding the adhesin intimin, a type III secretion system, a number of secreted (Esp) proteins, and the translocated intimin receptor named Tir (Nataro and Kaper, 1998). After entering the gastrointestinal tract, EPEC adhere to the mucosa of the small and large intestine, and at least three steps for pathogenesis have been described (Sperandio *et al.*, 1998). The initial step includes adherence to the host cell mediated by the action of pili. After attachment, the type III secretion system is used to inject virulence factors in the host cell, including the translocation of Tir protein into the host cell membrane. Finally, intimate bacterial attachment is completed due to the interaction of intimin with Tir (Kenny *et al.*, 1997) and the formation of a pedestal caused by the cytoskeletal rearrangement in the enterocyte (Hu and Torres, 2015).

Currently, the EPEC pathotype is subdivided into typical EPEC (tEPEC) and atypical EPEC (aEPEC) strains. This classification is based on the presence of a plasmid known as EAF (EPEC adherence factor) (pEAF) in the tEPEC and its absence in the aEPEC (Trabulsi *et al.*, 2002). *bfp* and *per* are two important loci harboured by the EAF plasmid, with the former gene encoding a type IV bundle-forming pilus (BFP), which promotes bacterial microcolony formation (Bieber *et al.*, 1998). The *per* gene encodes a transcriptional activator known as the plasmid encoded regulator of the LEE pathogenicity island (Frankel *et al.*, 1998; Gomez-Duarte and Kaper, 1995). The major reservoir for tEPEC are humans, and they are rarely found in animals (Trabulsi *et al.*, 2002). On the other hand, aEPEC are common in humans and in both healthy and diseased animals (Hernandes *et al.*, 2009); they lack of the pEAF and in most cases are characterized by the presence of a large virulence plasmid typically present in certain STEC, which encodes the enterohaemolysin, a type II secretion system, and often carries the genes *katP* and *espP* encoding a catalase peroxidase and a serine protease (Lim *et al.*, 2010). Given the genomic arrangement similar to that of some STEC, aEPEC are also defined as *E. coli* positive for the presence of the LEE locus but not producing Stx (Trabulsi *et al.*, 2002).

The best tool for diagnosis of EPEC infections and strain identification is undoubtedly the detection of the intimin-coding gene *eae*, usually by conventional or real-time PCR (Gomes *et al.*, 2016). Such a strategy can be coupled with the detection of the *bfpA* gene present on the pEAF for the discrimination of the tEPEC from the aEPEC (Franke *et al.*, 1994; Gunzburg *et al.*, 1995). It has to be stressed, however, that multiple alleles of *bfpA* have been identified (Blank *et al.*, 2000), and that, based on the design of the primers, some PCR methods may fail to identify all *bfpA*-positive EPEC strains. The BFP have been also considered as targets for phenotypic assays for identification of tEPEC based on the detection of the protein using monoclonal or rabbit-polyclonal antibodies in immunofluorescence or immunoblotting tests (Giron *et al.*, 1995; Nara *et al.*, 2010). It has been shown that about 90% of tEPEC indeed produce BFP in Dulbecco's modified medium and on MacConkey agar plates (Giron *et al.*, 1995; Gismero-Ordoñez *et al.*, 2002).

Diagnostic and typing markers for Shiga toxin-producing *E. coli* (STEC)

Shiga toxin (Stx)-producing *Escherichia coli* (STEC) strains are rare zoonotic pathogens, with cattle being their major reservoir. These microorganisms can cause severe diseases such as haemorrhagic colitis (HC) and the life-threatening haemolytic uraemic syndrome (HUS) and are continuously implicated in either sporadic infections or large outbreaks worldwide (Caprioli *et al.*, 2005). Transmission of infections to humans occurs through consumption of food or water contaminated by faeces of ruminants or via direct contact with infected persons or animals and the environment (Caprioli *et al.*, 2005). Although a wide range of STEC serotypes have been implicated in human infections, the vast majority of sporadic cases and outbreaks of HC and HUS are caused by a limited number of serotypes including O157:H7, O26:H11, O103:H2, O111:H8, O145:H28 together with their non-motile (NM) variants (Beutin and Fach, 2014). Additionally, serotypes O45:H2/H- and O121:H19 are of public health concern, mainly in the US (Beutin and Fach, 2014; Brooks *et al.*, 2005).

STEC-induced disease can vary in symptomatology, ranging from mild diarrhoea to HC and HUS, usually affecting children (Lindsey *et al.*, 2016). A prompt and accurate diagnosis of STEC infections is thus pivotal for driving the appropriate management of infected patients and set-up of public health intervention strategies aimed at containing the risk of infection.

The major virulence factors of STEC are the Shiga toxins, which comprise a family of structurally related cytotoxins with similar biological activity. The Stx belong to two main groups, Stx1 and Stx2, but several subtypes of each group have been recognized. In particular, three subtypes for Stx1 and seven for Stx2 have been described (Scheutz *et al.*, 2012). Epidemiological studies have highlighted that Stx2 is more often associated with severe human disease than Stx1 (Boerlin *et al.*, 1999). Among the Stx2 subtypes, Stx2a and Stx2c are those frequently found in strains isolated from patients with HUS, whereas the others are rarely associated with the most severe forms of STEC infection (Friedrich *et al.*, 2002; Persson *et al.*, 2007).

As for the diagnosis, the isolation of STEC strains is an important goal, also for the purpose of strain typing. However, as culture-based methods are laborious and exhibit clear limits in sensitivity for STEC detection, it is recommended that laboratories supplement culture-based approaches with other assay types (Gould *et al.*, 2009). These consist in methods usually relying on molecular biology techniques, often used for an initial screening of the sample and for confirmation of the isolated colonies. The gold standard for the identification of STEC is based on the detection of the production of Shiga toxin. This was traditionally done by the observation of the characteristic cytopathic effect induced by inoculation of STEC culture supernatants onto a semi-confluent Vero cell monolayer (Konowalchuk *et al.*, 1977). This assay is also carried out by directly applying an extract of the faecal specimen onto the Vero cell monolayers. This test is quite laborious and not routinely used in the diagnosis of STEC, which can, on the other hand, rely on the application of molecular methods, such as PCR, for the detection of virulence-associated genes. The detection of Stx-coding genes is widely applied for STEC detection and identification, either as an initial screening of enrichment cultures or for the confirmation of isolated colonies. It has to be noted that the existence of different Stx subtypes reflects in different *stx*-genes sequences, and single PCR reagents may fail in detecting all the subtypes. On the other hand, the possibility of determining the presence of certain *stx* gene subtypes in an isolated STEC strain represents a valid tool for initial strain characterization, adding a prognostic value, given the known association of certain Stx2 subtypes with the clinical severity of STEC infection and the risk of HUS (Bielaszewska *et al.*, 2006; Chui *et al.*,

2015; Friedrich *et al.*, 2002; Haugum *et al.*, 2014; Persson *et al.*, 2007).

Nonetheless, Stx production alone seems to be not sufficient to cause illness, at least the most severe forms, which must be preceded by the efficacious colonization of the host gut. A genetic marker related with the adhesion to the host intestinal mucosa present in certain STEC strains is the intimin-coding *eae* gene. As described for EPEC, the *eae* gene is considered the hallmark of the presence of the LEE pathogenicity island, responsible for the attaching and effacing adhesion, a colonization mechanism frequently associated with STEC strains associated with HC and HUS (BIOHAZ, 2013; Caprioli *et al.*, 2005). Therefore, *stx*-positive samples are generally screened for the presence of the *eae* gene to predict their potential to cause attaching and effacing lesions. The intimin-coding *eae* gene can also be subtyped for further strain characterization. In fact, while the N-terminus of the intimin protein is highly conserved, the C-terminus, comprising the active receptor-binding site, shows a certain degree of variability (Frankel *et al.*, 1995). The polymorphism in the C-terminus of the intimin protein is due to heterogeneity in the 3′ region of the coding gene and has led to the identification of at least 18 intimin subtypes (Ito *et al.*, 2007). Among these, the *eae-γ1* subtype has commonly been found in STEC O157:H7 and O145:H28, whereas eae-β, eae-ε and eae-θ subtypes have commonly been detected in STEC O26:H11, O103:H2 and O111:H8, respectively (Madic *et al.*, 2010; Oswald *et al.*, 2000; Tarr and Whittam, 2002).

The same analytical approach of detecting the STEC virulence genes *stx* and *eae* is also widely used for the detection of these pathogens in the vehicles of infection, including food items. In the United States, the US Department of Agriculture Food Safety and Inspection Service (USDA-FSIS) has issued a method for the Detection and Isolation of non-O157 Shiga Toxin-Producing *Escherichia coli* (STEC) from Meat Products and Carcass and Environmental Sponges (FSIS, 2014a). The method involves the detection and isolation of STEC belonging to specific serogroups, namely O26, O45, O103, O111, O121 and O145 and altogether termed as the 'big six' STEC serogroups. The selection of these specific serogroups is based on the results of an investigation conducted at the Centers for Disease Control and Prevention (CDC), which showed that approximately 70% of non-O157 STEC infections occurring from 1983 to 2002 in the United States were caused by strains belonging to one of those six serogroups (Brooks *et al.*, 2005). Following the recognition of these peculiar STEC serogroups as an important public health issue they have been listed in 2011, together with the STEC O157, as the STEC serogroups considered as adulterants in certain beef products (FSIS, 2011). The methodological strategy for testing beef for non-O157 STEC includes an initial enrichment of the meat sample (325 ± 32.5g test portion) followed by DNA extraction and real-time PCR screening of *stx* and *eae* genes. Samples positive for the presence of *stx* and *eae* genes are then tested for the presence of genes specific for the 'top 6' serogroups by real-time PCR, by using the BAX® system kits. With detection of any of the top-6 serogroup-associated genes, the sample is subjected to immunomagnetic separation (IMS) to isolate the STEC strain. As for the detection of strains belonging to serogroup O157, the USDA FSIS employs the Microbiology Laboratory Guidebook (MLG) 5.09 method 'Detection, Isolation and Identification of *Escherichia coli* O157:H7 from Meat Products and Carcass and Environmental Sponges' (FSIS, 2015), which relies on the enrichment in a selective medium, application of a screening test based on the BAX® system (FSIS, 2014b), followed by IMS for positive samples and plating on a highly selective medium (Modified Rainbow® Agar).

In 2012, the International Standardization Organization (ISO) published the Technical Specification ISO TS 13136 'Microbiology of food and animal feed – Real-time polymerase chain reaction (PCR)-based method for the detection of food-borne pathogens – Horizontal method for the detection of Shiga-toxin-producing *Escherichia coli* (STEC) and the determination of O157, O111, O26, O103 and O145 serogroups' (ISO, 2012). The aim of the method is the detection and isolation of any STEC, and the identification of the STEC strains belonging to serogroups of major public health concern. Similarly to the USDA-FSIS MLG method, the ISO method is based on a stepwise approach involving initial screening using DNA extracted from food enrichment cultures for the genes coding for Stx1 and Stx2 by real-time PCR. If positive, the DNA is also tested for the presence of *eae* gene. The isolation of STEC is always attempted in samples positive for *stx* genes, and *eae*-positive samples are also assayed for the top 5 serogroup-associated genes, to aid in the isolation step applying IMS.

In a recent study, Delannoy and colleagues (2013) showed that combining the detection of the *espK* gene with either *espV*, *ureD*, or *Z2098* was a highly sensitive and specific approach for identifying the most clinically relevant STEC serogroups (O26, O45, O103, O111, O121, O145, and O157). These markers were shown to be preferentially associated with *E. coli* strains carrying *stx* and *eae* genes, and the authors proposed to use them in combination with *stx* and *eae* screening for a better identification of food samples that are more likely to contain a STEC of public health concern. A different approach has been proposed to identify *E. coli* isolates belonging to all seven clinically important serogroups targeted by US regulations (i.e. O157 and the 'big six' non-O157 STEC) based on interrogation of informative single nucleotide polymorphisms (SNPs) in the sequence of the *gnd* gene (Elder *et al.*, 2016), a locus encoding 6-phosphogluconate dehydrogenase adjacent to the O-antigen-encoding gene cluster (Gilmour *et al.*, 2007). The authors selected the most informative SNPs based on an alignment of sequences representing each of the 'big six' non-O157 STEC serogroups and O157. The total number of SNPs required to classify isolates belonging to the 'big six' non-O157 STEC and O157 were few and resulted in good candidates to differentiate them from other non-O157 STEC isolates, without the need to sequence the entire *gnd* gene, and resulting in easy interpretation of the multiplex SNP typing assay.

Methods for typing STEC based on phenotype, such as serotyping and phage-typing, have been applied for many years (Sabat *et al.*, 2013). However, the latter approach is applicable only to type STEC O157 and, given the great diversity of STEC, their characterization beyond serotype is needed for epidemiological tracking of infections, especially during outbreak investigations and for the assessment of the risks related with human infections. In early 2000, Karmali and colleagues proposed the classification of STEC into seropathotypes that considers the STEC serotype in association with the ability to cause outbreaks, and severity of the disease (Karmali *et al.*, 2003) in order to assess the clinical and public health risks associated with STEC. In the proposed scheme the seropathotype A comprised serotypes O157:H7 and O157:NM, which are common causes of outbreaks and HUS in most countries; seropathotype B included STEC strains associated with HUS and often involved in outbreaks, although less frequently than seropathotype A while those belonging to seropathotype C strains were meant to be associated with sporadic HUS but not epidemics. Finally, seropathotype D strains were described as associated with diarrhoea but not with outbreaks or severe disease; and seropathotype E comprised multiple STEC serotypes that had never been associated with human disease and appeared to be linked exclusively to the animal reservoir (Karmali *et al.*, 2003). The scheme proposed by Karmali and co-workers is acknowledged as being the most appropriate to classify STEC (BIOHAZ, 2013), but is also considered a dynamic scheme, which needs to be continuously refined as our knowledge on STEC progresses (BIOHAZ, 2013).

A useful approach for characterization of STEC strains is in the definition of their virulence gene profile, or the so-called 'virulome' (Imamovic *et al.*, 2010). This strategy has complemented the seropathotype concept to give a more comprehensive classification (Coombes *et al.*, 2008; Karmali *et al.*, 2003; Konczy *et al.*, 2008). The STEC virulome includes a high number of virulence-associated mobile genetic elements such as the phages carrying the Shiga toxins (Stx)-coding genes and others possessing virulence genes involved in the efficient colonization of STEC in the host gut (Imamovic *et al.*, 2010; McDaniel and Kaper, 1997; Morabito *et al.*, 2003; O'Brien *et al.*, 1984). The latter includes the well-known LEE pathogenicity island and the O-islands (OI) 122 (Karmali *et al.*, 2003; Morabito *et al.*, 2003) and OI-57 (Imamovic *et al.*, 2010). STEC possessing the LEE locus also contain a variable repertoire of genes encoding virulence determinants, including a collection of non-LEE-encoded effector (*nle*) genes that specify translocated substrates of the type III secretion system. One of these, *nleB*, has been proposed as a diagnostic tool to identify the most virulent STEC strains (Bugarel *et al.*, 2011). As far as the LEE-negative STEC isolates are concerned, they have been shown to belong to a heterogeneous class composed of different subgroups (Jenkins *et* al., 2003; Michelacci et al., 2013). Some of these are identified by the presence of different subtypes of the subtilase-coding gene, *subAB*, or associated with genes encoding specific adhesins, such as *saa* (Paton *et al.*, 2004) or invasins, such as the *tia* gene (Tozzoli *et al.*, 2010).

Characterization of STEC strains of serotype O26:H11 circulating in France (Delannoy *et al.*, 2015a,b) showed that some isolates did not possess the markers *espK, espV, ureD*, or *Z2098* that were proposed to be associated with the most virulent STEC (Delannoy *et al.*, 2013). In order to identify this particular STEC clone, the authors developed a strategy based on the detection of a CRISPR (clustered regularly interspaced short palindromic repeats) sequence (Rooney *et al.*, 2004), and particularly, one that they termed $CRISPR_{O26:H11}$ (Delannoy *et al.*, 2016). The detection of CRISPR sequences to identify STEC had already been described for the identification of the seven STEC serogroups most often involved in HUS, as well as the O104:H4 STEC clone responsible for the large European outbreak in 2011 (Delannoy *et al.*, 2012a,b; Miko *et al.*, 2013), making such an approach a promising field of investigation for STEC detection and typing.

References

Albert, M.J., Alam, K., Islam, M., Montanaro, J., Rahaman, A.S., Haider, K., Hossain, M.A., Kibriya, A.K., and Tzipori, S. (1991). *Hafnia alvei*, a probable cause of diarrhea in humans. Infect. Immun. 59, 1507–1513.

Albert, M.J., Faruque, S.M., Ansaruzzaman, M., Islam, M.M., Haider, K., Alam, K., Kabir, I., and Robins-Browne, R. (1992). Sharing of virulence-associated properties at the phenotypic and genetic levels between enteropathogenic *Escherichia coli* and *Hafnia alvei*. J. Med. Microbiol. 37, 310–314.

Andrade, F.B., Gomes, T.A., and Elias, W.P. (2014). A sensitive and specific molecular tool for detection of both typical and atypical enteroaggregative *Escherichia coli*. J. Microbiol. Methods. *106*, 16–18.

Aranda, K.R., Fabbricotti, S.H., Fagundes-Neto, U., and Scaletsky, I.C (2007). Single multiplex assay to identify simultaneously enteropathogenic, enteroaggregative, enterotoxigenic, enteroinvasive and Shiga toxin-producing *Escherichia coli* strains in Brazilian children. FEMS Microbiol. Lett. *267*, 145–150.

Ashida, H., and Sasakawa, C. (2015). *Shigella* IpaH family effectors as a versatile model for studying pathogenic bacteria. Front Cell Infect. Microbiol. *5*, 100. https://doi.org/10.3389/fcimb.2015.00100

Baudry, B., Savarino, S.J., Vial, P., Kaper, J.B., and Levine, M.M. (1990). A sensitive and specific DNA probe to identify enteroaggregative *Escherichia coli*, a recently discovered diarrheal pathogen. J. Infect. Dis. *161*, 1249–1251.

Beutin, L., and Fach, P. (2014). Detection of Shiga toxin-producing *Escherichia coli* from nonhuman sources and strain typing. Microbiol Spectr *2*, https://doi.org/10.1128/microbiolspec.EHEC-0001-2013

Beutin, L., Gleier, K., Kontny, I., Echeverria, P., and Scheutz, F. (1997). Origin and characteristics of enteroinvasive strains of *Escherichia coli* (EIEC) isolated in Germany. Epidemiol. Infect. *118*, 199–205.

Bieber, D., Ramer, S.W., Wu, C.Y., Murray, W.J., Tobe, T., Fernandez, R., and Schoolnik, G.K. (1998). Type IV pili, transient bacterial aggregates, and virulence of enteropathogenic *Escherichia coli*. Science *280*, 2114–2118.

Bielaszewska, M., Friedrich, A.W., Aldick, T., Schurk-Bulgrin, R., and Karch, H. (2006). Shiga toxin activatable by intestinal mucus in *Escherichia coli* isolated from humans: predictor for a severe clinical outcome. Clin. Infect. Dis. *43*, 1160–1167.

BIOHAZ. (2013). EFSA Panel on Biological Hazards Scientific Opinion on VTEC-seropathotype and scientific criteria regarding pathogenicity assessment. EFSA Journal. *11*, 3138. doi: 10.2903/j.efsa.2013.3138.

Blanco, J.E., Blanco, M., and Blanco, J. (1995). [Enterotoxigenic, verotoxigenic, and necrotoxigenic *Escherichia coli* in food and clinical samples. Role of animals as reservoirs of strains pathogenic for humans]. Microbiologia *11*, 97–110.

Blank, T.E., Zhong, H., Bell, A.L., Whittam, T.S., and Donnenberg, M.S. (2000). Molecular variation among type IV pilin (*bfpA*) genes from diverse enteropathogenic *Escherichia coli* strains. Infect. Immun. *68*, 7028–7038.

Boerlin, P., McEwen, S.A., Boerlin-Petzold, F., Wilson, J.B., Johnson, R.P., and Gyles, C.L. (1999). Associations between virulence factors of Shiga toxin-producing *Escherichia coli* and disease in humans. J. Clin. Microbiol. *37*, 497–503.

Boisen, N., Scheutz, F., Rasko, D.A., Redman, J.C., Persson, S., Simon, J., Kotloff, K.L., Levine, M.M., Sow, S., Tamboura, B., *et al.* (2012). Genomic characterization of enteroaggregative *Escherichia coli* from children in Mali. J. Infect. Dis. *205*, 431–444. https://doi.org/10.1093/infdis/jir757

Brandal, L.T., Lindstedt, B.A., Aas, L., Stavnes, T.L., Lassen, J., and Kapperud, G. (2007). Octaplex PCR and fluorescence-based capillary electrophoresis for identification of human diarrheagenic *Escherichia coli* and *Shigella* spp. J. Microbiol. Methods. *68*, 331–341.

Brooks, J.T., Sowers, E.G., Wells, J.G., Greene, K.D., Griffin, P.M., Hoekstra, R.M., and Strockbine, N.A. (2005). Non-O157 Shiga toxin-producing *Escherichia coli* infections in the United States, 1983-2002. J. Infect. Dis. *192*, 1422–1429.

Bugarel, M., Martin, A., Fach, P., and Beutin, L. (2011). Virulence gene profiling of enterohemorrhagic (EHEC) and enteropathogenic (EPEC) *Escherichia coli* strains: a basis for molecular risk assessment of typical and atypical EPEC strains. BMC Microbiol. *11*, 142. https://doi.org/10.1186/1471-2180-11-142

Caprioli, A., Morabito, S., Brugère, H., and Oswald, E. (2005). Enterohaemorrhagic *Escherichia coli*: emerging issues on virulence and modes of transmission. Vet. Res. *36*, 289–311. https://doi.org/10.1051/vetres:2005002

Cheasty, T., and Rowe, B. (1983). Antigenic relationships between the enteroinvasive *Escherichia coli* O antigens O28ac, O112ac, O124, O136, O143, O144, O152, and O164 and Shigella O antigens. J. Clin. Microbiol. *17*, 681–684.

Cho, S.H., Kim, J., Oh, K.H., Hu, J.K., Seo, J., Oh, S.S., Hur, M.J., Choi, Y.H., Youn, S.K., Chung, G.T., *et al.* (2014). Outbreak of enterotoxigenic *Escherichia coli* O169 enteritis in schoolchildren associated with consumption of kimchi, Republic of Korea, 2012. Epidemiol. Infect. *142*, 616–623. https://doi.org/10.1017/S0950268813001477

Chui, L., Li, V., Fach, P., Delannoy, S., Malejczyk, K., Patterson-Fortin, L., Poon, A., King, R., Simmonds, K., Scott, A.N., *et al.* (2015). Molecular profiling of *Escherichia coli* O157:H7 and non-O157 strains isolated from humans and cattle in Alberta, Canada. J. Clin. Microbiol. *53*, 986–990. https://doi.org/10.1128/JCM.03321-14

Coombes, B.K., Wickham, M.E., Mascarenhas, M., Gruenheid, S., Finlay, B.B., and Karmali, M.A. (2008). Molecular analysis as an aid to assess the public health risk of non-O157 Shiga toxin-producing *Escherichia coli* strains. Appl. Environ. Microbiol. *74*, 2153–2160.

Cryan, B. (1990). Comparison of three assay systems for detection of enterotoxigenic *Escherichia coli* heat-stable enterotoxin. J. Clin. Microbiol. *28*, 792–794.

Curtis, V., Cairncross, S., and Yonli, R. (2000). Domestic hygiene and diarrhoea – pinpointing the problem. Trop. Med. Int. Health *5*, 22–32.

Dalton, C.B., Mintz, E.D., Wells, J.G., Bopp, C.A., and Tauxe, R.V. (1999). Outbreaks of enterotoxigenic *Escherichia coli* infection in American adults: a clinical and epidemiologic profile. Epidemiol. Infect. *123*, 9–16.

Delannoy, S., Beutin, L., Burgos, Y., and Fach, P. (2012a). Specific detection of enteroaggregative hemorrhagic *Escherichia coli* O104:H4 strains by use of the CRISPR locus as a target for a diagnostic real-time PCR. J. Clin. Microbiol. *50*, 3485–3492.

Delannoy, S., Beutin, L., and Fach, P. (2012b). Use of clustered regularly interspaced short palindromic repeat sequence polymorphisms for specific detection of enterohemorrhagic *Escherichia coli* strains of serotypes O26:H11, O45:H2, O103:H2, O111:H8, O121:H19, O145:H28, and O157:H7 by real-time PCR. J. Clin. Microbiol. *50*, 4035–4040.

Delannoy, S., Beutin, L., and Fach, P. (2013). Discrimination of enterohemorrhagic *Escherichia coli* (EHEC) from non-EHEC strains based on detection of various combinations of type III effector genes. J. Clin. Microbiol. *51*, 3257–3262. https://doi.org/10.1128/JCM.01471-13

Delannoy, S., Chaves, B.D., Ison, S.A., Webb, H.E., Beutin, L., Delaval, J., Billet, I., and Fach, P. (2016). Revisiting the STEC testing approach: using *espK* and *espV* to make enterohemorrhagic *Escherichia coli* (EHEC) detection more reliable in beef. Front Microbiol 7, 1. https://doi.org/10.3389/fmicb.2016.00001

Delannoy, S., Mariani-Kurkdjian, P., Bonacorsi, S., Liguori, S., and Fach, P. (2015a). Characteristics of emerging human-pathogenic *Escherichia coli* O26:H11 strains isolated in France between 2010 and 2013 and carrying the stx2d gene only. J. Clin. Microbiol. *53*, 486–492.

Delannoy, S., Mariani-Kurkdjian, P., Bonacorsi, S., Liguori, S., Ison, S.A., and Fach, P. (2015b). Draft genome sequences of human-pathogenic *Escherichia coli* O26:H11 strains carrying the stx2 gene only and circulating in France. Genome Announc. *3*, e00852–15. doi: 10.1128/genomeA.00852-15.

Donta, S.T., Moon, H.W., and Whipp, S.C. (1974). Detection of heat-labile *Escherichia coli* enterotoxin with the use of adrenal cells in tissue culture. Science *183*, 334–336.

Dudley, E.G., Thomson, N.R., Parkhill, J., Morin, N.P., and Nataro, J.P. (2006). Proteomic and microarray characterization of the AggR regulon identifies a *pheU* pathogenicity island in enteroaggregative *Escherichia coli*. Mol. Microbiol. *61*, 1267–1282.

DuPont, H.L., Formal, S.B., Hornick, R.B., Snyder, M.J., Libonati, J.P., Sheahan, D.G., LaBrec, E.H., and Kalas, J.P. (1971). Pathogenesis of *Escherichia coli* diarrhea. N. Engl. J. Med. *285*, 1–9. https://doi.org/10.1056/NEJM197107012850101

EC (2013). COMMISSION REGULATION (EU) No 209/2013 of 11 March 2013 amending Regulation (EC) No 2073/2005 as regards microbiological criteria for sprouts and the sampling rules for poultry carcases and fresh poultry meat, E. Commission, ed. Official Journal of the European Union, 5 pp.

Elder, J.R., Bugarel, M., den Bakker, H.C., Loneragan, G.H., and Nightingale, K.K. (2016). Interrogation of single nucleotide polymorphisms in *gnd* provides a novel method for molecular serogrouping of clinically important Shiga toxin producing *Escherichia coli* (STEC) targeted by regulation in the United States, including the 'big six' non-O157 STEC and STEC O157. J. Microbiol. Methods. *129*, 85–93.

Escher, M., Scavia, G., Morabito, S., Tozzoli, R., Maugliani, A., Cantoni, S., Fracchia, S., Bettati, A., Casa, R., Gesu, G.P., *et al.* (2014). A severe food-borne outbreak of diarrhoea linked to a canteen in Italy caused by enteroinvasive *Escherichia coli*, an uncommon agent. Epidemiol. Infect. *142*, 2559–2566.

Ethelberg, S., Lisby, M., Bottiger, B., Schultz, A.C., Villif, A., Jensen, T., Olsen, K.E., Scheutz, F., Kjelso, C., and Muller, L. (2010). Outbreaks of gastroenteritis linked to lettuce, Denmark, January 2010. Euro Surveill. *15*, 19484.

Evans, D.G., Evans, D.J. Jr., and Pierce, N.F. (1973). Differences in the response of rabbit small intestine to heat-labile and heat-stable enterotoxins of *Escherichia coli*. Infect. Immun. *7*, 873–880.

Fang, G.D., Lima, A.A., Martins, C.V., Nataro, J.P., and Guerrant, R.L. (1995). Etiology and epidemiology of persistent diarrhea in northeastern Brazil: a hospital-based, prospective, case-control study. J. Pediatr. Gastroenterol. Nutr. *21*, 137–144.

Fischer Walker, C.L., Sack, D., and Black, R.E. (2010). Etiology of diarrhea in older children, adolescents and adults: a systematic review. PLOS Negl Trop Dis *4*, e768. https://doi.org/10.1371/journal.pntd.0000768

Fleckenstein, J.M., Hardwidge, P.R., Munson, G.P., Rasko, D.A., Sommerfelt, H., and Steinsland, H. (2010). Molecular mechanisms of enterotoxigenic *Escherichia coli* infection. Microbes Infect. *12*, 89–98. https://doi.org/10.1016/j.micinf.2009.10.002

Formal, S.B., Hale, T.L., and Sansonetti, P.J. (1983). Invasive enteric pathogens. Rev. Infect. Dis. 5 (Suppl. 4), S702–707.

Franke, J., Franke, S., Schmidt, H., Schwarzkopf, A., Wieler, L.H., Baljer, G., Beutin, L., and Karch, H. (1994). Nucleotide sequence analysis of enteropathogenic *Escherichia coli* (EPEC) adherence factor probe and development of PCR for rapid detection of EPEC harboring virulence plasmids. J. Clin. Microbiol. 32, 2460–2463.

Frankel, G., Candy, D.C., Fabiani, E., Adu-Bobie, J., Gil, S., Novakova, M., Phillips, A.D., and Dougan, G. (1995). Molecular characterization of a carboxy-terminal eukaryotic-cell-binding domain of intimin from enteropathogenic *Escherichia coli*. Infect. Immun. *63*, 4323–4328.

Frankel, G., Phillips, A.D., Rosenshine, I., Dougan, G., Kaper, J.B., and Knutton, S. (1998). Enteropathogenic and enterohaemorrhagic *Escherichia coli*: more subversive elements. Mol. Microbiol. *30*, 911–921.

Friedrich, A.W., Bielaszewska, M., Zhang, W.L., Pulz, M., Kuczius, T., Ammon, A., and Karch, H. (2002). *Escherichia coli* harboring Shiga toxin 2 gene variants: frequency and association with clinical symptoms. J. Infect. Dis. *185*, 74–84.

FSIS (2012). Shiga toxin-producing *Escherichia coli* in certain raw beef products. Federal Register *77*, 31975–31981.

FSIS (2014a). Detection and isolation of non-O157 Shiga toxin-producing *Escherichia coli* (STEC) from meat products and carcass and environmental sponges. USDA FSIS Microbiology Laboratory Guidebook *MLG 5B.05*.

FSIS (2014b). FSIS Procedure for the use of *Escherichia coli* O157:H7 screening tests for meat products and carcass and environmental sponges. USDA FSIS Microbiology Laboratory Guidebook *MLG 5A.04*.

FSIS (2015). Detection, isolation and identification of *Escherichia coli* O157:H7 from meat products and carcass and environmental sponges. USDA FSIS Microbiology Laboratory Guidebook *MLG 5.09*.

Gaastra, W., and Svennerholm, A.M. (1996). Colonization factors of human enterotoxigenic *Escherichia coli* (ETEC). Trends Microbiol. *4*, 444–452.

Giannella, R.A. (1976). Suckling mouse model for detection of heat-stable *Escherichia coli* enterotoxin: characteristics of the model. Infect. Immun. *14*, 95–99.

Giannella, R.A., Drake, K.W., and Luttrell, M. (1981). Development of a radioimmunoassay for *Escherichia coli* heat-stable enterotoxin: comparison with the suckling mouse bioassay. Infect. Immun. 33, 186–192.

Gilmour, M.W., Olson, A.B., Andrysiak, A.K., Ng, L.K., and Chui, L. (2007). Sequence-based typing of genetic targets encoded outside of the O-antigen gene cluster is indicative of Shiga toxin-producing *Escherichia coli* serogroup lineages. J. Med. Microbiol. *56*, 620–628.

Giron, J.A., Qadri, F., Azim, T., Jarvis, K.J., Kaper, J.B., and Albert, M.J. (1995). Monoclonal antibodies specific for the bundle-forming pilus of enteropathogenic *Escherichia coli*. Infect. Immun. *63*, 4949–4952.

Gismero-Ordoñez, J., Dall'Agnol, M., Trabulsi, L.R., and Girón, J.A. (2002). Expression of the bundle-forming pilus by enteropathogenic *Escherichia coli* strains of heterologous serotypes. J. Clin. Microbiol. *40*, 2291–2296.

Gomes, T.A., Elias, W.P., Scaletsky, I.C., Guth, B.E., Rodrigues, J.F., Piazza, R.M., Ferreira, L.C., and Martinez, M.B. (2016). Diarrheagenic *Escherichia coli*. Braz. J. Microbiol. *47* (Suppl. 1), 3–30.

Gomez-Duarte, O.G., and Kaper, J.B. (1995). A plasmid-encoded regulatory region activates chromosomal *eaeA* expression in enteropathogenic *Escherichia coli*. Infect. Immun. *63*, 1767-1776.

Gonzales-Siles, L., and Sjöling, Å. (2016). The different ecological niches of enterotoxigenic *Escherichia coli*. Environ. Microbiol. *18*, 741–751. https://doi.org/10.1111/1462-2920.13106

Gould, L.H., Demma, L., Jones, T.F., Hurd, S., Vugia, D.J., Smith, K., Shiferaw, B., Segler, S., Palmer, A., Zansky, S., *et al.* (2009). Hemolytic uremic syndrome and death in persons with *Escherichia coli* O157:H7 infection, food-borne diseases active surveillance network sites, 2000-2006. Clin. Infect. Dis. *49*, 1480–1485.

Gross, R.J., Thomas, L.V., Cheasty, T., Day, N.P., Rowe, B., Toledo, M.R., and Trabulsi, L.R. (1983). Enterotoxigenic and enteroinvasive *Escherichia coli* strains belonging to a new O group, O167. J. Clin. Microbiol. *17*, 521–523.

Guerrant, R.L., Brunton, L.L., Schnaitman, T.C., Rebhun, L.I., and Gilman, A.G. (1974). Cyclic adenosine monophosphate and alteration of Chinese hamster ovary cell morphology: a rapid, sensitive in vitro assay for the enterotoxins of *Vibrio cholerae* and *Escherichia coli*. Infect. Immun. *10*, 320–327.

Gunzburg, S.T., Tornieporth, N.G., and Riley, L.W. (1995). Identification of enteropathogenic *Escherichia coli* by PCR-based detection of the bundle-forming pilus gene. J. Clin. Microbiol. 33, 1375–1377.

Guth, B.E., Silva, R.M., Toledo, M.R., Lima, T.M., and Trabulsi, L.R. (1989). Virulence factors and biochemical characteristics of serotypes of *Escherichia coli* serogroup O29. J. Clin. Microbiol. 27, 2161–2164.

Harris, J.R., Wachsmuth, I.K., Davis, B.R., and Cohen, M.L. (1982). High-molecular-weight plasmid correlates with *Escherichia coli* enteroinvasiveness. Infect. Immun. *37*, 1295–1298.

Haugum, K., Brandal, L.T., Lindstedt, B.A., Wester, A.L., Bergh, K., and Afset, J.E. (2014). PCR-based detection and molecular characterization of Shiga toxin-producing *Escherichia coli* strains in a routine microbiology laboratory over 16 years. J. Clin. Microbiol. *52*, 3156–3163.

Hernandes, R.T., Elias, W.P., Vieira, M.A., and Gomes, T.A. (2009). An overview of atypical enteropathogenic *Escherichia coli*. FEMS Microbiol. Lett. *297*, 137–149. https://doi.org/10.1111/j.1574-6968.2009.01664.x

Hu, J., and Torres, A.G. (2015). Enteropathogenic *Escherichia coli*: foe or innocent bystander? Clin. Microbiol. Infect. *21*, 729–734. https://doi.org/10.1016/j.cmi.2015.01.015

Huang, D.B., Okhuysen, P.C., Jiang, Z.D., and DuPont, H.L. (2004). Enteroaggregative *Escherichia coli*: an emerging enteric pathogen. Am. J. Gastroenterol. *99*, 383–389.

Huerta, M., Grotto, I., Gdalevich, M., Mimouni, D., Gavrieli, B., Yavzori, M., Cohen, D., and Shpilberg, O. (2000). A waterborne outbreak of gastroenteritis in the Golan

Heights due to enterotoxigenic *Escherichia coli*. Infection *28*, 267–271.

Imamovic, L., Tozzoli, R., Michelacci, V., Minelli, F., Marziano, M.L., Caprioli, A., and Morabito, S. (2010). OI-57, a genomic island of *Escherichia coli* O157, is present in other seropathotypes of Shiga toxin-producing *E. coli* associated with severe human disease. Infect. Immun. *78*, 4697–4704.

ISO (2012). Microbiology of food and animal feed — Real-time polymerase chain reaction (PCR)-based method for the detection of food-borne pathogens — Horizontal method for the detection of Shiga toxin-producing *Escherichia coli* (STEC) and the determination of O157, O111, O26, O103 and O145 serogroups. ISO/TS 13136:2012.

Ito, K., Iida, M., Yamazaki, M., Moriya, K., Moroishi, S., Yatsuyanagi, J., Kurazono, T., Hiruta, N., and Ratchtrachenchai, O.A. (2007). Intimin types determined by heteroduplex mobility assay of intimin gene (*eae*)-positive *Escherichia coli* strains. J. Clin. Microbiol. *45*, 1038–1041.

Itoh, Y., Nagano, I., Kunishima, M., and Ezaki, T. (1997). Laboratory investigation of enteroaggregative *Escherichia coli* O untypeable:H10 associated with a massive outbreak of gastrointestinal illness. J. Clin. Microbiol. *35*, 2546–2550.

Jenkins, C., Perry, N.T., Cheasty, T., Shaw, D.J., Frankel, G., Dougan, G., Gunn, G.J., Smith, H.R., Paton, A.W., and Paton, J.C. (2003). Distribution of the saa gene in strains of Shiga toxin-producing *Escherichia coli* of human and bovine origins. J. Clin. Microbiol. *41*, 1775–1778.

Kang, G., Mathan, M.M., and Mathan, V.I. (1995). Evaluation of a simplified HEp-2 cell adherence assay for *Escherichia coli* isolated from south Indian children with acute diarrhea and controls. J. Clin. Microbiol. *33*, 2204–2205.

Karmali, M.A., Mascarenhas, M., Shen, S., Ziebell, K., Johnson, S., Reid-Smith, R., Isaac-Renton, J., Clark, C., Rahn, K., and Kaper, J.B. (2003). Association of genomic O island 122 of *Escherichia coli* EDL 933 with verocytotoxin-producing *Escherichia coli* seropathotypes that are linked to epidemic and/or serious disease. J. Clin. Microbiol. *41*, 4930–4940.

Kenny, B., DeVinney, R., Stein, M., Reinscheid, D.J., Frey, E.A., and Finlay, B.B. (1997). Enteropathogenic *E. coli* (EPEC) transfers its receptor for intimate adherence into mammalian cells. Cell *91*, 511–520.

Konczy, P., Ziebell, K., Mascarenhas, M., Choi, A., Michaud, C., Kropinski, A.M., Whittam, T.S., Wickham, M., Finlay, B., and Karmali, M.A. (2008). Genomic O island 122, locus for enterocyte effacement, and the evolution of virulent verocytotoxin-producing *Escherichia coli*. J. Bacteriol. *190*, 5832–5840.

Konishi, N., Obata, H., Monma, C., Nakama, A., Kai, A., and Tsuji, T. (2011). Bacteriological and epidemiological characteristics of enterotoxigenic *Escherichia coli* isolated in Tokyo, Japan, between 1966 and 2009. J. Clin. Microbiol. *49*, 3348–3351.

Konowalchuk, J., Speirs, J.I., and Stavric, S. (1977). Vero response to a cytotoxin of *Escherichia coli*. Infect. Immun. *18*, 775–779.

Kotloff, K.L., Nataro, J.P., Blackwelder, W.C., Nasrin, D., Farag, T.H., Panchalingam, S., Wu, Y., Sow, S.O., Sur, D., Breiman, R.F., *et al.* (2013). Burden and aetiology of diarrhoeal disease in infants and young children in developing countries (the Global Enteric Multicenter Study, GEMS): a prospective, case-control study. Lancet. *382*, 209–222.

Lamberti, L.M., Bourgeois, A.L., Fischer Walker, C.L., Black, R.E., and Sack, D. (2014). Estimating diarrheal illness and deaths attributable to Shigellae and enterotoxigenic *Escherichia coli* among older children, adolescents, and adults in South Asia and Africa. PLOS Negl. Trop. Dis. *8*, e2705.

Lan, R., Alles, M.C., Donohoe, K., Martinez, M.B., and Reeves, P.R. (2004). Molecular evolutionary relationships of enteroinvasive *Escherichia coli* and *Shigella* spp. Infect. Immun. *72*, 5080–5088.

Levine, M.M., Ferreccio, C., Prado, V., Cayazzo, M., Abrego, P., Martinez, J., Maggi, L., Baldini, M.M., Martin, W., Maneval, D., *et al.* (1993). Epidemiologic studies of *Escherichia coli* diarrheal infections in a low socioeconomic level peri-urban community in Santiago, Chile. Am. J. Epidemiol. *138*, 849–869.

Lim, J.Y., Yoon, J., and Hovde, C.J. (2010). A brief overview of *Escherichia coli* O157:H7 and its plasmid O157. J. Microbiol. Biotechnol. *20*, 5–14.

Lindsey, R.L., Pouseele, H., Chen, J.C., Strockbine, N.A., and Carleton, H.A. (2016). Implementation of Whole Genome Sequencing (WGS) for Identification and Characterization of Shiga Toxin-Producing *Escherichia coli* (STEC) in the United States. Front. Microbiol. *7*, 766.

Liu, J., Gratz, J., Amour, C., Kibiki, G., Becker, S., Janaki, L., Verweij, J.J., Taniuchi, M., Sobuz, S.U., Haque, R., *et al.* (2013). A laboratory-developed TaqMan Array Card for simultaneous detection of 19 enteropathogens. J. Clin. Microbiol. *51*, 472–480. https://doi.org/10.1128/JCM.02658-12

Liu, L., Johnson, H.L., Cousens, S., Perin, J., Scott, S., Lawn, J.E., Rudan, I., Campbell, H., Cibulskis, R., Li, M., *et al.* (2012). Global, regional, and national causes of child mortality: an updated systematic analysis for 2010 with time trends since 2000. Lancet. *379*, 2151–2161.

MacDonald, E., Moller, K.E., Wester, A.L., Dahle, U.R., Hermansen, N.O., Jenum, P.A., Thoresen, L., and Vold, L. (2015). An outbreak of enterotoxigenic *Escherichia coli* (ETEC) infection in Norway, 2012: a reminder to consider uncommon pathogens in outbreaks involving imported products. Epidemiol. Infect. *143*, 486–493.

Madic, J., Peytavin de Garam, C., Vingadassalon, N., Oswald, E., Fach, P., Jamet, E., and Auvray, F. (2010). Simplex and multiplex real-time PCR assays for the detection of flagellar (H-antigen) *fliC* alleles and intimin (*eae*) variants associated with enterohaemorrhagic *Escherichia coli* (EHEC) serotypes O26:H11, O103:H2, O111:H8, O145:H28 and O157:H7. J. Appl. Microbiol. *109*, 1696–1705.

Matsushita, S., Yamada, S., Kai, A., and Kudoh, Y. (1993). Invasive strains of *Escherichia coli* belonging to serotype O121:NM. J. Clin. Microbiol. *31*, 3034–3035.

Mauricio, R.P., Jeffries, C.M., Svergun, D.I., and Deane, J.E. (2017). The *Shigella* virulence factor IcsA relieves

N-WASP autoinhibition by displacing the verprolin homology/cofilin/acidic (VCA) domain. J. Biol. Chem. *292*, 134–145. https://doi.org/10.1074/jbc.M116.758003

McDaniel, T.K., Jarvis, K.G., Donnenberg, M.S., and Kaper, J.B. (1995). A genetic locus of enterocyte effacement conserved among diverse enterobacterial pathogens. Proc. Natl. Acad. Sci. U.S.A. *92*, 1664–1668.

McDaniel, T.K., and Kaper, J.B. (1997). A cloned pathogenicity island from enteropathogenic *Escherichia coli* confers the attaching and effacing phenotype on *E. coli* K-12. Mol. Microbiol. *23*, 399–407.

Michelacci, V., Prosseda, G., Maugliani, A., Tozzoli, R., Sanchez, S., Herrera-León, S., Dallman, T., Jenkins, C., Caprioli, A., and Morabito, S. (2016). Characterization of an emergent clone of enteroinvasive *Escherichia coli* circulating in Europe. Clin. Microbiol. Infect. *22*, 287.e11–9. https://doi.org/10.1016/j.cmi.2015.10.025

Michelacci, V., Tozzoli, R., Caprioli, A., Martinez, R., Scheutz, F., Grande, L., Sanchez, S., and Morabito, S. (2013). A new pathogenicity island carrying an allelic variant of the subtilase cytotoxin is common among Shiga toxin producing *Escherichia coli* of human and ovine origin. Clin. Microbiol. Infect. *19*, E149–156.

Miko, A., Delannoy, S., Fach, P., Strockbine, N.A., Lindstedt, B.A., Mariani-Kurkdjian, P., Reetz, J., and Beutin, L. (2013). Genotypes and virulence characteristics of Shiga toxin-producing *Escherichia coli* O104 strains from different origins and sources. Int. J. Med. Microbiol. *303*, 410–421.

Morabito, S., Tozzoli, R., Oswald, E., and Caprioli, A. (2003). A mosaic pathogenicity island made up of the locus of enterocyte effacement and a pathogenicity island of *Escherichia coli* O157:H7 is frequently present in attaching and effacing E. *coli*. Infect. Immun. *71*, 3343–3348.

Morin, N., Santiago, A.E., Ernst, R.K., Guillot, S.J., and Nataro, J.P. (2013). Characterization of the AggR regulon in enteroaggregative *Escherichia coli*. Infect. Immun. *81*, 122–132. https://doi.org/10.1128/IAI.00676-12

Nara, J.M., Cianciarullo, A.M., Culler, H.F., Bueris, V., Horton, D.S., Menezes, M.A., Franzolin, M.R., Elias, W.P., and Piazza, R.M. (2010). Differentiation of typical and atypical enteropathogenic *Escherichia coli* using colony immunoblot for detection of bundle-forming pilus expression. J. Appl. Microbiol. *109*, 35–43.

Nataro, J.P., Deng, Y., Cookson, S., Cravioto, A., Savarino, S.J., Guers, L.D., Levine, M.M., and Tacket, C.O. (1995). Heterogeneity of enteroaggregative *Escherichia coli* virulence demonstrated in volunteers. J. Infect. Dis. *171*, 465–468.

Nataro, J.P., and Kaper, J.B. (1998). Diarrheagenic *Escherichia coli*. Clin. Microbiol. Rev. *11*, 142–201.

Nataro, J.P., Kaper, J.B., Robins-Browne, R., Prado, V., Vial, P., and Levine, M.M. (1987). Patterns of adherence of diarrheagenic *Escherichia coli* to HEp-2 cells. Pediatr. Infect. Dis. J. *6*, 829–831.

Nataro, J.P., Yikang, D., Yingkang, D., and Walker, K. (1994). AggR, a transcriptional activator of aggregative adherence fimbria I expression in enteroaggregative *Escherichia coli*. J. Bacteriol. *176*, 4691–4699.

Navarro-Garcia, F., and Elias, W.P. (2011). Autotransporters and virulence of enteroaggregative *E. coli*. Gut Microbes *2*, 13–24. https://doi.org/10.4161/gmic.2.1.14933

Nishi, J., Sheikh, J., Mizuguchi, K., Luisi, B., Burland, V., Boutin, A., Rose, D.J., Blattner, F.R., and Nataro, J.P. (2003). The export of coat protein from enteroaggregative *Escherichia coli* by a specific ATP-binding cassette transporter system. J. Biol. Chem. *278*, 45680–45689.

O'Brien, A.D., Newland, J.W., Miller, S.F., Holmes, R.K., Smith, H.W., and Formal, S.B. (1984). Shiga-like toxin-converting phages from *Escherichia coli* strains that cause hemorrhagic colitis or infantile diarrhea. Science *226*, 694–696.

Olsvik, O., Wasteson, Y., Lund, A., and Hornes, E. (1991). Pathogenic *Escherichia coli* found in food. Int. J. Food Microbiol. *12*, 103–113.

Orskov, I., Wachsmuth, I.K., Taylor, D.N., Echeverria, P., Rowe, B., Sakazaki, R., and Orskov, F. (1991). Two new *Escherichia coli* O groups: O172 from 'Shiga-like' toxin II-producing strains (EHEC) and O173 from enteroinvasive *E. coli* (EIEC). APMIS *99*, 30–32.

Oswald, E., Schmidt, H., Morabito, S., Karch, H., Marches, O., and Caprioli, A. (2000). Typing of intimin genes in human and animal enterohemorrhagic and enteropathogenic *Escherichia coli*: characterization of a new intimin variant. Infect. Immun. *68*, 64–71.

Pai, M., Kang, G., Ramakrishna, B.S., Venkataraman, A., and Muliyil, J. (1997). An epidemic of diarrhoea in south India caused by enteroaggregative *Escherichia coli*. Indian J. Med. Res. *106*, 7–12.

Paton, A.W., Srimanote, P., Talbot, U.M., Wang, H., and Paton, J.C. (2004). A new family of potent AB(5) cytotoxins produced by Shiga toxigenic *Escherichia coli*. J. Exp. Med. *200*, 35–46. https://doi.org/10.1084/jem.20040392

Pavlovic, M., Luze, A., Konrad, R., Berger, A., Sing, A., Busch, U., and Huber, I. (2011). Development of a duplex real-time PCR for differentiation between *E. coli* and *Shigella* spp. J. Appl. Microbiol. *110*, 1245–1251. https://doi.org/10.1111/j.1365-2672.2011.04973.x

Persson, S., Olsen, K.E., Ethelberg, S., and Scheutz, F. (2007). Subtyping method for *Escherichia coli* shiga toxin (verocytotoxin) 2 variants and correlations to clinical manifestations. J. Clin. Microbiol. *45*, 2020–2024.

Qadri, F., Svennerholm, A.M., Faruque, A.S., and Sack, R.B. (2005). Enterotoxigenic *Escherichia coli* in developing countries: epidemiology, microbiology, clinical features, treatment, and prevention. Clin. Microbiol. Rev *18*, 465–483.

Regua-Mangia, A.H., Gomes, T.A., Vieira, M.A., Andrade, J.R., Irino, K., and Teixeira, L.M. (2004). Frequency and characteristics of diarrhoeagenic *Escherichia coli* strains isolated from children with and without diarrhoea in Rio de Janeiro, Brazil. J. Infect. *48*, 161–167.

Reis, M.H., Guth, B.E., Gomes, T.A., Murahovschi, J., and Trabulsi, L.R. (1982). Frequency of *Escherichia coli* strains producing heat-labile toxin or heat-stable toxin or both in children with and without diarrhea in Sao Paulo. J. Clin. Microbiol. *15*, 1062–1064.

Rooney, R.M., Bartram, J.K., Cramer, E.H., Mantha, S., Nichols, G., Suraj, R., and Todd, E.C. (2004). A review of outbreaks of waterborne disease associated with ships:

evidence for risk management. Public Health Rep. *119*, 435–442. https://doi.org/10.1016/j.phr.2004.05.008

Sabat, A.J., Budimir, A., Nashev, D., Sa-Leao, R., van Dijl, J., Laurent, F., Grundmann, H., and Friedrich, A.W. (2013). Overview of molecular typing methods for outbreak detection and epidemiological surveillance. Euro Surveill. *18*, 20380.

Sack, R.B. (2011). The discovery of cholera-like enterotoxins produced by *Escherichia coli* causing secretory diarrhoea in humans. Indian J. Med. Res. *133*, 171–180.

Sankaran, K., Banerjee, S., Pavankumar, A.R., Jesudason, M., Reissbrodt, R., and Williams, P.H. (2009). Apyrase-based colorimetric test for detection of *Shigella* and enteroinvasive *Escherichia coli* in stool. Diagn. Microbiol. Infect. Dis. *63*, 243–250. https://doi.org/10.1016/j.diagmicrobio.2008.10.014

Sansonetti, P.J., Kopecko, D.J., and Formal, S.B. (1982). Involvement of a plasmid in the invasive ability of *Shigella flexneri*. Infect. Immun. *35*, 852–860.

Sarantuya, J., Nishi, J., Wakimoto, N., Erdene, S., Nataro, J.P., Sheikh, J., Iwashita, M., Manago, K., Tokuda, K., Yoshinaga, M., *et al.* (2004). Typical enteroaggregative *Escherichia coli* is the most prevalent pathotype among *E. coli* strains causing diarrhea in Mongolian children. J. Clin. Microbiol. *42*, 133–139.

Scavia, G., Staffolani, M., Fisichella, S., Striano, G., Colletta, S., Ferri, G., Escher, M., Minelli, F., and Caprioli, A. (2008). Enteroaggregative *Escherichia coli* associated with a food-borne outbreak of gastroenteritis. J. Med. Microbiol. *57*, 1141–1146.

Schauer, D.B., and Falkow, S. (1993). Attaching and effacing locus of a *Citrobacter freundii* biotype that causes transmissible murine colonic hyperplasia. Infect. Immun. *61*, 2486–2492.

Scheutz, F., Nielsen, E.M., Frimodt-Moller, J., Boisen, N., Morabito, S., Tozzoli, R., Nataro, J.P., and Caprioli, A. (2011). Characteristics of the enteroaggregative Shiga toxin/verotoxin-producing *Escherichia coli* O104:H4 strain causing the outbreak of haemolytic uraemic syndrome in Germany, May to June 2011. Euro Surveill. *16*, 19889.

Scheutz, F., Teel, L.D., Beutin, L., Piérard, D., Buvens, G., Karch, H., Mellmann, A., Caprioli, A., Tozzoli, R., Morabito, S., *et al.* (2012). Multicenter evaluation of a sequence-based protocol for subtyping Shiga toxins and standardizing Stx nomenclature. J. Clin. Microbiol. *50*, 2951–2963. https://doi.org/10.1128/JCM.00860-12

Schmidt, H., Knop, C., Franke, S., Aleksic, S., Heesemann, J., and Karch, H. (1995). Development of PCR for screening of enteroaggregative *Escherichia coli*. J. Clin. Microbiol. 33, 701–705.

Sheikh, J., Czeczulin, J.R., Harrington, S., Hicks, S., Henderson, I.R., Le Bouguénec, C., Gounon, P., Phillips, A., and Nataro, J.P. (2002). A novel dispersin protein in enteroaggregative *Escherichia coli*. J. Clin. Invest. *110*, 1329–1337. https://doi.org/10.1172/JCI16172

Silva, R.M., Toledo, M.R., and Trabulsi, L.R. (1980). Biochemical and cultural characteristics of invasive *Escherichia coli*. J. Clin. Microbiol. *11*, 441–444.

Small, P.L., Isberg, R.R., and Falkow, S. (1987). Comparison of the ability of enteroinvasive *Escherichia coli*, *Salmonella typhimurium*, *Yersinia pseudotuberculosis*, and *Yersinia enterocolitica* to enter and replicate within HEp-2 cells. Infect Immun. *55*, 1674–1679.

Sousa, M.A., Mendes, E.N., Collares, G.B., Peret-Filho, L.A., Penna, F.J., and Magalhaes, P.P. (2013). *Shigella* in Brazilian children with acute diarrhoea: prevalence, antimicrobial resistance and virulence genes. Memorias do Instituto Oswaldo Cruz *108*, 30–35.

Spangler, B.D. (1992). Structure and function of cholera toxin and the related *Escherichia coli* heat-labile enterotoxin. Microbiol. Rev. *56*, 622–647.

Sperandio, V., Kaper, J.B., Bortolini, M.R., Neves, B.C., Keller, R., and Trabulsi, L.R. (1998). Characterization of the locus of enterocyte effacement (LEE) in different enteropathogenic *Escherichia coli* (EPEC) and Shiga-toxin producing *Escherichia coli* (STEC) serotypes. FEMS Microbiol. Lett. *164*, 133–139.

Stacy-Phipps, S., Mecca, J.J., and Weiss, J.B. (1995). Multiplex PCR assay and simple preparation method for stool specimens detect enterotoxigenic *Escherichia coli* DNA during course of infection. J. Clin. Microbiol. *33*, 1054–1059.

Tarr, C.L., and Whittam, T.S. (2002). Molecular evolution of the intimin gene in O111 clones of pathogenic *Escherichia coli*. J. Bacteriol. *184*, 479–487.

Toma, C., Lu, Y., Higa, N., Nakasone, N., Chinen, I., Baschkier, A., Rivas, M., and Iwanaga, M. (2003). Multiplex PCR assay for identification of human diarrheagenic *Escherichia coli*. J. Clin. Microbiol. *41*, 2669–2671.

Torres, M.E., Pirez, M.C., Schelotto, F., Varela, G., Parodi, V., Allende, F., Falconi, E., Dell'Acqua, L., Gaione, P., Mendez, M.V., *et al.* (2001). Etiology of children's diarrhea in Montevideo, Uruguay: associated pathogens and unusual isolates. J. Clin. Microbiol. *39*, 2134–2139.

Tozzoli, R., Caprioli, A., Cappannella, S., Michelacci, V., Marziano, M.L., and Morabito, S. (2010). Production of the subtilase AB5 cytotoxin by Shiga toxin-negative *Escherichia coli*. J. Clin. Microbiol. *48*, 178–183. https://doi.org/10.1128/JCM.01648-09

Trabulsi, L.R., Keller, R., and Tardelli Gomes, T.A. (2002). Typical and atypical enteropathogenic *Escherichia coli*. Emerging Infect. Dis. *8*, 508–513.

Tsai, C.C., Chen, S.Y., and Tsen, H.Y. (2003). Screening the enteroaggregative *Escherichia coli* activity and detection of the aggA, aafA, and astA genes with novel PCR primers for the *Escherichia coli* isolates from diarrhea cases in Taiwan. Diagn. Microbiol. Infect. Dis. *46*, 159–165.

Vial, P.A., Robins-Browne, R., Lior, H., Prado, V., Kaper, J.B., Nataro, J.P., Maneval, D., Elsayed, A., and Levine, M.M. (1988). Characterization of enteroadherent-aggregative *Escherichia coli*, a putative agent of diarrheal disease. J. Infect. Dis. *158*, 70–79.

Vidal, R., Vidal, M., Lagos, R., Levine, M., and Prado, V. (2004). Multiplex PCR for diagnosis of enteric infections associated with diarrheagenic *Escherichia coli*. J. Clin. Microbiol. *42*, 1787–1789.

Voeroes, S., Redey, B., and Csizmazia, F. (1964). Antigenic structure of a new enteropathogenic *E. coli* strain. Acta Microbiol. Acad. Sci. Hung. *11*, 125–129.

Werber, D., Krause, G., Frank, C., Fruth, A., Flieger, A., Mielke, M., Schaade, L., and Stark, K. (2012). Outbreaks of virulent diarrheagenic *Escherichia coli*

– are we in control? BMC Med. *10,* 11. https://doi.org/10.1186/1741-7015-10-11

Youmans, B.P., Ajami, N.J., Jiang, Z.D., Petrosino, J.F., DuPont, H.L., and Highlander, S.K. (2014). Development and accuracy of quantitative real-time polymerase chain reaction assays for detection and quantification of enterotoxigenic *Escherichia coli* (ETEC) heat labile and heat stable toxin genes in travelers' diarrhea samples. Am. J. Trop. Med. Hyg. *90,* 124–132.

Extra-intestinal Pathogenic *Escherichia coli* (ExPEC): Characteristics, Virulence Genes, Detection and Control

4

Jeroen Geurtsen* and Jan T. Poolman

Bacterial Vaccine Discovery and Early Development, Janssen Vaccines & Prevention B.V., Leiden, The Netherlands.

*Corresponding author E-mail: jgeurtse@its.jnj.com

https://doi.org/10.21775/9781910190777.04

Abstract

Extraintestinal pathogenic *Escherichia coli* (ExPEC) are major human pathogens that carry combinations of genes encoding colonization factors and toxins that allow infection outside of the gastrointestinal tract, and that act to suppress host immune responses. Subgroups within the ExPEC family cause urinary tract infections, urosepsis, bacteraemia, and neonatal meningitis, and they may be associated with inflammatory bowel disease. However, clinical distinctions are blurred because ExPEC strains may cause disease in more than one site, and the genomic and phenotypic diversity amongst ExPEC has so far eluded clear-cut classification. A virulent, antibiotic-resistant clonal group called ST131 emerged and spread globally after 2000, and is now the most common cause of multiresistant ExPEC infections worldwide. Antibiotic resistant ExPEC and the looming threat of a pan-resistant ExPEC strain represent a worldwide threat and a major challenge facing infection control in the 21st century. The availability of *E. coli* vaccines could reduce the prevalence of targeted strains and make positive inroads on antibiotic resistance levels. Advances in ExPEC infection control require a better understanding of the interactions between pathogen and host and factors that drive fitness, transmissibility, and virulence.

Introduction

Escherichia coli account for only a small percentage of the overall gut flora, yet they are ubiquitous in the human gastrointestinal tract. *E. coli* display enormous diversity in phenotype and genotype, and their ability to colonize the human gut and cause human disease is equally as varied. *E. coli* are loosely divided into three families, although high genome plasticity and considerable genetic overlap between and within the three families makes classification of individual strains problematic. The largest family group are the commensal strains, which are normal residents of the gastrointestinal tract and do not usually cause disease. The second large family cause diarrhoeal intestinal disease (enteropathogenic, enterotoxigenic, enterohaemorrhagic, enteroaggregative, enteroinvasive and diffusely adherent *E. coli*). The third group is the extra-intestinal pathogenic *E. coli* or ExPEC strains. ExPEC carry combinations of genes that encode colonization factors and toxins that make these strains capable of causing disease outside of the gastrointestinal tract, predominantly urinary tract infections (UTIs), but also bacteraemia (Kaper *et al.*, 2004). Only around 20% of coding sequences are shared by all strains, and approximately 64% of sequences are shared between commensal and ExPEC strains (Lukjancenko *et al.*, 2010).

Within the ExPEC family are subgroups that cause, but are not restricted to, infections of the urinary tract (uropathogenic *E. coli* or UPEC), neonatal meningitis, and septicaemia (known as neonatal meningitis *E. coli* or NMEC), and strains that appear to be associated with Crohn's disease (adherent invasive *E. coli* or AIEC). These clinical distinctions are blurred because many ExPEC strains overlap in the clinical symptoms they produce. Furthermore, there are no single virulence traits or groups of traits that are pathognomonic of a particular ExPEC strain or of a single clinical profile. Multiple gene combinations can cause the same clinical profile, and conversely, different clinical profiles can be caused by the same strain. This has hindered attempts to classify specific *E. coli* pathotypes by the range of clinical symptoms they produce, and by the presence or absence of key virulence traits. Comparative genomics is contributing to this field, and has demonstrated substantial genetic diversity within pathovars, indicating that the distinction between pathovars is considerably more complex than originally thought. Nevertheless, the term 'ExPEC' distinguishes *E. coli* strains from commensal and diarrhoeagenic strains based on the presence of a number of house-keeping genes. The distinctions between ExPEC strains reflected by the bacterial nomenclature (UPEC, AIEC, etc) reflect clinically meaningful differences in disease potential.

It is worth noting that the clinical outcome of an ExPEC infection is not only determined by microbiological factors related to the infecting strain, such as the expression of virulence traits. Host factors are equally important (Johnson *et al.*, 2015a). Clinical disease is strongly influenced by age, the presence of underlying pathology such as immunosuppression, the route of bacterial entry, for example the urinary tract or an invasive surgical procedure, venous catheter, or indwelling urinary catheter, and previous exposure to antibiotics which predisposes to selection of antibiotic-resistant strains (Costelloe *et al.*, 2010; Roubaud Baudron *et al.*, 2014).

ExPEC-associated clinical diseases

E. coli is a major human pathogen and can cause infection in any body system. The frequency, severity and clinical significance of ExPEC as a cause of community- and hospital-acquired infections may be underestimated by the public and health care professionals alike. Given that disease coding in hospital databases may be incomplete, with bacteraemia coded without the causative organism either identified or recorded, *E. coli* bacteraemia/sepsis may be significantly underreported. It is perhaps not recognized that the incidence of ExPEC bacteraemia in older adults exceeds that due to pneumococcal bacteraemia (Jackson *et al.*, 2005); that *E. coli* is the most common cause of neonatal meningitis in preterm and low birth weight infants (Simonsen *et al.*, 2014); and that *E. coli* infections caused by antibiotic resistant strains account for one-half of the estimated global burden associated with antibiotic resistant infections (O'Neill, 2014).

ExPEC cause soft tissue infections, infections of the respiratory tract including pneumonia, skin infections, post-operative wound infections, prostatitis, intra-abdominal infections including peritonitis, infections of the biliary tract, and pelvic inflammatory disease, and they have been implicated in the pathogenesis of inflammatory bowel disease (Petersen *et al.*, 2015). Any of these infections can lead to bacteraemia and/or sepsis. However, the most common ExPEC infections are lower and upper UTIs, with severe upper UTIs leading to robust inflammatory responses and (multi) organ failure/urosepsis (often without a positive blood culture), as well as neonatal meningitis and bacteraemia. More recently, a potential role for ExPEC in chronic inflammatory diseases such as Crohn's disease was suggested and the implications for prevention and treatment are under investigation.

Most ExPEC infections are community-acquired, including uncomplicated cystitis, urosepsis and bacteraemia. In the hospital setting, *E. coli* are the most common cause of nosocomial infections overall, causing infections of central venous access lines and indwelling urinary catheters, (abdominal) surgical site infections, and ventilator-associated pneumonia (Weiner *et al.*, 2016). Compared to community-acquired infections, nosocomial ExPEC infections are associated with higher rates of antibiotic resistance, higher rates of admission to intensive care, and higher mortality (Bou-Antoun *et al.*, 2016; Søgaard *et al.*, 2011; Williamson *et al.*, 2013a). In addition to specific strategies for infection control and management in hospitals, targeted population-based strategies are needed at the community level to prevent *E. coli* infections.

Urinary tract infection (UTI)

Urinary tract infections are some of the most common bacterial infections; the major risk factors being female, having had a previous UTI, sexual activity, invasive procedures (e.g. catheterization), and underlying diseases such as diabetes (Foxman, 2014). UPEC cause up to 95% of uncomplicated cases of cystitis, 90% of pyelonephritis, and approximately one-third of catheter-associated UTI (Russo and Johnson, 2003). The lifetime risk for UTI among healthy women is 60%, and the risk of recurrence is 30–50% annually (Foxman, 2014). In 2009, UTIs accounted for approximately 0.9% of all outpatient and emergency department visits in the United States (US) (Foxman, 2014). In the same year, there were 22.57 cases of hospitalized *E. coli* UTI per 1000 hospitalizations, a figure which was 50% higher than in the year 2000, representing an increase driven by increasing rates of antibiotic resistance and resulting outpatient treatment failures (Zilberberg and Shorr, 2013). The health care burden of *E. coli* UTI in terms of outpatient consultations, diagnosis, antibiotic use, and hospitalization is immense.

UPEC are a highly genetically and phylogenically diverse group found in the gastrointestinal tract as commensals, but cause disease when they gain entry to the urinary tract. UPEC that cause cystitis appear to be more closely related to commensal strains than to strains that cause pyelonephritis (Johnson *et al.*, 2005). Unlike commensal strains, UPEC carry pathogenicity islands that are distinct regions of genomic DNA absent in non-pathogenic strains. As a consequence, UPEC display numerous virulence traits that allow their survival in the urinary tract and promote colonization of uroepithelium.

Neonatal meningitis

After group B streptococcus, NMEC is the leading cause of neonatal meningitis (Gaschignard *et al.*, 2011; Okike *et al.*, 2014). NMEC is the most frequent cause of neonatal meningitis in a subset of pre-term and low-birthweight infants, and may be increasing in term neonates due to extensive maternal antibiotic prophylaxis measures taken against group B streptococcus infection (Simonsen *et al.*, 2014). *E. coli* may also cause meningitis in older age groups following trauma, or in patients with liver cirrhosis (Cabellos *et al.*, 2008).

Full genome sequencing of NMEC strains is in its early stages (Lu *et al.*, 2011; Wijetunge *et al.*, 2015), and virulence traits and genomic traits considered typical of NMEC are not well characterized (Logue *et al.*, 2012). NMEC appear to resemble faecal commensal strains in terms of phylotype and O-serotype, but may be able to be differentiated by the presence of pathogenicity islands (Logue *et al.*, 2012). Additionally, characteristics such as lactose fermentation, haemolytic activity, and antibiotic resistance profile, do not differentiate between commensal and NMEC strains.

Bacteraemia

The most frequent cause of bacteraemia in all ages combined is ExPEC (Fig. 4.1). The incidence of *E. coli* bacteraemia is strongly age dependent, and although the majority of cases occur in older adults, the incidence of *E. coli* bacteraemia in infants less than 1 year of age is higher than in 1- to 64-year-olds (Bou-Antoun *et al.*, 2016). Any ExPEC infection may be complicated by bacteraemia and/or sepsis. Approximately three-quarters of cases of bacteraemia are community acquired, although this figure includes the subgroup of patients with recent health care exposures (Bou-Antoun *et al.*, 2016; Søgaard *et al.*, 2011; Williamson *et al.*, 2013a). The most common portal of entry by which *E. coli* reaches the blood stream in adults is the urinary tract (Hernández *et al.*, 2015; Jackson *et al.*, 2005; Roubaud Baudron *et al.*, 2014), including community-acquired bacteraemia, of which 35–45% have a primary focus in this compartment (Bou-Antoun *et al.*, 2016; Hernández *et al.*, 2015; Roubaud Baudron *et al.*, 2014).

In long-term residential care facilities, the risk of bacteraemia is around 40 times higher when an indwelling catheter is *in situ* (Rudman *et al.*, 1988). Up to 6% of patients undergoing trans-rectal prostatic biopsy develop complications including bacteraemia and/or urosepsis, despite antibiotic prophylaxis (Rudzinski and Kawakami, 2014; Williamson *et al.*, 2013b). Bacteraemia is also a common complication of ExPEC infections that occur at any surgical site, of which the most common is after abdominal surgery (Weiner *et al.*, 2016).

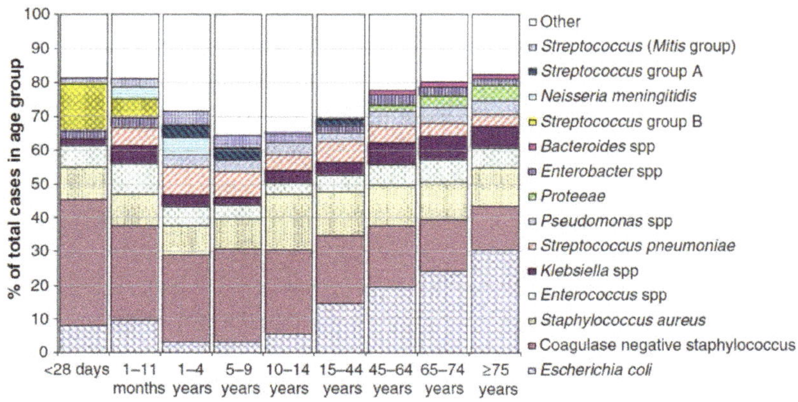

Age group	No of episodes of	% of total	Rate per 100,000
<1 year	4598	4.8	688.7
1–4 years	2099	2.2	85.3
5–9 years	859	0.9	30.1
10–14 years	720	0.7	23.6
15–44 years	14098	14.6	66.3
45–64 years	22176	23.0	172.6
65–74 years	17491	18.1	409.3
75+ years	34382	35.7	857.1
Total	**96423***		**187.4***

*The total number of isolates and the overall rate for the combined age groups is slightly lower in this analysis due to records without age data being excluded.

Figure 4.1 Isolates causing bacteraemia in England by age group in 2008 – data from 167 laboratories. Reproduced with permission from Wilson *et al.* (Wilson *et al.*, 2011).

The disease burden due to bacteraemia is strongly age dependent. While the overall incidence of *E. coli* bacteraemia is between 30/100,000 and 50/100,000 (Al-Hasan *et al.*, 2009; Laupland and Church, 2014), it is 150/100,000 among adults ≥ 65 years of age, and 452/100,000 in those aged ≥ 85 years (1998–2001, US data) (Jackson *et al.*, 2005). The case fatality rate is high at 13–19% (Ani *et al.*, 2015; Roubaud Baudron *et al.*, 2014), but increases to as much as 60% in elderly patients with nosocomial ExPEC infections (Roubaud Baudron *et al.*, 2014).

The combination of an ageing population and increasing antibiotic resistance has driven an increase in ExPEC bacteraemia. In the US and Europe between 2000 and 2008, the occurrence of ExPEC bacteraemia increased by around 6% to 8% annually, with an increasing proportion due to multidrug resistant ExPEC (Bou-Antoun *et al.*, 2016; Elixhauser *et al.*, 2011; de Kraker *et al.*, 2013; Søgaard *et al.*, 2011; Wilson *et al.*, 2011). Increases in the rates of ExPEC bacteraemia have also been reported from New Zealand, in studies from 2005 to 2011 (Williamson *et al.*, 2013a). Infections caused by antibiotic resistant *E. coli* are more likely to result in death, and they consume substantially more health care resources than infections by antibiotic-susceptible strains (WHO, 2014). Septicaemia was the most costly diagnostic category in all US hospitals in 2013, representing 3.6% of hospital stays, and 6.2% of national hospital aggregate costs (USD 23,663 million) (Torio and Moore, 2016).

Inflammatory bowel disease

A potential role for AIEC in inflammatory bowel disease was first suggested in 1999 when Boudeau *et al.* (1999) reported an invasive strain of *E. coli* (LF82) from a patient with Crohn's disease. Genome analysis provided the unexpected finding that AIEC strains show genetic similarities to UPEC and NMEC strains, although the clinical disease is not extra-intestinal *per-se* (Martinez-Medina *et al.*, 2009; Nash *et al.*, 2010). AIEC contain unique genes and distinct genomic islands which may contribute to their specific pathogenicity. By adhering to and invading intestinal epithelial cells, AIEC are able to replicate intracellularly, triggering host cell death and the release of tumour-necrosis factor. Their role in Crohn's disease is thought to be via promotion of pro-inflammatory granulomatous immune responses, although it is currently unclear whether a direct causal relationship exists.

Virulence traits

Individual ExPEC strains encode numerous different systems of virulence either within pathogenicity islands on the bacterial chromosome, or on mobile elements such as plasmids and phages. Virulence traits contribute to the ability to colonize, invade, and replicate intracellularly, produce biofilms or intracellular bacterial communities, retrieve iron necessary for optimal survival, and cause direct cellular damage (Table 4.1).

O and K surface antigens

The 'O' (*ohne hauch* referring to non-spreading growth on agar) and 'K' (*kapsel* – polysaccharide capsule) surface antigens were first described by Kauffman in the 1940s and are used for classical serotyping of *E. coli* (Orskov and Orskov, 1992) (Fig. 4.2). The third surface antigen 'H' (*hauch* – flagellum) used for classical serotyping, is discussed under 'motility factors' below. This leads to the O:K:H classification, allowing an almost endless combination of surface characteristics.

The O-antigen contributes to bacterial survival in human serum (serum resistance) by inhibiting phagocytosis and complement-mediated killing, possibly by shielding binding sites on the bacterial surface (Burns and Hull, 1998, 1999; Miajlovic and Smith, 2014; Phan *et al.*, 2013). The serum resistance capability seems to differ between individual O-antigens and is associated with their pathogenicity (Phan *et al.*, 2013; Sarkar *et al.*, 2014). Thus, while there are more than 180 distinct O-antigens described (Stenutz *et al.*, 2006), approximately 90% of meningitis isolates and more than 60% of bacteraemia isolates are caused by only 10 to 12 O-antigen types (Cross *et al.*, 1994), such as O1, O6, and O18 that are implicated in UTI, bacteraemia, and meningitis, and O25, O75, and others that are implicated in UTI and bacteraemia (Orskov and Orskov, 1985; Poolman and Wacker, 2015) (Fig. 4.3). Serotype O25, and to a lesser extent serotype O16, include the ST131 clonal group that has grown to be the most important cause of antibiotic resistant ExPEC infections worldwide (Johnson *et al.*, 2014). It is not known whether the vast array of different O-polysaccharide structures just reflects immunological variation, or whether underlying relationships between structure and function such as serum resistance exist. Other Gram-negative bacterial species such as *Pseudomonas aeruginosa* and *Salmonella* Typhimurium appear to regulate the length of the O-polysaccharide by expressing different O-antigen length regulators in response to external factors such as growth phase and local environment (Kintz and Goldberg, 2008; Murray *et al.*, 2005). For these species, different O-antigen lengths may be associated with increased serum resistance and virulence in animal models. *E. coli* is known to encode different O-antigen length regulators (Wzz, FepE), but whether modulation of O-antigen length occurs in response to environmental conditions is not currently known (Osawa *et al.*, 2013).

The K capsule antigen provides a physical barrier against host defences including the deposition of complement and contributes to serum resistance. Some capsules also mimic host structures to avoid recognition by host immune mechanisms. For example the K1 capsule resembles sialylated glycoproteins found in human neural tissue (Söderström *et al.*, 1984). There are more than 80 distinct K-antigens types known but relatively few are associated with pathogenicity (Whitfield, 2006). The genomes of strains with the K1 capsule which predominates amongst NMEC, have genes that are not found in commensal strains (such as K12 strains). The K1 capsule appears to contribute to survival after crossing the blood–brain barrier (Kaper *et al.*, 2004). K2 and K5 capsules protect UPEC from complement-mediated killing and are important virulence traits in UTI (Subashchandrabose and Mobley, 2015). Both the type and combination of O and K antigens contribute to the virulence of individual strains (Miajlovic and Smith, 2014; Phan *et al.*, 2013).

Motility structures

Flagella are considered a critical virulence trait in UTIs because they allow ascent through the urinary tract providing access to the kidneys (Lane *et al.*, 2007a; Subashchandrabose and Mobley, 2015). Flagellin, the main component of the flagellum, binds to toll-like receptor 5, stimulating the production of chemokines and cytokines and promotion of invasion of bacteria across cell membranes within the renal medulla and brain (Bens *et al.*, 2014; Haiko and Westerlund-Wikström, 2013). Individual H-antigens display differences in their ability to induce cytokine production. For example, the H4 antigen frequently expressed in ST131

Table 4.1 Representative virulence traits of extra-intestinal *Escherichia coli*

Virulence trait	Gene	Pathotype	Function
Surface antigens			
O antigen (lipopolysaccharide)	Between *galF* and *gnd*	UPEC	Resists opsonophagocytosis and complement-mediated killing
K1 (capsule)	*kpsM*	UPEC, NMEC	Anti-phagocytic, serum resistance, contributes to survival in the brain
H antigen (flagellin component)	*fliC*	All	Allows bacterial motility, induces cytokine expression, promotes cellular invasion
Adhesins			
Type 1 fimbriae (FimH tip protein)	*fimH*	UPEC, AIEC	Adherence, facilitates invasion, intracellular replication and formation of intracellular bacterial communities
P fimbriae	*papA/C/E/F/G*	UPEC, NMEC (presumed)	Adherence, binding to human brain microvascular endothelial cells
S fimbriae	*sfa*	UPEC, MNEC	Adherence, binding to sialyl galactosides
F1C fimbriae	*foc*	UPEC	Adherence to renal collecting ducts and tubules
Dr-binding adhesins	*afa/draBC*	UPEC, NMEC	Adherence to kidney cells
Long polar fimbriae	*lpfA2*	AIEC (presumed)	Possibly enhances attachment and microcolony formation
Curli	*csg*	UPEC	Adherence to matrix and plasma proteins, biofilm formation, bacterial aggregation
Ag43 (autotransporter)	*flu*	UPEC	Adherence, bacterial aggregation, biofilm formation
UpaB (autotransporter)	*upaB*	UPEC	Binding to extracellular matrix proteins
UpaG (autotransporter)	*upaG*	UPEC	Possibly adherence, bacterial aggregation, biofilm formation
UpaH (autotransporter)	*upaH*	UPEC	Binding to extracellular matrix proteins, biofilm formation
FdeC (autotransporter)	*fdeC*	UPEC	Adherence
Hemaglobin protease	*hbp*	UPEC, NMEC	Hemagglutination
Heat-resistant haemagglutinin	*hra*		
Iha	*iha*	UPEC	Adherence and iron uptake
TosA	*TosA*	UPEC	Adherence to kidney cells, fitness during infection
ExPEC adhesin 1	*yqi*	UPEC, NMEC	Initial colonization
Toxins			
Invasion of brain endothelium	*ibeA*	UPEC, NMEC (presumed), AIEC (presumed)	Contributes to invasion of the blood–brain barrier
Alpha-haemolysin	*hlyCABD*	UPEC	Adherence, cellular toxicity, lysis of erythrocytes
Vacuolating autotransporter toxin	*vat*	UPEC	Toxin that may improve fitness during infection, causes vacuole formation uroepithelial cells
Secreted autotransporter toxin	*sat*	UPEC	Damages renal tubular cells
Serine protease	*Pic*	UPEC	Targets cytoskeletal proteins
Cytotoxic necrotizing factor	*Cnf1*	UPEC, NMEC	Contributes to invasion of epithelium and the blood–brain barrier, disrupts neutrophil function

Table 4.1 Continued

Virulence trait	Gene	Pathotype	Function
Siderophores			
Siderophore receptor	*ireA*	UPEC	Iron uptake
Salmochelin receptor	*iroN*	UPEC	Iron uptake
Peri-plasmic iron binding receptor	*fecB*		Iron uptake
Chu outer membrane receptor protein	*chuA*	UPEC, MNEC	Haem binding
Aerobactin receptor	*iutA*	UPEC	Iron uptake
Yersiniabactin receptor	*fyuA*	UPEC	Iron uptake

UPEC: Uropathogenic *E. coli*, NMEC: Neonatal meningitis-associated *E. coli*, AIEC: Adherent invasive *E. coli*

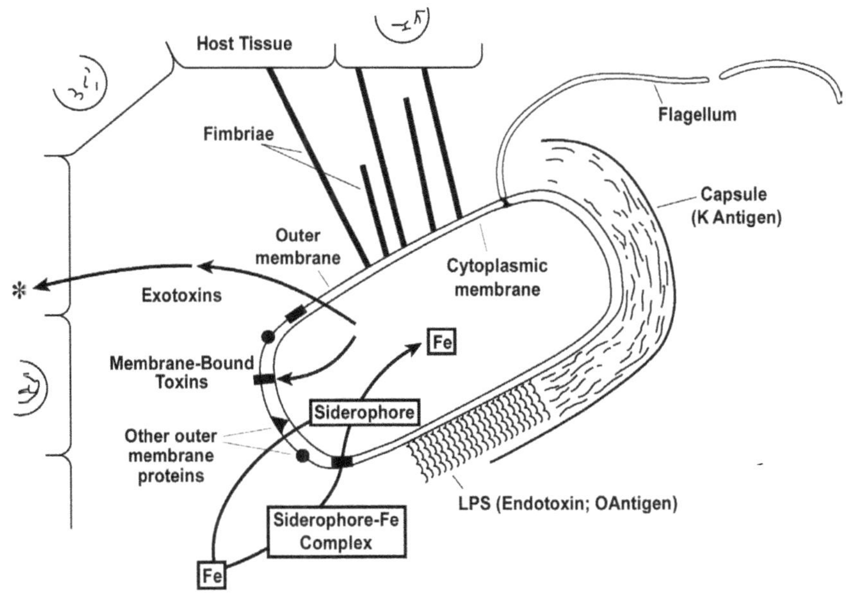

Figure 4.2 *Escherichia coli* cell surface structures. Reproduced with permission from Dr Barry Eisenstein (Eisenstein and Jones, 1988).

strains induces higher amounts of IL-10 release than H1 and H7 antigens, possibly contributing to the success of ST131 as a cause of UTI (Kakkanat *et al.*, 2015).

As well as contributing to motility within the urinary tract, epithelial invasion and access to the bloodstream, flagella promote adhesion, and may facilitate biofilm formation (Zhou *et al.*, 2015). There are more than 50 types of the H-antigen and more than one H-type can be present in an individual pathovar (Geue *et al.*, 2014).

Adhesins and invasins

Adhesive proteins on the bacterial surface mediate receptor binding, trigger host and bacterial cell signalling pathways, and allow colonization by ExPEC of epithelial surfaces outside of the environment of the large intestine. Adhesins recognize specific receptors on the surfaces of host cells or of other bacteria, and as a consequence, they vary considerably in their structure and function, ranging from curli and fimbrial structures (or pili) of variable size and shape that extend out from the bacterial cell surface and non-fimbrial surface proteins that are located directly on the cell surface.

Fimbriae allow contact between surfaces that occur at a distance, allowing the surfaces to remain separated. Their structure is such that fimbriae provide strong and stable adhesive anchorage that is resistant to dissociation. Fimbriae are assembled

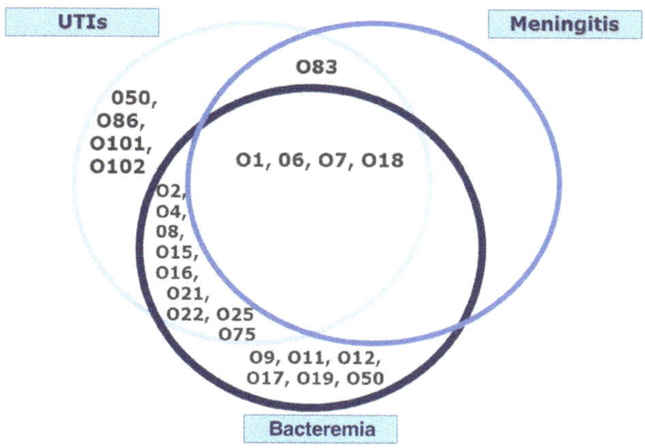

Figure 4.3 *Escherichia coli* O serotypes causing extra-intestinal infections. Reproduced with permission from Poolman *et al*. (Poolman and Wacker, 2015). Data from the article by Orskov *et al*. (Orskov and Orskov, 1985).

by the chaperone-usher pathway, which may be present in multiple copies in a single *E. coli* strain and capable of producing different adherence structures. The expression of chaperone-usher fimbriae pili is in direct response to the environmental context, shows phase variation, and is highly regulated. This allows expression of the correct fimbriae specific to the local environment (Lane *et al*., 2007b; Snyder *et al*., 2005). There are multiple possible fimbrial types, such as Afa/Dr, F9, S, F1C, type 3, Auf, Yad, and Ygi fimbriae (Antão *et al*., 2009; Chahales and Thanassi, 2015), but Type 1 fimbriae and P fimbriae are the best described.

Type 1 fimbriae are made up of multiple copies of the FimA building block that shows antigenic variation and forms the pilus rod and a distal tip composed of a single copy of FimF, FimG, and FimH subunits (Fig. 4.4). The distal FimH subunit is antigenically conserved and binds to alpha-D-manosylated proteins on uroplaktin, a major component of the bladder epithelium, triggering a host signalling mechanism that leads to invasion of the host cell and formation of intracellular bacterial communities (Abraham *et al*., 1988; Chahales and Thanassi, 2015). FimH contributes to surface motility and increased adhesive capacity under mechanical stress, which is important during colonization of the urinary tract (Sauer *et al*., 2016). FimH uses a 'catch-bond' mechanism for binding. A recent study established that the FimH catch-bond structure exists in different states, allowing rapid release, flagellar motility, and colonization of new tissues under some conditions, and tight adhesive binding in settings of shear mechanical stress, such as during urination (Sauer *et al*., 2016). FimH is

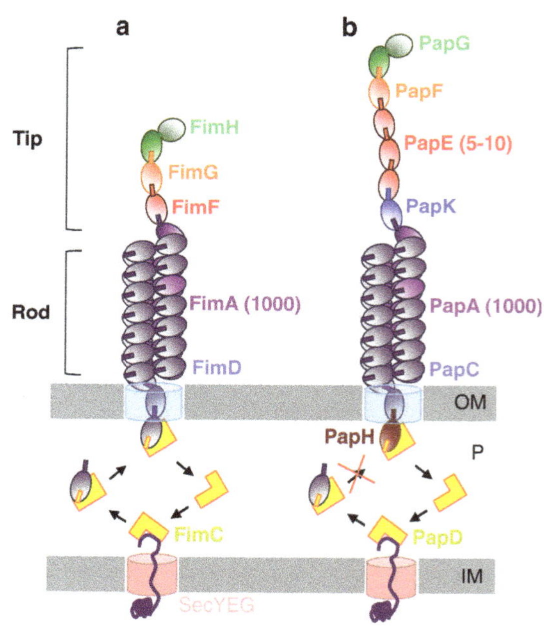

Figure 4.4 *Escherichia coli* type 1 (a) and P pili (b). Reproduced with permission from Lillington *et al*. 2014. Subunits are located at the periplasm (P) via the Sec general secretory pathway across the inner membrane (IM), where they are folded and transported to the usher with the aid of a chaperone. Here, they undergo donor-strand exchange before translocation beyond the outer membrane (OM) to form a pilus with a well regulated ordering of subunits, which may be divided into a 'tip' and a helically wound 'rod'.

thought to be a necessary requirement for colonization of the urinary tract and a precursor for UTI (Langermann *et al.*, 1997, 2000), and as such, an attractive candidate for an *E. coli* vaccine. A subunit FimH and the chaperone FimC (FIMCH) vaccine candidate reduced bladder and kidney colonization by UPEC in mice and non-human primates (Langermann *et al.*, 1997, 2000). However, the vaccine was reported to show insufficient efficacy when tested in the clinic, and the development was put on hold (Brumbaugh and Mobley, 2012). Other studies indicate that the regulation and expression of FimH by UPEC, and the effect of specific anti-FimH antibodies on binding, is more complex than initially thought (Brumbaugh and Mobley, 2012). For example, there is evidence that the conformation of the FimH adhesin (low-binding versus high-binding conformation) affects exposure and access of antibodies to binding epitopes, and thus potentially the immunogenicity and effectiveness of anti-FimH vaccines (Kisiela *et al.*, 2013). The P fimbriae are longer and more flexible than Type 1 fimbriae and are made up of six subunit proteins that terminate in the PapG adhesin (Fig. 4.4). The P fimbriae *pap* genes are found on pathogenicity islands in 80% of ExPEC strains that caused pyelonephritis. The terminal PapG adhesin binds to galabiose receptors in the kidney and is considered a key virulence trait in pyelonephritis in terms of adhesion and promotion of local inflammation and tissue damage.

The Dr-family of adhesins are thin flexible fibres with a distal invasin that promotes bacterial uptake within host cells. Binding to C55 (decay accelerating factor) on the host cell activates surface expression of the major histocompatibility complex (MHC) class I-related molecule. Dr adhesins agglutinate erythrocytes and bind to the basement membrane of renal tubular epithelium (Subashchandrabose and Mobley, 2015).

Curli are stable amyloid-like fibres made up of repeating CsgA protein. They are assembled directly on the cell surface and appear as a dense tangled mass on the cell surface (Chahales and Thanassi, 2015). Unlike most bacteria that express curli at 30 degrees Celsius, many EXPEC strains produce curli at 37°C (Chahales and Thanassi, 2015). Curli appear to have diverse roles in early colonization of the urinary tract, fostering bacterial aggregation and the formation of biofilms.

The invasion of brain endothelium invasin A identified in NMEC strains mediates translocation of the bacteria across the blood–brain barrier. The same invasin has been found in AIEC and appears to have an important role in the invasion of intestinal epithelium, and may promote survival of AIEC in macrophages (Cieza *et al.*, 2015).

Autotransporters

Autotransporters are a family of secreted proteins that are transported to the cell surface using the translocator that spans the outer membrane. Because autotransporters are referred to by the transport mechanism and not by their function, they consist of proteins with diverse functions including adhesins, invasins, toxins, and proteases. Multiple autotransporters are present in a single genome and the AIDA-I autotransporter family are the most frequently identified in UPEC strains (Chahales and Thanassi, 2015). Autotransporters may be directly involved in adhesion to cell surfaces, and can promote aggregation and biofilm formation is settings where type 1 pili are not expressed.

Siderophores and iron acquisition

Obtaining iron from the local environment is a survival requirement for *E. coli* and many other pathogens, and multiple mechanisms exist to ensure iron availability and provide redundancy (Garcia *et al.*, 2011). The nutrient-poor environment of the urinary tract stimulates the activation of iron sequestration mechanisms such as siderophores, iron transporters and outer membrane haem receptors (Subashchandrabose and Mobley, 2015).

Siderophores are iron scavenging molecules released by bacteria to bind and transport ferric iron from outside the host back into the cell cytosol. Siderophores are secreted molecules that have a high affinity for the ferric iron necessary for growth and persistence. Once bound, iron-laden siderophores are detected by bacterial receptors that signal transportation using the TonB complex located in the inner cell membrane (Subashchandrabose and Mobley, 2015). Once in the periplasm, ABC transporters effect translocation into the cell cytosol where the iron is removed. Siderophore genes are found on pathogenicity islands and multiple systems are usually present in ExPEC strains (Wiles *et al.*, 2008).

Haem receptors (ChuA and Hma) located on the outer membrane bind haem and facilitate its transport into the periplasm via the TonB-inner membrane complex (Subashchandrabose and Mobley, 2015). Ferrous iron is bound and transported across the cell membrane via the Sit system of proteins.

Toxins

Numerous secreted toxins are capable of affecting a wide variety of cellular processes. However, toxins play a greater role in the pathogenesis of intra-intestinal *E. coli* infections than in ExPEC infections, where strategies to adhere, invade, and survive in biofilms or intracellular communities are more necessary. Toxins secreted by ExPEC include alpha haemolysin, cytotoxic necrotizing factor, several autotransporter toxins, and the invasion of brain endothelium invasin.

Alpha-haemolysin haemolyses erythrocytes, thereby releasing their iron for uptake by siderophores. Haemolysin also inhibits host immune mechanisms and is directly toxic to host cells and immune cells, contributing to inflammation and tissue injury (Subashchandrabose and Mobley, 2015). Haemolysin is encoded by the *hly* gene, of which multiple copies may exist in a single strain, frequently with other virulence traits such as P fimbriae (Johnson, 1991). Haemolysin is produced by *E. coli* associated with pyelonephritis, and to a lesser extent, by strains associated with UTI and bacteraemia (Johnson, 1991).

The family of autotransporter toxins (serine-protease autotransporter toxins of Enterobacteriaceae or SPATE) include the secreted autotransporter toxin (Sat), which is found in 68% of pyelonephritis ExPEC isolates and 14% of faecal isolates (Welch, 2016). Vacuolating autotransporter toxin and cytotoxic necrotizing factor contribute to UPEC fitness during infection (Nichols *et al.*, 2016; Welch, 2016). Cytotoxic necrotizing factor influences actin remodelling and promotes access to deeper epithelial tissues by disrupting the epithelial barrier.

Toxins produced by ExPEC reach their target via numerous mechanisms that include secretory systems that deliver them across the cell membrane. The Type I secretory system is composed of three proteins: an outer membrane pore-forming protein, a membrane fusion protein, and an inner membrane binding cassette, which act to deliver alpha-haemolysin into the plasma membrane of host cells, leading to host cell disruption (Henderson *et al.*, 2004). Type V secreted toxins are the autotransporters that are exported through the cytoplasmic membrane via the Sec inner membrane secretion pathway. Both systems form β-barrel structures that embed in the outer membrane, allowing the passage of proteins through the outer membrane to the cell surface or extracellular environment (Henderson *et al.*, 2004).

More recently, the Type 6 secretory system has been characterized in detail (Journet and Cascales, 2016). This syringe-like structure consists of 13 subunits that are assembled within the cell membrane and loaded with effector toxins. On contact with a target bacterial or human host cell, an inner tube is propelled towards the target, puncturing the target cell membrane and delivering the effector molecules. Type 6 secretory systems in *E. coli* may deliver anti-bacterial toxins, or toxins that contribute to pathogenesis and virulence within the host (Journet and Cascales, 2016).

Evasion of host immune responses

ExPEC actively promote an immune-suppressive environment to improve survival outside of the gastrointestinal tract. Within the urinary tract, UPEC promote secretion of anti-inflammatory cytokines such as IL-10 that reduce Th-1 responses to infection. IL-10 is secreted in large amounts by mast cells early on during episodes of cystitis, a mechanism thought to attenuate the inflammatory response and allow regeneration of bladder epithelium (Abraham and Miao, 2015; Chan *et al.*, 2013). As a result, while the innate immune response to UPEC bladder infections is initially robust, downstream adaptive immunity and antibody production do not occur, potentially contributed to by the added immunosuppressive activities of IL-10 secretion induced by UPEC (Abraham and Miao, 2015). In a mouse model of epididymitis, the inflammatory response to an UPEC strain was markedly lower than that observed following infection with a strain of non-pathogenic *E. coli*, suggesting pathogen specific mechanisms capable of modulating immune responses (Lang *et al.*, 2014). Invasion of the uroepithelium enables the development of intracellular bacterial communities, which are shielded from host defence mechanisms, as well as the action of antibiotics. These reservoirs of bacteria reside within membrane-bound compartments in transitional cells, and are capable of remaining viable for long periods. The dispersal of intracellular bacterial communities promotes the spread of bacteria to new cells.

Within the blood stream, the O and K antigens play a critical role in inhibiting phagocytosis and complement-mediated killing (Sarkar *et al.*, 2014). O and K antigens also modulate and reduce neutrophil chemotaxis and the release of oxygen radicals. *E. coli* serum resistance is also enhanced by the presence of immunoglobulin-binding proteins on the cell surface that bind the Fc-region of immunoglobulin G in a non-immune manner (Leo and Goldman, 2009; Sandt and Hill, 2000). Immunoglobulin binding may prevent phagocytosis and physically protect the bacterium from the effects of complement.

Challenges in predicting virulence of individual ExPEC strains

Attempts to link a panel of virulence genes to clinical source, host characteristics and lethality in a mouse model have served to emphasize the highly complex interactions between bacterial and host characteristics in determining virulence (Johnson *et al.*, 2006, 2015a).

The assessment of the virulence potential of ExPEC strains has traditionally been done by either investigating one or more bacterial traits from a single strain in an animal model (Table 4.2), or investigation of multiple strains isolated from a defined source (UTI isolates) compared to faecal strains. Both approaches have significant limitations which have so far prevented identification of definitive links between the presence of virulence traits in individual strains and clinical disease. Confounding these associations is the close relationship between host characteristics and susceptibility to ExPEC infections (Johnson *et al.*, 2015a). Virulence is inversely proportional to underlying health status, whereby elderly, immunocompromised individuals, or individuals with an impaired urinary tract may be infected by strains of relatively low-virulence, while more virulent strains are required to infect younger (healthier) individuals, particularly those with an intact renal system (Johnson *et al.*, 2015a). Furthermore, the intestine is usually the reservoir for many or most ExPEC infections, which means that faecal strains are not always by definition, non-virulent.

The investigation of preventative and treatment modalities in animal models makes invaluable contributions to understanding the pathogenesis, prevention, and treatment of human diseases. Nevertheless, there are no animal models that are completely representative of the human setting. Inter-special differences in specific receptors influence the pathogenicity and the physiological response to infection. The phenotype of bacteria harvested from culture conditions may differ from those occurring outside the laboratory, altering their pathogenicity. The resulting phenotype may require a higher dose of pathogen to predictably induce infection in healthy animals, and animal models are unable to completely account for the effects of underlying chronic disease or age on susceptibility and the immune response. The route of administration of pathogen may not replicate bacterial entry into a human host. Experiments of *E. coli* strains in animal models are therefore limited by the uncertainty around their direct applicability to humans and because the presence or absence of virulence traits does not guarantee their expression or their role in infection.

The antibody response to ExPEC infection is predominantly against the O-antigen (Miajlovic and Smith, 2014). Studies in animal models indicate that O-specific and K-specific antibodies promote phagocytosis and confer protection against homologous lethal challenge (Kaijser and Ahlstedt, 1977). Therefore, *in vitro* tests of functional killing can provide useful information about virulence, and will be important in the development of, for instance, *E. coli* vaccines. Given that immunity to *E. coli* is mediated via complement-mediated lysis and opsonophagocytic killing (OPK) (Burns and Hull, 1999; Miajlovic and Smith, 2014), it is possible that correlates of protection using functional antibody responses in opsonophagocytic killing assays (OPKA) can be identified. An OPKA developed and qualified for measuring functional antibodies to multiple ExPEC strains has been used to assess the immunogenicity of a 4-valent bioconjugate ExPEC vaccine administered to humans (Abbanat *et al.*, 2016).

Table 4.2 In vivo test systems for characterization of ExPEC: advantages and limitations

Model	Advantages	Disadvantages
Murine subcutaneous sepsis model (Johnson *et al.*, 2006, 2015a)	Experimental virulence corresponds with bacterial traits, host characteristics, source	Mammalian model requiring specialized facilities Direct applicability to specific human systems (urinary tract, meninges, blood stream) not known
Mouse ascending urinary tract infection model (Johnson and Brown, 1996)	More specific for investigation of virulence of UPEC than the sepsis model Experimental virulence corresponds with bacterial traits and host characteristics	Mammalian model requiring specialized facilities Risk of vesico-ureteric reflux/damage to kidneys resulting in complicated UTIs Differences exist between mouse and human bladder in the immune response
Rat model of pneumonia (Russo *et al.*, 2003)	Clinical disease is similar to that observed in humans Able to assess contribution of different virulence traits and host factors on lung injury	Mammalian model requiring specialized facilities Applicability to human response to pneumonia not known
Rat model of meningitis (Tivendale *et al.*, 2010)	Able to assess contribution of different virulence traits and host factors on meningitis	Mammalian model requiring specialized facilities Applicability to human response to meningitis not known
Wax moth *Galleria mellonella* larvae model of virulence (Johnson *et al.*, 2015a; Williamson *et al.*, 2014)	Technically simple invertebrate model Dose-dependent effect of inoculum on larval survival Correlation between *in vitro* antibiotic resistance and larval survival Correlation between virulence (based on a virulence gene score) and larval killing	Not all virulent strains led to larval killing. Possibly due to lack of susceptibility to lave to certain virulence traits, or low/absent expression of the virulence trait in this model. Direct applicability to specific human systems (urinary tract, meninges, blood stream) not known Poor correlation with mouse model in terms of illness severity or lethality
Zebrafish embryo model of sepsis (Barber *et al.*, 2016)	Technically simple vertebrate model Strain-dependent lethality Different strains with similar lethality produced different morphological changes Differential impact of antibiotic treatment according to strain Infected embryos produced transcriptional responses that differ between strains and that resemble human responses to sepsis.	Less responsive to lipopolysaccharide than humans Direct applicability to specific human systems (urinary tract, meninges, blood stream) not known

Multidrug resistance amongst ExPEC

As is the case for other virulence traits, there are a large range of antibiotic resistance genes available to ExPEC strains (Table 4.3). Much of this genetic material is readily transferable via mobile genetic elements, and is continuously evolving. The World Health Organization has identified *E. coli* as one of seven pathogens with antibiotic resistance levels that are of international concern (WHO, 2014). Resistance mechanisms to all major antibiotic classes are present among *E. coli* strains. However, increasing resistance to fluoroquinolones and third generation cephalosporins is of major clinical importance. Fluoroquinolones are oral antibiotics commonly used for the treatment of UTI. Resistance to fluoroquinolones occurs via plasmid or chromosomally mediated production of transferases, or other mechanisms that act to prevent antimicrobial uptake into the cell (Allocati *et al.*, 2013). Virtually all *E. coli* were sensitive to fluoroquinolone antibiotics when they first became available in the 1980s. By contrast, in 2014 the WHO recorded that fluoroquinolone resistance levels were more than 50% in all WHO regions except for the Europe region, where it was up to 48% (2014).

Resistance to β-lactam antibiotics is mediated via plasmid-encoded β-lactamases which include the extended-spectrum β-lactamases (ESBL) capable

Table 4.3 Multiple ExPEC antibiotic resistance mechanisms

Category	Source	Type	Mechanism	Spectrum of activity
Extended spectrum β-lactamases	Plasmid	TEM, SHV	Enzyme inactivation	Penicillins, cephalosporins, monobactams
	Plasmid	CTX-M-14, -15		Penicillins, cephalosporins
	Plasmid	KPC		Penicillins, cephalosporins, monobactams, carbapenems
	Plasmid	AmpC		Penicillins, cephalosporins, monobactams
	Plasmid	NDM Metallo-β-lactamase		Penicillins, cephalosporins, carbapenems
	Plasmid	OXA β-lactamases		Penicillins, carbapenems
	Plasmid or Chromosome	TMQR	DNA gyrase and topoisomerase IV mutations	Quinolones
Methyltransferases	Mobile genetic element	*armA*	Reduce affinity of antibiotic for target binding site	Aminoglycosides
Phosphoethanolamine transferase	Plasmid	*mcr-1*	Reduce colistin affinity for lipid A	Colistin
Changes in the target enzymes	Plasmid	*dfrA7*, *sul1*, *sul2*	Inhibition of enzyme synthesis	Trimethoprim-sulphamethoxazole

of inactivating most β-lactam drugs, including third generation cephalosporins. Currently CTX-M-15 is the most prevalent of the CTX-M-encoding genes that have predominated among *E. coli* since 2000 (Livermore *et al.*, 2007; Mathers *et al.*, 2015). *E. coli* harbouring CTX-M-producing enzymes frequently carry genes conveying resistance to other antibiotics, such as fluoroquinolones, co-trimoxazole, and aminoglycosides.

Resistance genes present on transferable genetic material such as plasmids pose a significant risk, because these plasmids are readily capable of crossing within and between species. Plasmid-encoded carbapenemases (zinc site metallo-β-lactamases, $bla_{nd\text{-}5}$ and OXA β-lactamases), initially identified in *Klebsiella pneumoniae* around 2000, are now found in *E. coli* isolated mainly from nosocomial infections and in outbreaks of *E. coli* infection in health care facilities (Tzouvelekis *et al.*, 2012). Treatment options are limited and treatment failure rates and mortality are high in these populations. One of the few antibiotics remaining to treat carbapenem resistant ExPEC is colistin, for which a plasmid-encoded resistance gene has recently been identified.

In 2015, the first reports of colistin resistance conferred by a plasmid were reported in China (Liu *et al.*, 2016). Colistin is a polymyxin antibiotic whose use has increased in recent years in parallel with increasing resistance of bacteria to other antibiotics. Colistin resistance through chromosomally encoded alleles was previously documented. The plasmid-based colistin resistance gene *mcr-1* has subsequently been detected in *E. coli* isolates from five continents (Schwarz and Johnson, 2016). Further investigation showed that transferable colistin resistance was present in isolates from the 1980s, but it appears that these isolates had minimal clinical impact as they were not detected at the time (Schwarz and Johnson, 2016). The detection of colistin-resistant *E. coli* in meat sources and in humans suggests expansion of *mcr-1* due to selective pressure caused by the use of antibiotics in animals. The potential for transmission of this plasmid amongst *E. coli* is major global concern and heralds the possibility of a pan-resistant *E. coli* strain.

Successful clones

Changes to the bacterial genome through the acquisition of mobile genetic elements, such as plasmids through horizontal transfer, recombination events, and point-mutations can be favourable or unfavourable in terms of their effect on bacterial stability and survival. *Escherichia coli* strains that are resistant to antibiotics are generally less virulent than antibiotic-susceptible strains, suggesting that there is a fitness cost for resistance to antibiotics (Johnson *et al.*, 2004). Antibiotic resistance is frequently mediated via the horizontal transfer of plasmids such as FIA, FIB, and FII replicon types which belong to incompatibility group F (IncF) (Carattoli, 2009). Horizontal transfer of such genetic material spreads antibiotic resistance within the bacterial population without affecting the distribution of strains. By contrast, successful bacterial clones have acquired genetic components that render them stable and promote their survival through antibiotic-resistance mechanisms, and they are also capable of efficient transmission in populations. In the case of the ST131 clone, the unusual combination of high virulence combined with multidrug resistance has driven its rapid and worldwide dissemination (Johnson *et al.*, 2010; Olesen *et al.*, 2013).

The ST131 clonal group

The ST131 clonal group was first identified in 2008 by groups investigating the ESBL antibiotic resistance gene CTX-M-15 (Coque *et al.*, 2008; Nicolas-Chanoine *et al.*, 2008). ST131 is defined by seven house-keeping genes (sequence types or STs) used for multi-locus sequence typing (MLST) (Tartof *et al.*, 2013). ST131 belongs to the phylogenetic group B2, and most isolates exhibit serotype O25b:H4, with smaller subsets being O16 or non-typeable (Johnson *et al.*, 2014).

Genomic sequencing data indicate that ST131 emerged from a single progenitor before 2000, which has since successfully achieved global dissemination due to a unique combination of antibiotic resistance, high virulence, and transmissibility (Petty *et al.*, 2014). The success of this clone is attributed to the type 1 fimbriae FimH complex 30 (H30, not to be confused with flagellin H typing) lineage and nested sub-clones (H30-R and H30-Rx) (Price *et al.*, 2013). H30-R are fluoroquinolone-resistant strains that display a unique mutation in the *gyrA*/*parC* alleles (Johnson *et al.*, 2013; Petty *et al.*, 2014; Price *et al.*, 2013). Today, the fluoroquinolone-resistant, CTX-M-15-positive H30 clade C sub-lineage (H30-Rx) dominates among antibiotic resistant ExPEC worldwide. The mechanisms behind the success of H30 ST131 are not clear, but they could be a function of improved transmissibility, longer duration of colonization as well as virulence and antimicrobial resistance (Banerjee and Johnson, 2014). ST131 H30 appears to be highly virulent, being more prevalent than other clones in patients with recurrent or persistent UTI and in patients with sepsis (Price *et al.*, 2013; Tchesnokova *et al.*, 2013).

The health care-related impact and cost of ST131 infections is incalculable. ST131 constitutes the bulk of *E. coli* multidrug resistant clinical isolates and has been linked to rising numbers of hospitalizations for UTI due to outpatient treatment failures (Kudinha *et al.*, 2013; Olesen *et al.*, 2013). In 2007, 70% of fluoroquinolone-resistant strains and more than 50% of multidrug resistant isolates in the United States were ST131 (Johnson *et al.*, 2010). Between 2002 and 2008 the prevalence of *E. coli* fluoroquinolone resistance was 16% in Europe (de Kraker *et al.*, 2013). Between 2007 and 2010, 31.3% of *E. coli* isolates from US hospitals were ST131 (Edelsberg *et al.*, 2014).

Resistance of ST131 H30 to aminoglycosides, trimethoprim–sulfamethoxazole and, more recently, cabapenems (Accogli *et al.*, 2014; Cai *et al.*, 2014; Johnson *et al.*, 2015b) significantly limits antibiotic treatment options and raises the spectre of the development of a pan-resistant bacterium with the capability for rapid worldwide dissemination.

Other successful clones

CTX-M enzyme production is not limited to ST131 but has been identified in other successful pandemic clones such as ST38, ST405, and ST648. The strains all contribute to the global exchange and transmission of genetic material conveying antibiotic resistance. These clones appear to have compensated for the fitness cost associated with antibiotic resistance, and are identified in an increasing number of countries (Naseer and Sundsfjord, 2011; Schaufler *et al.*, 2016).

The ST405 clonal group consists predominantly of O102:H6 strains and belongs to phylogroup D. ST405 strains have been identified worldwide and are causes of community-acquired UTI (Naseer and Sundsfjord, 2011). As well as CTX-M-15, ST405 strains have been shown to harbour genes for carbapenemases (NDM beta-lactamases) and more recently, *mcr-1*, which conveys colistin resistance (Mediavilla *et al.*, 2016; Pitout, 2012). The first US isolate known to harbour both colistin and carbapenem resistance was identified by Mediavilla *et al.* (2016). This strain was isolated in 2014 from an elderly man with recurrent UTI, suggesting that such strains may already have been circulating for several years.

ST38 are O86:H18 strains from phylogroup D. ST38 produce CTX-M-9 and CTX-M-14 and are associated with NDM and OXA-48 beta-lactamases. ST648 is also associated with CTX-M-15 and NDM beta-lactamases. MCR-1 in conjunction with ESBL and NDM-1 has been detected in ST648 strains causing diarrhoea, raising concerns about the development of pan-resistant ST648 strains (Zhang *et al.*, 2016). Some ExPEC strains (for example serogroups O25 and O15) show outbreak potential and have been associated with community and hospital-based outbreaks of UTI, pyelonephritis, and bacteraemia (George and Manges, 2010). Why some specific *E. coli* types are predisposed to cause outbreaks of disease is not well understood.

Control of ExPEC infections

The control of ExPEC infections centres on their ubiquity, their frequency, and the problem of antibiotic resistance. Given that around 75% of ExPEC bacteremias and virtually all uncomplicated UTIs are community-acquired, population-based prevention is needed to impact on the frequency of infection (Bou-Antoun *et al.*, 2016). The global significance of antibiotic resistance and its impact on human health is the subject of growing international efforts to curtail the spread of resistant strains and reduce antibiotic pressure by limiting antibiotic use. The World Health Organization conducted a global survey of antibiotic resistance in 2014 and a public survey of knowledge and attitudes to antibiotic use in approximately 10,000 individuals from 12 countries (WHO, 2014, 2015a). A 2014 United Kingdom Prime Ministerial review committee called for a response to the threats posed by antibiotic resistance (O'Neill, 2014). In 2015 the WHO released a Global Action Plan on antibiotic resistance with the intention of reducing infection and inappropriate antibiotic use and promoting research and development of new medicines and vaccines (WHO, 2015b).

A key objective in the Global Action Plan is the prevention of infection through improved hygiene and infection control practices in hospitals and health care facilities and by implementing preventative strategies such as vaccination. Vaccination against pneumococcal infection has proved effective in reducing infections by antibiotic-resistant strains (Tomczyk *et al.*, 2016). Currently no effective vaccine against *E. coli* is available for widespread use (Poolman and Wacker, 2015). However, a multivalent bioconjugate vaccine using O-antigen has been tested in a phase I study in healthy women with recurrent UTI (Poolman and Wacker, 2015). Initial results suggest that the candidate vaccine induced functional antibody responses and decreased the number of single-pathogen *E. coli* UTI episodes, warranting further investigation. Importantly, an effective ExPEC vaccine is one of few strategies that could have a sustainable impact in reversing current upwards trends in antibiotic resistance.

Detection and characterization

Antibody-based O-serotyping is the traditional method for identification of *E. coli* clones but is laborious and time consuming. Because of the large number of known serotypes and the need to test each serotype individually, O-typing is used for outbreak investigations but has historically not been suitable for routine clinical application. Genetic subtyping was first proposed in the 1980s, leading to the development of PCR assays to detect housekeeping genes typical of the four described phylogroups; A, B1, B2 and D (*chuA*, *yjaA*, TspE4.C2). MLST extended this classification by identifying phylogroups C, E, F as well as *Escherichia* clade 1 (Clermont *et al.*, 2013). The prevalence of individual phylogroups varies regionally. For example, phylogroup B1 is more prevalent in developing countries compared to industrialized countries, possibly linked to differences in diet (Gordon, 2013). Phylogroups also differ in terms of their transmissibility, survival in the environment and in their distribution in human clinical disease syndromes, although the mechanisms underlying these differences are not understood. ExPEC strains are more likely to be phylogroups B2 or D. Faecal *E. coli* strains and ExPEC strains that cause cystitis are more likely to belong to phylogroup A, B1, and D, whereas strains causing pyelonephritis are more often phylogroup B2 or D (Johnson *et al.*, 2005).

Molecular clonal subtyping methods in use include pulsed-field gel electrophoresis (PFGE), which relies on the visual identification of banding patterns using gel electrophoresis, and MLST (Mathers *et al.*, 2015; Tartof *et al.*, 2013). More than 1000 STs have been identified. Both PFGE and MLST are costly and their use is limited to epidemiological or research settings rather than clinical practice. Other molecular approaches being explored include PCR methods to detect specific clades (Johnson *et al.*, 2014; Li *et al.*, 2010), and oligonucleotide-based microarray testing which could allow rapid testing for all known O-serotypes (Geue *et al.*, 2014).

Matrix-assisted laser desorption ionization-time of flight mass spectrometry (MALDI-TOF MS) allows rapid mass fingerprinting of bacterial colonies, and it is a tool being increasingly used in clinical laboratories (Karger, 2016). To date, identification of *E. coli* to the level of strain has not been achieved using MALDI-TOF MS (Dallagassa *et al.*, 2014). More recently, mass spectrometry shows potential in predicting antibiotic susceptibility and could provide advantages in terms of rapidity of results in the clinical setting (Charretier and Schrenzel, 2016).

Whole-genome sequencing is a highly discriminatory tool and may become the standard method for bacterial subtyping in the future, although tools to collate, assimilate, and interpret the vast numbers of data that arise from this process are first needed (Fratamico *et al.*, 2016). Currently the clinical application of whole-genome sequencing is limited to investigation of outbreak settings (Dallman *et al.*, 2015).

Successful control and treatment of ExPEC will probably be possible only if detection methods can be improved. Rational antibiotic use would be aided by the availability of rapid, low-cost tools for identifying specific strains that are highly likely to be resistant to empirical antibiotic therapies. Antimicrobial testing of bacterial isolates takes several days using traditional culture-based methods, by which time seriously ill patients may have succumbed if inappropriate antibiotics have been administered. Attempts have been made to correlate PCR-identified clonal types identified from the primary clinical sample to patterns of antimicrobial susceptibility suggest that rapid identification of clones likely to be resistant to antibiotics is possible in the clinical setting, thereby reducing the risk of inappropriate antibiotic administration and improving clinical outcomes (Tchesnokova *et al.*, 2013). Progress in this regard is being made using PCR with high sensitivity and specificity to identify four common ST lineages that cause the majority of cases of bacteraemia (Doumith *et al.*, 2015). However, routine application of such assays and their widespread availability in clinical settings has yet to be achieved.

Other potential methods for rapid antimicrobial testing use single cell analyses. These techniques track the growth rate or characterize changes in the morphology or architecture of a single bacterial cell in response to different antibiotic conditions, and can be used to predict antibiotic susceptibility within hours, rather than days (Besant *et al.*, 2015; Choi *et al.*, 2014; Quach *et al.*, 2016).

Molecular methods such as MLST that detect combinations of housekeeping genes may be insufficient compared to typing of O, H, and K antigenic surface structures, which are directly relevant for *E. coli* pathogenicity. Rapid and inexpensive methods that allow simultaneous testing for multiple O-types will provide clinically relevant information with implications for treatment and outcome. Aside from clinical use, rapid O-serotype identification methods such as these could allow widespread epidemiological assessment of prevalent *E. coli* strains, which is currently lacking, and could contribute important strain-specific information to support the introduction of an O-antigen based *E. coli* vaccine.

Future trends and challenges

Limiting advances in our understanding of ExPEC infections and how to prevent them is the absence of world-wide epidemiological surveillance data stratified by age, location, antibiotic pre-treatment, and clinical disease syndrome. These data are needed to begin to investigate links between clinical presentation, course and outcome of disease, and to gain a more detailed understanding of *E. coli* transmission patterns in different settings, such as the general community and environment, and in hospitals, and nursing homes. Current experimental systems do not allow full investigation of interactions between host characteristics and pathogenicity, nor do they replicate local infection environments (such as the urinary tract) that drive the expression of specific virulence traits contributing to survival and replication in these environments.

Determination of the potential impact of preventative measures such as vaccination requires infectious disease models capable of testing the impact of different assumptions and approaches in the diverse patient groups at risk of ExPEC infection. The very large differences in the underlying and immunological characteristics of populations at risk of ExPEC infections (healthy young women with UTI, elderly with bacteraemia and/or urosepsis, neonates with meningitis, individuals prior to surgical or uro-invasive procedures), make this particularly challenging.

Antibiotic resistant ExPEC and the looming threat of a pan-resistant ExPEC strain represent a worldwide threat and the major challenge facing control of ExPEC infections in the 21st century. The genomic and phenotypic diversity amongst ExPEC has so far eluded clear-cut classification, which has complicated efforts to identify and treat specific strains associated with high virulence and/or multidrug resistance.

Unfortunately, without intervention, the proportion of multidrug-resistant ExPEC infections will continue rising, and the burden of disease, particularly in the growing population of elderly, will increase in parallel. Reversals will be possible with improved rapid diagnostics to facilitate early strain identification and guide rational antibiotic choice. Novel antibiotics are likely to provide a temporary reprieve should they become available. However, new antibiotics are likely to come at substantial cost, and will be reserved for clinical situations where no other option exists to prevent early development of resistance. The return on investment for new antibiotics is therefore potentially very difficult. The availability of *E. coli* vaccines for use amongst risk groups (e.g. the elderly, those with a compromised urinary tract, prior to surgery) and potentially in the general population (prevention of bacteraemia and UTI) could reduce the prevalence of targeted strains and make positive inroads on antibiotic resistance levels.

Underlying these advances is the need to better understand the interactions between pathogen and host, and why certain ExPEC clones are successful and others are not. In this respect, close examination of the ST131 strain could provide important insights. Could we have predicted the success of ST131 earlier? Armed with this knowledge, could we have prevented or significantly curtailed its spread? Would this information allow early identification of new, potentially hypervirulent and/or multiresistant strains, and prevention of their transmission? In view of the plasticity of the *E. coli* genome, an understanding of the factors that drive fitness, transmissibility and virulence will contribute to current ExPEC control efforts, and could allow early intervention to prevent widespread dissemination of future pathogenic clones.

References

Abbanat, D., Kae, K., Davies, T., Amsler, K., He, W., Contractor, K., Alaimo, C., Dreyer, V., Gambillara, V., Poolman, J., *et al.* (2016). Antibody response measured by ELISA and opsonophagocytic killing assays elicited by a vaccine targeting extraintestinal pathogenic *Escherichia coli*. [abstract]. Interscience Conference of Antimicrobial Agents and Chemotherapy, Jun 16-20, 2016. Boston, USA. Available at http://files.abstractsonline.com/SUPT/1/4060/AllPosterAbstracts.pdf

Abraham, S.N., and Miao, Y. (2015). The nature of immune responses to urinary tract infections. Nat. Rev. Immunol. *15*, 655–663. https://doi.org/10.1038/nri3887

Abraham, S.N., Sun, D., Dale, J.B., and Beachey, E.H. (1988). Conservation of the D-mannose-adhesion protein among type 1 fimbriated members of the family Enterobacteriaceae. Nature. *336*, 682–684. https://doi.org/10.1038/336682a0

Accogli, M., Giani, T., Monaco, M., Giufrè, M., García-Fernández, A., Conte, V., D'Ancona, F., Pantosti, A., Rossolini, G.M., and Cerquetti, M. (2014). Emergence of *Escherichia coli* ST131 sub-clone H30 producing VIM-1 and KPC-3 carbapenemases, Italy. J. Antimicrob. Chemother. *69*, 2293–2296. https://doi.org/10.1093/jac/dku132

Al-Hasan, M.N., Lahr, B.D., Eckel-Passow, J.E., and Baddour, L.M. (2009). Antimicrobial resistance trends of *Escherichia coli* bloodstream isolates: a population-based study, 1998-2007. J. Antimicrob. Chemother. *64*, 169–174. https://doi.org/10.1093/jac/dkp162

Allocati, N., Masulli, M., Alexeyev, M.F., and Di Ilio, C. (2013). Escherichia coli in Europe: an overview. Int. J. Environ. Res. Public Health *10*, 6235–6254. https://doi.org/10.3390/ijerph10126235

Ani, C., Farshidpanah, S., Bellinghausen Stewart, A., and Nguyen, H.B. (2015). Variations in organism-specific severe sepsis mortality in the United States: 1999-2008. Crit. Care Med. 43, 65–77. https://doi.org/10.1097/CCM.0000000000000555

Antão, E.M., Wieler, L.H., and Ewers, C. (2009). Adhesive threads of extraintestinal pathogenic *Escherichia coli*. Gut Pathog. *1*, 22. https://doi.org/10.1186/1757-4749-1-22

Banerjee, R., and Johnson, J.R. (2014). A new clone sweeps clean: the enigmatic emergence of *Escherichia coli* sequence type 131. Antimicrob. Agents Chemother. *58*, 4997–5004. https://doi.org/10.1128/AAC.02824-14

Barber, A.E., Fleming, B.A., and Mulvey, M.A. (2016). Similarly lethal strains of extraintestinal pathogenic *Escherichia coli* trigger markedly diverse host responses in a zebrafish model of sepsis. mSphere. *1, pii: e00062-16.* doi: 10.1128/mSphere.00062-16.

Bens, M., Vimont, S., Ben Mkaddem, S., Chassin, C., Goujon, J.-M., Balloy, V., Chignard, M., Werts, C., and Vandewalle, A. (2014). Flagellin/TLR5 signalling activates renal collecting duct cells and facilitates invasion and cellular translocation of uropathogenic *Escherichia coli*. Cell. Microbiol. *16*, 1503–1517. https://doi.org/10.1111/cmi.12306

Besant, J.D., Sargent, E.H., and Kelley, S.O. (2015). Rapid electrochemical phenotypic profiling of antibiotic-resistant bacteria. Lab Chip. *15*, 2799–2807. https://doi.org/10.1039/c5lc00375j

Bou-Antoun, S., Davies, J., Guy, R., Johnson, A.P., Sheridan, E.A., and Hope, R.J. (2016). Descriptive epidemiology of *Escherichia coli* bacteraemia in England, April 2012 to March 2014. Euro Surveill. *21*. https://doi.org/10.2807/1560-7917.ES.2016.21.35.30329

Boudeau, J., Glasser, A.L., Masseret, E., Joly, B., and Darfeuille-Michaud, A. (1999). Invasive ability of an *Escherichia coli* strain isolated from the ileal mucosa of a patient with Crohn's disease. Infect. Immun. *67*, 4499–4509.

Brumbaugh, A.R., and Mobley, H.L. (2012). Preventing urinary tract infection: progress toward an effective *Escherichia coli* vaccine. Expert Rev. Vaccines *11*, 663–676. https://doi.org/10.1586/erv.12.36

Burns, S.M., and Hull, S.I. (1998). Comparison of loss of serum resistance by defined lipopolysaccharide mutants and an acapsular mutant of uropathogenic *Escherichia coli* O75:K5. Infect. Immun. *66*, 4244–4253.

Burns, S.M., and Hull, S.I. (1999). Loss of resistance to ingestion and phagocytic killing by O(-) and K(-) mutants of a uropathogenic *Escherichia coli* O75:K5 strain. Infect. Immun. *67*, 3757–3762.

Cabellos, C., Viladrich, P.F., Ariza, J., Maiques, J.M., Verdaguer, R., and Gudiol, F. (2008). Community-acquired bacterial meningitis in cirrhotic patients. Clin. Microbiol. Infect. *14*, 35–40. https://doi.org/10.1111/j.1469-0691.2007.01839.x

Cai, J.C., Zhang, R., Hu, Y.Y., Zhou, H.W., and Chen, G.X. (2014). Emergence of *Escherichia coli* sequence type 131 isolates producing KPC-2 carbapenemase in China. Antimicrob. Agents Chemother. *58*, 1146–1152. https://doi.org/10.1128/AAC.00912-13

Carattoli, A. (2009). Resistance plasmid families in Enterobacteriaceae. Antimicrob. Agents Chemother. *53*, 2227–2238. https://doi.org/10.1128/AAC.01707-08

Chahales, P., and Thanassi, D.G. (2015). Structure, function, and assembly of adhesive organelles by uropathogenic bacteria. Microbiol. Spectr. 3. https://doi.org/10.1128/microbiolspec.UTI-0018-2013

Chan, C.Y., St John, A.L., and Abraham, S.N. (2013). Mast cell interleukin-10 drives localized tolerance in chronic bladder infection. Immunity *38*, 349–359. https://doi.org/10.1016/j.immuni.2012.10.019

Charretier, Y., and Schrenzel, J. (2016). Mass spectrometry methods for predicting antibiotic resistance. Proteomics Clin. Appl. *10*, 964–981. https://doi.org/10.1002/prca.201600041

Choi, J., Yoo, J., Lee, M., Kim, E.G., Lee, J.S., Lee, S., Joo, S., Song, S.H., Kim, E.C., Lee, J.C., *et al.* (2014). A rapid antimicrobial susceptibility test based on single-cell morphological analysis. Sci. Transl. Med. *6*, 267ra174. https://doi.org/10.1126/scitranslmed.3009650

Cieza, R.J., Hu, J., Ross, B.N., Sbrana, E., and Torres, A.G. (2015). The IbeA invasin of adherent-invasive *Escherichia coli* mediates interaction with intestinal epithelia and macrophages. Infect. Immun. *83*, 1904–1918. https://doi.org/10.1128/IAI.03003-14

Clermont, O., Christenson, J.K., Denamur, E., and Gordon, D.M. (2013). The Clermont *Escherichia coli* phylo-typing method revisited: improvement of specificity and detection of new phylo-groups. Environ. Microbiol. Rep. *5*, 58–65. https://doi.org/10.1111/1758-2229.12019

Coque, T.M., Novais, A., Carattoli, A., Poirel, L., Pitout, J., Peixe, L., Baquero, F., Cantón, R., and Nordmann, P. (2008). Dissemination of clonally related *Escherichia coli* strains expressing extended-spectrum β-lactamase CTX-M-15. Emerging Infect. Dis. *14*, 195–200. https://doi.org/10.3201/eid1402.070350

Costelloe, C., Metcalfe, C., Lovering, A., Mant, D., and Hay, A.D. (2010). Effect of antibiotic prescribing in primary care on antimicrobial resistance in individual patients: systematic review and meta-analysis. BMJ. *340*, c2096. https://doi.org/10.1136/bmj.c2096

Cross, A., Artenstein, A., Que, J., Fredeking, T., Furer, E., Sadoff, J.C., and Cryz, S.J. (1994). Safety and immunogenicity of a polyvalent *Escherichia coli* vaccine in human volunteers. J. Infect. Dis. *170*, 834–840.

Dallagassa, C.B., Huergo, L.F., Stets, M.I., Pedrosa, F.O., Souza, E.M., Cruz, L.M., Assis, F.E.A., Wolf, S., Volanski, W., Picheth, G., *et al.* (2014). Matrix-assisted laser desorption ionization-time of flight mass spectrometry analysis of *Escherichia coli* categories. Genet. Mol. Res. *13*, 716–722. https://doi.org/10.4238/2014

Dallman, T.J., Byrne, L., Launders, N., Glen, K., Grant, K.A., and Jenkins, C. (2015). The utility and public health implications of PCR and whole genome sequencing for the detection and investigation of an outbreak of Shiga toxin-producing *Escherichia coli* serogroup O26:H11. Epidemiol. Infect. *143*, 1672–1680. https://doi.org/10.1017/S0950268814002696

Doumith, M., Day, M., Ciesielczuk, H., Hope, R., Underwood, A., Reynolds, R., Wain, J., Livermore, D.M., and Woodford, N. (2015). Rapid identification of major *Escherichia coli* sequence types causing urinary tract and bloodstream infections. J. Clin. Microbiol. *53*, 160–166. https://doi.org/10.1128/JCM.02562-14

Edelsberg, J., Weycker, D., Barron, R., Li, X., Wu, H., Oster, G., Badre, S., Langeberg, W.J., and Weber, D.J. (2014). Prevalence of antibiotic resistance in US hospitals. Diagn. Microbiol. Infect. Dis. *78*, 255–262. https://doi.org/10.1016/j.diagmicrobio.2013.11.011

Eisenstein, B.I., and Jones, G.W. (1988). The spectrum of infections and pathogenic mechanisms of *Escherichia coli*. Adv. Intern. Med. 33, 231–252.

Elixhauser, A., Friedman, B., and Stranges, E. (2011). Septicemia in U.S. Hospitals, 2009: Statistical Brief

#122. In Health care Cost and Utilization Project (HCUP) Statistical Briefs, (Rockville (MD): Agency for Health Care Policy and Research (US)). [cited 21 Jan 2015] Available at http://www.ncbi.nlm.nih.gov/books/NBK65391/

Foxman, B. (2014). Urinary tract infection syndromes: occurrence, recurrence, bacteriology, risk factors, and disease burden. Infect. Dis. Clin. North Am. *28*, 1–13. https://doi.org/10.1016/j.idc.2013.09.003

Fratamico, P.M., DebRoy, C., Liu, Y., Needleman, D.S., Baranzoni, G.M., and Feng, P. (2016). Advances in molecular serotyping and subtyping of *Escherichia coli*. Front. Microbiol. *7*, 644. https://doi.org/10.3389/fmicb.2016.00644

Garcia, E.C., Brumbaugh, A.R., and Mobley, H.L. (2011). Redundancy and specificity of *Escherichia coli* iron acquisition systems during urinary tract infection. Infect. Immun. *79*, 1225–1235. https://doi.org/10.1128/IAI.01222-10

Gaschignard, J., Levy, C., Romain, O., Cohen, R., Bingen, E., Aujard, Y., and Boileau, P. (2011). Neonatal Bacterial Meningitis: 444 Cases in 7 Years. Pediatr. Infect. Dis. J. *30*, 212–217.

George, D.B., and Manges, A.R. (2010). A systematic review of outbreak and non-outbreak studies of extraintestinal pathogenic *Escherichia coli* causing community-acquired infections. Epidemiol. Infect. *138*, 1679–1690. https://doi.org/10.1017/S0950268810001639

Geue, L., Monecke, S., Engelmann, I., Braun, S., Slickers, P., and Ehricht, R. (2014). Rapid microarray-based DNA genoserotyping of *Escherichia coli*. Microbiol. Immunol. *58*, 77–86. https://doi.org/10.1111/1348-0421.12120

Gordon, D. (2013). The ecology of *Escherichia coli*. In *Escherichia coli* (Second Edition) Pathotypes and Principles of Pathogenesis, M.l Donnenberg, ed. (Academic Press), p. 1-20Haiko, J., and Westerlund-Wikström, B. (2013). The role of the bacterial flagellum in adhesion and virulence. Biology 2, 1242–1267.

Henderson, I.R., Navarro-Garcia, F., Desvaux, M., Fernandez, R.C., and Ala'Aldeen, D. (2004). Type V protein secretion pathway: the autotransporter story. Microbiol. Mol. Biol. Rev. *68*, 692–744. https://doi.org/10.1128/MMBR.68.4.692-744.2004

Hernández, C., Fehér, C., Soriano, A., Marco, F., Almela, M., Cobos-Trigueros, N., De La Calle, C., Morata, L., Mensa, J., and Martínez, J.A. (2015). Clinical characteristics and outcome of elderly patients with community-onset bacteremia. J. Infect. *70*, 135–143. https://doi.org/10.1016/j.jinf.2014.09.002

Jackson, L.A., Benson, P., Neuzil, K.M., Grandjean, M., and Marino, J.L. (2005). Burden of community-onset *Escherichia coli* bacteremia in seniors. J. Infect. Dis. *191*, 1523–1529. https://doi.org/10.1086/429344

Johnson, J.R. (1991). Virulence factors in *Escherichia coli* urinary tract infection. Clin. Microbiol. Rev. *4*, 80–128.

Johnson, J.R., and Brown, J.J. (1996). Defining inoculation conditions for the mouse model of ascending urinary tract infection that avoid immediate vesicoureteral reflux yet produce renal and bladder infection. J. Infect. Dis. *173*, 746–749.

Johnson, J.R., Kuskowski, M.A., Gajewski, A., Sahm, D.F., and Karlowsky, J.A. (2004). Virulence characteristics and phylogenetic background of multidrug-resistant and antimicrobial-susceptible clinical isolates of *Escherichia coli* from across the United States, 2000-2001. J. Infect. Dis. *190*, 1739–1744. https://doi.org/10.1086/425018

Johnson, J.R., Owens, K., Gajewski, A., and Kuskowski, M.A. (2005). Bacterial characteristics in relation to clinical source of *Escherichia coli* isolates from women with acute cystitis or pyelonephritis and uninfected women. J. Clin. Microbiol. *43*, 6064–6072. https://doi.org/10.1128/JCM.43.12.6064-6072.2005

Johnson, J.R., Clermont, O., Menard, M., Kuskowski, M.A., Picard, B., and Denamur, E. (2006). Experimental mouse lethality of *Escherichia coli* isolates, in relation to accessory traits, phylogenetic group, and ecological source. J. Infect. Dis. *194*, 1141–1150. https://doi.org/10.1086/507305

Johnson, J.R., Johnston, B., Clabots, C., Kuskowski, M.A., and Castanheira, M. (2010). *Escherichia coli* sequence type ST131 as the major cause of serious multidrug-resistant *E. coli* infections in the United States. Clin. Infect. Dis. *51*, 286–294. https://doi.org/10.1086/653932

Johnson, J.R., Tchesnokova, V., Johnston, B., Clabots, C., Roberts, P.L., Billig, M., Riddell, K., Rogers, P., Qin, X., Butler-Wu, S., *et al.* (2013). Abrupt emergence of a single dominant multidrug-resistant strain of *Escherichia coli*. J. Infect. Dis. *207*, 919–928. https://doi.org/10.1093/infdis/jis933

Johnson, J.R., Clermont, O., Johnston, B., Clabots, C., Tchesnokova, V., Sokurenko, E., Junka, A.F., Maczynska, B., and Denamur, E. (2014). Rapid and specific detection, molecular epidemiology, and experimental virulence of the O16 subgroup within *Escherichia coli* sequence type 131. J. Clin. Microbiol. *52*, 1358–1365. https://doi.org/10.1128/JCM.03502-13

Johnson, J.R., Porter, S., Johnston, B., Kuskowski, M.A., Spurbeck, R.R., Mobley, H.L.T., and Williamson, D.A. (2015a). Host characteristics and bacterial traits predict experimental virulence for *Escherichia coli* bloodstream isolates from patients with urosepsis. Open Forum Infect. Dis. *2*, ofv083. doi: 10.1093/ofid/ofv083.

Johnson, T.J., Hargreaves, M., Shaw, K., Snippes, P., Lynfield, R., Aziz, M., and Price, L.B. (2015b). Complete genome sequence of a carbapenem-resistant extraintestinal pathogenic *Escherichia coli* strain belonging to the sequence type 131 H30R subclade. Genome Announc. 3. https://doi.org/10.1128/genomeA.00272-15

Journet, L., and Cascales, E. (2016). The Type VI Secretion System in *Escherichia coli* and Related Species. EcoSal Plus 7. https://doi.org/10.1128/ecosalplus.ESP-0009-2015

Kaijser, B., and Ahlstedt, S. (1977). Protective capacity of antibodies against *Escherichia coli* and K antigens. Infect. Immun. *17*, 286–289.

Kakkanat, A., Totsika, M., Schaale, K., Duell, B.L., Lo, A.W., Phan, M.D., Moriel, D.G., Beatson, S.A., Sweet, M.J., Ulett, G.C., *et al.* (2015). The role of H4 flagella in *Escherichia coli* ST131 virulence. Sci. Rep. *5*, 16149. https://doi.org/10.1038/srep16149

Kaper, J.B., Nataro, J.P., and Mobley, H.L. (2004). Pathogenic *Escherichia coli*. Nat. Rev. Microbiol. *2*, 123–140. https://doi.org/10.1038/nrmicro818

Karger, A. (2016). Current developments to use linear MALDI-TOF spectra for the identification and typing of bacteria and the characterization of other cells/organisms related to infectious diseases. Proteomics Clin. Appl. *10*, 982–993. https://doi.org/10.1002/prca.201600038

Kintz, E., and Goldberg, J.B. (2008). Regulation of lipopolysaccharide O antigen expression in *Pseudomonas aeruginosa*. Future Microbiol. *3*, 191–203. https://doi.org/10.2217/17460913.3.2.191

Kisiela, D.I., Rodriguez, V.B., Tchesnokova, V., Avagyan, H., Aprikian, P., Liu, Y., Wu, X.-R., Thomas, W.E., and Sokurenko, E.V. (2013). Conformational inactivation induces immunogenicity of the receptor-binding pocket of a bacterial adhesin. Proc. Natl. Acad. Sci. U.S.A. *110*, 19089–19094. https://doi.org/0.1073/pnas.1314395110

de Kraker, M.E.A., Jarlier, V., Monen, J.C.M., Heuer, O.E., van de Sande, N., and Grundmann, H. (2013). The changing epidemiology of bacteraemias in Europe: trends from the European Antimicrobial Resistance Surveillance System. Clin. Microbiol. Infect. *19*, 860–868. https://doi.org/ 10.1111/1469-0691.12028

Kudinha, T., Johnson, J.R., Andrew, S.D., Kong, F., Anderson, P., and Gilbert, G.L. (2013). *Escherichia coli* sequence type 131 as a prominent cause of antibiotic resistance among urinary *Escherichia coli* isolates from reproductive-age women. J. Clin. Microbiol. *51*, 3270–3276. https://doi.org/10.1128/JCM.01315-13

Lane, M.C., Alteri, C.J., Smith, S.N., and Mobley, H.L. (2007a). Expression of flagella is coincident with uropathogenic *Escherichia coli* ascension to the upper urinary tract. Proc. Natl. Acad. Sci. U.S.A. *104*, 16669–16674. https://doi.org/10.1073/pnas.0607898104

Lane, M.C., Simms, A.N., and Mobley, H.L.T. (2007b). complex interplay between type 1 fimbrial expression and flagellum-mediated motility of uropathogenic *Escherichia coli*. J. Bacteriol. *189*, 5523–5533. https://doi.org/ 10.1128/JB.00434-07

Lang, T., Hudemann, C., Tchatalbachev, S., Stammler, A., Michel, V., Aslani, F., Bhushan, S., Chakraborty, T., Renz, H., and Meinhardt, A. (2014). Uropathogenic *Escherichia coli* modulates innate immunity to suppress Th1-mediated inflammatory responses during infectious epididymitis. Infect. Immun. *82*, 1104–1111. https://doi.org/10.1128/IAI.01373-13

Langermann, S., Palaszynski, S., Barnhart, M., Auguste, G., Pinkner, J.S., Burlein, J., Barren, P., Koenig, S., Leath, S., Jones, C.H., *et al.* (1997). Prevention of mucosal *Escherichia coli* infection by FimH-adhesin-based systemic vaccination. Science *276*, 607–611.

Langermann, S., Möllby, R., Burlein, J.E., Palaszynski, S.R., Auguste, C.G., DeFusco, A., Strouse, R., Schenerman, M.A., Hultgren, S.J., Pinkner, J.S., *et al.* (2000). Vaccination with FimH adhesin protects cynomolgus monkeys from colonization and infection by uropathogenic *Escherichia coli*. J. Infect. Dis. *181*, 774–778. https://doi.org/10.1086/315258

Laupland, K.B., and Church, D.L. (2014). Population-based epidemiology and microbiology of community-onset bloodstream infections. Clin. Microbiol. Rev. *27*, 647–664. https://doi.org/10.1128/CMR.00002-14

Leo, J.C., and Goldman, A. (2009). The immunoglobulin-binding Eib proteins from *Escherichia coli* are receptors for IgG Fc. Mol. Immunol. *46*, 1860–1866. https://doi.org/10.1016/j.molimm.2009.02.024

Li, D., Liu, B., Chen, M., Guo, D., Guo, X., Liu, F., Feng, L., and Wang, L. (2010). A multiplex PCR method to detect 14 *Escherichia coli* serogroups associated with urinary tract infections. J. Microbiol. Methods *82*, 71–77. https://doi.org/10.1016/j.mimet.2010.04.008

Lillington, J., Geibel, S., and Waksman, G. (2014) Biogenesis and adhesion of type 1 nad P pili. Biochim. Biophys. Acta *1840*, 2783–2793. https://doi.org/10.1016/j.bbagen.2014.04.021

Liu, Y.-Y., Wang, Y., Walsh, T.R., Yi, L.-X., Zhang, R., Spencer, J., Doi, Y., Tian, G., Dong, B., Huang, X., *et al.* (2016). Emergence of plasmid-mediated colistin resistance mechanism MCR-1 in animals and human beings in China: a microbiological and molecular biological study. Lancet Infect. Dis. *16*, 161–168. https://doi.org/10.1016/S1473-3099(15)00424-7

Livermore, D.M., Canton, R., Gniadkowski, M., Nordmann, P., Rossolini, G.M., Arlet, G., Ayala, J., Coque, T.M., Kern-Zdanowicz, I., Luzzaro, F., *et al.* (2007). CTX-M: changing the face of ESBLs in Europe. J. Antimicrob. Chemother. *59*, 165–174. https://doi.org/10.1093/jac/dkl483

Logue, C.M., Doetkott, C., Mangiamele, P., Wannemuehler, Y.M., Johnson, T.J., Tivendale, K.A., Li, G., Sherwood, J.S., and Nolan, L.K. (2012). Genotypic and phenotypic traits that distinguish neonatal meningitis-associated *Escherichia coli* from fecal *E. coli* isolates of healthy human hosts. Appl. Environ. Microbiol. 78, 5824–5830. https://doi.org/10.1128/AEM.07869-11

Lu, S., Zhang, X., Zhu, Y., Kim, K.S., Yang, J., and Jin, Q. (2011). Complete genome sequence of the neonatal-meningitis-associated *Escherichia coli* strain CE10. J. Bacteriol. *193*, 7005. https://doi.org/10.1128/JB.06284-11

Lukjancenko, O., Wassenaar, T.M., and Ussery, D.W. (2010). Comparison of 61 sequenced *Escherichia coli* genomes. Microb. Ecol. *60*, 708–720. https://doi.org/10.1007/s00248-010-9717-3

Martinez-Medina, M., Mora, A., Blanco, M., López, C., Alonso, M.P., Bonacorsi, S., Nicolas-Chanoine, M.-H., Darfeuille-Michaud, A., Garcia-Gil, J., and Blanco, J. (2009). Similarity and divergence among adherent-invasive *Escherichia coli* and extraintestinal pathogenic *E. coli* strains. J. Clin. Microbiol. *47*, 3968–3979. https://doi.org/10.1128/JCM.01484-09

Mathers, A.J., Peirano, G., and Pitout, J.D. (2015). *Escherichia coli* ST131: The quintessential example of an international multiresistant high-risk clone. Adv. Appl. Microbiol. *90*, 109–154. https://doi.org/10.1016/bs.aambs.2014.09.002

Mediavilla, J.R., Patrawalla, A., Chen, L., Chavda, K.D., Mathema, B., Vinnard, C., Dever, L.L., and Kreiswirth, B.N. (2016). Colistin- and carbapenem-resistant *Escherichia coli* harboring *mcr-1* and blaNDM-5, causing a complicated urinary tract infection in a patient from the United States. MBio. 7. https://doi.org/10.1128/mBio.01191-16.

Miajlovic, H., and Smith, S.G. (2014). Bacterial self-defence: how *Escherichia coli* evades serum killing. FEMS Microbiol. Lett. *354*, 1–9. https://doi.org/10.1111/1574-6968.12419

Murray, G.L., Attridge, S.R., and Morona, R. (2005). Inducible serum resistance in *Salmonella typhimurium* is dependent on *wzz*(*fepE*)-regulated very long O antigen chains. Microbes Infect. 7, 1296–1304.

Naseer, U., and Sundsfjord, A. (2011). The CTX-M conundrum: dissemination of plasmids and *Escherichia coli* clones. Microb. Drug Resist. *17*, 83–97. https://doi.org/10.1089/mdr.2010.0132

Nash, J.H., Villegas, A., Kropinski, A.M., Aguilar-Valenzuela, R., Konczy, P., Mascarenhas, M., Ziebell, K., Torres, A.G., Karmali, M.A., and Coombes, B.K. (2010). Genome sequence of adherent-invasive *Escherichia coli* and comparative genomic analysis with other *E. coli* pathotypes. BMC Genomics. *11*, 667. https://doi.org/10.1186/1471-2164-11-667

Nichols, K.B., Totsika, M., Moriel, D.G., Lo, A.W., Yang, J., Wurpel, D.J., Rossiter, A.E., Strugnell, R.A., Henderson, I.R., Ulett, G.C., *et al.* (2016). Molecular Characterization of the Vacuolating Autotransporter Toxin in Uropathogenic *Escherichia coli*. J. Bacteriol. *198*, 1487–1498. https://doi.org/10.1128/JB.00791-15

Nicolas-Chanoine, M.H., Blanco, J., Leflon-Guibout, V., Demarty, R., Alonso, M.P., Caniça, M.M., Park, Y.J., Lavigne, J.P., Pitout, J., and Johnson, J.R. (2008). Intercontinental emergence of *Escherichia coli* clone O25:H4-ST131 producing CTX-M-15. J. Antimicrob. Chemother. *61*, 273–281. https://doi.org/10.1093/jac/dkm464

Okike, I.O., Johnson, A.P., Henderson, K.L., Blackburn, R.M., Muller-Pebody, B., Ladhani, S.N., Anthony, M., Ninis, N., Heath, P.T., and neoMen Study Group (2014). Incidence, etiology, and outcome of bacterial meningitis in infants aged <90 days in the United kingdom and Republic of Ireland: prospective, enhanced, national population-based surveillance. Clin. Infect. Dis. *59*, e150-157. https://doi.org/10.1093/cid/ciu514

Olesen, B., Hansen, D.S., Nilsson, F., Frimodt-Møller, J., Leihof, R.F., Struve, C., Scheutz, F., Johnston, B., Krogfelt, K.A., and Johnson, J.R. (2013). Prevalence and characteristics of the epidemic multiresistant *Escherichia coli* ST131 clonal group among extended-spectrum beta-lactamase-producing E. coli isolates in Copenhagen, Denmark. J. Clin. Microbiol. *51*, 1779–1785. https://doi.org/10.1128/JCM.00346-13

O'Neill, J. (Chair) (2014). Antimicrobial resistance: tackling a crisis for the health and wealth of nations. The Review on Antimicrobial Resistance. [cited 03 May 2016]. Available at http://www.his.org.uk/files/4514/1829/6668/AMR_Review_Paper_-_Tackling_a_crisis_for_the_health_and_wealth_of_nations_1.pdf

Orskov, F., and Orskov, I. (1992). *Escherichia coli* serotyping and disease in man and animals. Can. J. Microbiol. *38*, 699–704.

Orskov, I., and Orskov, F. (1985). *Escherichia coli* in extra-intestinal infections. J Hyg 95, 551–575.

Osawa, K., Shigemura, K., Iguchi, A., Shirai, H., Imayama, T., Seto, K., Raharjo, D., Fujisawa, M., Osawa, R., and Shirakawa, T. (2013). Modulation of O-antigen chain length by the *wzz* gene in *Escherichia coli* O157 influences its sensitivities to serum complement. Microbiol. Immunol. 57, 616–623. https://doi.org/10.1111/1348-0421.12084

Petersen, A.M., Halkjær, S.I., and Gluud, L.L. (2015). Intestinal colonization with phylogenetic group B2 *Escherichia coli* related to inflammatory bowel disease: a systematic review and meta-analysis. Scand. J. Gastroenterol. *50*, 1199–1207. https://doi.org/10.3109/00365521.2015.1028993

Petty, N.K., Ben Zakour, N.L., Stanton-Cook, M., Skippington, E., Totsika, M., Forde, B.M., Phan, M.D., Gomes Moriel, D., Peters, K.M., Davies, M., *et al.* (2014). Global dissemination of a multidrug resistant *Escherichia coli* clone. Proc. Natl. Acad. Sci. U.S.A. *111*, 5694–5699. https://doi.org/10.1073/pnas.1322678111

Phan, M.-D., Peters, K.M., Sarkar, S., Lukowski, S.W., Allsopp, L.P., Gomes Moriel, D., Achard, M.E.S., Totsika, M., Marshall, V.M., Upton, M., *et al.* (2013). The serum resistome of a globally disseminated multidrug resistant uropathogenic *Escherichia coli* clone. PLOS Genet. *9*, e1003834. https://doi.org/10.1371/journal.pgen.1003834

Pitout, J.D.D. (2012). Extraintestinal Pathogenic *Escherichia coli*: A combination of virulence with antibiotic resistance. Front. Microbiol. 3, 9. doi: 10.3389/fmicb.2012.00009. https://doi.org/10.3389/fmicb.2012.00009

Poolman, J.T., and Wacker, M. (2015). Extraintestinal Pathogenic *Escherichia coli*, a common human pathogen: challenges for vaccine development and progress in the field. J. Infect. Dis. *213*, 6–13. https://doi.org/10.1093/infdis/jiv429

Price, L.B., Johnson, J.R., Aziz, M., Clabots, C., Johnston, B., Tchesnokova, V., Nordstrom, L., Billig, M., Chattopadhyay, S., Stegger, M., *et al.* (2013). The epidemic of extended-spectrum-β-lactamase-producing *Escherichia coli* ST131 is driven by a single highly pathogenic subclone, H30-Rx. MBio. *4*, e00377-00313. https://doi.org/10.1128/mBio.00377-13

Quach, D.T., Sakoulas, G., Nizet, V., Pogliano, J., and Pogliano, K. (2016). Bacterial cytological profiling (BCP) as a rapid and accurate antimicrobial susceptibility testing method for *Staphylococcus aureus*. EBioMedicine *4*, 95–103. https://doi.org/10.1016/j.ebiom.2016.01.020

Roubaud Baudron, C., Panhard, X., Clermont, O., Mentré, F., Fantin, B., Denamur, E., and Lefort, A. (2014). *Escherichia coli* bacteraemia in adults: age-related differences in clinical and bacteriological characteristics, and outcome. Epidemiol. Infect. *142*, 2672–2683. https://doi.org/10.1017/S0950268814000211

Rudman, D., Hontanosas, A., Cohen, Z., and Mattson, D.E. (1988). Clinical correlates of bacteremia in a Veterans Administration extended care facility. J. Am. Geriatr. Soc. 36, 726–732.

Rudzinski, J.K., and Kawakami, J. (2014). Incidence of infectious complications following transrectal ultrasound-guided prostate biopsy in Calgary, Alberta, Canada: A retrospective population-based analysis. Can. Urol. Assoc. J. *8*, E301-305. https://doi.org/10.5489/cuaj.1751

Russo, T.A., and Johnson, J.R. (2003). Medical and economic impact of extraintestinal infections due to *Escherichia coli*: focus on an increasingly important endemic problem. Microbes Infect. 5, 449–456.

Russo, T.A., Davidson, B.A., Carlino-MacDonald, U.B., Helinski, J.D., Priore, R.L., and Knight, P.R. (2003). The effects of *Escherichia coli* capsule, O-antigen, host neutrophils, and complement in a rat model of Gram-negative pneumonia. FEMS Microbiol. Lett. *226*, 355–361.

Sandt, C.H., and Hill, C.W. (2000). Four different genes responsible for nonimmune immunoglobulin-binding activities within a single strain of *Escherichia coli*. Infect. Immun. *68*, 2205–2214.

Sarkar, S., Ulett, G.C., Totsika, M., Phan, M.D., and Schembri, M.A. (2014). Role of capsule and O antigen in the virulence of uropathogenic *Escherichia coli*. PLOS ONE *9*, e94786. https://doi.org/10.1371/journal.pone.0094786

Sauer, M.M., Jakob, R.P., Eras, J., Baday, S., Eriş, D., Navarra, G., Bernèche, S., Ernst, B., Maier, T., and Glockshuber, R. (2016). Catch-bond mechanism of the bacterial adhesin FimH. Nat. Commun. *7*, 10738. https://doi.org/10.1038/ncomms10738

Schaufler, K., Semmler, T., Pickard, D.J., de Toro, M., de la Cruz, F., Wieler, L.H., Ewers, C., and Guenther, S. (2016). Carriage of extended-spectrum beta-lactamase-plasmids does not reduce fitness but enhances virulence in some strains of pandemic *E. coli* lineages. Front. Microbiol. *7*, 336. https://doi.org/10.3389/fmicb.2016.00336

Schwarz, S., and Johnson, A.P. (2016). Transferable resistance to colistin: a new but old threat. J. Antimicrob. Chemother. *71*, 2066–2070. https://doi.org/10.1093/jac/dkw274

Simonsen, K.A., Anderson-Berry, A.L., Delair, S.F., and Davies, H.D. (2014). Early-onset neonatal sepsis. Clin. Microbiol. Rev. *27*, 21–47. https://doi.org/10.1128/CMR.00031-13

Snyder, J.A., Haugen, B.J., Lockatell, C.V., Maroncle, N., Hagan, E.C., Johnson, D.E., Welch, R.A., and Mobley, H.L. (2005). Coordinate expression of fimbriae in uropathogenic *Escherichia coli*. Infect. Immun. *73*, 7588–7596. https://doi.org/10.1128/IAI.73.11.7588-7596.2005

Söderström, T., Hansson, G., and Larson, G. (1984). The *Escherichia coli* K1 capsule shares antigenic determinants with the human gangliosides GM3 and GD3. N. Engl. J. Med. *310*, 726–727. https://doi.org/10.1056/NEJM198403153101121

Søgaard, M., Nørgaard, M., Dethlefsen, C., and Schønheyder, H.C. (2011). Temporal changes in the incidence and 30-day mortality associated with bacteremia in hospitalized patients from 1992 through 2006: a population-based cohort study. Clin. Infect. Dis. *52*, 61–69. https://doi.org/10.1093/cid/ciq069

Stenutz, R., Weintraub, A., and Widmalm, G. (2006). The structures of *Escherichia coli* O-polysaccharide antigens. FEMS Microbiol. Rev. *30*, 382–403. https://doi.org/10.1111/j.1574-6976.2006.00016.x

Subashchandrabose, S., and Mobley, H.L. (2015). Virulence and Fitness Determinants of Uropathogenic *Escherichia coli*. Microbiol. Spectr. *3*, https://doi.org/10.1128/microbiolspec.UTI-0015-2012

Tartof, S.Y., Lewis, M., Kenyon, C., White, K., Osborn, A., Liko, J., Zell, E., Martin, S., Messonnier, N.E., Clark, T.A., *et al.* (2013). Waning immunity to pertussis following 5 doses of DTaP. Pediatrics *131*, e1047-1052.https://doi.org/10.1542/peds.2012-1928

Tchesnokova, V., Billig, M., Chattopadhyay, S., Linardopoulou, E., Aprikian, P., Roberts, P.L., Skrivankova, V., Johnston, B., Gileva, A., Igusheva, I., *et al.* (2013). Predictive diagnostics for *Escherichia coli* infections based on the clonal association of antimicrobial resistance and clinical outcome. J. Clin. Microbiol. *51*, 2991–2999. https://doi.org/10.1128/JCM.00984-13

Tivendale, K.A., Logue, C.M., Kariyawasam, S., Jordan, D., Hussein, A., Li, G., Wannemuehler, Y., and Nolan, L.K. (2010). Avian-pathogenic *Escherichia coli* strains are similar to neonatal meningitis *E. coli* strains and are able to cause meningitis in the rat model of human disease. Infect. Immun. *78*, 3412–3419. https://doi.org/10.1128/IAI.00347-10

Tomczyk, S., Lynfield, R., Schaffner, W., Reingold, A., Miller, L., Petit, S., Holtzman, C., Zansky, S.M., Thomas, A., Baumbach, J., *et al.* (2016). Prevention of antibiotic-nonsusceptible invasive pneumococcal disease with the 13-valent pneumococcal conjugate vaccine. Clin. Infect. Dis. *62*, 1119–1125. https://doi.org/10.1093/cid/ciw067

Torio, C., and Moore, B. (2016). National Inpatient Hospital Costs: The Most Expensive Conditions by Payer, 2013. Statistical brief #204. Health care Costs and Utilization Program. [cited 08 Sept 2016]. Available at https://www.hcup-us.ahrq.gov/reports/statbriefs/sb204-Most-Expensive-Hospital-Conditions.jsp

Tzouvelekis, L.S., Markogiannakis, A., Psichogiou, M., Tassios, P.T., and Daikos, G.L. (2012). Carbapenemases in *Klebsiella pneumoniae* and other Enterobacteriaceae: an evolving crisis of global dimensions. Clin. Microbiol. Rev. *25*, 682–707. https://doi.org/10.1128/CMR.05035-11

Weiner, L.M., Webb, A.K., Limbago, B., Dudeck, M.A., Patel, J., Kallen, A.J., Edwards, J.R., and Sievert, D.M. (2016). Antimicrobial-resistant pathogens associated with health care-associated infections: summary of data reported to the National Health care Safety Network at the Centers for Disease Control and Prevention, 2011-2014. Infect. Control Hosp. Epidemiol. *37*, 1–14. https://doi.org/10.1017/ice.2016.174

Welch, R.A. (2016). Uropathogenic *Escherichia coli*-Associated Exotoxins. Microbiol. Spectr. *4*. https://doi.org/10.1128/microbiolspec.UTI-0011-2012

Whitfield, C. (2006). Biosynthesis and assembly of capsular polysaccharides in *Escherichia coli*. Annu. Rev. Biochem. *75*, 39–68. https://doi.org/10.1146/annurev.biochem.75.103004.142545

Wijetunge, D.S., Katani, R., Kapur, V., and Kariyawasam, S. (2015). Complete Genome Sequence of *Escherichia coli* Strain RS218 (O18:H7:K1), Associated with Neonatal Meningitis. Genome Announc. 3, e00804–15. https://doi.org/10.1128/genomeA.00804-15

Wiles, T.J., Kulesus, R.R., and Mulvey, M.A. (2008). Origins and virulence mechanisms of uropathogenic *Escherichia coli*. Exp. Mol. Pathol. *85*, 11–19. https://doi.org/10.1016/j.yexmp.2008.03.007

Williamson, D.A., Lim, A., Wiles, S., Roberts, S.A., and Freeman, J.T. (2013a). Population-based incidence and comparative demographics of community-associated and health care-associated *Escherichia coli* bloodstream infection in Auckland, New Zealand, 2005–2011. BMC Infect. Dis. *13*, 385. https://doi.org/10.1186/1471-2334-13-385

Williamson, D.A., Barrett, L.K., Rogers, B.A., Freeman, J.T., Hadway, P., and Paterson, D.L. (2013b). Infectious complications following transrectal ultrasound-guided prostate biopsy: new challenges in the era of multidrug-resistant *Escherichia coli*. Clin. Infect. Dis. *57*, 267–274. https://doi.org/10.1093/cid/cit193

Williamson, D.A., Mills, G., Johnson, J.R., Porter, S., and Wiles, S. (2014). *In vivo* correlates of molecularly inferred virulence among extraintestinal pathogenic *Escherichia coli* (ExPEC) in the wax moth *Galleria mellonella* model system. Virulence 5, 388–393. https://doi.org/10.4161/viru.27912

Wilson, J., Elgohari, S., Livermore, D.M., Cookson, B., Johnson, A., Lamagni, T., Chronias, A., and Sheridan, E. (2011). Trends among pathogens reported as causing bacteraemia in England, 2004-2008. Clin. Microbiol. Infect. *17*, 451–458. https://doi.org/10.1111/j.1469-0691.2010.03262.x

Zhang, H., Seward, C.H., Wu, Z., Ye, H., and Feng, Y. (2016). Genomic insights into the ESBL and MCR-1-producing ST648 *Escherichia coli* with multidrug resistance. Sci. Bull. *61*, 875–878. https://doi.org/10.1007/s11434-016-1086-y

Zhou, M., Yang, Y., Chen, P., Hu, H., Hardwidge, P.R., and Zhu, G. (2015). More than a locomotive organelle: flagella in *Escherichia coli*. Appl. Microbiol. Biotechnol. *99*, 8883–8890. https://doi.org/10.1007/s00253-015-6946-x

Zilberberg, M.D., and Shorr, A.F. (2013). Secular trends in gram-negative resistance among urinary tract infection hospitalizations in the United States, 2000-2009. Infect. Control Hosp. Epidemiol. *34*, 940–946. https://doi.org/10.1086/671740

World Health Organization. (2014). Antimicrobial resistance: global report on surveillance. [cited 21 Jan 2015] Available at http://apps.who.int/iris/bitstream/10665/112642/1/9789241564748_eng.pdf?ua=1

World Health Organization. (2015a). Antibiotic resistance: Multi-country public awareness survey. [cited 21 Jan 2015] Available at http://www.who.int/drugresistance/documents/baselinesurveynov2015/en/

World Health Organization. (2015b). Global action plan on antimicrobial resistance. [cited 21 Jan 2015] Available at http://www.who.int/drugresistance/global_action_plan/en/

Bacteriophage-based Strategies to Control Pathogenic *Escherichia coli* in Humans and Animals

5

Michca Gordon, Brigitte Cadieux and Lawrence D. Goodridge*

Department of Food Science and Agricultural Chemistry, McGill University, Ste Anne de Bellevue, QC, Canada.

*Correspondence: lawrence.goodridge@mcgill.ca

https://doi.org/10.21775/9781910190777.05

Abstract

Escherichia coli is a bacterial species with the ability to cause a wide spectrum of diseases among humans and other animals. A great deal of research has been conducted to find ways to control this diffuse group of pathogens. Such research has accelerated recently with the understanding that the overuse of antibiotics has contributed to a rapid increase in the spread of multidrug resistant bacteria. Additionally, in some cases (i.e. treatment of Shiga-toxin-producing *E. coli* infections), antibiotic treatment is contraindicated, necessitating the development of alternative treatment strategies. This review explores the use of bacteriophages (phages) as a biocontrol strategy, and their potential to treat *E. coli* infections in humans and food animals.

Introduction

Escherichia coli is one of the most frequent causes of bacterial infections in humans and animals, and causes one of the widest spectrum of diseases of any bacterial species. As a zoonotic agent, both humans and animals act as a reservoir for this pathogen. While the large majority of *E. coli* isolates are regarded as commensals in humans, pathogenic strains are strongly associated with diarrhoeal illness, causing mortality worldwide, particularly in children under the age of five (Croxen *et al.*, 2013). Many *E. coli* isolates that cause illness in humans are considered part of the natural microflora of food animals [for example, Shiga-toxin-producing *E. coli* (STEC) in cattle], but *E. coli* can still cause diarrhoea in food-producing animals, such as swine and poultry, as well as in companion animals.

Food and waterborne *E. coli* are a group of pathogens of great public health significance because these microorganisms are highly virulent and ingestion of a concentration as low as ten cells is able to cause infection (Russell *et al.*, 2000). STEC are food associated *E. coli* that have been consistently implicated in food-borne outbreaks. However, there are at least seven pathotypes of *E. coli* capable of causing diarrhoea in humans, including STEC, enterohemorrhagic *E. coli* (EHEC) (a subset of STEC), enterotoxigenic *E. coli* (ETEC), enteropathogenic *E. coli* (EPEC), enteroaggregative *E. coli* (EAEC), enteroinvasive *E. coli* (EIEC), diffusely adherent *E. coli* (DAEC), and adherent invasive *E. coli* (AIEC) (Croxen *et al.*, 2013; CDC, 2017). Many of these *E. coli* pathotypes have been associated with food and water, and may be regarded as continuing to emerge as important causes of food-borne disease. Food animals are a major reservoir for *E. coli*. Transmission to humans primarily occurs through the consumption of animal-derived foods, including contaminated, undercooked or raw ground meat products, and raw milk, as well as consumption of fresh produce that has been contaminated by the faeces of domesticated or wild animals. Therefore, a major focus has been placed on controlling the presence of STEC in food-producing animals.

Therapeutic treatment in humans and animals is threatened by the emergence of antimicrobial resistance (Allocati *et al.*, 2013). The prevalence of multidrug resistant *E. coli* strains is increasing globally due to the overuse of antibiotics and the spread of mobile genetic elements, such as plasmids (Allocati *et al.*, 2013) and bacteriophage like plasmids (Colavecchio *et al.*, 2017), that carry antibiotic resistance determinants. The continued and rapid spread of multidrug resistant bacteria has led to a restriction on the use of antibiotics to control diseases in farm animals, leading to an acute need to develop alternative approaches to reduce the presence and spread of pathogenic *E. coli* in animals used to produce food.

Bacteriophages (phages) were co-discovered in 1915 by Frederick Twort, a British pathologist, and in 1917 by Felix d'Herelle, a French-Canadian microbiologist (Fruciano and Bourne, 2007). Phages were identified as possible biocontrol tools to reduce the presence of pathogenic bacteria immediately upon their discovery (Fruciano and Bourne, 2007). Their use in the treatment of disease was hampered by a lack of knowledge regarding the physico-chemical nature of phages, including host range and various growth characteristics. The discovery of antibiotics in the 1940s effectively halted all phage biocontrol research in the western world, though much research continued in the former Soviet Union. In the 1980s, concern regarding the emergence of multidrug resistant bacteria led to a resurgence in phage research, and there is now a renewed interest in the use of phages to eliminate pathogenic *E. coli* in both animals and humans. A number of phage therapy studies to control *E. coli* have been conducted in food-producing animals, including cattle, ruminants, swine, and poultry (reviewed in Brüssow, 2005 and Johnson *et al.*, 2008). In addition, phages are being evaluated in several human trials to combat pathogenic *E. coli*. This chapter will discuss the latest studies investigating the potential use of phage therapy to control pathogenic *E. coli* in food animals and humans.

Therapeutic use of bacteriophages

Phages have been observed to be ubiquitous in nature, representing the most abundant biological entity in the biosphere with an estimated 10^{31} virions, outnumbering bacteria by a factor of ten (Brüssow *et al.*, 2004). The ubiquitous nature of phages, and the fact that they have been identified for most zoonotic bacterial pathogens, has led to many efforts aimed at using phages to treat infections in humans and food-producing animals. In order to be effective at controlling disease-causing bacteria, phages should have several characteristics including:

1. They should have high specificity for their target bacterial host, leaving the remaining microbiota untouched.
2. Phages used for therapeutic approaches should be capable of rapid multiplication to high concentrations at the site of treatment.
3. Phages used in phage therapy should be capable of continuously adapting to bacteriophage insensitive mutants (BIMs), as phage resistance mechanisms are developed by the bacteria in order to resist phage attack.
4. Phage preparations should be sufficiently purified to ensure the lack of toxic substances, such as lipopolysaccharide, which could potentially harm the host (i.e. human, animal).
5. Phages used in phage therapy should be capable of withstanding environmental stresses.
6. Phages should have a prolonged shelf life.

Pre-harvest strategies to reduce pathogenic *E. coli*

STEC is commonly found in the gastrointestinal tract (GIT) and faeces of farm animals. During harvest, meat may become contaminated with intestinal contents or faecal material from the hide, leading to transmission of the food-borne pathogen to humans. A number of research studies have investigated the use of phages as a pre-harvest strategy to control STEC in food-producing animals. These studies involve administering phages directly to livestock as a way to reduce the incidence of STEC in food producing animals, thereby preventing the pathogen from entering the food chain and improving the safety of animal derived foods (Hynes and Wachsmuth, 2000). Central to these control studies has been a focus on the most effective methods to deliver the phages to the GIT, which is the site of *E. coli* colonization (Ricke, 2012). However, the key for successful pre-harvest use of phages as a control measure is not only effective delivery of phages, but also survival of the phages until they arrive at the GIT. Details of research studies on the control of STEC by phage in different food-producing animals, as well as some of the challenges associated with this approach, will be presented in the following sections.

Ruminants

Cattle and other ruminants are considered to be the natural reservoirs of STEC (Zhao *et al.*, 2005). The control of STEC in animals is difficult because carriage in ruminants is asymptomatic and shedding can be intermittent (Matthews *et al.*, 2006). As such, the scientific literature reports both failure and varied success in the use of phage therapy to control the presence of STEC in these animals (Bach *et al.*, 2003; 2009; Callaway *et al.*, 2003; 2006; 2008; Raya *et al.*, 2006; 2011; Rivas *et al.*, 2010; Rozema *et al.*, 2009; Sheng *et al.*, 2006). Initial studies, in which oral administration was used to deliver phages, did not prove to be very effective, in part because ruminant physiology poses unique challenges. Specifically, the distinct microbial population in the large caecum is responsible for the secondary fermentation of animal feed components, leading to acidic conditions, which can have a negative impact on the survival of phages and lead to treatment failure (Watanabe *et al.*, 2007).

Other studies demonstrated that phage therapy was successful when high numbers of STEC were present, but unsuccessful in the presence of low numbers of food-borne pathogens, probably due to the lack of optimum phage-target bacterial interactions (Ricke, 2012). For example, unlike bacterial infections where bacterial numbers are high, the use of phage to eliminate food-borne pathogens in the GIT, where total bacterial populations are high and food-borne pathogens are only a minor component, presents challenges. In such conditions, it can be difficult for the phages to physically come in contact with the target bacteria, which are often hidden within the intestinal epithelium, or may be present in biofilms (Ricke, 2012). Therefore, a potential challenge for effective phage control of STEC in ruminants is the requirement for high numbers of bacteria for the phages to interact with. In fact, the concentration of STEC O157:H7 (the principal STEC serotype associated with cattle) in bovine faeces has been reported to range from 10^1 to 10^2 colony forming units (CFU)/g, but some cattle, known as super shedders, excrete as high as1 0^7 CFU/g of faeces and maintain the population of this pathogen within the herd (Fegan *et al.*, 2004; Widiasih *et al.*, 2004).

Cattle

Many studies aimed at reducing STEC in ruminants have focused on cattle. Rivas *et al.* (2010) examined the ability of phages to control STEC O157:H7 in the rumen and the bovine GIT. First, the authors used an *ex vivo* rumen model to study the efficacy of phages e11/2 and e4/1c against STEC O157:H7. Rumen samples were inoculated with either 10^3 or 10^6 CFU/ml inoculum of STEC O157:H7 and challenged separately with each phage. In the presence of phage e11/2, the numbers of bacteria were significantly ($P < 0.05$) reduced to below the limit of detection within 1 hour. Phage e4/1c significantly ($P < 0.05$) reduced STEC O157:H7 numbers within 2 hours of incubation, but the number of surviving bacteria then remained unchanged over a further 22 hour incubation period. Next, both phages were combined into a cocktail and evaluated for reduction of faecal shedding of STEC O157:H7 in experimentally inoculated cattle. In one study, adult fistulated cows were orally inoculated with 10^{10} CFU of STEC O157:H7. Animals were dosed daily with a phage cocktail of 10^{11} plaque forming units (PFU) for 3 days post inoculation. *STEC* O157:H7 and phage numbers in faecal and/or rumen samples were monitored over 7 days post inoculation. STEC O157:H7 numbers rapidly declined in all animals within 24 to 48 hours; however, there was no significant difference ($P > 0.05$) between the numbers of STEC O157:H7 shed by the phage-treated or control animals. The authors concluded that while the results from the rumen model suggest that phages are effective in the rumen, further research is required to improve the antimicrobial effectiveness of phages for the elimination of STEC O157:H7 *in vivo*.

The rectal anal junction is recognized as the primary site of colonization of STEC O157:H7 in cattle, leading to strategies to apply phages directly to the rectal-anal mucosa (Ricke, 2012). In one study, steers that received phage by the rectal route had similar phage faecal titres as the untreated control, and significantly lower titres than the steers which received phage orally (Rozema *et al.*, 2009).

Additionally, the levels of *E. coli* O157:H7 shedding were 0.5 log higher in steers inoculated rectally, compared to those untreated, and there was no difference in the number of *E. coli* O157-positive samples among treatments. The authors rationalized that the oral administration of phage may have permitted a longer period of retention of phage within the digestive tract and increased the opportunity for the phage to interact with the target host cells. Another example of rectal administration of phage was observed in a study conducted by Sheng *et al.* (2006). The combination of oral/rectal delivery of a cocktail of phages, KH1 and SH1, was able to reduce the levels of pathogenic *E. coli* O157:H7 by 1.5 log on day 1, and 1.3 log 7 days after phage treatment, when compared to control steers, which had not been administered the phage cocktail. However, while the combination treatment using the phage cocktail was successful in reduction of *E. coli* O157:H7 levels, total eradication of the pathogen from cattle was not achieved.

Another factor influencing the effectiveness of phage therapy in cattle (and other animals) is the source from which the phages were isolated. For example, Callaway *et al.* (2008) has stipulated that phages used to control the presence of STEC in the bovine GIT should ideally be isolated from the same environment, since the fitness of most organisms is typically greatest in their natural environment. This could potentially explain why some studies involving experimentally inoculated animals were not successful.

Sheep

A number of studies aimed at reducing STEC in ruminants have also been conducted on sheep, likely because they are smaller and thus easier to handle, and are also more cost-effective than adult bovines. As with the bovine studies, phages have been successful at reducing STEC O157:H7 during *ex vivo* studies (Bach *et al.*, 2003). In contrast, the use of phages to reduce STEC O157:H7 shedding *in vivo* has resulted in varying success (Bach *et al.*, 2003; 2009; Callaway *et al.*, 2003; 2006; 2008; Raya *et al.*, 2006; 2011). Raya *et al.* (2006) isolated phage CEV1 from sheep that were resistant to STEC O157:H7 colonization and used this phage to reduce populations of STEC O157:H7 in sheep. *In vitro*, CEV1 efficiently infected STEC O157:H7 grown both aerobically and anaerobically. In an *in vivo* study, four sheep were treated once orally with 10^{11} PFU of phage CEV1, 3 days after challenge with STEC O157:H7. Sheep receiving a single oral dose of CEV1 showed a 2–3 log-unit reduction in caecal and rectal STEC O157:H7 levels within two days compared to levels in control animals, although rumen concentrations of STEC O157:H7 remained unchanged.

Other studies have not been as successful. For example, Callaway *et al.* (2003) showed that while STEC O157 specific phages caused a decrease in the concentration of STEC O157:H7 throughout the ovine GIT, the differences were not statistically significant. Furthermore, a study by Bach *et al.* (2003) evaluated the ability of a single phage (DC22) to eliminate STEC O157:H7 in an artificial rumen system and in experimentally inoculated sheep. Results from this study indicated phage DC22, with a phage to bacteria ratio (multiplicity of infection, MOI) of 10^5 was successful at eliminating STEC O157:H7 in the artificial rumen. However, after inoculation of the sheep with the same MOI, phage DC22 had no effect on faecal shedding of STEC O157:H7. This might have been due to an insufficient MOI in the sheep compared to the high MOI in the enclosed fermentation system. Also, some non-specific binding of the phage to food particles and other debris might have occurred in the rumen, thereby decreasing the number of phages reaching the lower GIT where STEC O157:H7 resides. Finally, the inability of a single phage to control STEC O157:H7 in this study strengthens the idea that a cocktail of phages might be more effective at controlling STEC O157:H7 in livestock.

Thus, other research groups have evaluated the use of phage cocktails to decrease STEC in sheep during brief time periods. Bach *et al.* (2009) reported that STEC O157:H7 phage mixtures administered orally to Canadian Arcott rams reduced STEC O157:H7 shedding when compared to controls not receiving phage, but the administered phage also declined rapidly over a 21-day period causing the authors to suspect that phage inactivation occurred in the GIT.

Callaway *et al.* (2006) developed a 21-phage cocktail, which was used to inoculate sheep artificially contaminated with STEC O157:H7. Following introduction of the cocktail, there was a statistically significant reduction in populations of STEC O157:H7 in the caecum and rectum.

When sheep were dosed with 10^5, 10^6, or 10^7 PFU of the phage cocktail, the 10^5 PFU (MOI of 10) concentration was the most effective at reducing STEC O157:H7 populations throughout the GIT, indicating that properly selected phages can reduce STEC O157:H7 in ovines. These authors followed with another study, with an eight-phage cocktail orally administered to sheep (Callaway *et al.*, 2008). Bacterial reductions were observed throughout the GIT with the greatest reductions occurring in the cecum. Reduced rectal STEC O157:H7 levels were also observed, but complete elimination was never detected. Similar results were reported by Raya *et al.* (2011), whose results showed that a cocktail of CEV1 and CEV2 phages led to more than a three log reduction of *E. coli* in the ovine GIT.

Swine

Swine have been shown to be a reservoir for STEC and pork products have been associated with outbreaks involving STEC O157 and O111:H- (Tseng *et al.*, 2014). Still, most phage therapy applications in swine have been directed at controlling the presence of porcine specific ETEC, which causes diarrhoea in piglets. ETEC serotype O149:H10:F4 is the dominant ETEC serotype that causes disease in swine (Jamalludeen *et al.*, 2007). One potential limitation of phage therapy in food animals is the fact that phages lose activity at low pH, which is one reason why oral administration of phages to ruminants, where the phages must traverse four individual compartments of the stomach, has not been very successful in reducing STEC (Bach *et al.*, 2009). Since swine are monogastric animals, phage transit time through the low pH environment of the stomach is significantly reduced compared to ruminant animals. This means that approaches used to protect the phages in the low pH environment such as encapsulation are likely to be more effective in swine, and phages are potentially more likely to reach the lower GIT where the *E. coli* colonization occurs. In fact, studies on the use of phage to control ETEC in swine have reported promising results.

Smith and Huggins (1983) were the first to study phage therapy to control *E. coli* in porcine in the modern era. These authors investigated the efficacy of a two-phage cocktail against infection induced by the ETEC strain P433 in neonatal pigs. Seven piglets were infected with *E. coli* P433, and then treated at the onset of diarrhoea (13–16 hours later) with an orally administered cocktail of phage P433/1 and phage P433/2. Seven piglets infected with ETEC strain P433, but not treated with the phage cocktail, served as controls. All control piglets became very ill, and four died. In contrast, apart from the diarrhoea, which ceased shortly after introduction of the phage cocktail, the treated piglets showed no other symptoms and all seven piglets survived.

Jamalludeen *et al.* (2007) isolated phages from the sewage of 38 swine farms, and evaluated their host ranges on various *E. coli* isolates. Six phages (GJ1-GJ6) that lysed an ETEC O149:H10:F4 isolate and three phages (GJ7-GJ9) that lysed an ETEC O149:H43:F4 isolate were assessed for ability to lyse various ETEC O149 isolates *in vitro*. Phages GJ1-GJ6 lysed 99–100% of 85 O149:H10:F4 ETEC, 0–12% of 42 O149:H43:F4 ETEC, 3–35% of 37 non-O149 porcine ETEC, and 6–68% of the 72 strains of the ECOR collection. Phages GJ7-GJ9 lysed 86–98% of the O149:H43:F4 ETEC, 2–53% of the O149:H10:F4 ETEC, and 24–41% of the non-O149 porcine ETEC. The phages were unaffected by exposure to pH 5–9 for 16 hours, and bacterial resistance to one phage was generally not accompanied by resistance to other phages, demonstrating that bacterial resistance was not cross-protective, and indicating that the phages were potentially useful in a cocktail to treat ETEC induced diarrhoea (Jamalludeen *et al.*, 2007).

A subsequent study by the same authors (Jamalludeen *et al.*, 2009) was conducted, during which phages GJ1-GJ7 were individually evaluated for prophylactic and therapeutic treatment of an experimental ETEC O149:H10:F4 infection in weaned pigs. For prophylaxis, the phages were administered orally shortly after challenge, and for therapeutic use, the phages were given 24 hours after challenge, following the onset of diarrhoea. The six (GJ1–GJ6) phages that were tested individually using the prophylactic approach successfully reduced diarrhoea in the treated animals, but fecal counts of the phages were low and did not exceed 10^3 PFU/g of faeces. Another attempt was conducted in which the pigs were given florfenicol prior to the bacterial challenge to reduce the competing intestinal flora and improve the colonization chances of the ETEC isolate (Jamalludeen *et al.*, 2009). Additionally, the pigs were administered sodium bicarbonate prior to the ETEC challenge in an attempt to protect the phages from exposure to low pH in the stomach. Oral administration of sodium bicarbonate prior to the ETEC challenge and phage administration resulted in high levels of phages in the faeces. In the therapeutic model of infection, a combination of three (GJ1, GJ2, and GJ7) phages, previously used in the prophylactic study, significantly reduced the severity of diarrhoea in the piglets. Enumeration of phages in faeces after treatment indicated that the phages were replicating to high concentrations in the intestinal tract of ETEC infected pigs within one to two days of administration, before declining progressively. The authors concluded that the selected phages were effective in reducing the duration of experimental O149:H10:F4 ETEC diarrhoea in weaned pigs when given prophylactically or therapeutically.

Poultry

In poultry, phage therapy has mostly focused on reducing *Salmonella* and *Campylobacter*, which are two of the main pathogens involved in poultry associated food-borne outbreaks (FAO and WHO, 2009). However, avian pathogenic *Escherichia coli* (APEC) is another major poultry pathogen, causing a great variety of diseases in birds, including infection of the lower respiratory tract, blood and intestine, collectively referred to as colibacillosis. Being the most common bacterial disease affecting poultry, colibacillosis accounts for huge economic losses to the poultry industry (La Ragione and Woodward, 2002). Farms affected by APEC can lose up to 8% of their hens. Direct economic losses are a consequence of mortality, cost of antimicrobial treatments, and condemnations at slaughter, while indirect losses are attributed to poor performance and stunted growth of surviving birds (Barnes *et al.*, 2003).

Phage therapy has been investigated as a treatment for controlling APEC in poultry. The literature demonstrates mixed results of these studies. For example, in a large study, Lau *et al.* (2010) evaluated the efficacy of phage EC1, against APEC O78:K80 in poultry. A total of 480 1-day-old birds were randomly assigned to four treatments groups (I–IV). Birds from the control groups (groups I and II) received buffer or 10^{10} PFU of phage EC1, respectively, by intratracheal inoculation. Group III consisted of birds challenged (by intratracheal inoculation) with 10^8 CFU of APEC O78:K80, followed by treatment with 10^{10} of phage EC1 at 2 hours post infection. Birds in group IV were challenged with 10^8 CFU of APEC O78:K80 only, and served as positive controls. The infection was found to be less severe in the treated APEC challenged group. Mean total viable cell counts of *E. coli* in the lungs were significantly lower in treated, APEC-challenged birds than in untreated, APEC-challenged birds on days one and two post infection. Additionally, no APEC was detectable in blood samples on any sampling day. The body weight of birds from the APEC-challenged group treated with phage EC1 was not significantly different from that of birds from both control groups but was 15.4% higher than that of the untreated, APEC-challenged group on day 21 post infection. The total mortality rate of birds during the 3-week experimental period decreased from 83.3% in the untreated, APEC-challenged group (IV) to 13.3% in birds treated with phage EC1 (group III). These results suggest that phage EC1 is an effective *in vivo* antimicrobial and could be used to treat colibacillosis in chickens.

Oliveira *et al.* (2010) investigated the performance of a three-phage cocktail in treating severe respiratory *E. coli* infections in experimentally contaminated birds and naturally infected flocks. Three lytic coliphages, phi F78E (Myoviridae), phi F258E (Siphoviridae) and phi F61E (Myoviridae) were combined in a cocktail at a concentration of 5.0×10^7 PFU/ml, and used to treat naturally APEC infected flocks. In addition, experimentally infected birds were treated with phage phi F78E at two different titres (10^7 PFU/ml and 10^9 PFU/ml). Phage administration was performed orally and by spray (inhalation), in a single application. The morbidity, mortality, and pathology scores of the treated birds were compared with control birds that had not received the phage cocktail. The results revealed that the efficacy of phage therapy in the experimental infections was dosage dependent, with 10^7 PFU/ml not enough to treat the infected chickens, while a concentration of 10^9 PFU/ml of phi F78E allowed a decrease of 25% and 43% in mortality and morbidity, respectively. In naturally infected flocks, the lower concentration of the phage cocktail (10^7 PFU/ml) effectively decreased mortality to levels below 0.5% within 3 weeks. Collectively, these studies demonstrate that phage treatment is a valuable alternative to control APEC infections in poultry.

On the other hand, Tsonos *et al.* (2014) developed a cocktail of four different APEC-specific phages, which was then evaluated for its ability to control APEC in experimentally infected chickens. Chickens were intratracheally infected with APEC strain CH2 (serogroup O78) and the phage cocktail was administered intratracheally, intra-oesophageally, or via drinking water, 2 hours after the infection. The results showed that the treated groups did not show a significant decrease in mortality, lesion scores or weight loss compared to untreated groups, and this result was not attributed to the development of BIMs, since re-isolated bacteria from infected chickens remained sensitive to the phage cocktail. The authors concluded that the efficiency of the phage cocktail used in treating CH2-infected chickens *in vivo* was negligible. It is possible that this result was due to the lack of a high enough concentration of phages reaching the bacteria.

Many variables influence the ability of phages to effectively find and kill their host cells, and failure to control these variables often leads to treatment failure. In poultry, as in other animals, one such variable is the method used to administer the phage(s). Oliveira *et al.* (2009) evaluated the effect of the route of administration, and the dosage, on the systemic dissemination of three coliphages in chickens. *In vivo* trials were conducted by inoculating chickens orally, by spray, and intramuscularly with 10^6, 10^7 and 10^8 PFU/ml suspensions of three lytic phages including phi F78E, phi F258E, and phi F61E. Birds were sacrificed at 3, 10 and 24 hours after inoculation and the phage titre was measured in the lungs, air sac membranes, liver, duodenum, and spleen. When administered by spray, the phages reached the respiratory tract within 3 hours. Oral administration also allowed all phages to be recovered in lungs, but only phi F78E was recovered from the duodenum, the liver, and the spleen. It is possible that phage phi F78E was able to replicate in commensal *E. coli* strains present in the chicken gut, thus leading to a higher concentration of this phage in the intestines that resulted in systemic circulation of the phage with consequent recovery in the organs. When phages were administered intramuscularly, they were found in all of the sampled organs, but this approach is not a practical way of protecting a large number of birds in a poultry unit. The overall results of this work suggest that oral and spray administration allowed phages to reach and remain active in the respiratory tract, and these delivery methods seem to be promising administration routes to treat respiratory *E. coli* infections in poultry. Huff *et al.* (2013) also investigated whether different phage administration routes lead to different colibacillosis treatment outcomes. Two studies were conducted with six treatments, including an untreated control, birds treated by spray administration of phage and not challenged, birds administered phage intratracheally and not challenged, birds not treated and challenged intratracheally with *E. coli*, birds sprayed with phage and intratracheally challenged with *E. coli*, and birds administered phage intratracheally and challenged intratracheally with *E. coli*. In both studies, neither a coarse nor a fine spray protected the birds from an intratracheal *E. coli* challenge. However, when phages were administered intratracheally, there was complete protection. This research demonstrates the importance of proper administration routes to deliver high phage titres to the site of a bacterial infection, and the lack of constant delivery of high enough phage concentrations to the site of active infection is likely the reason why many phage therapy approaches are unsuccessful in animals.

Another reason for phage treatment failure is rapid clearance of phages by the host immune system. Huff *et al.* (2010) evaluated whether prior exposure to phage limits the ability of the same phage to treat colibacillosis. In a series of experiments, birds were either first treated with a phage followed by challenge with *E. coli*, and re-treatment with the same phage, or challenged with *E. coli*, and then treated with the phage. Mortality in the birds challenged with *E. coli* and not treated with phage was 55%, and phage therapy significantly reduced mortality to 8%, which was not significantly different from non-*E. coli*-challenged controls (3%). However, mortality in the birds administered phage prior to challenge with *E. coli* and treated with phage (33%) was not significantly different from the birds that were challenged with *E. coli* and untreated (55%), suggesting rapid clearance of the phage by the immune system. An *in vitro* assay of phage activity (infection kinetics of *E. coli* host cells) showed that serum from birds pretreated with phage inhibited phage activity. Additionally, the IgG levels to the phage in serum from birds pre-treated with phage were significantly higher at all dilutions compared with control serum. This study shows that prior exposure to phage will limit the use of the same (or similar) phage as a therapeutic, and suggests that the reduced efficacy is due to an immune response to phage.

Human trials

It is striking to note that the first demonstration of phage therapy occurred in humans, when, shortly after discovering phages, d'Herelle demonstrated the efficacy of phage therapy by curing four patients of shigellosis, and followed that with several subsequent studies confirming the ability of phages to reduce *Shigella*-based infection (Summers, 1999). *Shigella* and *E. coli* are highly related as demonstrated by the fact that the shigellae and *E. coli* share > 75% nucleotide similarity. *Shigella* spp. appear to be metabolically inactive biogroups of *E. coli*, and both are considered to be a single species, based on DNA homology (Brenner, 1984).

While d'Herelle's early work in treating disease causing bacteria with phages was successful (d'Herelle, 1917), early human phage therapy trials were impeded by limited knowledge of phage biology, poor experimental design, and in some cases, a lack of understanding of the causative nature of the illness that was being treated. Interest in phage therapy waned in the Western world with the discovery of antibiotics. However, phage therapy research continued in the former Soviet Union, and a great deal of insight into the potential for phages to treat human disease can be learned from these studies. For example, in 1963, a total of 30,769 children (ranging in age from 6 months to 7 years old) were enrolled in a phage therapy prophylaxis trial against bacterial dysentery. In the study, children living on one side of the streets in the geographical study area received a weekly oral inoculation of *Shigella* phages in a pressed tablet form, while children on the other side of the streets received a placebo. The children were followed for 109 days. Phage administration was associated with a 3.8-fold decrease in dysentery incidence (1.8 vs 6.7 episodes per 1000 children from the treatment and placebo groups, respectively). Phage exposure also decreased the incidence of diarrhoea (15 vs 45 episodes per 1000 children, 6–12 months old, in treatment and placebo group, respectively). These results suggest a protective effect of the anti-*Shigella* phage preparation against serotypes of *E. coli*. Protective effects were most pronounced in children younger than three years (Babalova *et al.*, 1968).

Interest in phage therapy was rekindled in the 1980s in the Western world with the realization that overuse of antibiotics was contributing to an increase in multidrug resistant bacterial pathogens. Phage therapy research had continued throughout the years in the former Eastern bloc nations, but the studies were generally insufficiently documented in the published Russian literature for a rigorous scientific evaluation to be accomplished. Thus, rudimentary studies aimed at evaluating the safety of phage therapy as a treatment in humans had to be repeated. One such study was conducted by Bruttin and Brüssow (2005), who evaluated the presence and safety of the *E. coli* phage T4 in 15 healthy adult volunteers, who each received T4 phage in their drinking water. The volunteers received either a lower phage T4 dose (10^3 PFU/ml), a higher phage dose (10^5 PFU/ml), or a placebo. All volunteers receiving the higher phage dose showed T4 presence in faecal samples one day following exposure, while only 50% of volunteers who received the lower phage dose shed phage T4 in their faeces. No faecal phage were detectable a week after a 2-day course of oral phage application. Oral phage application did not cause a decrease in total faecal *E. coli* counts. In addition, no substantial phage T4 replication on the commensal *E. coli* population was observed. No adverse events related to phage application were reported. Serum transaminase levels remained in the normal range (indicative of the absence of liver toxicity), and neither phage T4 nor T4-specific antibodies were observed in the serum of the subjects at the end of the study. This study indicated that concentrated doses of phage could be orally administered to humans without adverse health effects. In contrast, Bocian *et al.* (2016) tested the effects of two phages (T4 and A3/R), as well as phage-generated bacterial lysates, on differentiation of human myeloid dendritic cells (DCs) from monocytes, which can serve as an indication of the activation of the immune response. Neither of the phages significantly reduced the expression of markers associated with differentiation of DCs and their role in the activation of T cells and phagocytosis receptors (CD64 and DEC-205). However,

Phage T4 bacterial lysate significantly decreased the percentages of DEC-205- and CD1c-positive cells. The percentage of DEC-205-positive cells was also significantly reduced in DCs differentiated in the presence of phage A3/R lysate. Therefore, while phages do not substantially affect differentiation of DCs, some products of phage-induced bacterial lysis may influence the differentiation and potentially also some functions of DCs. The authors concluded that their results have important implications for phage therapy of bacterial infections, because during infections, monocytes recruited to the site of inflammation are an important source of inflammatory DCs.

Notwithstanding the above studies, it is evident that the safety of human (and other animal) phage therapeutics will be determined through carefully developed clinical trials. Several phage products have undergone clinical trials, or are currently being evaluated in clinical trials. These trials have focused on phage products active against wound infections. Treating exposed wounds eliminates many of the limitations of using phages in animals including rapid clearance by the immune system, inactivation due to low pH of the stomach, and problems associated with effective delivery of the phage product to the site of the infection.

Rhoads *et al.* (2009) described a phase 1 trial to evaluate the safety of a phage-based preparation for leg ulcers. Forty-two patients with chronic venous leg ulcers were enrolled in the trial, and 39 patients completed the trial. The ulcers were treated for 12 weeks with either a saline control or phages targeted against *Pseudomonas aeruginosa, Staphylococcus aureus,* and *E. coli.* All patients were followed until week 24. No significant differences ($P > 0.05$) were observed between the test and control groups for frequency of adverse events, rate of healing, or frequency of healing. It should be noted that this trial only evaluated the safety of the phage treatment. The efficacy of the preparation was not evaluated. To date, no other clinical trials on this treatment have been conducted.

An ongoing clinical trial is evaluating the efficacy, safety, and pharmacodynamics of Phagoburn, which consists of two therapeutic phage cocktails to treat either *E. coli* or *P. aeruginosa* burn wound infections (Unknown, 2017). The Phagoburn clinical trial began in July 2015, and is the first international clinical study on phages that meets international standards in clinical evaluation. This randomized and monitored phase I/II single-blind trial is simultaneously being conducted in France, Switzerland and Belgium in 11 different burn wound treatment centres acting as clinical centres. Two hundred and twenty patients have been enrolled in the trial. The study includes two arms of 110 patients each: one arm for burn wounds infected by *E. coli,* and the other arm for burn wounds infected by *P. aeruginosa.* The effect of the phage cocktails is compared to a reference antibiotic treatment (i.e. silver sulfadiazine). The primary outcome measure is the time for bacteria reduction adjusted on antibiotic treatment; the secondary outcome measures are the assessment of treatment tolerance, the incidence on delay of infection reduction with different bacterial species from the targets, and the number of sites cured.

One phage product currently on the market is Pyophage, a phage cocktail that is directly available to the public without prescription in Georgia and Russia, and contains phages targeting *Staphylococcus, Streptococcus, Pseudomonas, Proteus* and *E. coli.* As with the phage products that have undergone clinical trials, *Pyophage* is used to treat skin or wound infections (Abedon *et al.*, 2011).

Future trends

The many phage therapy studies conducted in animals and humans have highlighted the advantages of using phages prophylactically or therapeutically. Such advantages include the ubiquitous nature of phages, their relatively narrow host specificity, and low inherent toxicity. Still, such advantages are limited by the inherent disadvantages of employing phages to treat disease, such as the need to optimize phage delivery, and the need to overcome the fact that a given phage tends to only infect a narrow range of related bacterial strains, and can promote the development of BIMs. Future trends directed at addressing these challenges are described below.

Optimization of phage delivery

A major barrier to the successful use of phages for treatment of disease is their delivery to the site of infection. In animals, the most practical approach continues to be oral delivery. However, this necessitates the transit of the phages through low pH organs such as the stomach. Several approaches including encapsulation of the phage cocktail have been proposed, but such approaches have been met with mixed results with respect to survival and viability of the phages. Nobrega *et al.* (2016) report a simple and efficient alternative, consisting of genetic modification of phages to display lipids on their surfaces. In their work, *E. coli* phage T7 was used as a model phage and the *E. coli* PhoE signal peptide was genetically fused to the phage T7 major capsid protein, enabling phospholipid attachment to the phage capsid. The stability of the genetically modified phages was analysed in simulated GIT conditions, demonstrating improved stability of the mutant phages with survival rates ranging from 10^2–10^7 PFU/ml higher than wild-type phages. The authors concluded that their work demonstrated that genetic phage modification is a good strategy to improve phage tolerance to acidic GIT conditions, increasing the prospects of oral phage administration in veterinary medicine.

Improving phage host range

Genetic modification of phages may also be employed to solve the issue of phage host range, which is another challenge that must be overcome for effective phage therapy deployment. One much described advantage of using phages to treat disease is their narrow host range, meaning that, unlike antibiotics, the gut microflora will not be affected by phage therapy. On the other hand, having narrow host range is also a disadvantage as there are many diverse strains of bacteria within a given species or even serotype that cause disease, and in order to be effective, a phage product would have to contain phages that can infect all strains within the target bacterial species. With these hurdles in mind, a company called Synthetic Genomics is employing an alternative strategy to the traditional phage cocktail approach to solve host range challenges. Synthetic Genomics is attempting to engineer a synthetic phage that combines the properties of multiple phages into a single genome, by using bioinformatics and viral engineering methods to understand the natural host ranges of various phages, and then inserting genes from other phages to expand the host ranges to include other subtypes of the target bacteria (Madhusoodanan, 2016). Such genetically engineered phages are potentially simpler to manufacture, but may have additional regulatory hurdles to clear than cocktails of phages.

The idea of expanding the host range of phages for use in therapeutic approaches is not a new one. Instead of genetically modifying a phage to expand its host range, one can take advantage of phages that can use more than one receptor on the bacterial cell surface. There are many examples of dual specificity phages described in the literature. For example, members of the T-even group of phages (including the type phage, T4) bind to two different receptors (outer membrane proteins and lipopolysaccharide) on the bacterial cell surface in a sequential manner. Moreno and Wandersman (1980) observed that both OmpC and LamB could serve as receptors for host range mutants of the T-even phage, Tula, which employs OmpF as its receptor. The researchers described an elegant method to isolate the mutants, which initially involved plating the wild-type phage Tula on an *E. coli* strain that produced very low amounts of the OmpF and LamB proteins. The resulting phage (TP2) used either OmpF or OmpC as its receptor. When Tula was plated on an OmpF mutant, phage TP5 was created, which has tri-specificity and utilized OmpF, OmpC, or LamB as a receptor. Plating Tula on other mutant strains produced other phages capable of using OmpF or LamB to adsorb to the host cell (phage TP1), or the OmpC protein as a receptor (TP6). Such studies suggest that the natural ability of T-even phages to expand their host ranges could be exploited to create phages with expanded host ranges using a directed receptor approach, addressing the challenge of bacteria within the same species that express varied cell surface proteins.

Conclusion

Several challenges of phage therapy as a method to control bacterial infections in humans and animals must be overcome before routine use can be realized. Efficient and cost effective delivery methods must be developed to ensure that the appropriate phage concentrations are consistently introduced to the site of infection. In the case of *E. coli*, in which multiple pathotypes that cause disease have been discovered, phage products must be designed to ensure adequate coverage of all bacterial strains that cause disease within a given pathotype. Additional criteria such as selection approaches for phages and methods of phage preparation must be standardized. The regulatory process for phage-based antimicrobials used prophylactically or as therapeutics must be defined. Finally, carefully developed clinical trials assessing the safety and efficacy of phage products must expand. Despite the enormous amount of research and development still to occur, the increasing emergence of multi-drug resistant bacterial pathogens, and the lack of effective antibiotics to treat the diseases they cause ensure that phage therapy will remain as a potential treatment alternative for years to come.

Bacteriophage therapy web resources

Phage Therapy: http://www.phage-therapy.org/

Eliava Phage Therapy Center: http://eliavaphagetherapy.com/

Phage Therapy Center: http://www.phagetherapycenter.com/pii/PatientServlet?command=static_home

Institute of Immunology and Experimental Therapy Polish Academy of Sciences: https://www.iitd.pan.wroc.pl/en/Phages/

Evergreen Phage Lab: http://blogs.evergreen.edu/phage/therapy-news/

Phages for Human Applications Group Europe (P.H.A.G.E.): http://www.p-h-a-g-e.org/

The DSMZ biological resource center phage therapy page: https://www.dsmz.de/home/news-and-events/faq-phage-therapy.html

References

Abedon, S.T., Kuhl, S.J., Blasdel, B.G., and Kutter, E.M. (2011). Phage treatment of human infections. Bacteriophage *1*, 66–85. https://doi.org/10.4161/bact.1.2.15845

Allocati, N., Masulli, M., Alexeyev, M.F., and Di Ilio, C. (2013). *Escherichia coli* in Europe: an overview. Int. J. Environ. Res. Public Health *10*, 6235–6254. https://doi.org/10.3390/ijerph10126235

Babalova, E.G., Katsitadze, K.T., Sakvarelidze, L.A., Imnaishvili, N.S.h., Sharashidze, T.G., Badashvili, V.A., Kiknadze, G.P., Meĭpariani, A.N., Gendzekhadze, N.D., Machavariani, E.V., *et al.* (1968). [Preventive value of dried dysentery bacteriophage.] Zh. Mikrobiol. Epidemiol. Immunobiol. 45, 143–145.

Bach, S.J., Johnson, R.P., Stanford, K., and McAllister, T.A. (2009). Bacteriophages reduce *Escherichia coli* O157:H7 levels in experimentally inoculated sheep. Can. J. Anim. Sci. *89*, 285-293.

Bach, S.J., McAllister, T.A., Veira, D.M., Gannon, V.P., and Holley, R.A. (2003). Effect of bacteriophage DC22 on *Escherichia coli* O157: H7 in an artificial rumen system (Rusitec) and inoculated sheep. Anim. Res. 52, 89–101.

Barnes, H.J., Vaillancourt, J.P., and Gross, W.B. (2003). Colibacillosis. In Diseases of Poultry, 11th edition, H.J. Barnes, A.M. Fadly, J.R. Glisson, L.R. McDougald, D.E. Swayne, and Y.M. Saif, ed. (Ames: Iowa State University Press), pp. 631–656.

Bocian, K., Borysowski, J., Zarzycki, M., Pacek, M., Weber-Dąbrowska, B., Machcińska, M., Korczak-Kowalska, G., and Górski, A. (2016). The Effects of T4 and A3/R Bacteriophages on Differentiation of Human Myeloid Dendritic Cells. Front. Microbiol. 7, 1267.

Brenner, D.J. (1984). Family I. Enterobacteriaceae. In Bergey's Manual of Systematic Bacteriology, vol. 1, N.R. Krieg and J.G. Holt, ed. (Baltimore: Williams and Wilkins), pp. 408–420.

Brüssow, H. (2005). Phage therapy: the *Escherichia coli* experience. Microbiology. *151*, 2133-2140.

Brüssow, H., Canchaya, C., and Hardt, W.D. (2004). Phages and the evolution of bacterial pathogens: from genomic rearrangements to lysogenic conversion. Microbiol. Mol. Biol. Rev. *68*, 560–602. https://doi.org/10.1128/MMBR.68.3.560-602.2004

Bruttin, A., and Brüssow, H. (2005). Human volunteers receiving *Escherichia coli* phage T4 orally: a safety test of phage therapy. Antimicrob. Agents Chemother. *49*, 2874–2878.

Callaway, T.R., Edrington, T.S., Anderson, R.C., Jung, Y.S., Genovese, K.J., Elder, R.O., and Nisbet, D.J. (2003). Isolation of naturally-occurring bacteriophage from sheep that reduce populations of *E. coli* O157:H7 *in vitro* and *in vivo*. Proc. 5th Int. Symp. on Shiga Toxin-Producing *Escherichia coli* Infections, p. 25.

Callaway, T.R., Edrington, T.S., Brabban, A.D., Anderson, R.C., Rossman, M.L., Engler, M.J., Carr, M.A., Genovese, K.J., Keen, J.E., Looper, M.L., *et al.* (2008). Bacteriophage isolated from feedlot cattle can reduce *Escherichia coli* O157: H7 populations in ruminant gastrointestinal tracts. Food-borne Pathog. Dis. *5*, 183–191.

Callaway, T.R., Edrington, T.S., Brabban, A.D., Kutter, E.S., Anderson, R.C., and Nisbet, D.J. (2006). Isolation and use of bacteriophage to reduce *Escherichia coli* O157:H7 populations in ruminants. Proc. Int. Conf. Perspect. Bacteriophage Preparation, pp. 72–83.

Center for Disease Control and Prevention. (2017). *E. coli* (*Escherichia coli*). Available at http://www.cdc.gov/ecoli/general/index.html.

Colavecchio, A., Jeukens, J., Freschi, L., Edmond Rheault, J.-G., Kukavica-Ibrulj, I., Levesque, R.C., LeJeune, J., and Goodridge, L.G. (2017). Complete genome sequences of two phage-like plasmids carrying the CTX-M-15 extended-spectrum β-lactamase gene. Genome Announc. 5, e00102-17.

Croxen, M.A., Law, R.J., Scholz, R., Keeney, K.M., Wlodarska, M., and Finlay, B.B. (2013). Recent advances in understanding enteric pathogenic *Escherichia coli*. Clin. Microbiol. Rev. *26*, 822–880. https://doi.org/10.1128/CMR.00022-13

d'Hérelle, F. (1917). [Sur un microbe invisible antagoniste des bacilles dysentérique]. Acad. Sci. Paris. *165*, 373–375.

Fegan, N., Vanderlinde, P., Higgs, G., and Desmarchelier, P. (2004). The prevalence and concentration of *Escherichia coli* O157 in faeces of cattle from different production systems at slaughter. J. Appl. Microbiol. *97*, 362–370. https://doi.org/10.1111/j.1365-2672.2004.02300.x

Food and Agriculture Organization of the United Nations and World Health Organization. (2009). Microbiological Risk Assessment Series. *Salmonella* and *Campylobacter* in chicken meat. Available at http://www.who.int/foodsafety/publications/micro/MRA19.pdf.

Fruciano, D.E., and Bourne, S. (2007). Phage as an antimicrobial agent: d'Herelle's heretical theories and their role in the decline of phage prophylaxis in the West. Can. J. Infect. Dis. Med. Microbiol. *18*, 19–26.

Huff, W.E., Huff, G.R., Rath, N.C., and Donoghue, A.M. (2010). Immune interference of bacteriophage efficacy when treating colibacillosis in poultry. Poult. Sci. *89*, 895–900. https://doi.org/10.3382/ps.2009-00528

Huff, W.E., Huff, G.R., Rath, N.C., and Donoghue, A.M. (2013). Method of administration affects the ability of bacteriophage to prevent colibacillosis in 1-day-old broiler chickens. Poult. Sci. *92*, 930–934. https://doi.org/10.3382/ps.2012-02916

Hynes, N., and Wachsmuth, I. (2000). *Escherichia coli* O157:H7 risk assessment in ground beef: a public health tool. Proc. 4th Int. Symp. on Shiga Toxin-Producing *Escherichia coli* Infections, p. 46.

Jamalludeen, N., Johnson, R.P., Friendship, R., Kropinski, A.M., Lingohr, E.J., and Gyles, C.L. (2007). Isolation and characterization of nine bacteriophages that lyse O149 enterotoxigenic *Escherichia coli*. Vet. Microbiol. *124*, 47–57.

Jamalludeen, N., Johnson, R.P., Shewen, P.E., and Gyles, C.L. (2009). Evaluation of bacteriophages for prevention and treatment of diarrhea due to experimental enterotoxigenic *Escherichia coli* O149 infection of pigs. Vet. Microbiol. *136*, 135–141. https://doi.org/10.1016/j.vetmic.2008.10.021

Johnson, R.P., Gyles, C.L., Huff, W.E., Ojha, S., Huff, G.R., Rath, N.C., and Donoghue, A.M. (2008). Bacteriophages for prophylaxis and therapy in cattle, poultry and pigs. Anim. Health. Res. Rev. *9*, 201–215. https://doi.org/10.1017/S1466252308001576

La Ragione, R.M., and Woodward, M.J. (2002). Virulence factors of *Escherichia coli* serotypes associated with avian colisepticaemia. Res. Vet. Sci. *73*, 27–35.

Lau, G.L., Sieo, C.C., Tan, W.S., Hair-Bejo, M., Jalila, A., and Ho, Y.W. (2010). Efficacy of a bacteriophage isolated from chickens as a therapeutic agent for colibacillosis in broiler chickens. Poult. Sci. *89*, 2589–2596. https://doi.org/10.3382/ps.2010-00904

Madhusoodanan, J. (2016). Viral Soldiers. The Scientist. Available at http://www.the-scientist.com/?articles.view/articleNo/44785/title/Viral-Soldiers.

Matthews, L., Low, J.C., Gally, D.L., Pearce, M.C., Mellor, D.J., Heesterbeek, J.A., Chase-Topping, M., Naylor, S.W., Shaw, D.J., Reid, S.W., *et al.* (2006). Heterogeneous shedding of *Escherichia coli* O157 in cattle and its implications for control. Proc. Natl. Acad. Sci. U.S.A. *103*, 547–552.

Moreno, F., and Wandersman, C. (1980). OmpC and LamB proteins can serve as substitute receptors for host range mutants of coliphage TuIa. J. Bacteriol. *144*, 1182–1185.

Nobrega, F.L., Costa, A.R., Santos, J.F., Siliakus, M.F., van Lent, J.W., Kengen, S.W., Azeredo, J., and Kluskens, L.D. (2016). Genetically manipulated phages with improved pH resistance for oral administration in veterinary medicine. Sci. Rep. *6*, 39235. https://doi.org/10.1038/srep39235

Oliveira, A., Sereno, R., and Azeredo, J. (2010). *In vivo* efficiency evaluation of a phage cocktail in controlling severe colibacillosis in confined conditions and experimental poultry houses. Vet. Microbiol. *146*, 303–308.

Oliveira, A., Sereno, R., Nicolau, A., and Azeredo, J. (2009). The influence of the mode of administration in the dissemination of three coliphages in chickens. Poult. Sci. *88*, 728–733. https://doi.org/10.3382/ps.2008-00378

Raya, R.R., Oot, R.A., Moore-Maley, B., Wieland, S., Callaway, T.R., Kutter, E.M., and Brabban, A.D. (2011). Naturally resident and exogenously applied T4-like and T5-like bacteriophages can reduce *Escherichia coli* O157:H7 levels in sheep guts. Bacteriophage *1*, 15–24. https://doi.org/10.4161/bact.1.1.14175

Raya, R.R., Varey, P., Oot, R.A., Dyen, M.R., Callaway, T.R., Edrington, T.S., Kutter, E.M., and Brabban, A.D. (2006). Isolation and characterization of a new T-even bacteriophage, CEV1, and determination of its potential to reduce *Escherichia coli* O157:H7 levels in sheep. Appl. Environ. Microbiol. *72*, 6405–6410.

Rhoads, D.D., Wolcott, R.D., Kuskowski, M.A., Wolcott, B.M., Ward, L.S., and Sulakvelidze, A. (2009). Bacteriophage therapy of venous leg ulcers in humans: results of a phase I safety trial. J. Wound Care *18*, 237–238, 240–243. https://doi.org/10.12968/jowc.2009.18.6.42801

Ricke, S.C., Van Loo, E.J., Johnson, M.G., and O'Bryan, C.A. (2012). Organic meat production and processing (Oxford, UK: Wiley-Blackwell).

Rivas, L., Coffey, B., McAuliffe, O., McDonnell, M.J., Burgess, C.M., Coffey, A., Ross, R.P., and Duffy, G. (2010). *In vivo* and *ex vivo* evaluations of bacteriophages e11/2 and e4/1c for use in the control of *Escherichia coli* O157:H7. Appl. Environ. Microbiol. *76*, 7210–7216. https://doi.org/10.1128/AEM.01530-10

Rozema, E.A., Stephens, T.P., Bach, S.J., Okine, E.K., Johnson, R.P., Stanford, K., and McAllister, T.A. (2009). Oral and rectal administration of bacteriophages for control of *Escherichia coli* O157:H7 in feedlot cattle. J. Food Prot. *72*, 241–250.

Russell, J.B., Diez-Gonzalez, F., and Jarvis, G.N. (2000). Invited review: effects of diet shifts on *Escherichia coli* in cattle. J. Dairy Sci. *83*, 863–873.

Sheng, H., Knecht, H.J., Kudva, I.T., and Hovde, C.J. (2006). Application of bacteriophages to control intestinal *Escherichia coli* O157:H7 levels in ruminants. Appl. Environ. Microbiol. *72*, 5359–5366.

Smith, H.W., and Huggins, M.B. (1983). Effectiveness of phages in treating experimental *Escherichia coli* diarrhoea in calves, piglets and lambs. J. Gen. Microbiol. *129*, 2659–2675. https://doi.org/10.1099/00221287-129-8-2659

Summers, W.C. (1999). Felix d'Herelle and the origins of molecular biology (New Haven, CT: Yale University Press).

Tseng, M., Fratamico, P.M., Manning, S.D., and Funk, J.A. (2014). Shiga toxin-producing *Escherichia coli* in swine: the public health perspective. Anim. Health Res. Rev. *15*, 63–75. https://doi.org/10.1017/S1466252313000170

Tsonos, J., Oosterik, L.H., Tuntufye, H.N., Klumpp, J., Butaye, P., De Greve, H., Hernalsteens, J.P., Lavigne, R., and Goddeeris, B.M. (2014). A cocktail of *in vitro* efficient phages is not a guarantee for *in vivo* therapeutic results against avian colibacillosis. Vet. Microbiol. *171*, 470–479. https://doi.org/10.1016/j.vetmic.2013.10.021

Unknown. (2017). Evaluation of Phage Therapy for the Treatment of *Escherichia coli* and *Pseudomonas aeruginosa* Wound Infections in Burned Patients (PHAGOBURN). ClinicalTrials.gov. Available at https://clinicaltrials.gov/ct2/show/NCT02116010.

Watanabe, R., Matsumoto, T., Sano, G., Ishii, Y., Tateda, K., Sumiyama, Y., Uchiyama, J., Sakurai, S., Matsuzaki, S., Imai, S., *et al.* (2007). Efficacy of bacteriophage therapy against gut-derived sepsis caused by *Pseudomonas aeruginosa* in mice. Antimicrob. Agents Chemother. *51*, 446–452.

Widiasih, D.A., Ido, N., Omoe, K., Sugii, S., and Shinagawa, K. (2004). Duration and magnitude of faecal shedding of Shiga toxin-producing *Escherichia coli* from naturally infected cattle. Epidemiol. Infect. *132*, 67–75.

Zhao, S., Maurer, J.J., Hubert, S., De Villena, J.F., McDermott, P.F., Meng, J., Ayers, S., English, L., and White, D.G. (2005). Antimicrobial susceptibility and molecular characterization of avian pathogenic *Escherichia coli* isolates. Vet. Microbiol. *107*, 215–224.

New Developments in Detection Technologies for *Escherichia coli* and Other Pathogenic Organisms

6

Wen Ren[1], Renjie Wang[1], Lei Ouyang[1] and Joseph Irudayaraj[2*]

[1]Department of Agricultural and Biological Engineering, Purdue University, West Lafayette, IN, USA.
[2]Department of Bioengineering, University of Illinois at Urbana–Champaign, Urbana, IL, USA.

*Correspondence: jirudaya@illinois.edu

https://doi.org/10.21775/9781910190777.06

Abstract

Escherichia coli O157:H7 and other Shiga-toxin-producing *E. coli* (STEC), among other pathogens, are a serious threat to public safety and health. A gap between the zero tolerance regulatory requirement for O157:H7 and several other STEC serogroups, the challenges related to diverse sample conditions, and the analytical methods currently being used still exists, requiring additional research efforts. Due to advances in material science, instrumentation, and bioanalytical techniques, significant progress has been made on all aspects of technology development, including sample preparation; however, challenges remain. In this chapter, we review the unique strategies proposed for pathogen detection with a primary focus on technology development, with potential implications for monitoring for STEC.

E. coli O157:H7 is a Shiga toxin-producing pathogen that causes serious illness, including life-threatening haemorrhagic colitis with symptoms of bloody diarrhoea and thrombotic thrombocytopenic purpura (Buchanan and Doyle, 1997; Jay, 2012), resulting in up to 40% mortality (Chen and Frankel, 2005). It has been reported that 100 or fewer Shiga toxin-producing *E. coli* (STEC) O157:H7 cells can induce illness (Mead and Griffin, 1998; Nataro and Kaper, 1998). Therefore, there is a zero tolerance regulatory requirement on this pathogen and, thus, there is a strong demand for technologies to detect *E. coli* O157:H7 with an ultra-high sensitivity. Extensive efforts were undertaken to improve the sensitivity of *E. coli* O157:H7 detection technologies by incorporating a variety of analytical techniques such as colorimetry, spectrometry, microfluidics, and paper and chip-based methods. However, there is still a gap between the proposed analytical methods and expected sensitivity. Meanwhile, challenges exist to detect *E. coli* O157:H7 in complex samples such as food matrix or other environmental systems, including soil, plants, and water where the presence of other materials can influence detection. Hence, to detect *E. coli* O157:H7 in real samples, additional steps are commonly needed to separate and enrich the target pathogen of interest. To bring analytical methods from lab to on-site application, improvements in technologies that will deliver high sensitivity and rapid results with minimum involvement of operator time are necessary. In this section, we will discuss detection methods for *E. coli* O157:H7 in light of some of the latest and promising analytical techniques for pathogen detection.

Briefly, the essence of detection is the transfer of signals generated due to the presence of the target to some form of proportional detectable signal. One common route is to detect whole cell, while it is also possible to detect based on the pathogen's DNA or RNA, which can be amplified utilizing the polymerase chain reaction (PCR), loop-mediated isothermal amplification (LAMP) or other method, resulting in an increase in nucleic acid target of many orders of magnitude, resulting in high sensitivity. An alternative route is to monitor the consumption or generation of particular chemicals due to the physiological activity of target pathogens or the response of other bio-systems to the pathogens. No matter which route is undertaken for the detection of pathogens, the purpose is to provide a reliable technique with high sensitivity that is rapid, as well as cost-effective.

Colorimetric detection methods

Among the various analytical techniques, colorimetric methods promise to be the most simple for on-site applications, since it is possible to detect the presence of target pathogens by the naked eye without instrumentation. Further, colorimetric detection methods provide an easy and cost-effective choice when only semi-quantitative judgement is required to report on a target for regulatory purposes. Detection based on a colorimetric signal enables the direct recognition of pathogens or evaluation of their indicators such as DNA or RNA. Meanwhile, the generation and change in colour of chemicals or nanoparticles employed in the assay can be used for detection. In general, colorimetric routes provide versatile strategies to recognize the presence and determine the concentration of target pathogens including *E. coli* O157:H7. Therefore, it is not surprising that currently the number of publications on colorimetric detection of pathogens is on the rise. In the following sections, we will introduce new developments in colorimetric detection methods for *E. coli* O157:H7, as well as methods for other pathogens, which could be easily applied for *E. coli* O157:H7.

Colorimetry is a direct way to detect target pathogens in samples. To identify the target pathogens from samples, antibody is the most commonly used recognition ligand for specific capture for colorimetric determination. For instance, Wang *et al.* (2010) synthesized oval shaped gold nanoparticles (GNPs) modified with antibody against *Salmonella*, which will specifically attach to *Salmonella* strains. The interaction of antibody-modified GNPs on *Salmonella* Typhimurium resulted in a colour change from red to purple GNP aggregation on the surface of *S.* Typhimurium with good specificity, while the limit of detection (LOD) at10^4 cells/ml was not satisfactory. Cao *et al.* (2011) proposed a dual enlargement technique for antibody-modified GNP-based detection of pathogens. On a solid substrate, the captured target bacteria were labelled with modified GNPs. To enhance the signal, $HAuCl_4$ based growth followed by a silver enhancement increased the colorimetric signal from the nanoparticles, resulting in a dual enhancement for the detection of *Campylobacter jejuni* with a LOD down to 10^3 CFU/ml. Shim *et al.* (2014) used antibody-modified metallic nanoparticles (MNPs) to separate *S.* Typhimurium and after filtration, the brown colour from MNPs could be used for the determination of *S.* Typhimurium concentration, which provided a LOD down to 20 cells/ml and 100 cells/ml in vegetable samples. Using bifunctional nanostructures of gold and magnetic material, a significant improvement in capture efficiency for *S. aureus* was demonstrated both in PBS buffer and milk samples (Sung *et al.*, 2013). In addition to the colour from the nanoparticles themselves, nanoparticles are also a good platform to integrate the recognition ligand and enzyme, which can catalyse a colorimetric reaction for improved sensitivity. The functionalized nanoparticles as a platform could be utilized to fabricate unique sensors for pathogen detection. Cho *et al* (2013b) used magnetic beads to concentrate pathogens and GNP network enzyme-linked immunosorbent assay (ELISA) to amplify the signal to detect as low as 10 CFU/ml. For instance, Wu *et al.* (2014) used GNPs to load antibody and horseradish peroxidase (HRP) to label *S.* Typhimurium for ELISA detection assisted with aptamer modified MNPs separation. Compared with the direct labelling of antibodies by HRP, the GNP platform increased the number of HRP molecules linked to the target, which further improved the colorimetric signal from the HRP-catalysed 3,3,5,5-tetramethylbenzidine (TMB) reaction, resulting in a LOD of 10^3 CFU/ml, a 10-fold lower LOD with HRP-based chemiluminescent detection. Ren *et al* (2017) combined enzyme-amplified colorimetric detection with a concentration strategy termed as 'net fishing' utilizing porous nitrocellulose (NC) membrane modified with antibodies to capture *E. coli* O157:H7. The captured pathogens were then labelled by GNP probes where HRP was immobilized to generate a colorimetric signal with TMB to achieve a LOD of 100 CFU/ml.

A novel strategy known as plasmonic ELISA based on colorimetry was described (de la Rica and Stevens, 2012). In this strategy, antibody-conjugated HRPs catalysed H_2O_2 consumption, and then addition of gold ions resulted in GNP formation and aggregation for colorimetric detection of target concentration. A modified plasmonic ELISA was proposed in which the product of the reaction catalysed by the antibody-modified enzyme neutralized the surface charge of nanoparticles, resulting in aggregation and colour change (Nie *et al.*, 2014). These two protocols were introduced to improve pathogen detection performance. For instance, Chen *et al.* (2015) utilized SiO_2 nanoparticles as a platform modified with multiple poly(acrylic acid) (PAA) to which numerous catalases were linked. The modified SiO_2 nanoparticles linked with antibodies could specifically recognize the target, *Listeria monocytogenes*. The catalase consumes H_2O_2, and the concentration determines GNP formation and aggregation; and the corresponding colour, red and purple, could be used for the detection of target with a LOD as low as 80 CFU/ml in PBS and 80 CFU/g in lettuce samples. As shown in Fig. 6.1, Liu *et al.* (2013) modified magnetic beads with antibody-conjugated acetylcholinesterase as probes for enterovirus 71 detection to detect 10^4 copies/ml of target by the naked eye. Rather than enzyme, Bui *et al.* (2015) introduced an enzyme-free protocol for a plasmonic immunoassay. By conjugating liposomes loaded with cysteine to the target pathogen with an appropriate antibody, the released cysteine due to the breakdown of liposome will trigger the aggregation of GNPs. With this procedure, as low as 6.7 attomolar of target antibody could be recognized and the determination of 30 ± 8 CFU/ml of *E. coli* O157:H7 was enabled by the naked eye. On the other hand, the immunological reaction between antibody and antigen could be influenced by other factors in real samples and the lack of stability could reduce the specificity of this antibody-based assay.

Besides antibody, other recognition ligands have also been used for the detection of pathogens. Aptamers can specifically recognize target pathogens and be used for colorimetric detection. For instance, Lavu *et al.* (2016) selected aptamers against *S.* Typhimurium and proposed a colorimetric detection technique. In their work, the aptamer served to protect GNPs against aggregation triggered by salt; when the aptamer preferentially interacted with the target pathogen, the presence of target pathogen resulted in the aggregation of GNPs with added salt, along with a colour change for the detection of pathogens. With a biotin-labelled

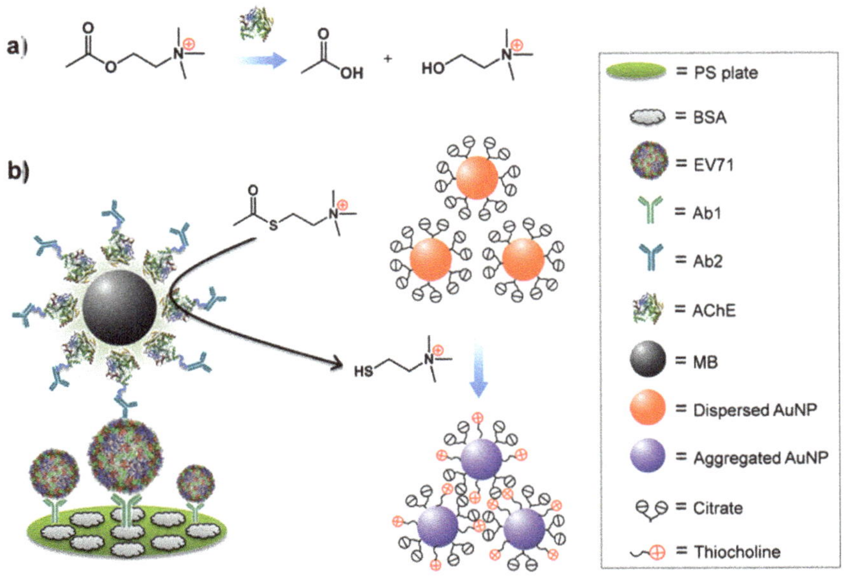

Figure 6.1 Plasmonic ELISA concept applied for pathogen detection (Liu *et al.*, 2013). Copyright Wiley-VCH Verlag GmbH & Co. KGaA. Reproduced with permission.

aptamer, Moon *et al.* (2014) prepared GNPs modified with aptamers for *S.* Typhimurium detection. In their design, the target pathogens were linked to GNPs and therefore increased the stability of GNPs against salt; while in the absence of target pathogens, the short sequence aptamer modification could not protect GNPs against aggregation due to salt. Besides aptamer, Liu *et al.* (2016) selected a specific fusion protein from phage for selective biosensing of *S. aureus*. The fusion protein selected from phage against *S. aureus* could specifically bind to the target. After modification with fusion proteins, cysteamine-stabilized GNPs bound to *S. aureus*, and thus reduced the colour of the solution. With this approach, detection as low as 19 CFU/ml *S. aureus* was possible in 30 minutes.

A different technique termed as chemical nose was proposed for pathogen detection based on the difference in electrostatic aggregation behaviour of the positively charged nanostructure on the surface of various pathogens. Verma *et al.* (2014) used optimized nanostars for *S. aureus* detection with a clear colour contrast in the presence of 5×10^5 CFU/ml; the method exhibited good specificity compared with polystyrene and liposome. Later, this strategy was extended to detect *S. aureus, Achromobacter xylosoxidans, Delftia acidovorans and Stenotrophomonas maltophilia* (Verma *et al.*, 2014). The distribution of surface charge on bacteria due to teichoic acid for Gram-positive bacteria or lipopolysaccharides and phospholipids for Gram-negative bacteria (Berry *et al.*, 2005; Berry and Saraf, 2005) caused a specific interaction between gold nanostars and bacteria, resulting in a unique response in colour and absorbance spectra. A chemical nose system was used to identify various bacteria in polymicrobial mixtures (Verma *et al.*, 2016). Similarly, Su *et al.* (2013) synthesized hexadecyl trimethyl ammonium chloride (CTAC) and 4-mercaptophenylboronic acid (MPBA) capped Au@Pt NPs to recognize target *E. coli* O157:H7. The Au@Pt NPs catalysed the oxidization of TMB as a peroxidase mimetic for colorimetric detection of *E. coli* O157:H7. Miranda *et al.* (2011) utilized cationic GNPs, which bound to β-galactosidase to inhibit its catalytic activity; while the target pathogen, *E. coli* (XL1), interacted with cationic GNPs and releasing β-galactosidase, which restored the catalytic activity and hydrolysed β-D-galactopyranoside (CPRG) for a colour change from pale yellow to red for pathogen detection with a LOD at 100 CFU/ml.

Besides NPs, vesicles could also be used for pathogen determination. Pires *et al.* (2011) synthesized 10,12-pentacosadyinoic acid (PCDA) +*N*-[(2-tetradecanamide)-ethyl]-ribonamide (TDER) vesicles to detect bacteria based on the colour response of the vesicles. Substrates from bacteria in the supernatant induced a colour change in the vesicles for colorimetric detection. The authors compared the colorimetric response of vesicles to *E. coli* and *S. aureus* and tested these bacteria in apple juice. A vesicle composite of PCDA/dimyristoyl-sn-glycero-3-phosphatidylcholine (DMPC) and PCDA/Sphingomyelin (SPH)/Cholesterol (CHO) was improved for colorimetric detection of *S. aureus, Salmonella* Choleraesuis, and *E. coli* (de Oliveira et al., 2013) at a LOD between 1 and 10 cells in buffer and chicken meat (de Oliveira *et al.* 2015).

Genetic engineering can be used for developing bacteriophage for colorimetric detection of bacteria. Burnham *et al.* (2014) engineered wild-type T4 bacteriophage and recombinant reporter *lacZ* T4 bacteriophage to carry the β-galactosidase gene for *E. coli* colorimetric detection. During the bacteriophage-mediated cell lysis, β-galactosidase is released to catalyse the CPRG hydrolysis, resulting in a colour change for detection. Benefiting from the proliferation of bacteriophage to target bacteria, the method exhibited good sensitivity with a LOD down to 10 CFU/ml for *E. coli*. It is possible to engineer bacteriophage to carry various genes for different enzymes (Alcaine *et al.*, 2015). Generally speaking, the proliferation of bacteriophage could be utilized to amplify the signal for detection of bacteria.

The nucleic acid indicator could target the pathogens with high specificity in complex samples. For instance, Martos *et al.* (2011) used colorimetric dot-blot assay to detect *Phaeomoniella chlamydospora* by testing against indicator DNA from 17 non-target fungi associated with grapevine. DNA/RNA-based determination could be an alternative for pathogen recognition. Furthermore, PCR and LAMP enables dramatic amplification of the nucleic acid indicators to result in greatly improved sensitivity, making DNA/RNA-based techniques a good indicator for colorimetric detection, which often suffer from unsatisfactory LOD. It should be noted that in recent years, reports of DNA/RNA-based colorimetric detection of pathogens has drawn more attention, suggesting that it is a promising future direction in this field.

Some chemicals and commercial reagents can reveal the presence of amplicons from PCR, LAMP, or other DNA/RNA amplification methods. Based on the specific hybridization between primer and indicator nucleic acid sequence chosen from the target pathogen, simple colorimetric detection methods could be developed. Safavieh *et al.* (2014) used naphthol blue (HNB) and calcein to measure the amplicons resulting from LAMP targeting the *tuf* gene of *E. coli* and the *mcat* gene of *S. aureus* for colorimetric detection of pathogens. Based on the colour change of HNB, a LOD at 200 CFU/ml was achieved with a cassette device facilitating the application of point-of-care (POC) diagnosis. Bearinger *et al.* (2011) developed a disposable POC testing device for the detection of *S. aureus* by integrating nucleic acid extraction and LAMP or reverse transcriptase-LAMP. Oh *et al.* (2016a) proposed an integrated centrifugal microfluidic device termed as a lab-on-a-disc to test for the presence of four pathogens, *E. coli* O157:H7, *S.* Typhimurium, *Vibrio parahaemolyticus*, and *Listeria monocytogenes* based on LAMP with a reagent, Eriochrome Black T (EBT), for multiplex colorimetric determination. Detection could be completed in an automated manner with a lab-on-a-disc and a miniaturized rotary instrument. A LOD of ten bacterial cells in 65 minutes was demonstrated in milk. They also proposed a similar device with a different design based on the lab-on-a-disc concept for an automatic multiplex detection of *E. coli* O157:H7, *S.* Typhimurium, and *Vibrio parahaemolyticus* (Oh *et al.*, 2016b).

To measure PCR amplicons, HRP was used to amplify the colorimetric signal. Qi *et al.* (2012) proposed a strategy based on a HRP-catalysed reaction for *S.* Typhimurium detection with its feature DNA sequence. A complementary sequence of the biotinylated amplicon was used to capture target amplicons on slides, and HRP was linked to the biotinylated amplicons. The chemical 3,3′-diaminobenzidine (DAB) was catalysed by HRP, resulting in a deposition of the DAB modified GNPs followed by a silver enhancement to generate a visible dot for colorimetric determination. This strategy provided a LOD down to 10^3 CFU/ml for *S.* Typhimurium detection. Based on the biotinylated amplicons, Zhang *et al.* (2015) chose feature genes, *rfbE* gene (*E. coli* O157:H7), *invA* gene (*Salmonella enterica*), and the *inlA* gene (*L. monocytogenes*) to detect the target bacteria. The biotinylated amplicons were anchored in a well with the complementary capture sequence. With biotinylated antibody against HRP, a network of biotin-streptavidin-HRP was constructed on the biotinylated amplicons for improved sensitivity. As low as 5 copies could be distinguished by the naked eye. In addition to slides or wells, MNPs could also be used to capture and separate biotinylated amplicons. Jansaento *et al.* (2016) linked capture DNA on MNPs to hybridize with target DNA selected from *Campylobacter*. Then a biotinylated reverse primer was linked to the other end of the captured target DNA. After PCR amplification, biotin was conjugated to the MNPs and then streptavidin-HRP was linked to the MNPs for colorimetric detection of *Campylobacter* based on the HRP-catalysed reaction. Even without PCR, the HRP enabled sensitive detection of pathogens with the DNA signature. Paniel and Baudart (2013) proposed a sandwich-like strategy to capture DNA extracted from *E. coli* with a probe sequence modified with digoxigenin and antibody-modified HRP. The HRP-catalysed TMB reaction was utilized for colorimetric determination of DNA indicator from *E. coli*. Assisted with magnetic separation, the proposed method could detect *E. coli* between 2 CFU and 20 CFU per 100 ml samples. A G-quadruplex DNA structure (a specific DNA secondary structure which could selectively interact with corresponding target molecules to provide function) bound with haemin worked as a functional DNAzyme to catalyse the oxidation of 2,2′-azino-bis(3-ethyl-benzthiazoline)-6-sulfonic acid (ABTS) in the presence of H_2O_2, resulting in a colour change from colorless to green. Based on this phenomenon, Gomez *et al.* (2014) combined a peptide nucleic acid (PNA) technique and rolling circle amplification (RCA) to transfer the gene from *E. coli*, *S.* Typhimurium, and *Clostridium difficile* to the G-quadruplex to construct a DNAzyme with haemin. PNA was used to recognize the specific target DNA and the resulting PNA/DNA padlock probes circularized and incorporated G-quadruplex structure flanked by nicking enzyme recognition sites. During the following RCA procedure, short single-stranded DNA (ssDNA) sequences, which could fold into G-quadruplex structures were produced. Then the formation of G-quadruplex structures and binding with haemin resulted in functional DNAzymes to catalyse ABTS oxidation for colorimetric detection of gene targets. Duy *et al.* (2015) proposed a colorimetric detection method with PNA. A segment of the 18S ribosomal RNA gene from *Synchytrium endobioticum* was chosen as an indicator and the PNA sequence was designed to selectively recognize the target RNA. The mismatched non-target nucleic acid sequence was digested with micrococcal nuclease while the hybridization of target RNA and PNA probe remained in the presence of micrococcal nuclease. Cyanine dye 3,3′-diethylthiadicarbocyanine iodide ($DiSC_2$ (5)) was used to measure the hybridized double-stranded RNA/PNA sequences that induced the colour change of $DiSC_2$ (5) from blue to purple. The proposed method could recognize *Synchytrium endobioticum* from potato wart samples (Duy *et al.*, 2015).

From the initial work on colorimetric detection of DNA with GNPs in 1998 (Storhoff *et al.*, 1998), the detection performance of NP-based colorimetric detection of nucleic acids, such as DNA/RNA indicators targeting specific pathogens, was consecutively improved, meanwhile there is strong potential for on-site application and POC testing benefiting from its simplicity. Based on different nanostructures, various strategies were reported to transfer the amount of target DNA/RNA to the colorimetric signal. In these strategies, the detection procedure with label-free GNPs should be a simple route. Prasad *et al.* (2011) proposed the colorimetric detection of *Salmonella* spp. with non-functionalized GNPs. The amplicon complementary ssDNA was used as probes, which could protect non-functionalized GNPs against aggregation due to the addition of salt. In presence of the target amplicons, the hybridized double-strand DNA (dsDNA) could not protect GNPs and the colour of the solution changed from red to blue due to the aggregation of GNPs triggered by salt. The method exhibited good sensitivity (89.15%) and specificity (99.04%) for detection in food samples. Based on 23S-16S rRNA intergenic spacer sequences, Khalil *et al.* (2014) proposed the colorimetric route to identify *Acinetobacter baumannii*, which was usually misrecognized. In addition to GNPs, Wee *et al.* (2015) expanded the label-free detection strategy to MNPs and reported a bridging flocculation procedure to determine target DNA from pathogens. Isothermal recombinase polymerase amplification was used to amplify long DNA products, which link to MNPs and flocculate the MNPs rather than re-dispersion after magnetic capture, and thus the resulting solution will be colourless. The absence of target DNA will retain the short primer and will not flocculate MNPs, resulting in a brown-coloured solution.

The stability of the GNPs modified with dsDNA against salt is better than that modified with ssDNA. The protection difference between ssDNA and dsDNA was utilized for colorimetric sensing to identify pathogens. Veigas *et al.* (2010) prepared GNPs modified with complementary DNA sequence of the DNA indicator to detect single base mutations/single nucleotide polymorphisms. The hybridized dsDNA of the PCR-amplified DNA indicator and the complementary DNA modified onto GNPs will protect GNPs against aggregation with the added salt. The proposed method could recognize a single base pair mismatch with high specificity (100%) and was used for multiplex detection of *Mycobacterium tuberculosis* complex and *Plasmodium* sp. with capture DNA sequences on GNPs for different target DNAs (Veigas *et al.* 2015). To simplify the GNP probe preparation, Jung *et al.* (2010b) developed a different route for colorimetric detection of target DNA from *Chlamydia trachomatis*. PCR thiolated primer was used to produce amplicons, and the generated thiol-labelled amplicons in dsDNA could link to GNPs against aggregation due to salt addition. This procedure could detect as low as 100 copies of target DNA.

With the capture DNA sequence and modified NPs, a sandwich-like structure could be constructed to form a NP/DNA network and prevent aggregation triggered by salt. Jyoti *et al.* (2010) modified GNPs with two short thiolated DNA sequences to capture the Shiga toxin 2 (*stx2*) gene for detection of STEC. The results demonstrated that the increase in amount of target DNA improved the stability of GNPs against salt due to network formation. Quintela *et al.* (2015) utilized asymmetric PCR to amplify the *stx1* and *stx2* gene to detect STEC strains. A LOD as low as 10 CFU/g was achieved with high specificity (100%).

Besides the above strategies, DNAzymes, which can catalyse the cleavage of nucleic acid molecules were also used for NP-based colorimetric detection (Santoro and Joyce, 1997; Schlosser and Li, 2010). Zagorovsky and Chan (2013) utilized DNAzyme and DNA-modified GNPs to detect AF-1, *N. gonorrhoeae* and *T. pallidum*, malarial parasite *P. falciparum*, and HBV virus. The modified GNPs form a network conjugated by DNA linkers, which includes a substrate sequence able to be cleaved by a multicomponent nucleic acid enzyme (MNAzyme). The presence of target DNA will assemble the MNAzyme subunit, which includes the complementary sequence, and the assembled MNAzyme cleaves the linker, and thus no GNP/DNA network was formed, resulting in the aggregation of GNPs with added salt. The proposed method exhibited a LOD at 50 pM and could be used for the detection of various pathogens.

In summary, the simplicity of colorimetric detection makes it an attractive technique for POC testing and on-site detection of pathogens. Various nanostructures, particularly GNPs and MNPs, have been used to improve the specificity and sensitivity of colorimetric detection. Meanwhile, with the nucleic acid sequence selected from target pathogens as the indicator, it is possible to further improve the LOD with amplification techniques such as PCR, LAMP, RCA and so on, while the high specificity of a nucleic acid sequence as the target enables better selectivity and serotype recognition. In the future, research should focus on further improvements in the sensitivity and specificity of colorimetric detection methods, which should be coupled with improved sample preparation methods for complex samples.

Spectroscopic detection methods

Spectroscopic approaches combined with nanotechnology are promising candidates for developing pathogen detection methods with high sensitivity, specificity, and the capacity to multiplex. Approaches such as fluorescence, surface plasmon resonance (SPR), infrared (IR) spectroscopy, and surface-enhanced Raman spectroscopy (SERS) have gained prominence in monitoring pathogenic organisms because of their appeal in improving sensitivity and in non-destructive analysis requiring minimum sample pretreatment. These approaches could also be integrated with other sample preparation or separation technologies, such as microfluidic approaches or membrane or magnetic separation for practical application, which show promise in POC or *in situ* monitoring for food safety and environmental protection.

Fluorescence analysis is one of the common methods in pathogen sensor development such as in fluorescence immunoassays. Traditional fluorescence methods based on organic dyes usually suffer from poor photostability and producing a broad spectral band, which might lead to poor reproducibility and anti-interference ability. Fluorescent nanoparticles have been introduced to circumvent such disadvantages. For example, quantum dots may serve as effective fluorescence tracers for detection. (Wang *et al.*, 2016) Furthermore, and approach based on *in-situ* immunomagnetic beads with enhanced fluorescence was developed for the detection of *E. coli* O157:H7, *S.* Typhimurium, and *Listeria monocytogenes*, with a LOD down to 5 CFU/ml (Cho *et al.*, 2014). When the fluorophore molecules were located in the intense plasmon electric field of the noble metal nanostructures, the surface enhanced fluorescence (SEF) signal could be further improved to address sensitivity. Such a strategy was used for the detection of the *E. coli* O157:H7 *eaeA* gene with a LOD at 3.33×10^{-18} M (Sun *et al.*, 2015). Functional fluorescent conjugated polymers were developed as sensors for the detection of bacteria within 10–15 minutes, and the specific binding to carbohydrates of different bacteria enabled detection of different pathogens (Disney *et al.*, 2004). With a cell-phone camera with an additional accessory, the fluorescent signal was observed, facilitating a cell-phone based *E. coli* detection platform for screening of liquid samples (Zhu *et al.*, 2012). A LOD down to 5–10 CFU/ml achieved from this work revealed the promising potential for fast, cost-effective and *in situ* detection.

Vibrational spectroscopy such as IR absorption provides molecular structural information from pathogens. The mid-IR spectra (400–4000/cm) of pathogens can be used to characterize their composition based on proteins, fatty acids, nucleic acids, lipopolysaccharides, etc. for classification. The composition of pathogens varies according to species and strains, thus it is possible to identify each pathogen based on its own characteristic spectrum, or 'fingerprint' spectrum. Fourier-transform infrared spectroscopy (FTIR) has been used for pathogen fingerprinting. The difference in the spectra among matrices contaminated with pathogens and the blank could be used to indicate the presence of targets. For example, based on this concept eight bacteria were quantified in juice, which was consistent with standard plate counts (Yu *et al.*, 2004). However, in many cases, the complex background spectra interfere with spectral pathogen detection. Indirect methods with steps to reduce the background signal were useful, for example, filtration and magnetic separation were the most common and effective techniques (Davis *et al.*, 2010).

IR spectra also enable the differentiation of live and dead pathogens with fingerprint information from pathogens in different health states. Differences in the spectra between the live and dead pathogenic cells, mainly in the amide and nucleic acid regions were applied to differentiate live and heat-killed *E. coli* O157:H7 in beef (Davis *et al.*, 2010). The FTIR spectra of pathogens may also change during the growth process (Yu *et al.*, 2006; Yu and Irudayaraj, 2005). FTIR exhibits unique advantages to identify and classify pathogenic organisms based on the vibrational modes of the molecular chemical groups. Sensitivity and the interference from matrices were compounding factors in developing a robust platform.

Raman scattering spectroscopy has also been used for detection and classification based on the molecular fingerprint information from the analytes with low interference from water. But the inherent weak scattering provides low sensitivity for detection. SERS discovered in the 1970s could amplify the Raman signal up to 10^{10} in the electromagnetic field of the plasmonic structures (Albrecht and Creighton, 1977; Fleischmann *et al.*, 1974). Localized surface plasmon resonance (LSPR) of the plasmonic nanostructures, known as SERS substrates, creates an amplified electric field near the surface to greatly enhance the Raman scattering of the molecules in the evanescent field (Granger *et al.*, 2016). SERS demonstrates extremely high sensitivity, even down to single molecule detection (Lee and Irudayaraj, 2013; Wang and Irudayaraj, 2013). SERS has been utilized in a wide range of applications for analytical and biological science. For pathogen detection, high sensitivity of SERS could be obtained by employing suitable substrates compared to FTIR; the widths of the SERS spectra peaks are much narrower than that of fluorescence, which would minimize spectra overlap, which is useful for fingerprint identification, classification, and multiplex detection; combined with portable instrumentation technologies, SERS could be an exciting approach for fast and *in situ* detection.

In a SERS-based method, the targets could be selectively isolated from the matrix with capture elements, such as target-specific antibody, aptamers, DNA, proteins, or phage. The obtained SERS spectrum could be exceptionally specific, making it possible to differentiate species, strains, and even live or dead cells. With an *in-situ* coating strategy, detection of bacteria down to 2.5×10^2 cell/ml was demonstrated, as well as discrimination of three strains of *E. coli* according to the fingerprint spectra (Zhou *et al.*, 2014). By employing assembled silver nanospheres as the SERS substrate, detection of pathogenic bacteria was performed with a concentration as low as 10 CFU/ml, while the differentiation of *E. coli* O157, *S.* Typhimurium., and *S. aureus*, including live and dead cells was also accomplished (Guo *et al.* 2012a). The unique spectral characteristics of molecules embedded in the membrane of pathogenic organisms enable direct identification by using a specific SERS fingerprint.

The analysis of SERS spectra with chemometrics such as Principle Component Analysis (PCA) has been found to be effective in classification based on the fingerprint spectra (Gupta *et al.*, 2004; Sivakesava *et al.*, 2004; Yu and Irudayaraj, 2006). Better immobilization and separation methods have been used to overcome interference from complex matrices to enhance sensitivity based on magnetic particle-assisted capture and separation (Zhang *et al.*, 2012) and a functionalized filtration process (Chen *et al.* 2014). Recently, electrokinetic method was proposed for the isolation and concentration of pathogens from human blood samples, which further expands the application of SERS in food safety and the POC area (Cheng *et al.*, 2013; Kwon *et al.*, 2012). The idea of a smaller device (Knauer *et al.*, 2012) and easier signal readout (such as imaging) (Yang *et al.*, 2016) also represents a developing direction.

Label-free SERS methods have been proposed for differentiation and classification of pathogenic organisms, but the minor spectral difference and the weak Raman response requires highly optimized substrates. In these approaches, the lack of reproducibility needs to be addressed.

By employing SERS-active reporters and target specific antibodies or aptamers as recognition elements to fabricate nanoprobes, multiple pathogens have been identified based on specific SERS spectra from reporters. A SERS-based multiplex pathogen detection method was developed based on this strategy (Ravindranath *et al.*, 2011). Ag, Au, and Ag/Au core/shell nanoparticles with unique Raman labels were used as probes to complete the immune reaction specific to targets. Highly selective and sensitive SERS detection was used to identify species (*E. coli* O157:H7 vs. *S.* Typhimurium) and strains (*E. coli* O157:H7 vs. *E. coli* K12) at a LOD down to 10^2 CFU/ml. The simple and robust approach exhibited excellent potential for multiplex detection using an indirect strategy. Magnetic or filter assisted separation was used to identify multiple viral antigens or pathogens (Cho *et al.*, 2015a; Neng *et al.*, 2013). A typical magnetic separation process combined with SERS detection is illustrated in Fig. 6.2. Recent efforts have also utilized multifunctional substrates for optical labelling and simultaneous photothermal ablation with gold nanorods (Wang and Irudayaraj, 2010), capture and detection with inactivation (Wang *et al.* 2015a), as well as concentration and detection with kinetic monitoring by dielectrophoresis (Madiyar *et al.*, 2015), which enables sensitive and specific discrimination.

SPR is a charge-density oscillation at the interface of two media such as metal and dielectric, measuring the refractive index change upon target binding. It is a rapid, label-free and real-time monitoring technique, but for the detection of pathogenic organisms with low sensitivity due to small refractive index change, the slow diffusion-driven mass transfer still needs to be addressed. Huang *et al.* (2014) investigated the sensitivity of an SPR biosensor for pathogen detection as a function of flow conditions and probing depth, showing the importance of accurate control of these parameters to maximize the response and the match of the target size with the surface-plasmon evanescent field by using LSPR. LSPR band position intensity of GNRs can be affected by the change in the dielectric constant around the vicinity; more

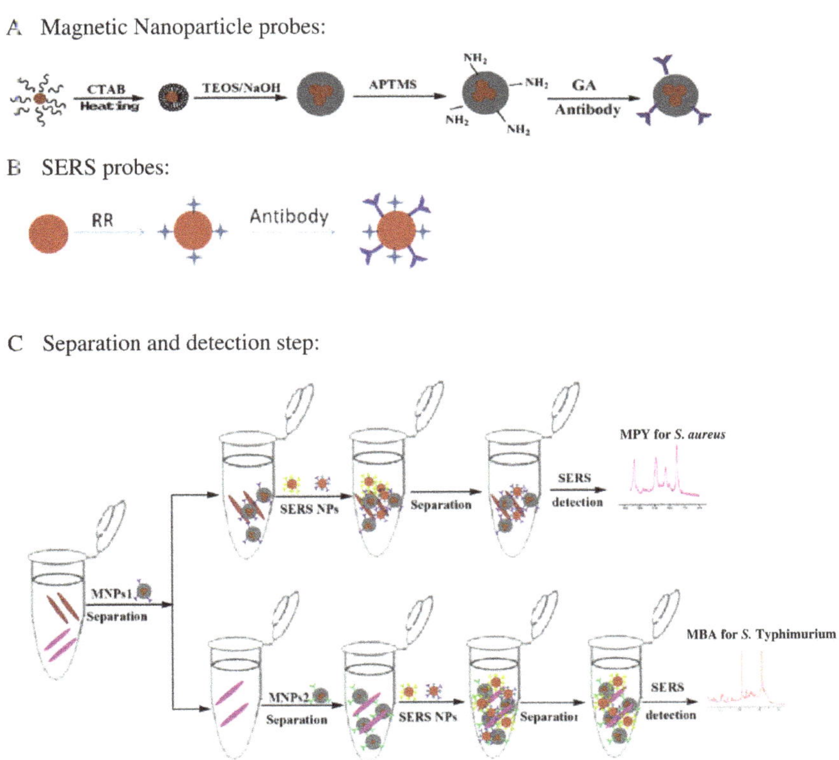

Figure 6.2 Magnetic separation and SERS detection of pathogens with Raman tag labelled probes Reprinted by permission from Springer Nature: Analytical and Biochemical Chemistry, Wang *et al.*, 2011. Copyright 2011.

importantly, their aspect ratios could be easily controlled, and their unique simple 'multiplexing' advantage could be realized. By replacing the toxic hexadecyltrimethylammonium bromide (CTAB) bilayer with cystamine on the GNR surface, functionalized GNRs with different ratios were used as novel optical labels for the SPR detection of *E. coli* and *S.* Typhimurium at a LOD of 10^2 CFU/ml within 30 minutes (Wang and Irudayaraj, 2008).

Although the SPR approach could extend the LOD (Subramanian *et al.*, 2006; Waswa *et al.*, 2006), the interfering effects from the matrix has been a constraint to extend this technology for the evaluation of complex samples such as food matrices. Functional coating and specific capture element modification would suppress this influence. For example, low fouling coatings such as poly (carboxybetaine acrylamide) (pCBAA) brushes were applied as a unique sensing surface for multi-step SPR detection of pathogens in crude hamburger samples (Vaisocherová-Lísalová *et al.*, 2016). Other surface modification, including avidin–biotin arrays, primary and secondary antibody sandwich assays, and self-assembled monolayer-based SPR assays were also proposed for the detection of pathogenic organisms (Torun *et al.*, 2012; Wang *et al.*, 2013). Besides common antibody-based capture methods, other recognition strategies, such as bacterial phage and whole cell imprinting, could also improve SPR detection (Yilmaz *et al.*, 2015).

In addition to the methods above, an optical scattering technology (BARDOT) was developed to obtain two-dimensional scattering patterns from the pathogen colonies, which could be used for both identification and detection of bacteria (Rajwa *et al.*, 2010). Similar to BARDOT, by monitoring scattering information, dark field imaging associated with gold nanoparticles has been used to detect *E. coli* as low as 10^4 CFU/ml within 30 minutes (Xu *et al.*, 2012). Detection of a single bacterium *in situ* from a complex environment was demonstrated by hyperspectral coherent anti-Stoke Raman scattering (CARS) for *in situ* imaging of pathogens (Hong *et al.*, 2016).

Sensing strategies that require no labelling probes for analytes are preferred because of their ability for real-time high-throughput imaging. Spectroscopic methods could match this demand with good sensitivity. On the other hand, it should be noted that a signal readout instrument would be a limitation for on-site application, which relies on the development of a portable sensing platform and instrument.

Paper-based detection methods

Until now, detection strategies such as PCR, ELISA, and spectroscopic methods were used to detect pathogens in different samples, and these exhibited good detection performance. However, the required instruments, trained operators, and a well-equipped laboratory limit the application of these methods in a resource-limited setting. Furthermore, the strict regulatory requirements related to food quality and safety monitoring demand sensitive and specific determination of pathogens. The assays or sensors based on paper, a ubiquitous material based on cellulose or nitrocellulose with porous structure, provide the possibility to match the above demands. The unique attributes of paper-based devices that make them an excellent tool for pathogen monitoring are (i) the material (paper/membrane) is inexpensive and available at different porosity; (ii) can be functionalized with antibodies, aptamers and targets for the recognition of organisms or their indicators (iii) paper could be functionalized with various chemical groups, antibodies, or nucleic acid sequences; (iv) ease of patterning to construct channels for the fabrication of devices to control sample flow or multiplex detection; (v) samples could utilize the capillary flow action for transport without an additional instrument; (vi) ease of disposal and generation of less biohazardous wastes with incineration (Ju *et al.*, 2016; Kong and Hu, 2012; Srinivasan and Tung, 2015). Besides, paper provides a wet environment using water or buffer, enabling biochemical reactions such as antibody–antigen interactions, nucleic acid sequence hybridization, and enzyme catalysed reactions. Paper-based assays or sensors for pathogen detection in different routes were carried out with fluorescent (Chandan *et al*,. 2016) and colorimetric signals, and lateral flow assays (LFA) have been a tremendous success in this field, primarily because of their simplicity and on-site detection ability. In this segment we will focus on colorimetric paper-based sensors.

Using a patterned paper substrate by wax printing, Jokerst *et al.* (2012) developed a microspot assay based on a colorimetric reaction catalysed by bacteria-specific enzymes to detect three bacteria, *E. coli* O157:H7, *S.* Typhimurium., and *Listeria monocytogenes*. The bacteria-specific enzymes, β-galactosidase for *E. coli*, phosphatidylinositol-specific phospholipase C (PI-PLC) for *Listeria monocytogenes*, and esterase for *Salmonella enterica*, were used to catalyse the colorimetric reaction, resulting in colour signals from different substrates to determine the concentration of target bacteria. The proposed paper-based assay could detect target bacteria as low as 10 CFU/cm^2 in ready-to-eat meat after enrichment. Paper could also be used to directly anchor pathogens for subsequent detection. Shih *et al.* (2015) simply dropped samples on paper and performed a paper-based ELISA to detect *E. coli*. The *E. coli* bacteria in the sample were anchored on paper and then labelled with primary and secondary antibodies modified with HRP. The colour change due to HRP-catalysed TMB reaction resulted in blue spots for quantitative assessment of *E. coli*. With the paper-based ELISA, *E. coli* at 10^5 CFU/ml could be detected. Specific nucleic acid indicators could also be detected with the paper-based method. Saikrishnan *et al.* (2014) used cellulose paper as a substrate to capture pathogenic DNA and HRP was use to amplify the colorimetric signal. Synthesis of a biotinylated segment of IS6110 DNA from *Mycobacterium tuberculosis* was chosen as the model. The complementary sequence was covalently conjugated to the paper substrate to capture the target DNA sequence, while streptavidin-conjugated HRP was used to label the captured target to generate a colorimetric signal. The investigation demonstrated a possible detection route assisted with DNA amplification techniques.

Colorimetric detection could also be accomplished utilizing phage with the paper-based device. Khan *et al.* (2015) conjugated antibodies against T7 bacteriophage as an indicator to complete a paper-based sandwich ELISA assay. The target phage anchored on paper was then labelled by an antibody modified with enzyme to catalyse DAB oxidation to generate a brownish colour.

Lateral flow assay (LFA) is one of the most common paper-based detection methods. A typical lateral flow device is constructed on a NC membrane where the recognition ligands such as antibody, aptamer, or ssDNA are conjugated to capture the target. Meanwhile, the sample pad is assembled at the end of the strip for sample application. Here the conjugated pad links sample pad and NC membrane where the as-prepared probes would label the target as the sample flows through the conjugate pad. At the other end, the absorbent pad would extract and retain all of the sample solutions. The assembled strip has a long shelf life. As a simple, facile, cost-effective, and practical analytical technique, LFA exhibits strong potential for onsite monitoring of pathogens. It is a simple one-step detection process and could be completed in 15–30 minutes. When the colorimetric strategy is employed, signal readout can be performed by the naked eye, making LFA suitable for on-site detection. A limiting factor of LFA platforms is its poor sensitivity. For pathogen detection, LFA can be performed with antibodies to specifically recognize pathogens, as well as feature DNA/RNA chosen from pathogens as indicators. The method based on the determination of pathogens is simple without the need for bacterial cell lysis and DNA/RNA extraction procedures, while detection utilizing DNA/RNA as indicators accompanied with amplification could provide high sensitivity.

To generate a signal in the detection zone on the NC membrane, various materials have been utilized to prepare probes. The most common probes are GNPs, which could be easily observed as a red colour benefiting from their unique optical property. Meanwhile, the biocompatibility of GNPs facilitates the modification of antibody or probe DNA sequence. Braun-Kiewnick *et al.* (2011) used a typical LFA concept to detect *Erwinia amylovora*, a pathogen implicated in fire blight disease. Good specificity was demonstrated with 39 *E. amylovora* strains out of 100 bacterial strains tested with a LOD at $10^{5.7}$ CFU/ml based on the red colour generated from GNPs. This strategy could also be applied with GNP aggregates rather than monodispersed GNPs to generate the signal (Shen *et al.*, 2013), or can be further improved with silver enhancement (Panferov *et al.*, 2016). In addition to the colour of NPs, Jiang *et al.* (2016) utilized the catalytic activity of enzyme mimetic Pt-Au bimetal NPs in a LFA for colorimetric signal amplification (Jiang *et al.* 2016). Instead of GNPs, Pt-Au NPs were used as probes, which can catalyse the TMB reaction using peroxidase to improve the sensitivity of *E. coli* O157:H7 detection by more than 1000-fold.

GNP probes can also be used for feature nucleic acid sequence detection. Similar to the antibody-based LFA, complementary DNA sequences were used to capture and label the feature DNA/RNA or its amplicons as the indicator for the presence of target pathogens. GNP probes could be labelled on the captured target sequence according to the antibody against antigen conjugated to the antigen-terminated probe sequence. Meanwhile, the probe sequence could be directly linked to GNPs for colorimetric detection. Rastogi *et al.* (2012) used locked nucleic acid (LNA) to construct GNP probes to detect target DNA from *E. coli* O157:H7. The LNA exhibited better enzymatic stability and stronger affinity to target DNA, thus resulting in an improved detection performance. Interestingly, since the aptamer sequence could be amplified, Fang *et al.* (2014) proposed a unique LFA-based strategy to detect *Salmonella* Enteritidis. Two types of aptamers were used to interact with target bacteria: one was labelled by biotin, which could link with streptavidin coated magnetic beads for immunomagnetic enrichment, while the other was released from bacteria after the enrichment and amplified by a strand displacement amplification. The obtained amplicons were tested as indicators by LFA for the detection of bacteria. A LOD at 10 CFU/ml was demonstrated and good specificity was shown for *S.* Typhimurium, *E. coli* O157, *Pseudomonas aeruginosa*, and *Citrobacter freundii*.

By patterning multiple antibodies in the detection zone, it is easy to develop a multiplex LFA strategy. Song *et al.* (2016a) conjugated antibodies specific to *Shigella boydii* and *E. coli* O157:H7 at two test lines on a single LF strip and established a process for simultaneous detection of two pathogens with GNP probes modified with corresponding antibodies. Rather than a test line, Safenkova *et al.* (2016) spotted antibodies on a LF strip for multiplex detection of several targets on a single strip. With the test points modified with different antibodies in the detection zone and corresponding GNP probes, the proposed LFA could detect eight potato pathogens in a single test. As shown in Fig. 6.3A, Cho *et al.* (2015b) integrated HRP and antibody on GNPs as probes to develop an enzymatically amplified colorimetric LFA sensor for the detection of *E. coli* O157:H7. Cross-flow

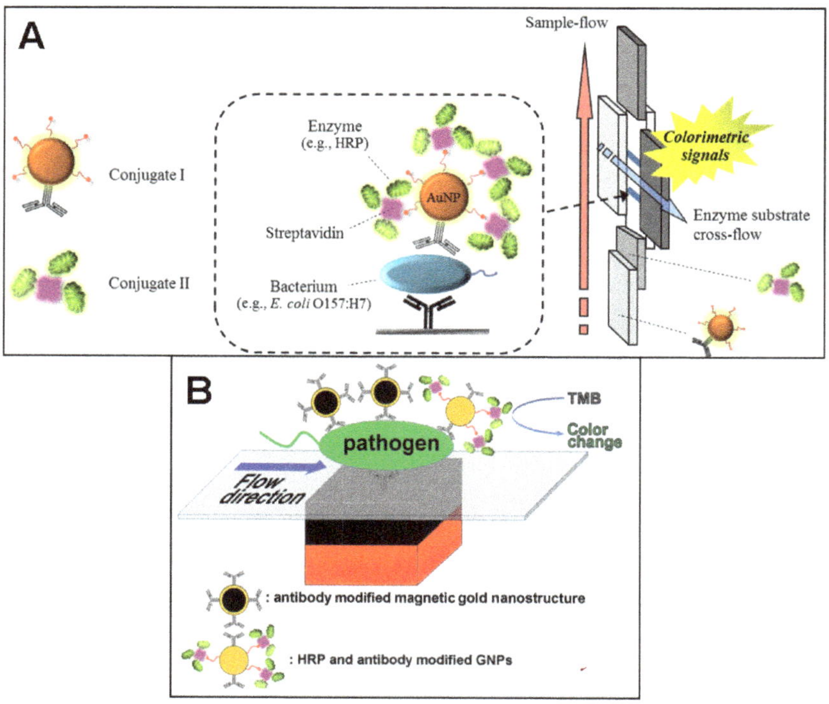

Figure 6.3 Scheme of HRP enhanced LFA with GNP probes (A) (reprinted from International Journal of Food Microbiology, 206, Il-Hoon Cho, Arun Bhunia, Joseph Irudayaraj, Rapid pathogen detection by lateral-flow immunochromatographic assay with gold nanoparticle-assisted enzyme signal amplification, 60–66, Copyright 2015, with permission from Elsevier) and magnetic focus enhanced LFA with HRP catalysed signal amplification (B) (reproduced from reference Ren *et al.*, 2016 with permission from The Royal Society of Chemistry).

was employed to wash and remove the unbound probes and HRP, as well as to add TMB. The colour was generated due to the catalysed TMB oxidation with HRP to indicate the concentration of *E. coli* O157:H7. Compared with the conventional LFA, sensitivity was improved 1000-fold with a LOD as low as 100 CFU/ml.

In addition to GNPs, other nanostructures such as latex nanoparticles and polymer nanoparticles, both of which can exhibit various colours, were used for multiplex detection. For instance, with latex beads loaded with a blue chemical, Imamura *et al.* (2015) performed LFA to detect *Porphyromonas gingivalis* in subgingival plaque. Similar to the GNP-based strategy, latex beads were modified with antibodies to label the target, and then the colorimetric signal from the blue colour loaded by latex beads was used for assessment. Lee *et al.* (2016b) modified latex beads in blue and red colour with dengue virus (DENV) and Chikungunya viruses (CHIKV), respectively. With the blue and red colour at different detection zones, it was possible to detect multiple targets simultaneously. Polymer nanoparticles are another cost-effective option for the construction of probes for LFA detection of pathogens or feature nucleic acid sequences (Cox *et al.* 2015).

Magnetic NPs enable separation and enhancement for LFA, which is a unique approach to improve LFA. Cho *et al.* (2013a) combined immunomagnetic concentration with LFA for the detection of *Listeria monocytogenes*. The MNPs were modified with antibody and HRP for signal enhancement. Benefiting from the immunomagnetic separation and concentration, as well as the enzymatic signal amplification, the detection sensitivity was improved to around 100 CFU/ml in buffer and 2% milk samples. In addition to magnetic pre-concentration, the magnetic force could also be utilized to slow down the movement of the probe-labelled target during the lateral flow process. Ren *et al.* (2016) proposed a magnetic focus enhanced LFA method. Fe_3O_4/Au core/shell nanostructures were synthesized for the preparation of magnetic probes while GNPs were modified with antibody and HRP to generate an enzymatic signal amplification with TMB. Rather than the immunomagnetic concentration, the magnetic probes were used to control the movement of the labelled target in LF strips as shown in Fig. 6.3B. During LFA, all the elements, such as the labelled target, are driven by the flow, resulting in a short reaction time for the capture of antibody in the detection zone of the target, which caused a low capture rate thus limiting sensitivity. With a simple magnet, the movement of magnetic probe-labelled targets could be slowed down in the strip to achieve a higher capture rate, resulting in better sensitivity.

A fluorescent signal, which requires as an additive readout instrument could provide better sensitivity and more precise quantitation compared to the colorimetric determination. Until now various fluorescent probes have been employed in LFA, including fluorescein isothiocyanate (FITC), FITC encapsulated NPs, quantum dots (QDs), and upconversion NPs for the detection of pathogens (Alcaine *et al.*, 2016; Hua *et al.*, 2015; Rajendran *et al.*, 2014; Song *et al.*, 2016b; Zhang *et al.*, 2014). The basic concept of LFA with these fluorescent beacons is similar to that of colorimetric LFA. The beacons were conjugated to recognition ligands to label the targets captured on the LF strip. Instead of this strategy, Morales-Narváez *et al.* (2015) used fluorescent QDs conjugated with antibody on a LF strip to capture target *E. coli*. Then graphene oxide (GO) sheets were employed to quench the fluorescence upon direct contact with QDs; while the *E. coli* cells captured by the antibody modified on QDs would separate QDs from the GO sheets to maintain the fluorescent signal from QDs. Therefore the 'on' status of fluorescence indicated the presence of the target. The proposed LFA method exhibited a good sensitivity with a LOD at 10 CFU/ml in buffer.

Besides LFA strips, various microdevices were developed to improve LFA. The simplest concept is to integrate multiple strips to detect different target pathogens. Compared with multiplex detection with multiple detection zones on a single strip, this strategy facilitates easier LF strip preparation to detect multiple targets. Numerous device designs have been proposed that combine LFA with nucleic acid amplification methods such as LAMP or PCR (Kim *et al.*, 2012; Lee *et al.*, 2016a) for use in real samples, for example oral fluids (Chen *et al.*, 2013). The integrated device could greatly simplify nucleic acid-based LFA approaches.

Paper-based assays and sensors are promising in the field of food security monitoring. Among the proposed paper-based devices, LFA shows a strong potential for onsite monitoring and screening of samples. However, it should also be noted that the sensitivity, reliability, and quantitation of LFA still require additional improvement, especially related to food matrices. Owing to the complexity of food matrices and the varying nature of food systems and ingredients, a one for all solution is not possible. However, it is conceivable that devices for a select set of food matrices could be developed with extensive sensitivity and specificity evaluations.

Chip-based detection methods

The microfluidic lab-on-a-chip technique exhibits a strong potential as a portable, sensitive, and real-time bacteria detection sensor, benefiting from the integration of chemical and biological functional units in a small platform and advantages such as easy operation and potential automation. Microfluidic chips can be used to quantify bacteria by integrating different detection techniques. In this process, bacteria separation and enrichment are critical in achieving high specificity and sensitivity with a microfluidic chip.

The highly selective interaction between antigen and antibody has been widely utilized to identify microorganisms and for capturing target bacteria. In a microfluidic chip platform, the antigen–antibody interaction is one of the most effective ways to capture and enrich target bacteria. Both physical and chemical binding processes can immobilize antibody molecules onto a glass or plastic surface of a functional area on a microchip (Rozhok *et al.*, 2005). Antibody-based microfluidic chip methods have been developed by integrating with different biosensor techniques for detection of bacteria with a high signal to noise ratio (Garcia-Aljaro *et al.*, 2010; Jung *et al.*, 2010a).

Growth of the bacteria during the detection process is a way to increase pathogen number to allow detection. However, this can be time-consuming and require a complex operation. In order to decrease operation time and simplify the procedure, many alternative technologies were proposed. Bouguelia *et al.* (2013) reported a culture–capture–measure (CCM) approach based on SPR assisted with an immuno-recognition microchip system. They used the antibody to specifically capture target pathogenic bacteria such as *E. coli* O157:H7, *S.* Enteritidis, and *Streptococcus pneumoniae* for 'on-chip' microbial culture. Specific detection was achieved with a LOD around 19.6 CFU/ml.

Integrated with immuno-capture, optical tools were used to improve LOD. Wang *et al.* (2015b) established an *in-situ* detection method for *S.* Typhimurium with high sensitivity and reliability by combining fluorescence detection based on QDs and antibody capture with a multichannel microfluidic chip. For *S.* Typhimurium enrichment, primary antibody was immobilized at the signal generating zone to capture *S.* Typhimurium in the flow. SERS can be utilized with microfluidic devices to provide highly resolved chemical fingerprints of pathogens (Efrima and Zeiri, 2009). The integration of SERS detection and immunoreaction on the microfluidic chip can achieve specific bacteria capture with higher reproducibility in flow cells (Guven *et al.*, 2011). Consequently, the advantages of immunorecognition in flow-through systems decreased the non-specific interference and improved sensitivity.

Antibody immunoreaction was integrated with an impedance-based microchip sensor. Currently, the impedance technique is widely used for biological and biomedical detection because of the advantages of high efficiency and ease of integration (Seidel and Niessner, 2008). A simple and portable microfluidic chip sensor, including monoclonal antibody enrichment and impedance measurement was developed by Boehm and colleagues (2007). Benefiting from the capture of target bacteria with antibodies modified on the surface of microchip chamber, the bacteria concentration within the chamber was increased to provide higher sensitivity.

Since the limited contact between suspended bacteria and antibody modified on the planar surface, immunoreaction-based enrichment suffers from low capture efficiency. For instance, the capture efficiency of an anti-*E. coli* antibody functionalized indium tin oxide electrode surface was only 16% for *E. coli* O157:H7 (Ruan *et al.*, 2002). Micro-beads coated with antibody were introduced to improve capture efficiency in a microchip system for target bacteria detection due to their high surface-to-volume ratio and suspension in samples. Guan *et al.* (2010) integrated *E. coli* O157:H7 antibody-coated glass beads into a microfluidic chip for *E. coli* O157:H7 enrichment and detection with bioluminescence technology. The capture efficiency was improved to 91.75–95.62% in the presence of glass beads coated with antibodies, which filled the microchamber in a microfluidic chip. Magnetic particles were also used for enrichment, enabling the concentration of target bacteria from a mixture or a large volume sample assisted with an external magnetic field. Applying this strategy, Sen *et al.* (2011) developed a microsystem to extract, capture, and detect *E. coli* O157:H7 in soil samples. In the microsystem, the magnetic beads were used to extract bacteria from soil samples to a confined space, resulting in less interference and better detection sensitivity. A collection of bacterial cells with magnetic particles onto the surface of microelectrodes can also be applied to improve the impedance detection in the microchip. Varshney *et al.* (2007) presented an impedance biosensor with the embedded gold interdigitated array microelectrode and magnetic nanoparticle (MNP)–antibody conjugates for detecting *E. coli* O157:H7 in ground beef samples. MNP–antibody conjugates were able to separate and concentrate target bacteria in a microfluidic flow cell with a magnetic field before impedance measurement.

Besides antibody immunoreaction-based methods, dielectrophoresis (DEP) is another widely useful approach for efficient manipulation and enrichment of bacteria. In the presence of a non-uniform electric field and polarization effects, dielectric particles can express dielectrophoretic activity and move towards high (positive DEP) or low (negative DEP) electric fields (Pohl and Pohl, 1978). Integration of DEP with a microfluidic chip has been shown to enrich targets at a designed location in a non-destructive and non-invasive manner. Moreover, a different mode of DEP manipulation can be realized by adjusting the structure and relative position of the microelectrodes in a microfluidic chip. For example, Cheng *et al.* (2010) developed a three-dimensional DEP by positioning microelectrodes at the bottom and top of the microfluidic chip cell. Applying this three-dimensional DEP microchip, the efficiency for concentrating different bacteria was higher than that with 2D microelectrodes.

Cells with different dielectric properties undergo different DEP force under the same electronic condition, which can be used for selective interrogation of bacteria. Suehiro *et al.* (2006) demonstrated a selective DEP impedance method based on different dielectric properties between viable and non-viable *E. coli*. Results showed that the heat-treated *E. coli* did not experience positive DEP when the field frequency was higher than 1 MHz due to the decrease in cytoplasmic conductivity of heat-treated cells caused by membrane perforation. Viable *E. coli* and heat-treated *E. coli* could be separated by applying a field frequency higher than 1 MHz for impedance detection. It should be noted that the drag force induced by the balance between the DEP force and liquid flow requires optimization to avoid non-specific capture. Selective DEP could enrich specific bacteria from mixed suspensions with non-target bacteria. In addition, a new method was developed to increase the sensitivity of DEP impedance detection by integrating electropermeabilization (Suehiro *et al.*, 2003). After the bacteria were trapped by the positive DEP force, electropermeabilization was performed by using the high alternating electric field to release the intracellular ion from trapped bacteria to enhance the performance by two orders of magnitude.

DEP enrichment could also be combined with fluorescence detection. He *et al.* (2013) used fluorescent nanoparticle labels and positive DEP enrichment to detect *S.* Typhimurium with high efficiency and sensitivity. After incubating *S.* Typhimurium with antibody-fluorescent nanoparticle conjugates, the mixture was injected into the microfluidic chip directly. *S.* Typhimurium. labelled with fluorescent nanoparticles were retained in the microelectrode region by positive DEP force, while the antibody-fluorescent nanoparticles conjugates flowed away. The positive DEP enrichment and fluorescent nanoparticle labelling significantly decreased the LOD down to 56 CFU/ml.

In order to capture bacteria from the moving fluid, the strength of DEP had to overcome the hydrodynamic drag force. The capture efficiency of bacteria from the suspension was also limited by the position of microelectrodes. Syed *et al.* (2011) developed a novel microelectrode configuration for better DEP enrichment and impedance detection with 'points-and-lid' geometry. The microelectrodes were composed of a vertical nanoelectrode array made of aligned carbon nanofiber and a macroscopic indium-tin-oxide (ITO) transparent electrode. This microelectrode configuration enhanced the electric field gradient, and the ability to overcome large hydrodynamic drag force to improve capture efficiency.

Antibody immunoreaction and DEP based bacteria enrichment methods on a microchip platform are two of the most commonly adopted approaches. In addition to these two strategies, various novel materials and technologies have been applied for bacteria capture and enrichment on a microchip. For capture, antibodies have been widely applied in the immunoreaction field; however, disadvantages including high cost and cross-reactivity is inevitable. Therefore, other recognition strategies have been explored for microchip based detection methods.

Zuo *et al.* (2013) developed an aptamer-based hybrid microfluidic system, in which aptamer-functionalized graphene oxide (GO) was used to capture and detect *Lactobacillus acidophilus*. As shown in Fig. 6.4, when fluorescent–aptamer conjugates adsorbed on the GO surface, the fluorescence was quenched (fluorescence 'OFF'). After fluorescent labelled aptamers were bound to target bacteria, the fluorescent–aptamer conjugates were released from the GO surface and the quenching effect can be reversed (fluorescence 'ON'). This aptamer-based microchip system enabled 96 tests in single measurement within 10 minutes, providing a high throughput and efficient analysis. Zinc-coordinated bis(dipicolylamine) (bis-Zn-DPA) is a synthetic ligand, which provides strong affinity towards both Gram-negative and Gram-positive bacteria, while very little binding to mammalian cells or negatively charged proteins (Ngo *et al.*, 2012). Lee *et al.* (2014) developed a magnetic microfluidic device with bis-Zn-DPA modified MNPs that can capture and isolate bacteria from the bloodstream. Using this microfluidic device, the capture and cleaning rate of *E. coli* from bovine whole blood can be almost 100%. Mannoor *et al.* (2010) developed a portable biosensor using micro-capacitive electrode arrays functionalized with the antimicrobial peptide as the recognition ligand to detect pathogens.

The biophysical properties of pathogenic bacteria can also be exploited for enrichment. Based on the negative charge of bacteria surfaces due to the carboxylate groups and ionized phosphoryl, Balasubramanian *et al.* (2007) presented a microfluidic device to capture and enrich bacteria from water by using electrostatic trapping and an electrophoretic transport technique without chemical modification. Due to the negative surface charges, bacteria moved towards the anode and were concentrated

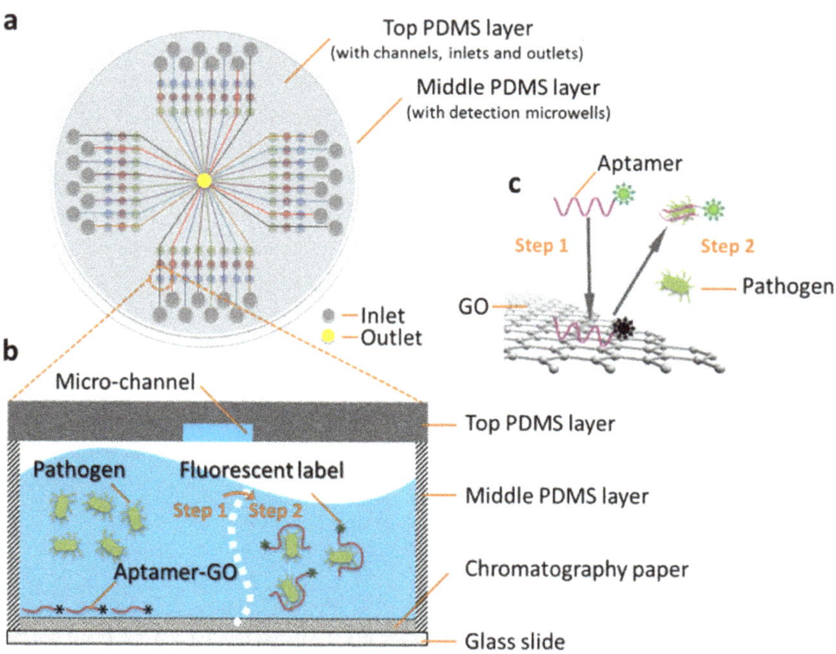

Figure 6.4 GO-based fluorescence quenching strategy integrated in a microfluidic system for pathogen detection (reproduced from reference Zuo *et al.*, 2013 with permission from The Royal Society of Chemistry).

within an electric field. The capture efficiencies for common pathogenic bacteria, such as *E. coli*, *Salmonella*, and *Pseudomonas* were up to 90–99%. Xin *et al.* (2013) reported that the manipulation of *E. coli* cells was achieved by integrating optical trapping with the microfluidic chip. *E. coli* cells can be stably trapped in the tip of an abruptly tapered optical fibre in a microfluidic channel due to the highly concentrated energy density at the abruptly tapered tip.

Filtration is a simple and effective method for concentration of bacteria and can selectively capture specific targets after functional modification. Guo *et al.* (2012b) applied filtration in a microfluidic device to capture and release bacteria. Zhang *et al.* (2010) reported an enrichment approach termed as evaporation microfluidics to separate *E. coli* from the fluid flow by exploiting evaporation microfluidics. This microfluidic sample preparation device consisted of a liquid sample flow layer made of poly(dimethylsiloxane), a metal airflow layer, and a porous PTFE (Teflon™) membrane between the liquid and air layers. Upon evaporation, bacteria were concentrated via water volume reduction. This microfluidic device can quickly concentrate 100 ml of a solution containing bacteria into 500 nl in 15 minutes, and the recovery concentration was above 85%. Microfluidic chips have a strong potential for use as a portable and real-time detection technology. However, culture enrichment is the key element to improve detection sensitivity on a microchip platform. It could be expected that the development of nanotechnology and aided-devices would further improve the microfluidic chip-based pathogen detection.

In the chapter, recent advances in pathogen detection methods not only for *E. coli* O157:H7 but also for other pathogenic organisms were summarized. It should be noted that the stricter regulatory limit for *E. coli* O157:H7 (zero tolerance) and six other non-O157 STEC serogroups (O26, O45, O103, O111, O121, and O145) demands higher sensitivity of the detection method, while differences in the application and functional environment play a role in reliability, simplicity, portability, recognition of serotype, detection speed, cost, and mass production. Spectroscopic methods are promising for their high throughput and specificity for detection, while colorimetric methods and paper-based methods are attractive for potential on-site monitoring due their simplicity and rapidity of detection at low cost. Chip-based methods with microfluidic devices could be integrated into various analytical techniques with pathogen enrichment, enabling portable and sensitive detection. The combination of these methods with other enrichment or analytical techniques would further improve the detection performance for *E. coli* O157:H7 or other pathogens and bring to the limelight a methodology that is practical and product specific.

Acknowledgements

Funding from the US Department of Agriculture, Agricultural Research Service, under Agreement No. 8072-42000-077 with the Center for Food Safety Engineering at Purdue University is appreciated. The authors thank Dr Pina Fratamico for critical feedback and insightful discussions.

References

Albrecht, M.G., and Creighton, J.A. (1977). Anomalously intense Raman spectra of pyridine at a silver electrode. J. Am. Chem. Soc. *99*, 5215–5217. https://doi.org/10.1021/ja00457a071

Alcaine, S.D., Law, K., Ho, S., Kinchla, A.J., Sela, D.A., and Nugen, S.R. (2016). Bioengineering bacteriophages to enhance the sensitivity of phage amplification-based paper fluidic detection of bacteria. Biosens. Bioelectron. *82*, 14–19. https://doi.org/10.1016/j.bios.2016.03.047

Alcaine, S.D., Pacitto, D., Sela, D.A., and Nugen, S.R. (2015). Phage & phosphatase: a novel phage-based probe for rapid, multi-platform detection of bacteria. Analyst *140*, 7629–7636. https://doi.org/10.1039/c5an01181g

Balasubramanian, A.K., Soni, K.A., Beskok, A., and Pillai, S.D. (2007). A microfluidic device for continuous capture and concentration of microorganisms from potable water. Lab Chip *7*, 1315–1321. https://doi.org/10.1039/b706559k

Bearinger, J.P., Dugan, L.C., Baker, B.R., Hall, S.B., Ebert, K., Mioulet, V., Madi, M., and King, D.P. (2011). Development and initial results of a low cost, disposable, point-of-care testing device for pathogen detection. IEEE Trans. Biomed. Eng. *58*, 805–808. https://doi.org/10.1109/TBME.2010.2089054

Berry, V., Gole, A., Kundu, S., Murphy, C.J., and Saraf, R.F. (2005). Deposition of CTAB-terminated nanorods on bacteria to form highly conducting hybrid systems. J. Am. Chem. Soc. *127*, 17600–17601. https://doi.org/10.1021/ja056428l

Berry, V., and Saraf, R.F. (2005). Self-assembly of nanoparticles on live bacterium: an avenue to fabricate electronic devices. Angew. Chem. Int. Ed. Engl. *44*, 6668–6673. https://doi.org/10.1002/anie.200501711

Boehm, D.A., Gottlieb, P.A., and Hua, S.Z. (2007). On-chip microfluidic biosensor for bacterial detection and identification. Sens. Actuator B-Chem. *126*, 508–514. https://doi.org/10.1016/j.snb.2007.03.043

Bouguelia, S., Roupioz, Y., Slimani, S., Mondani, L., Casabona, M.G., Durmort, C., Vernet, T., Calemczuk, R., and Livache, T. (2013). On-chip microbial culture for the specific detection of very low levels of bacteria. Lab Chip *13*, 4024–4032. https://doi.org/10.1039/c3lc50473e

Braun-Kiewnick, A., Altenbach, D., Oberhansli, T., Bitterlin, W., and Duffy, B. (2011). A rapid lateral-flow immunoassay for phytosanitary detection of *Erwinia amylovora* and on-site fire blight diagnosis. J. Microbiol. Methods *87*, 1–9. https://doi.org/10.1016/j.mimet.2011.06.015

Buchanan, R.L., and Doyle, M.P. (1997). Food-borne disease significance of *Escherichia coli* O157:H7 and other enterohemorrhagic *E. coli*. Food Technol. *51*, 69–76.

Bui, M.P.N., Ahmed, S., and Abbas, A. (2015). Single-digit pathogen and attomolar detection with the naked eye using liposome-amplified plasmonic immunoassay. Nano Lett. *15*, 6239–6246. https://doi.org/10.1021/acs.nanolett.5b02837

Burnham, S., Hu, J., Anany, H., Brovko, L., Deiss, F., Derda, R., and Griffiths, M.W. (2014). Towards rapid on-site phage-mediated detection of generic *Escherichia coli* in water using luminescent and visual readout. Anal. Bioanal. Chem. *406*, 5685–5693. https://doi.org/10.1007/s00216-014-7985-3

Cao, C., Gontard, L.C., Thuy Tram, l.e.L., Wolff, A., and Bang, D.D. (2011). Dual enlargement of gold nanoparticles: from mechanism to scanometric detection of pathogenic bacteria. Small 7, 1701–1708. https://doi.org/10.1002/smll.201100294

Chandan, H.R., Venkataramana, M., Kurkuri, M.D., and Balakrishna, R.G. (2016). Simple quantum dot bioprobe/label for sensitive detection of *Staphylococcus aureus* TNase. Sens. Actuator B-Chem. *222*, 1201–1208. https://doi.org/10.1016/j.snb.2015.07.121

Chen, H.D., and Frankel, G. (2005). Enteropathogenic *Escherichia coli*: unravelling pathogenesis. FEMS Microbiol. Rev. *29*, 83–98. https://doi.org/10.1016/j.femsre.2004.07.002

Chen, J., Wu, X., Huang, Y.-W., and Zhao, Y. (2014). Detection of *E. coli* using SERS active filters with silver nanorod array. Sens. Actuator B-Chem. *191*, 485–490. https://doi.org/10.1016/j.snb.2013.10.038

Chen, R., Huang, X.L., Xu, H.Y., Xiong, Y.H., and Li, Y.B. (2015). Plasmonic enzyme-linked immunosorbent assay using nanospherical brushes as a catalase container for colorimetric detection of ultralow concentrations of *Listeria monocytogenes*. ACS Appl. Mater. Interfaces *7*, 28632–28639. https://doi.org/10.1021/acsami.5b10181

Chen, Z.Y., Abrams, W.R., Geva, E., de Dood, C.J., Gonzalez, J.M., Tanke, H.J., Niedbala, R.S., Zhou, P., Malamud, D., and Corstjens, P. (2013). Development of a generic microfluidic device for simultaneous detection of antibodies and nucleic acids in oral fluids. Biomed. Res. Int. https://doi.org/10.1155/2013/543294

Cheng, I.-F., Chang, H.-C., Chen, T.-Y., Hu, C., and Yang, F.-L. (2013). Rapid (< 5 min) identification of pathogen in human blood by electrokinetic concentration and surface-enhanced Raman spectroscopy. Sci. Rep. U.K. 3, 2365. https://doi.org/10.1038/srep02365

Cheng, I.F., Lin, C.C., Lin, D.Y., and Chang, H.C. (2010). A dielectrophoretic chip with a roughened metal surface for on-chip surface-enhanced Raman scattering analysis of bacteria. Biomicrofluidics. *4*, 34104. https://doi.org/10.1063/1.3474638

Cho, I.-H., Bhandari, P., Patel, P., and Irudayaraj, J. (2015a). Membrane filter-assisted surface enhanced Raman spectroscopy for the rapid detection of *E. coli* O157:H7 in ground beef. Biosens. Bioelectron. *64*, 171–176. https://doi.org/10.1016/j.bios.2014.08.063

Cho, I.-H., Bhunia, A., and Irudayaraj, J. (2015b). Rapid pathogen detection by lateral-flow immunochromatographic assay with gold nanoparticle-assisted enzyme signal amplification. Int. J. Food Microbiol. *206*, 60–66. https://doi.org/10.1016/j.ijfoodmicro.2015.04.032

Cho, I.-H., Mauer, L., and Irudayaraj, J. (2014). In-situ fluorescent immunomagnetic multiplex detection of food-borne pathogens in very low numbers. Biosens. Bioelectron. *57*, 143–148. https://doi.org/10.1016/j.bios.2014.02.012

Cho, I.H., and Irudayaraj, J. (2013a). Lateral-flow enzyme immunoconcentration for rapid detection of *Listeria monocytogenes*. Anal. Bioanal. Chem. *405*, 3313–3319. https://doi.org/10.1007/s00216-013-6742-3

Cho, I.H., and Irudayaraj, J. (2013b). In-situ immuno-gold nanoparticle network ELISA biosensors for pathogen detection. Int. J. Food Microbiol. *164*, 70–75. https://doi.org/10.1016/j.ijfoodmicro.2013.02.025

Cox, C.R., Jensen, K.R., Mondesire, R.R., and Voorhees, K.J. (2015). Rapid detection of *Bacillus anthracis* by gamma phage amplification and lateral flow immunochromatography. J. Microbiol. Methods. *118*, 51–56. https://doi.org/10.1016/j.mimet.2015.08.011

Davis, R., Irudayaraj, J., Reuhs, B.L., and Mauer, L.J. (2010). Detection of *E. coli* O157:H7 from ground beef using Fourier transform infrared (FT-IR) spectroscopy and chemometrics. J. Food Sci. *75*, M340–346. https://doi.org/10.1111/j.1750-3841.2010.01686.x

de la Rica, R., and Stevens, M.M. (2012). Plasmonic ELISA for the ultrasensitive detection of disease biomarkers with the naked eye. Nat. Nanotechnol. *7*, 821–824. https://doi.org/10.1038/nnano.2012.186

de Oliveira, T.V., Soares, N.d.F.F., Silva, D.J., de Andrade, N.J., Medeiros, E.A.A., and Badaro, A.T. (2013). Development of PDA/phospholipids/lysine vesicles to detect pathogenic bacteria. Sens. Actuator B-Chem. *188*, 385–392. https://doi.org/10.1016/j.snb.2013.06.022

de Oliveira, T.V., Soares, N.d.e.F., de Andrade, N.J., Silva, D.J., Medeiros, E.A., and Badaró, A.T. (2015). Application of PCDA/SPH/CHO/Lysine vesicles to detect pathogenic bacteria in chicken. Food Chem. *172*, 428–432. https://doi.org/10.1016/j.foodchem.2014.09.055

Disney, M.D., Zheng, J., Swager, T.M., and Seeberger, P.H. (2004). Detection of bacteria with carbohydrate-functionalized fluorescent polymers. J. Am. Chem. Soc. *126*, 13343–13346. https://doi.org/10.1021/ja047936i

Duy, J., Smith, R.L., Collins, S.D., and Connell, L.B. (2015). Rapid colorimetric detection of the fungal phytopathogen *Synchytrium endobioticum* using cyanine dye-indicated PNA hybridization. Am. J. Potato Res. *92*, 398–409. https://doi.org/10.1007/s12230-015-9450-z

Efrima, S., and Zeiri, L. (2009). Understanding SERS of bacteria. J. Raman Spectrosc. *40*, 277–288. https://doi.org/10.1002/jrs.2121

Fang, Z.Y., Wu, W., Lu, X.W., and Zeng, L.W. (2014). Lateral flow biosensor for DNA extraction-free detection of *Salmonella* based on aptamer mediated strand displacement amplification. Biosens. Bioelectron. *56*, 192–197. https://doi.org/10.1016/j.bios.2014.01.015

Fleischmann, M., Hendra, P.J., and McQuillan, A.J. (1974). Raman spectra of pyridine adsorbed at a silver electrode. Chem. Phys. Lett. *26*, 163–166. https://doi.org/10.1016/0009-2614(74)85388-1

Garcia-Aljaro, C., Bangar, M.A., Baldrich, E., Munoz, F.J., and Mulchandani, A. (2010). Conducting polymer nanowire-based chemiresistive biosensor for the detection of bacterial spores. Biosens. Bioelectron. *25*, 2309–2312. https://doi.org/10.1016/j.bios.2010.03.021

Gomez, A., Miller, N.S., and Smolina, I. (2014). Visual detection of bacterial pathogens via PNA-based padlock probe assembly and isothermal amplification of DNAzymes. Anal. Chem. *86*, 11992–11998. https://doi.org/10.1021/ac5018748

Granger, J.H., Schlotter, N.E., Crawford, A.C., and Porter, M.D. (2016). Prospects for point-of-care pathogen diagnostics using surface-enhanced Raman scattering (SERS). Chem. Soc. Rev. *45*, 3865–3882. https://doi.org/10.1039/c5cs00828j

Guan, X., Zhang, H.J., Bi, Y.N., Zhang, L., and Hao, D.L. (2010). Rapid detection of pathogens using antibody-coated microbeads with bioluminescence in microfluidic chips. Biomed. Microdevices. *12*, 683–691. https://doi.org/10.1007/s10544-010-9421-6

Guo, H., Jiang, D., Li, H., Xu, S., and Xu, W. (2012a). Highly efficient construction of silver nanosphere dimers on poly (dimethylsiloxane) sheets for surface-enhanced Raman scattering. J. Phys. Chem. *117*, 564–570. https://doi.org/10.1021/jp309396x

Guo, P., Hall, E.W., Schirhagl, R., Mukaibo, H., Martin, C.R., and Zare, R.N. (2012b). Microfluidic capture and release of bacteria in a conical nanopore array. Lab Chip *12*, 558–561. https://doi.org/10.1039/c2lc21092d

Gupta, M.J., Irudayaraj, J., and Debroy, C. (2004). Spectroscopic quantification of bacteria using artificial neural networks. J. Food Prot. *67*, 2550–2554.

Guven, B., Basaran-Akgul, N., Temur, E., Tamer, U., and Boyaci, I.H. (2011). SERS-based sandwich immunoassay using antibody coated magnetic nanoparticles for *Escherichia coli* enumeration. Analyst. *136*, 740–748. https://doi.org/10.1039/c0an00473a

He, X., Hu, C., Guo, Q., Wang, K., Li, Y., and Shangguan, J. (2013). Rapid and ultrasensitive *Salmonella* Typhimurium quantification using positive dielectrophoresis driven on-line enrichment and fluorescent nanoparticles label. Biosens. Bioelectron. *42*, 460–466. https://doi.org/10.1016/j.bios.2012.11.020

Hong, W., Liao, C.S., Zhao, H., Younis, W., Zhang, Y., Seleem, M.N., and Cheng, J.X. (2016). In situ detection of a single bacterium in complex environment by hyperspectral CARS imaging. Chemistry Select. *1*, 513–517. https://doi.org/10.1002/slct.201600166

Hua, F., Zhang, P.P., Zhang, F.L., Zhao, Y., Li, C.F., Sun, C.Y., Wang, X.C., Yang, R.F., Wang, C.B., Yu, A.L., and Zhou, L. (2015). Development and evaluation of an up-converting phosphor technology-based lateral flow assay for rapid detection of *Francisella tularensis*. Sci. Rep. U.K. *5*, 17178. https://doi.org/10.1038/srep17178

Huang, C.J., Knoll, W., Sessitsch, A., and Dostalek, J. (2014). SPR bacterial pathogen biosensor: the importance of fluidic conditions and probing depth. Talanta *122*, 166–171. https://doi.org/10.1016/j.talanta.2014.01.022

Imamura, K., Takayama, S., Saito, A., Inoue, E., Nakayama, Y., Ogata, Y., Shirakawa, S., Nagano, T., Gomi, K., Morozumi, T., Akiishi, K., Watanabe, K., and Yoshie, H. (2015). Evaluation of a novel immunochromatographic device for rapid and accurate clinical detection of porphyromonas gingivalis in subgingival plaque. J. Microbiol. Methods. *117*, 4–10. https://doi.org/10.1016/j.mimet.2015.07.002

Jansaento, W., Jangpatarapongsa, K., Polpanich, D., and Wonglumsom, W. (2016). Detection of *Campylobacter* DNA using magnetic nanoparticles coupled with PCR and a colorimetric end-point system. Food Sci. Biotechnol. *25*, 193–198. https://doi.org/10.1007/s10068-016-0029-3

Jay, J.M. (2012). Modern food microbiology, (New York, US: Springer Science & Business Media). https://doi.org/10.1007/978-1-4615-7476-7

Jiang, T., Song, Y., Wei, T., Li, H., Du, D., Zhu, M.J., and Lin, Y. (2016). Sensitive detection of *Escherichia coli* O157:H7 using Pt-Au bimetal nanoparticles with peroxidase-like amplification. Biosens. Bioelectron. *77*, 687–694. https://doi.org/10.1016/j.bios.2015.10.017

Jokerst, J.C., Adkins, J.A., Bisha, B., Mentele, M.M., Goodridge, L.D., and Henry, C.S. (2012). Development of a paper-based analytical device for colorimetric detection of select food-borne pathogens. Anal. Chem. *84*, 2900–2907. https://doi.org/10.1021/ac203466y

Ju, Q., Noor, M.O., and Krull, U.J. (2016). Paper-based biodetection using luminescent nanoparticles. Analyst *141*, 2838–2860. https://doi.org/10.1039/c6an00129g

Jung, J.H., Cheon, D.S., Liu, F., Lee, K.B., and Seo, T.S. (2010a). A graphene oxide based immuno-biosensor for pathogen detection. Angew. Chem. Int. Ed. Engl. *49*, 5708–5711. https://doi.org/10.1002/anie.201001428

Jung, Y.L., Jung, C., Parab, H., Li, T., and Park, H.G. (2010b). Direct colorimetric diagnosis of pathogen infections by utilizing thiol-labeled PCR primers and unmodified gold nanoparticles. Biosens. Bioelectron. *25*, 1941–1946. https://doi.org/10.1016/j.bios.2010.01.010

Jyoti, A., Pandey, P., Singh, S.P., Jain, S.K., and Shanker, R. (2010). Colorimetric detection of nucleic acid signature of shiga toxin producing *Escherichia coli* using gold nanoparticles. J. Nanosci. Nanotechnol. *10*, 4154–4158. https://doi.org/10.1166/jnn.2010.2649

Khalil, M.A.F., Azzazy, H.M.E., Attia, A.S., and Hashem, A.G.M. (2014). A sensitive colorimetric assay for identification of *Acinetobacter baumannii* using unmodified gold nanoparticles. J. Appl. Microbiol. *117*, 465–471. https://doi.org/10.1111/jam.12546

Khan, M.S., Pande, T., and van de Ven, T.G.M. (2015). Qualitative and quantitative detection of T7 bacteriophages using paper based sandwich ELISA. Colloids Surf. B Biointerfaces. *132*, 264–270. https://doi.org/10.1016/j.colsurfb.2015.05.028

Kim, Y.T., Chen, Y., Choi, J.Y., Kim, W.J., Dae, H.M., Jung, J., and Seo, T.S. (2012). Integrated microdevice of reverse transcription-polymerase chain reaction with colorimetric immunochromatographic detection for rapid gene expression analysis of influenza A H1N1 virus. Biosens. Bioelectron. *33*, 88–94. https://doi.org/10.1016/j.bios.2011.12.024

Knauer, M., Ivleva, N.P., Niessner, R., and Haisch, C. (2012). A flow-through microarray cell for the online SERS detection of antibody-captured *E. coli* bacteria. Anal. Bioanal. Chem. *402*, 2663–2667. https://doi.org/10.1007/s00216-011-5398-0

Kong, F., and Hu, Y.F. (2012). Biomolecule immobilization techniques for bioactive paper fabrication. Anal. Bioanal. Chem. *403*, 7–13. https://doi.org/10.1007/s00216-012-5821-1

Kwon, J.S., Ravindranath, S.P., Kumar, A., Irudayaraj, J., and Wereley, S.T. (2012). Opto-electrokinetic manipulation for high-performance on-chip bioassays. Lab Chip *12*, 4955–4959. https://doi.org/10.1039/c2lc40662d

Lavu, P.S.R., Mondal, B., Ramlal, S., Murali, H.S., and Batra, H.V. (2016). Selection and characterization of aptamers using a modified whole cell bacterium SELEX for the detection of *Salmonella enterica* serovar Typhimurium. ACS. Comb. Sci. *18*, 292–301. https://doi.org/10.1021/acscombsci.5b00123

Lee, D., Kim, Y.T., Lee, J.W., Kim, D.H., and Seo, T.S. (2016a). An integrated direct loop-mediated isothermal amplification microdevice incorporated with an immunochromatographic strip for bacteria detection in human whole blood and milk without a sample preparation step. Biosens. Bioelectron. *79*, 273–279. https://doi.org/10.1016/j.bios.2015.12.044

Lee, J.J., Jeong, K.J., Hashimoto, M., Kwon, A.H., Rwei, A., Shankarappa, S.A., Tsui, J.H., and Kohane, D.S. (2014). Synthetic ligand-coated magnetic nanoparticles for microfluidic bacterial separation from blood. Nano Lett. *14*, 1–5. https://doi.org/10.1021/nl3047305

Lee, K., and Irudayaraj, J. (2013). Correct spectral conversion between surface-enhanced Raman and plasmon resonance scattering from nanoparticle dimers for single-molecule detection. Small. *9*, 1106–1115. https://doi.org/10.1002/smll.201201985

Lee, S., Mehta, S., and Erickson, D. (2016b). Two-color lateral flow assay for multiplex detection of causative agents behind acute febrile illnesses. Anal. Chem. *88*, 8359–8363. https://doi.org/10.1021/acs.analchem.6b01828

Liu, D.B., Wang, Z.T., Jin, A., Huang, X.L., Sun, X.L., Wang, F., Yan, Q., Ge, S.X., Xia, N.S., Niu, G., Liu, G., Walker, A.R.H., and Chen, X.Y. (2013). Acetylcholinesterase-catalyzed hydrolysis allows ultrasensitive detection of pathogens with the naked eye. Angew. Chem. Int. Ed. *52*, 14065–14069. https://doi.org/10.1002/anie.201307952

Liu, P., Han, L., Wang, F., Petrenko, V.A., and Liu, A.H. (2016). Gold nanoprobe functionalized with specific fusion protein selection from phage display and its application in rapid, selective and sensitive colorimetric biosensing of *Staphylococcus aureus*. Biosens. Bioelectron. *82*, 195–203. https://doi.org/10.1016/j.bios.2016.03.075

Madiyar, F.R., Bhana, S., Swisher, L.Z., Culbertson, C.T., Huang, X., and Li, J. (2015). Integration of a nanostructured dielectrophoretic device and a surface-enhanced Raman probe for highly sensitive rapid bacteria detection. Nanoscale. *7*, 3726–3736. https://doi.org/10.1039/C4NR07183B

Mannoor, M.S., Zhang, S., Link, A.J., and McAlpine, M.C. (2010). Electrical detection of pathogenic bacteria via immobilized antimicrobial peptides. Proc. Natl. Acad. Sci. U.S.A. *107*, 19207–19212. https://doi.org/10.1073/pnas.1008768107

Martos, S., Torres, E., El Bakali, M.A., Raposo, R., Gramaje, D., Armengol, J., and Luque, J. (2011). Co-operational PCR coupled with dot blot hybridization for the detection of *Phaeomoniella chlamydospora* on infected grapevine wood. J. Phytopathol. *159*, 247–254. https://doi.org/10.1111/j.1439-0434.2010.01758.x

Mead, P.S., and Griffin, P.M. (1998). Escherichia coli O157:H7. Lancet *352*, 1207–1212. https://doi.org/10.1016/s0140-6736(98)01267-7

Miranda, O.R., Li, X.N., Garcia-Gonzalez, L., Zhu, Z.J., Yan, B., Bunz, U.H.F., and Rotello, V.M. (2011). Colorimetric bacteria sensing using a supramolecular enzyme-nanoparticle Biosensor. J. Am. Chem. Soc. *133*, 9650–9653. https://doi.org/10.1021/ja2021729

Moon, J., Kim, G., and Park, S. (2014). Development of ssDNA aptamers for the capture and detection of *Salmonella* Typhimurium. Anal. Methods. *6*, 7442–7448. https://doi.org/10.1039/c4ay01035c

Morales-Narvaez, E., Naghdi, T., Zor, E., and Merkoci, A. (2015). Photo luminescent lateral-flow immunoassay revealed by graphene oxide: highly sensitive paper-based pathogen detection. Anal. Chem. *87*, 8573–8577. https://doi.org/10.1021/acs.analchem.5b02383

Nataro, J.P., and Kaper, J.B. (1998). Diarrheagenic *Escherichia coli*. Clin. Microbiol. Rev. *11*, 142–201

Neng, J., Harpster, M.H., Wilson, W.C., and Johnson, P.A. (2013). Surface-enhanced Raman scattering (SERS) detection of multiple viral antigens using magnetic capture of SERS-active nanoparticles. Biosens. Bioelectron. *41*, 316–321. https://doi.org/10.1016/j.bios.2012.08.048

Ngo, H.T., Liu, X., and Jolliffe, K.A. (2012). Anion recognition and sensing with Zn(II)-dipicolylamine complexes. Chem. Soc. Rev. *41*, 4928–4965. https://doi.org/10.1039/c2cs35087d

Nie, X.M., Huang, R., Dong, C.X., Tang, L.J., Gui, R., and Jiang, J.H. (2014). Plasmonic ELISA for the ultrasensitive detection of *Treponema pallidum*. Biosens. Bioelectron. *58*, 314–319. https://doi.org/10.1016/j.bios.2014.03.007

Oh, S.J., Park, B.H., Choi, G., Seo, J.H., Jung, J.H., Choi, J.S., Kim, D.H., and Seo, T.S. (2016a). Fully automated and colorimetric food-borne pathogen detection on an integrated centrifugal microfluidic device. Lab Chip. *16*, 1917–1926. https://doi.org/10.1039/c6lc00326e

Oh, S.J., Park, B.H., Jung, J.H., Choi, G., Lee, D.C., Kim, D.H., and Seo, T.S. (2016b). Centrifugal loop-mediated isothermal amplification microdevice for rapid, multiplex and colorimetric food-borne pathogen detection. Biosens. Bioelectron. *75*, 293–300. https://doi.org/10.1016/j.bios.2015.08.052

Panferov, V.G., Safenkova, I.V., Varitsev, Y.A., Drenova, N.V., Kornev, K.P., Zherdev, A.V., and Dzantiev, B.B. (2016). Development of the sensitive lateral flow immunoassay with silver enhancement for the detection of *Ralstonia solanacearum* in potato tubers. Talanta *152*, 521–530. https://doi.org/10.1016/j.talanta.2016.02.050

Paniel, N., and Baudart, J. (2013). Colorimetric and electrochemical genosensors for the detection of *Escherichia coli* DNA without amplification in seawater. Talanta *115*, 133–142. https://doi.org/10.1016/j.talanta.2013.04.050

Pires, A.C.D., Soares, N.D.F., da Silva, L.H.M., da Silva, M.D.H., De Almeida, M.V., Le Hyaric, M., de Andrade, N.J., Soares, R.F., Mageste, A.B., and Reis, S.G. (2011). A colorimetric biosensor for the detection of food-borne bacteria. Sens. Actuator B-Chem. *153*, 17–23. https://doi.org/10.1016/j.snb.2010.09.069

Pohl, H.A. (1978). Dielectrophoresis: The Behavior of Neutral Matter in Nonuniform Electric Fields. (New York: Cambridge University Press, Cambridge).

Prasad, D., Shankaracharya, and Vidyarthi, A.S. (2011). Gold nanoparticles-based colorimetric assay for rapid detection of *Salmonella* species in food samples. World J. Microbiol. Biotechnol. *27*, 2227–2230. https://doi.org/10.1007/s11274-011-0679-5

Qi, H.J., Chen, S.H., Hao, R.Z., Shi, H., Zhang, M.L., and Wang, S.Q. (2012). Introduction of nanogold-DAB as a HRP substrate for simplifying detection in visual DNA microarrays. Anal. Methods. *4*, 1178–1181. https://doi.org/10.1039/c2ay05854e

Quintela, I.A., de los Reyes, B.G., Lin, C.S., and Wu, V.C.H. (2015). Simultaneous direct detection of shiga-toxin producing *Escherichia coli* (STEC) strains by optical biosensing with oligonucleotide-functionalized gold. Nanoscale. *7*, 2417–2426. https://doi.org/10.1039/c4nr05869k

Rajendran, V.K., Bakthavathsalam, P., and Ali, B.M.J. (2014). Smartphone based bacterial detection using biofunctionalized fluorescent nanoparticles. Microchim. Acta. *181*, 1815–1821. https://doi.org/10.1007/s00604-014-1242-5

Rajwa, B., Dundar, M.M., Akova, F., Bettasso, A., Patsekin, V., Hirleman, E.D., Bhunia, A.K., and Robinson, J.P. (2010). Discovering the unknown: detection of emerging pathogens using a label-free light-scattering system. Cytometry A *77*, 1103–1112. https://doi.org/10.1002/cyto.a.20978

Rastogi, S.K., Gibson, C.M., Branen, J.R., Aston, D.E., Branen, A.L., and Hrdlicka, P.J. (2012). DNA detection on lateral flow test strips: enhanced signal sensitivity using LNA-conjugated gold nanoparticles. Chem. Commun. *48*, 7714–7716. https://doi.org/10.1039/c2cc33430e

Ravindranath, S.P., Wang, Y., and Irudayaraj, J. (2011). SERS driven cross-platform based multiplex pathogen detection. Sens. Actuator B-Chem. *152*, 183–190. https://doi.org/10.1016/j.snb.2010.12.005

Ren, W., Cho, H., Zhou, Z.W., and Irudayaraj, J. (2016). Ultrasensitive detection of microbial cells using magnetic focus enhanced lateral flow sensors. Chem. Commun. *52*, 4930–4933. https://doi.org/10.1039/c5cc10240e

Ren, W., Liu, W.J., and Irudayaraj, J. (2017). A net fishing enrichment strategy for colorimetric detection of *E. coli* O157: H7. Sens. Actuator B-Chem. *247*, 923–929. https://doi.org/10.1016/j.snb.2017.03.090

Rozhok, S., Shen, C.K., Littler, P.L., Fan, Z., Liu, C., Mirkin, C.A., and Holz, R.C. (2005). Methods for fabricating microarrays of motile bacteria. Small *1*, 445–451. https://doi.org/10.1002/smll.200400072

Ruan, C.M., Yang, L.J., and Li, Y.B. (2002). Immunobiosensor chips for detection of *Escherichia coli* O157: H7 using electrochemical impedance spectroscopy. Anal. Chem. *74*, 4818–4820. https://doi.org/10.1021/ac025647b

Safavieh, M., Ahmed, M.U., Sokullu, E., Ng, A., Braescuac, L., and Zourob, M. (2014). A simple cassette as point-of-care diagnostic device for naked-eye colorimetric bacteria detection. Analyst. *139*, 482–487. https://doi.org/10.1039/c3an01859h

Safenkova, I.V., Pankratova, G.K., Zaitsev, I.A., Varitsev, Y.A., Vengerov, Y.Y., Zherdev, A.V., and Dzantiev, B.B. (2016). Multiarray on a test strip (MATS): rapid multiplex immunodetection of priority potato pathogens. Anal. Bioanal. Chem. *408*, 6009–6017. https://doi.org/10.1007/s00216-016-9463-6

Saikrishnan, D., Goyal, M., Rossiter, S., and Kukol, A. (2014). A cellulose-based bioassay for the colorimetric detection of pathogen DNA. Anal. Bioanal. Chem. *406*, 7887–7898. https://doi.org/10.1007/s00216-014-8257-y

Santoro, S.W., and Joyce, G.F. (1997). A general purpose RNA-cleaving DNA enzyme. Proc. Natl. Acad. Sci. U.S.A. *94*, 4262–4266. https://doi.org/10.1073/pnas.94.9.4262

Schlosser, K., and Li, Y. (2010). A versatile endoribonuclease mimic made of DNA: characteristics and applications of the 8-17 RNA-cleaving DNAzyme. Chembiochem *11*, 866–879. https://doi.org/10.1002/cbic.200900786

Seidel, M., and Niessner, R. (2008). Automated analytical microarrays: a critical review. Anal. Bioanal. Chem. *391*, 1521–1544. https://doi.org/10.1007/s00216-008-2039-3

Sen, A., Harvey, T., and Clausen, J. (2011). A microsystem for extraction, capture and detection of *E-coli* O157:H7. Biomed. Microdevices *13*, 705–715. https://doi.org/10.1007/s10544-011-9540-8

Shen, G., Xu, H., Gurung, A.S., Yang, Y., and Liu, G. (2013). Lateral flow immunoassay with the signal enhanced by gold nanoparticle aggregates based on polyamidoamine dendrimer. Anal. Sci. *29*, 799–804

Shih, C.M., Chang, C.L., Hsu, M.Y., Lin, J.Y., Kuan, C.M., Wang, H.K., Huang, C.T., Chung, M.C., Huang, K.C., Hsu, C.E., *et al.* (2015). Paper-based ELISA to rapidly detect *Escherichia coli*. Talanta *145*, 2–5. https://doi.org/10.1016/j.talanta.2015.07.051

Shim, W.B., Song, J.E., Mun, H., Chung, D.H., and Kim, M.G. (2014). Rapid colorimetric detection of *Salmonella* Typhimurium using a selective filtration technique combined with antibody-magnetic nanoparticle nanocomposites. Anal. Bioanal. Chem. *406*, 859–866. https://doi.org/10.1007/s00216-013-7497-6

Sivakesava, S., Irudayaraj, J., and DebRoy, C. (2004). Differentiation of microorganisms by FTIR-ATR and NIR spectroscopy. T. ASAE *47*, 951–957. https://doi.org/10.13031/2013.16074

Song, C.M., Liu, C., Wu, S.Y., Li, H.L., Guo, H.Q., Yang, B., Qiu, S., Li, J.W., Liu, L., Zeng, H.J., Zhai, X.Z., and Liu, Q. (2016a). Development of a lateral flow colloidal gold immunoassay strip for the simultaneous detection of *Shigella boydii* and *Escherichia coli* O157:H7 in bread, milk and jelly samples. Food Control. *59,* 345–351. https://doi.org/10.1016/j.foodcont.2015.06.012

Song, C., Liu, J., Li, J., and Liu, Q. (2016b). Dual FITC lateral flow immunoassay for sensitive detection of *Escherichia coli* O157:H7 in food samples. Biosens. Bioelectron. *85,* 734–739

Srinivasan, B., and Tung, S. (2015). Development and applications of portable biosensors. J. Lab. Autom. *20,* 365–389. https://doi.org/10.1177/2211068215581349

Storhoff, J.J., Elghanian, R., Mucic, R.C., Mirkin, C.A., and Letsinger, R.L. (1998). One-pot colorimetric differentiation of polynucleotides with single base imperfections using gold nanoparticle probes. J. Am. Chem. Soc. *120,* 1959–1964. https://doi.org/10.1021/ja972332i

Su, H., Zhao, H., Qiao, F., Chen, L., Duan, R., and Ai, S. (2013). Colorimetric detection of *Escherichia coli* O157:H7 using functionalized Au@Pt nanoparticles as peroxidase mimetics. Analyst *138,* 3026–3031. https://doi.org/10.1039/c3an00026e

Subramanian, A., Irudayaraj, J., and Ryan, T. (2006). Mono and dithiol surfaces on surface plasmon resonance biosensors for detection of *Staphylococcus aureus.* Sens. Actuator B-Chem. *114,* 192–198. https://doi.org/10.1016/j.snb.2005.04.030

Suehiro, J., Ohtsubo, A., Hatano, T., and Hara, M. (2006). Selective detection of bacteria by a dielectrophoretic impedance measurement method using an antibody-immobilized electrode Chip. Sens. Actuator B-Chem. *119,* 319–326. https://doi.org/10.1016/j.snb.2005.12.027

Suehiro, J., Shutou, M., Hatano, T., and Hara, M. (2003). High sensitive detection of biological cells using dielectrophoretic impedance measurement method combined with electropermeabilization. Sens. Actuator B-Chem. *96,* 144–151. https://doi.org/10.1016/s0925-4005(03)00517-3

Sun, J., Ji, J., Sun, Y., Abdalhai, M.H., Zhang, Y., and Sun, X. (2015). DNA biosensor-based on fluorescence detection of *E. coli* O157:H7 by Au@Ag nanorods. Biosens. Bioelectron. *70,* 239–245. https://doi.org/10.1016/j.bios.2015.03.009

Sung, Y.J., Suk, H.J., Sung, H.Y., Li, T., Poo, H., and Kim, M.G. (2013). Novel antibody/gold nanoparticle/magnetic nanoparticle nanocomposites for immunomagnetic separation and rapid colorimetric detection of *Staphylococcus aureus* in milk. Biosens. Bioelectron. *43,* 432–439. https://doi.org/10.1016/j.bios.2012.12.052

Syed, L.U., Liu, J., Price, A.K., Li, Y.F., Culbertson, C.T., and Li, J. (2011). Dielectrophoretic capture of *E. coli* cells at micropatterned nanoelectrode arrays. Electrophoresis. *32,* 2358–2365. https://doi.org/10.1002/elps.201100020

Torun, O., Hakkı Boyacı, I., Temür, E., and Tamer, U. (2012). Comparison of sensing strategies in SPR biosensor for rapid and sensitive enumeration of bacteria. Biosens. Bioelectron. *37,* 53–60. https://doi.org/10.1016/j.bios.2012.04.034

Vaisocherová-Lísalová, H., Víšová, I., Ermini, M.L., Špringer, T., Song, X.C., Mrázek, J., Lamačová, J., Lynn, N.S., Šedivák, P., and Homola, J. (2016). Low-fouling surface plasmon resonance biosensor for multi-step detection of food-borne bacterial pathogens in complex food samples. Biosens. Bioelectron. *80,* 84–90. https://doi.org/10.1016/j.bios.2016.01.040

Varshney, M., Li, Y., Srinivasan, B., and Tung, S. (2007). A label-free, microfluidics and interdigitated array microelectrode-based impedance biosensor in combination with nanoparticles immunoseparation for detection of *Escherichia coli* O157:H7 in food samples. Sens. Actuator B-Chem. *128,* 99–107. https://doi.org/10.1016/j.snb.2007.03.045

Veigas, B., Machado, D., Perdigao, J., Portugal, I., Couto, I., Viveiros, M., and Baptista, P.V. (2010). Au-nanoprobes for detection of SNPs associated with antibiotic resistance in *Mycobacterium tuberculosis.* Nanotechnology *21,* 415101. https://doi.org/10.1088/0957-4484/21/41/415101

Veigas, B., Pedrosa, P., Carlos, F.F., Mancio-Silva, L., Grosso, A.R., Fortunato, E., Mota, M.M., and Baptista, P.V. (2015). One nanoprobe, two pathogens: gold nanoprobes multiplexing for point-of-care. J. Nanobiotechnol. *13,* 48. https://doi.org/10.1186/s12951-015-0109-1

Verma, M.S., Chen, P.Z., Jones, L., and Gu, F.X. (2014). 'Chemical nose' for the visual identification of emerging ocular pathogens using gold nanostars. Biosens. Bioelectron. *61,* 386–390. https://doi.org/10.1016/j.bios.2014.05.045

Verma, M.S., Wei, S.C., Rogowski, J.L., Tsuji, J.M., Chen, P.Z., Lin, C.W., Jones, L., and Gu, F.X. (2016). Interactions between bacterial surface and nanoparticles govern the performance of 'chemical nose' biosensors. Biosens. Bioelectron. *83,* 115–125. https://doi.org/10.1016/j.bios.2016.04.024

Wang, C., and Irudayaraj, J. (2008). Gold nanorod probes for the detection of multiple pathogens. Small *4,* 2204–2208. https://doi.org/10.1002/smll.200800309

Wang, C., and Irudayaraj, J. (2010). Multifunctional magnetic–optical nanoparticle probes for simultaneous detection, separation, and thermal ablation of multiple pathogens. Small. *6,* 283–289. https://doi.org/10.1002/smll.200901596

Wang, H., Zhou, Y., Jiang, X., Sun, B., Zhu, Y., Wang, H., Su, Y., and He, Y. (2015a). Simultaneous capture, detection, and inactivation of bacteria as enabled by a surface-enhanced Raman scattering multifunctional chip. Angew. Chem. Int. Ed. *54,* 5132–5136. https://doi.org/10.1002/anie.201412294

Wang, N., Wang, Y., Guo, T., Yang, T., Chen, M., and Wang, J. (2016). Green preparation of carbon dots with papaya as carbon source for effective fluorescent sensing of iron (III) and *Escherichia coli.* Biosens. Bioelectron. *85,* 68–75. https://doi.org/10.1016/j.bios.2016.04.089

Wang, R., Ni, Y., Xu, Y., Jiang, Y., Dong, C., and Chuan, N. (2015b). Immuno-capture and in situ detection of *Salmonella* typhimurium on a novel microfluidic chip. Anal. Chim. Acta *853,* 710–717. https://doi.org/10.1016/j.aca.2014.10.042

Wang, S.G., Singh A.K., Senapati, D., Neely, A., Yu, H.T., and Ray, P.C. (2010). Rapid colorimetric identification and targeted photothermal lysis of *Salmonella* bacteria by using bioconjugated oval-shaped gold nanoparticles. Chem. Eur. J. *16*, 5600–5606. https://doi.org/10.1002/chem.201000176

Wang, Y., and Irudayaraj, J. (2013). Surface-enhanced Raman spectroscopy at single-molecule scale and its implications in biology. Philos. Trans. R. Soc. Lond., B, Biol. Sci. *368*, 20120026. https://doi.org/10.1098/rstb.2012.0026

Wang, Y., Ravindranath, S., and Irudayaraj, J. (2011). Separation and detection of multiple pathogens in a food matrix by magnetic SERS nanoprobes. Anal Bioanal Chem *399*, 1271–1278. https://doi.org/10.1007/s00216-010-4453-6

Wang, Y., Ye, Z., Si, C., and Ying, Y. (2013). Monitoring of *Escherichia coli* O157:H7 in food samples using lectin based surface plasmon resonance biosensor. Food Chem. *136*, 1303–1308. https://doi.org/10.1016/j.foodchem.2012.09.069

Waswa, J., Debroy, C., and Irudayaraj, J. (2006). Rapid detection of *Salmonella* enteritidis and *Escherichia coli* using surface plasmon resonance biosensor. J. Food Process. Eng. *29*, 373–385. https://doi.org/10.1111/j.1745-4530.2006.00071.x

Wee, E.J., Lau, H.Y., Botella, J.R., and Trau, M. (2015). Re-purposing bridging flocculation for on-site, rapid, qualitative DNA detection in resource-poor settings. Chem. Commun. *51*, 5828–5831. https://doi.org/10.1039/c4cc10068a

Wu, W.H., Li, J., Pan, D., Song, S.P., Rong, M.G., Li, Z.X., Gao, J.M., and Lu, J.X. (2014). Gold nanoparticle-based enzyme-linked antibody-aptamer sandwich assay for detection of *Salmonella* Typhimurium. ACS Appl. Mater. Interfaces. *6*, 16974–16981. https://doi.org/10.1021/am5045828

Xin, H., Li, Y., Li, L., Xu, R., and Li, B. (2013). Optofluidic manipulation of *Escherichia coli* in a microfluidic channel using an abruptly tapered optical fiber. Appl. Phys. Lett. *103*, 033703. https://doi.org/10.1063/1.4813905

Xu, X., Chen, Y., Wei, H., Xia, B., Liu, F., and Li, N. (2012). Counting bacteria using functionalized gold nanoparticles as the light-scattering reporter. Anal. Chem. *84*, 9721–9728. https://doi.org/10.1021/ac302471c

Yang, D., Zhou, H., Haisch, C., Niessner, R., and Ying, Y. (2016). Reproducible *E. coli* detection based on label-free SERS and mapping. Talanta *146*, 457–463. https://doi.org/10.1016/j.talanta.2015.09.006

Yilmaz, E., Majidi, D., Ozgur, E., and Denizli, A. (2015). Whole cell imprinting based *Escherichia coli* sensors: a study for SPR and QCM. Sens. Actuator B-Chem. *209*, 714–721. https://doi.org/10.1016/j.snb.2014.12.032

Yu, C., Ganjoo, A., Jain, H., Pantano, C.G., and Irudayaraj, J. (2006). Mid-IR biosensor: detection and fingerprinting of pathogens on gold island functionalized chalcogenide films. Anal. Chem. *78*, 2500–2506. https://doi.org/10.1021/ac051828c

Yu, C., and Irudayaraj, J. (2005). Spectroscopic characterization of microorganisms by fourier transform infrared microspectroscopy. Biopolymers. *77*, 368–377. https://doi.org/10.1002/bip.20247

Yu, C., and Irudayaraj, J. (2006). Identification of pathogenic bacteria in mixed cultures by FTIR spectroscopy. T. ASABE *49*, 1623–1632. https://doi.org/10.13031/2013.22021

Yu, C., Irudayaraj, J., Debroy, C., Schmilovtich, Z.E., and Mizrach, A. (2004). Spectroscopic differentiation and quantification of microorganisms in apple juice. J. Food Sci. *69*, 268–272. https://doi.org/10.1111/j.1365-2621.2004.tb13627.x

Zagorovsky, K., and Chan, W.C. (2013). A plasmonic DNAzyme strategy for point-of-care genetic detection of infectious pathogens. Angew. Chem. Int. Ed. Engl. *52*, 3168–3171. https://doi.org/10.1002/anie.201208715

Zhang, H., Zhang, Y.L., Lin, Y.K., Liang, T.W., Chen, Z.H., Li, J.F., Yue, Z.F., Lv, J.Z., Jiang, Q., and Yi, C.Q. (2015). Ultrasensitive detection and rapid identification of multiple food-borne pathogens with the naked eyes. Biosens. Bioelectron. *71*, 186–193. https://doi.org/10.1016/j.bios.2015.04.034

Zhang, J.Y., Do, J., Premasiri, W.R., Ziegler, L.D., and Klapperich, C.M. (2010). Rapid point-of-care concentration of bacteria in a disposable microfluidic device using meniscus dragging effect. Lab Chip. *10*, 3265–3270. https://doi.org/10.1039/c0lc00051e

Zhang, L., Xu, J., Mi, L., Gong, H., Jiang, S., and Yu, Q. (2012). Multifunctional magnetic–plasmonic nanoparticles for fast concentration and sensitive detection of bacteria using SERS. Biosens. Bioelectron. *31*, 130–136. https://doi.org/10.1016/j.bios.2011.10.006

Zhang, P.P., Liu, X., Wang, C.B., Zhao, Y., Hua, F., Li, C.F., Yang, R.F., and Zhou, L. (2014). Evaluation of up-converting phosphor technology-based lateral flow strips for rapid detection of *Bacillus anthracis* spore, *Brucella* spp., and *Yersinia pestis*. PLOS ONE. *9*, e105305. https://doi.org/10.1371/journal.pone.0105305

Zhou, H., Yang, D., Ivleva, N.P., Mircescu, N.E., Niessner, R., and Haisch, C. (2014). SERS detection of bacteria in water by in situ coating with Ag nanoparticles. Anal. Chem. *86*, 1525–1533. https://doi.org/10.1021/ac402935p

Zhu, H., Sikora, U., and Ozcan, A. (2012). Quantum dot enabled detection of *Escherichia coli* using a cell-phone. Analyst *137*, 2541–2544. https://doi.org/10.1039/c2an35071h

Zuo, P., Li, X., Dominguez, D.C., and Ye, B.C. (2013). A PDMS/paper/glass hybrid microfluidic biochip integrated with aptamer-functionalized graphene oxide nano-biosensors for one-step multiplexed pathogen detection. Lab Chip. *13*, 3921–3928. https://doi.org/10.1039/c3lc50654a

Understanding Pathogenic *Escherichia coli* Through Whole-genome Sequencing

7

Valeria Michelacci[1] and Eelco Franz[2*]

[1] European Reference Laboratory for *E. coli*, Istituto Superiore di Sanità, Rome, Italy.
[2] Centre for Infectious Disease Control, National Institute for Public Health and the Environment, Bilthoven, The Netherlands.

*Correspondence: eelco.franz@rivm.nl

https://doi.org/10.21775/9781910190777.07

Abstract

Escherichia coli is a highly diverse microorganism with very different eco-evolutionary paths, ranging from commensalism to highly virulent intestinal and extra-intestinal pathovars. The recent broad access to whole-genome sequencing technology has expanded the understanding of the genomic basis of this diversity in an unprecedented way, but without undermining the prevailing views on *E. coli* diversity and phylogeny based on older low-resolution typing methods. Genomics indicate that in addition to recombination and mutation the acquisition and loss of genes is a major source for genetic variation in *E. coli*. The relatively small core genome compared with a large pan-genome led to the recognition of continued diversification by gene acquisition. Genomics provides insights into the plasticity of the pathovars concept and led to the current viewpoint of the apparent continuum of pathogenic *E. coli* rather than strictly separated pathovars. The chimaeric and dynamic genome of *E. coli* provides a challenge to the useful application of genomics into public health (i.e. diagnostics, risk assessment, surveillance and disease outbreak investigation). The combined study of the variation of different fractions of the genomes, which are subjected to different mutation rates such as the core and the accessory genome could eventually provide the best way forward to a consensus typing approach.

Introduction

Because of the increasing speed and decreasing operational cost of high-throughput sequencing, comparative genomic analysis of pathogens such as *E. coli* provides a promising public health benefit by increasing the understanding of pathogen ecology and epidemiology (Franz *et al.*, 2016). In addition, whole-genome sequencing (WGS) can provide evolutionary reconstructions of evolutionary relationships and transmission events at a level of resolution, which was not achievable by traditional (molecular) typing methods. In the current post-genomics era we have transitioned from sequencing archetypical isolates from a group of pathogens to sequencing a larger number of samples in an effort to understand the population structure and genomics plasticity of the pathogen associated with human and animal disease (Franz *et al.*, 2014). The wealth and continuously growing numbers of WGS information for different *E. coli* offer insight into the plasticity of the pathovars (Steyert *et al.*, 2012). The first part of this chapter deals with how genomics has provided a means to understand pathogenic *E. coli*.

The most significant advantage of WGS for pathogenic microorganisms is that their typing can be conducted at a much greater resolution than with traditional molecular typing methods, and therefore it has emerged as a powerful tool for public health applications such as surveillance and outbreak investigations. In addition, WGS can offer additional advantages such as the identification of serotypes, virulence markers, and antimicrobial resistance genes. However, the various methodologies by which WGS data can be analysed represents a significant hurdle in standardization and harmonization and in providing epidemiologists and decision makers with interpretable information for action (Franz *et al.*, 2016). In addition, in the second part of this chapter we will discuss the advances made in the area of WGS-based typing of pathogenic *E. coli*, with Shiga toxin-producing *E. coli* (STEC) in particular, and the difficulties involved with the application of different typing methods for useful public health application.

Genomics to understand the continuum of pathogenic *E. coli*

In general, *Escherichia coli* is considered to be roughly clonal with a level of recombination that allows building a robust phylogeny showing distinct phylogenetic groups based on MLEE (Selander *et al.*, 1987), MLST (LeCointre *et al.*, 1998, Reid *et al.*, 2000, Escobar-Páramo *et al.*, 2004, Wirth *et al.*, 2006), and whole-genome analysis (Touchon *et al.*, 2009). There is a huge phenotypic diversity across this phylogenetic tree, enabling a large degree of pathoadaptive niche differentiation. Based on disease potential in humans *E. coli* strains can broadly be classified in harmless commensals, intestinal pathogens, and those capable of infecting extraintestinal sites. Early work already showed that *E. coli* strains could differ considerably in their genome size but it was anticipated that pathogenic strains would differ from commensals by allelic variation of shared genes, different expression profiles, or the presence of a very limited number of virulence genes (Bergthorsson and Ochman, 1998). However, comparative genome scale analysis of the commensal K-12 and pathogenic O157:H7 EDL933 strain showed that over 30% of the genes in the pathogenic strain were unique to that strain as compared to the commensal (Perna *et al.*, 2001). This indicated that in addition to recombination and mutation, the acquisition and loss of genes is a major source for genetic variation in *E. coli*. This was further highlighted by extensive comparative genomics revealing that an individual *E. coli* strain on average has 4721 genes, with a core genome (i.e. genes present in all strains) of around 2000 genes, and a pan-genome (i.e. collection of all *E. coli* genes) of approximately of 18,000 genes (Hendrickson, 2009, Touchon *et al.*, 2009, Rasko *et al.*, 2008). This pan-genome is considered open since there is continuous addition of new genes with every newly sequenced strain (Rasko *et al.*, 2008). Studies comparing the phylogenetic history of the strains to the phylogenetic history of the virulence genes show that both histories are incongruent, which is indicative for multiple parallel acquisition of virulence genes (Reid 2000). However, the arrival of some virulence genes in only some specific clones or lineages of *E. coli* is suggestive for the role of the genetic background of the strain in the acquisition of those virulence factors (Escobar-Páramo *et al.*, 2004). This ongoing evolution by gene acquisition and diversification has large implications for the diversity of *E. coli* as a species and the emergence of pathogens..

An emergent clone of enteroinvasive *E. coli*

Enteroinvasive *E. coli* (EIEC) cause intestinal disease indistinguishable from that caused by *Shigella*. Both pathogens have a characteristic form of pathogenesis involving invasion of mucosal epithelial cells of the large intestine (Van Nhieu and Sansonetti, 2009). In contrast to the other pathogenic groups of *E. coli*, EIEC can be considered as the only non-opportunistic 'professional' pathotype. Early work on the phylogeny of *Shigella* and EIEC based on multi-locus analysis already showed that *Shigella* and EIEC are so closely related that they should be considered as two different steps of the same evolutionary pathway rather than two distinct species (Van Den Beld and Reubsaet, 2012, Michelacci *et al.*, 2016b). Based on genomic analysis a picture is emerging where *Shigella* and EIEC developed through acquisition of common virulence factors in multiple lineages of *E. coli*, including the virulence plasmid (pINV)

(Sahl *et al.*, 2015, Hazen *et al.*, 2016). In addition, *Shigella* and EIEC are known to have accumulated pathoadaptive mutations inactivating genes whose product is thought to be detrimental for survival in the intracellular environment. It was suggested that the strains of *Shigella* were derived from *E. coli* belonging to the ST-280 clonal complex (Wirth *et al.*, 2006) Although EIEC disease is rare in industrialized countries, two large outbreaks in Italy and the United Kingdom were reported in 2012 and 2014 (Escher *et al.*, 2014, Newitt *et al.*, 2016). Characterization of the strains from both outbreaks and a sporadic case from Spain in 2013 revealed that all of them belonged to serotype O96:H19 and ST-99 that does not belong to any known clonal complex and is only distantly related to the ST-280 clonal complex (Michelacci *et al.*, 2016b). In addition, the EIEC outbreak strains were all phylogroup B2 while all EIEC reference strains belong to phylogenetic group E. Interestingly, while the outbreak strains possessed all the virulence genes harboured on the large virulence plasmid they showed none of the pathoadaptive mutations that are characteristic for *Shigella* and EIEC. These genetic observations raised the hypothesis that the outbreak strains are closer to (commensal) *E. coli* than to the traditional *Shigella* and EIEC pathogens (Michelacci *et al.*, 2016b). This was further supported by phenotypic characterization, with the outbreak strains showing some motility (*Shigella* and EIEC are non-motile) and faster growth, which might explain the relation between the size of the outbreaks and the implicated salads as source of infection. Altogether these results suggest that the O96:H19 ST-99 associated outbreak strains emerged as the result of a recent acquisition of the invasion plasmid by a commensal *E. coli* strain (Michelacci *et al.*, 2016b). Based on comparative genomics it was recently concluded that EIEC isolates are closer related to other *E. coli* than to *Shigella* and that EIEC likely arose through the acquisition of the virulence plasmid by permissive recipient strains of multiple distinct lineages of *E. coli* (Hazen *et al.*, 2016). The insights from the O96:H19 ST-99 outbreak shows that emergence of new EIEC-like clones is ongoing and could, at least potentially, further evolve into EIEC and possibly *Shigella* with accumulation of pathoadaptive mutations and loss of motility (Michelacci *et al.*, 2016b).

Increased understanding of evolution and public health risks of attaching and effacing *E. coli* (AEEC)

Shiga toxin-producing *E. coli* (STEC) are a diverse group of enteric pathogens capable of causing severe gastrointestinal disease (haemorrhagic colitis), acute kidney failure (haemolytic uraemic syndrome: HUS), and chronic post-infection sequelae (e.g. irritable bowel syndrome and end-stage renal disease) or death (Tarr *et al.*, 2005). While large food-borne outbreaks and significant numbers of sporadic cases caused by STEC serotype O157:H7 have been reported since the early 1980s, infections and outbreaks caused by other serotypes (e.g. non-O157 STEC) with various combinations of (putative) virulence traits (i.e. virulence profiles) are increasingly recognized (Mathusa *et al.*, 2010, Gould *et al.*, 2013). Some of these non-O157 STEC strains cause outbreaks and severe disease, whereas others are associated with only mild sequelae or with no human disease at all. This observation resulted in the development of the STEC seropathotype (SPT) classification, which is based on the serotype association with human epidemics and HUS (Karmali *et al.*, 2003). Serotypes

responsible for haemorrhagic colitis (HC) and haemolytic uraemic syndrome (HUS), O157:H7 and O157:NM, were assigned to seropathotype A. Seropathotype B strains (O26:H11, O103:H2, O111:NM, O121:H19 and O145:NM) have a strong association with outbreaks and HUS, but less commonly than those of seropathotype A. Seropathotype C serotypes (O91:H21, O104:H21, O113:H21, O5:NM, O121:NM and O165:H25) are associated with sporadic HUS cases but not with epidemics. Seropathotype D serotypes are associated with diarrhoea but not with HUS and/or outbreaks. Seropathotype E serotypes comprise STEC that had never been associated with human disease and had been isolated only from animals. However, the seropathotype classification does not define pathogenic STEC nor does it provide an exhaustive list of pathogenic serotypes (Franz *et al.*, 2015). This relates primarily to the complexity of designating individual strains as pathogens due to the large variance in serotype and genetic content associated with (severe) disease. To classify the clinical and/or epidemic potential of a STEC isolate at an early stage it is crucial to identify virulence profiles of the strains of concern.

Despite considerable efforts, the mechanisms and evolutionary forces underlying the evolution and emergence of new STEC types are not well understood. Phylogenetic analysis revealed that the gain and loss of virulence elements, including the *locus for enterocyte effacement* (LEE) pathogenicity island, the plasmid-borne haemolysin, and phage-encoded Shiga toxins, has occurred several times, both in parallel and in separate lineages (Reid *et al.*, 2000). This convergent evolution suggests a selective advantage associated with the build-up of specific combinations of virulence factors. The LEE island harbours the genetic determinants responsible for the production of a typical histopathological lesion to the enterocyte termed 'attaching and effacing' (A/E), characterized by the effacement of the microvilli brush border and the presence of microfilamentous structures, known as pedestals, protruding from the cell surface and intimately accommodating the adhered bacteria (McDaniel and Kaper, 1997). LEE-positive strains are designated as attaching and effacing *E. coli* (AEEC), which in turn comprises typical and atypical enteropathogenic *E. coli* (EPEC), differentiating from the presence or absence of the gene encoding the bundle-forming pili, respectively, and, when combined with the presence of Shiga toxin genes, enterohaemorrhagic *E. coli* (EHEC). As a result, EHEC is a subgroup of STEC since the latter are defined only by the presence of Shiga toxin (Stx)-encoding genes. Several studies showed a strong correlation between the presence of LEE and other virulence markers like the occurrence of correlated *terB, toxB, etpD, adfO, ckf, efa1*, and almost all the so-called non-LEE-encoded effectors(*nle*) genes (Ju *et al.*, 2013, Franz *et al.*, 2015). This strong correlation between virulence factors indicates that these arrays are not a random collection but are rather specific sets (Franz *et al.*, 2015).

Comparative genomics consistently shows a high phylogenetic diversity of (LEE-positive and negative) STEC (Steyert *et al.*, 2012, Hazen *et al.*, 2016, Haugum *et al.*, 2014, Ferdous *et al.*, 2016a). Genomes of AEEC are represented in each of the *E. coli* phylogenetic groups (A, B1, B2, D, E, and F) (Hazen *et al.*, 2013). It appears that EPEC and EHEC lineages emerged in multiple lineages following the acquisition of mobile genetic elements (such as Stx converting phage and the EAF plasmid) and AEEC seem not to share commonalities in intimin-types, Stx-types and serotypes (Hazen *et al.*, 2013). AEEC and STEC isolates are scattered throughout the phylogenetic tree and with no correlation between serogenotype, MLST type and disease outcome (Ferdous *et al.*, 2016a; Haugum *et al.*, 2014). Because of the large heterogeneity of STEC and the role of mobile genetic elements, the interpretation of evolutionary origins of STEC based on WGS data should be done with caution. The heterogeneity in phylogenetic background of STEC also highlights the many opportunities for new STEC to evolve and emerge as potential public health threat.

Hybrid shigatoxigenic *E. coli*

Stx-carrying enteroaggregative *E. coli*

A particularly virulent Stx-producing clone emerged in 2011 as the causative agent of a large outbreak of severe infections across Europe, originating in Germany and France after the consumption of contaminated fenugreek sprouted seeds imported from Egypt (Werber *et al.*, 2012). The episode involved a total of 4,033 cases, comprising 901 HUS and 50 deaths (Karch *et al.*, 2012). The strain belonged to the unusual serotype O104:H4 and showed a rare combination of virulence genes, combining the ability to produce Stx2 with features typical of enteroaggregative *E. coli* (EAggEC). This gave origin to discussion about the correct nomenclature to be used to describe such a hybrid pathotype, with the two main proposals Stx-EAggEC and EAHEC (enteroaggregative haemorrhagic *E. coli*). The former will be used throughout this chapter. The genome of the outbreak strain was sequenced within two days from the report of the outbreak and showed that the outbreak strain was highly similar to the reference EAggEC O104:H4 strain 55989, with the major difference of the integration of a *stx2*-bacteriophage in *wrbA* insertion site (Rohde *et al.*, 2011). The retrospective revision of the literature and the analysis of the strain collections highlighted that in the time period 1992–2011 several other sporadic cases of HUS had already been caused by Stx-EAggEC strains and some others occurred in the 2 years immediately after the outbreak, all associated to five different serotypes: O111:H2, O86:HNM, O104:H4, O111:H21 and

O127:H4 (Scheutz *et al.*, 2011; Brzuszkiewicz *et al.*, 2011; Morabito *et al.*, 1998; Iyoda *et al.*, 2000; Scavia *et al.*, 2011; Dallman *et al.*, 2012; Tozzoli *et al.*, 2014). WGS was used to investigate the whole genome of Stx-EAggEC strains and of the bacteriophages harbouring the *stx* genes in such isolates, in order to shed light on the emergence of this newly described pathotype.

The analysis of the whole-genome sequence of the O104:H4 strain responsible for the German outbreak highlighted that it belonged to ST-678 and also harboured a plasmid encoding the genetic determinants for the resistance to expanded-spectrum beta-lactamases (ESBL), in addition to the *stx*-bacteriophage and the virulence determinants typical of EAggEC pathotype (Rohde *et al.*, 2011). The whole-genome sequences of three O104:H4 Stx-EAggEC strains isolated in 2013 in the Netherlands from two patients with a recent story of travel to Turkey were compared with those of the strain responsible for the German outbreak, and two other Stx-EAggEC O104:H4 strains isolated in 2009 in the Republic of Georgia. All the strains were typed as ST-678 and only the strains isolated from patients from the German outbreak of 2011 and one of those isolated from the Dutch patients harboured the ESBL plasmid, which could have been acquired and lost through horizontal gene transfer (Ferdous *et al.*, 2015). The core-genome phylogenetic analysis performed through core genome MLST allowed the hypothesis that Stx-EAggEC O104:H4 strains highly similar (27 alleles difference) to the 2011 outbreak clone were still circulating two years later (Ferdous *et al.*, 2015). This was confirmed by another comprehensive analysis of Stx-EAggEC O104:H4, also including strains isolated from the Republic of Georgia and Tunisia, which highlighted the existence of different clades of such pathogen, some of which deriving from a recent common ancestor shared by the outbreak clone (Zhou *et al.*, 2015). Moreover, differences in the content of mobile genetic elements, in particular of those involved in virulence such as the *stx2*-phages and the plasmids encoding AAF fimbriae, allowed hypothesizing a mosaic structure for the genome of Stx-EAggEC O104:H4 as an effect of the constraints exerted by multiple ecological niches (Zhou *et al.*, 2015).

Next generation sequencing (NGS) techniques proved useful for investigating the emergence of these hybrid strains also through their application in the study of *stx*-encoding bacteriophages. The genome of the *stx2*-phage integrated in the chromosome of the O104:H4 strain responsible for the German outbreak was shown to be nearly identical to that harbouring *stx2* in the in the genome of the O111:H2 strain responsible for an outbreak that occurred in France in 1992 (Grande *et al.*, 2014). Moreover, highly similar bacteriophages were shown to be responsible for Stx2 production in all the other O104:H4 Stx-EAggEC strains described, as well as in a Stx-EAggEC O127:H4 strain isolated in Italy in 2013 (Tozzoli *et al.*, 2014). On the other side, the sequence of the *stx2*-phage present in a Stx-EAggEC O111:H21 strain isolated from a small outbreak of HUS occurred in 2012 in Northern Ireland was shown to be largely different, with the only exception of the *stx2* genes and a portion of a gene encoding a tail fibre, which was hypothesized to play a role in the ability of such phages to infect EAggEC strains (Grande *et al.*, 2014). The finding of a conserved *stx2*-phage that infected different EAggEC serogroups over a 20-year time span, in spite of the high variability described for *stx*-phages (Smith *et al.*, 2012) allowed the formulation of the hypothesis that such *stx2*-phages are kept under a strong selective pressure (Grande *et al.*, 2014).

Stx-carrying enterotoxigenic *E. coli*

A similar *stx2*-bacteriophage was described to be present in the genome of two enterotoxigenic *E. coli* strains isolated from a case of HUS and one of diarrhoea in Finland (Nyholm *et al.*, 2015). The phage, which was described as being similar to the P13374 phage of the German outbreak strain, was integrated in both strains in the locus *wrbA* as in the O104:H4 EAggEC strains (Nyholm *et al.*, 2015). It is interesting to note that the ETEC strain isolated from the HUS case was also positive for the presence of the LEE locus and the plasmid-borne enterohaemolysin-coding gene, suggesting that the emergence of such a hybrid strain could have been due to the acquisition of the plasmid or of the plasmid-associated Heat Stable enterotoxin coding genes by a typical attaching and effacing STEC strain. Such a scenario could explain the severity of the symptoms induced in the patient.

Other two Stx-ETEC have been analysed in the attempt to identify the sequence of the *stx*-encoding bacteriophage, one from an uncomplicated diarrhoea case occurred in Denmark (Tozzoli *et al.*, 2014) and one isolated from bovine faeces in Finland (Nyholm *et al.*, 2015). In the mentioned studies the complete sequences of the *stx*-phages could not be determined or match those previously identified (Tozzoli *et al.*, 2014, Nyholm *et al.*, 2015). The strain from human diarrhoea was positive for the presence of the *stx2* genes and *elt* genes while the bovine isolate displayed the presence of *stx1* and *est* genes, suggesting a completely different genetic background.

A recent study compared 24 whole-genome sequences from hybrid Stx-ETEC strains from the sequences public archives and identified the existence of different clones scattered among the different *E. coli* phylogroups and displaying different combinations of virulence genes in different allelic forms, isolated from environmental, animal, and human clinical samples (Leonard *et al.*, 2016). The analysis of the plasmid sequences found in the Stx-ETEC strains also revealed extensive mosaicism. The sequences of the virulence plasmid identified in one isolate from cattle in Hong Kong showed the presence of two identical copies of the *est* gene in addition to the gene encoding the K88 colonization factor, typical of ETEC strains. In addition, the same plasmid carried the *katP*, *espP* and *ehx* genes (Burland *et al.*, 1998, Makino *et al.*, 1998) typically encoded on a virulence plasmid similar to those observed in STEC O157 and other highly pathogenic STEC (Losada *et al.*, 2016). A similar plasmid has been identified in other Stx-ETEC strains, sometimes lacking only the virulence genes typical of STEC (Leonard *et al.*, 2016). These findings highlight the importance of plasmid exchange and recombination in the emergence of hybrid strains featuring new combinations of virulence genes.

Stx2f-carrying *E. coli*

Other emerging STEC, investigated thanks to the availability of genomics technologies, are those producing a peculiar subtype of Shiga toxin, named Stx2f. The Stx2f subtype of Shiga-toxin was described in the late 1990s as being produced by *E. coli* strains circulating in birds, especially pigeons (Dell'Omo *et al.*, 1998) but was only associated with human cases of mild disease after 2003 (Prager *et al.*, 2009; Friesema *et al.*, 2014; Jenkins *et al.*, 2003). In the last few years four cases of haemolytic uraemic syndrome have been associated with infections with Stx2f-producing *E. coli* (Friesema *et al.*, 2015; Grande *et al.*, 2016). A study compared the whole genome of three *stx2f*-positive strains isolated from HUS and sets of isolates from diarrhoea cases and from pigeons (Grande *et al.*, 2016). This analysis allowed observing that the strains from the animal reservoir and from human diarrhoea shared a similar virulence genes profile, with the presence of the LEE locus and the absence of other important virulence associated MGEs as the OI-122 (Morabito *et al.*, 2003) and the large virulence plasmid of STEC (Makino *et al.*, 1998; Burland *et al.*, 1998). On the other hand, the strains from HUS showed the complete repertoire of virulence determinants additional to the Stx typically displayed by highly pathogenic STEC and atypical EPEC (aEPEC), which are thought to be derived from highly pathogenic STEC strains following the loss of *stx*-phages (Trabulsi *et al.*, 2002). These observations led to the conclusion that the population of STEC producing Stx2f present in pigeons may be directly associated with those causing human diarrhoea, identifying the pigeon as the natural reservoir of zoonotic STEC. In addition, it was observed that the strains from HUS, which harboured very similar *stx2f*-phages, but in a different genomic background, may have emerged following an event of phage acquisition by an *E. coli* with an aEPEC genomic scaffold (Grande *et al.*, 2016).

These examples show that WGS provides an effective mean for identifying rare combinations of virulence genes in strains isolated in routine diagnostics and allowed formulating hypothesis about the evolution of such pathogens. As a matter of fact, in the last few years WGS technologies have already allowed to identify the circulation of Stx-bacteriophages having broad spectra of infection, which have been shown to be kept under strong selective pressure and maintained in certain strains with a high conservation level after decades, representing a major finding to be kept in consideration for formulating a hypothesis on the emergence and circulation of hybrid STEC strains.

WGS-based typing of STEC for surveillance and outbreak investigation

WGS allows linked cases to be identified with unprecedented specificity and sensitivity that will facilitate targeted and appropriate public health investigations (Dallman *et al.*, 2015b). Because of the increasing speed and decreasing operational and acquisition cost of high-throughput sequencing, comparative genomic analysis of food-borne pathogens is increasingly integrated into surveillance and outbreak investigations (Franz *et al.*, 2016). It is expected that WGS will eventually become the sole standard method for genotyping pathogens like *E. coli* for public health purposes. WGS can be used in real time to provide the highest strain-level resolution for outbreak investigation. Moreover, WGS provides a reliable and robust one-step process for characterization of STEC since different types of information can be retrieved from the WGS data, including serotyping, virulence profiling, and antimicrobial resistance profiles. This enables the prediction of clinical outcome and monitoring of hypervirulent strains emergence (Chattaway *et al.*, 2016).

SNP- and allele-based typing

Despite the great potential demonstrated for WGS bacterial typing, the variety of methods by which WGS data can be analysed represents a hurdle in standardization and harmonization and in providing epidemiologists and decision makers with interpretable information for action. The harmonization of the typing methodologies and the use of a common nomenclature are crucial aspects for surveillance and outbreak investigation in order to obtain comparable results and to enable proper communication. Applying this concept to whole-genome sequencing is not an easy task. As the use of NGS technologies arose and spread during the last decade, the procedures for WGS data production and analysis have not been optimized and harmonized yet, hindering the complete shift towards the use of WGS in replacement of currently used standardized methodologies such as pulsed field gel electrophoresis and multilocus VNTR (variable number of tandem repeats) analysis.

Part of the current discussion about the implementation of WGS for surveillance and outbreak investigation is focused on the choice of the most appropriate analytical pipeline for deriving the phylogeny of a batch of test isolates, allowing inference of clonal correlation. In detail, discussion is ongoing about the advantages and disadvantages of whole genome single nucleotide polymorphisms (SNPs) pipelines and MLST schemes based on extended panels of genes, in view of their possible use for global surveillance and outbreak investigation programmes.

The first WGS-based typing approaches developed for comparing bacterial isolates were based on the study of the SNPs identified in the whole genome of test isolates when compared with the sequence of a reference isolate of the same species (Grad *et al.*, 2012; Underwood *et al.*, 2013). These approaches allowed the determination of phylogeny for all those bacterial sequences for which choosing an appropriate reference sequence was possible, including during epidemiological investigations of STEC infection (Dallman *et al.*, 2015b; Holmes *et al.*, 2015). However, the choice of a reference sequence affects the quality of typing data obtained by limiting the number of nucleotide positions analysed to those shared among the sequences of the test strains and the reference sequence. This choice could be particularly difficult for *E. coli*, which displays a high genetic diversity.

In addition to the reference-based whole genome SNPs analysis, other strategies are based on pipelines, which compare the presence of SNPs in short stretches of sequences (kmers) (Gardner *et al.*, 2015). Such an approach would allow comparing isolates' profiles without the need of selecting a reference sequence but at the same time it would introduce issues related with the reproducibility of the analysis. As a matter of fact, the addition of new test sequences to a previously analysed batch of sequences would change the topology of the resulting phylogenetic dendrogram.

Further to the difficulties in selecting the typing strategy, also defining a common nomenclature for the outputs generated by whole genome SNPs analyses is not straightforward. At present, the only proposal for a SNPs-based nomenclature is the 'SNP address', which is based on a hierarchical single linkage clustering on the pairwise SNP difference among the test strains at various distance thresholds (Δ250, Δ100, Δ50, Δ25, Δ10, Δ5, Δ0), resulting in unique keys assigned to the SNPs profiles (Dallman *et al.*, 2015a).

To develop a harmonized nomenclature for WGS typing data might be much easier for MLST-based analyses. The concept of a fixed loci scheme and allele nomenclature such as that employed in the seven-genes MLST schemes currently used for many bacteria (Maiden *et al.*, 1998; Wirth *et al.*, 2006) could be expanded to a wider panel of genes. Read mapping-based tools able to automatically detect genes, alleles and multi-locus sequence types from WGS are already available (Inouye *et al.*, 2014; Jünemann *et al.*, 2013) and proposals for typing schemes are being developed for several bacterial pathogens such as *Enterococcus faecium* (de Been *et al.*, 2015), *Legionella pneumophila* (Moran-Gilad *et al.*, 2015), *Listeria monocytogenes* (Ruppitsch *et al.*, 2015) and *Mycobacterium tuberculosis* (Kohl *et al.*, 2014). However, for these to be used in surveillance programmes, an agreed standard scheme and nomenclature has to be agreed upon and used. The majority of the studies using MLST-based approaches identify the allelic signatures of the genes composing the pathogen's core genome, which are defined as those shared among all the test isolates. This method is often referred to as cgMLST (Jünemann *et al.*, 2013), but it needs to be highlighted that in this case the 'core' genome does not represent a real core shared by all the strains representative of the species, but rather the genetic fraction shared by all the test strains, and for this reason it is sometimes also termed 'shared genome' (ECDC, 2015).

Several studies demonstrated the efficacy of such approaches to study the evolution of pathogenic *E. coli*. However, as in the case of SNPs-based typing strategies, not only the same algorithm and parameters must be applied, but also the same schemes and nomenclature for loci and alleles must be used in order to grant the comparability of the results obtained with different panels of strains and analysed by different laboratories (ECDC, 2015).

The main effort in the next few years can be foreseen to be represented by compiling a shared typing scheme, including the choice of the loci to be analysed and the definition of the different alleles for each locus. In particular, the selection of the loci under analysis still needs extensive discussion considering the large variability of the *E. coli* species, for which different portions of the genome could be exploited for different typing purposes. As a matter of fact, the core genome has been demonstrated to be useful for phylogenetic studies (Ferdous *et al.*, 2015, 2016b; Mellmann *et al.*, 2011), even though an agreed common scheme for cgMLST has not been constructed yet.

Core and accessory genome typing of STEC: different approaches for different questions

The variation of the accessory genome has proven to be sufficiently usable to subtype STEC strains at a deeper level than serotyping. In fact, accessory genome typing provided a better understanding of the eco-epidemiological relations between *stx2f* carrying isolates from different sources, as mentioned earlier in this chapter (Grande, 2016). Here, core-genome SNP analysis revealed that *stx2f*-carrying isolates from pigeons were more closely related to *stx2f* carrying HUS-related isolates and non-*stx2f* clinical isolates (including HUS) than to *stxf2* isolates from human cases of mild diarrhoea (Grande, 2016). However, clustering based on virulence profile showed that isolates from pigeons and mild human disease clustered more closely together, while the *stx2f* HUS-related isolates clustered more with the non-*stx2f* human clinical isolates (Grande 2016). Thus, in the case of STEC at least, the phylogeny may not reflect the present-day eco-epidemiological relations well enough in order to draw conclusions on source attribution. In addition, it suggests that the severity of the symptoms induced by infection may depend more on the ability to achieve a proficient colonization of the host gut mucosa rather than on the subtype of the produced toxin. A recent study analysed the allelic variation of 91 virulence genes of STEC, harboured on the three main pathogenicity islands, such as the LEE, the OI-122 and the OI-57 (Michelacci *et al.*, 2016a). The study was performed by real-time PCR followed by melting curves analysis, thus reducing the resolution power for the allelic determination to that of a sequence variation causing a significant shift of the melting temperature (Michelacci *et al.*, 2016a). Nevertheless, a number of alleles ranging from 2 to 10 were identified for each of the loci analysed and the allelic combination of the 91 loci has been used to define a unique allelic signature for each strain tested. Such signatures, used for cluster analyses, showed that the dendrograms identified groups of STEC strains segregating with their serogroup, but allowing at the same time the identification of subpopulations within the single serogroups and reaching a similar discriminatory power to that achieved through whole-genome SNPs analysis (Michelacci *et al.*, 2016a). This study could be considered as a proof of concept, as the same approach could be translated in tools extracting the allelic information on the same genes from WGS data, and performing an MLST-like analysis on an extended panel of highly variable genes. Such a typing scheme, when validated for the ability to cluster epidemiologically related isolates, would increase the possibilities to adopt a WGS-based tool for enhanced surveillance and monitoring of STEC. Additionally, such an 'accessory genome MLST' (agMLST), could take advantage of the available information on the STEC virulence genes already collected into a public database (Joensen *et al.*, 2014). A beta version of such a tool, using the mentioned database, is currently available for testing through the ARIES Galaxy curated and maintained by the European Reference Laboratory for *E. coli* (https://w3.iss.it/site/aries/).

The simultaneous study of the variation of different fractions of the genomes subjected to different mutations rates, such as the core and the accessory genome, could provide the best fitting approach, made up of high sensitivity and depth of investigation, to respond the request for a high discriminatory power and phylogenetic perspective needed for the study of rapidly evolving pathogenic bacteria, such as *E. coli*. The interplay among all the public health institutions involved in surveillance and outbreak investigations will be crucial to define a common approach to make the best possible use of NGS technologies in the fight against infectious disease.

References

Bergthorsson, U., and Ochman, H. (1998). Distribution of chromosome length variation in natural isolates of *Escherichia coli*. Mol. Biol. Evol. *15*, 6–16.

Brzuszkiewicz, E., Thurmer, A., Schuldes, J., Leimbach, A., Liesegang, H., Meyer, F.D., Boelter, J., Petersen, H., Gottschalk, G., and Daniel, R. (2011). Genome sequence analyses of two isolates from the recent *Escherichia coli* outbreak in Germany reveal the emergence of a new pathotype: entero-aggregative-haemorrhagic *Escherichia coli* (EAHEC). Arch. Microbiol. *193*, 883–891.

Burland, V., Shao, Y., Perna, N.T., Plunkett, G., Sofia, H.J., and Blattner, F.R. (1998). The complete DNA sequence and analysis of the large virulence plasmid of *Escherichia coli* O157:H7. Nucleic Acids Res. *26*, 4196–4204.

Chattaway, M.A., Dallman, T.J., Gentle, A., Wright, M.J., Long, S.E., Ashton, P.M., Perry, N.T., and Jenkins, C. (2016). Whole genome sequencing for public health surveillance of Shiga toxin-producing *Escherichia coli* other than serogroup O157. *Frontiers Microbiol.* 7, 258.

Dallman, T., Ashton, P.M., Byrne, L., Perry, N., Petrovska, L., Ellis, R.J., Allison, L., Hanson, M., Holmes, A., Gunn, G., Chase-Topping, M., Woolhouse, M., Grant, K., Gally, D., Wain, J., and Jenkins, C. (2015a). Applying phylogenomics to understand the emergence of Shiga-toxin-producing *Escherichia coli* O157:H7 strains causing severe human disease in the United Kingdom. Microb. Genom. *1*(3): e000029.

Dallman, T., Smith, G.P., O'brien, B., Chattaway, M.A., Finlay, D., Grant, K.A., and Jenkins, C. (2012). Characterization of a verocytotoxin-producing enteroaggregative *Escherichia coli* serogroup O111:H21 strain associated with a household outbreak in Northern Ireland. J. Clin. Microbiol. *50*, 4116–4119.

Dallman, T.J., Byrne, L., Ashton, P.M., Cowley, L.A., Perry, N.T., Adak, G., Petrovska, L., Ellis, R.J., Elson, R., Underwood, A., Green, J., Hanage, W.P., Jenkins, C., Grant, K., and Wain, J. 2015b. Whole-genome sequencing for national surveillance of Shiga toxin-producing *Escherichia coli* O157. Clin. Infect Dis. *61*, 305-312.

De Been, M., Pinholt, M., Top, J., Bletz, S., Mellmann, A., van Schaik, W., Brouwer, E., Rogers, M., Kraat, Y., Bonten, M., Corander, J., Westh, H., Harmsen, D., and Willems, R.J. (2015). Core genome multilocus sequence typing scheme for high-resolution typing of *Enterococcus faecium*. J. Clin Microbiol. *53*, 3788–3797.

Dell'Omo, G., Morabito, S., Quondam, R., Agrimi, U., Ciuchini, F., Macri, A., and Caprioli, A. (1998). Feral pigeons as a source of verocytotoxin-producing Escherichia *coli*. Vet Rec. *142*, 309–310.

ECDC (European Centre for Disease Prevention and Control). (2015). Expert Opinion on the introduction of next-generation typing methods for food- and waterborne diseases in the EU and EEA. Stockholm.

Escher, M., Scavia, G., Morabito, S., Tozzoli, R., Maugliani, A., Cantoni, S., Fracchia, S., Bettati, A., Casa, R., Gesu, G.P., Torresani, E., and Caprioli, A. (2014). A severe food-borne outbreak of diarrhoea linked to a canteen in Italy caused by enteroinvasive *Escherichia coli*, an uncommon agent. Epidemiol. Infect. *142*, 2559–2566.

Escobar-Páramo, P., Clermont, O., Blanc-Potard, A.B., Bui, H., Le Bouguénec, C., and Denamur, E. (2004). A specific genetic background is required for acquisition and expression of virulence factors in *Escherichia coli*. Mol. Biol. Evol. *21*, 1085–1094. https://doi.org/10.1093/molbev/msh118

Ferdous, M., Friedrich, A.W., Grundmann, H., De Boer, R.F., Croughs, P D., Islam, M.A., Kluytmans-Van Den Bergh, M.F.Q., Kooistra-Smid, A.M.D., and Rossen, J.W.A. (2016a). Molecular characterization and phylogeny of Shiga toxin–producing *Escherichia coli* isolates obtained from two Dutch regions using whole genome sequencing. Clin. Microbiol. Infect. *22*, 642.e1-642.e9.

Ferdous, M., Kooistra-Smid, A.M., Zhou, K., Rossen, J.W., and Friedrich, A.W. (2016b). Virulence, antimicrobial resistance properties and phylogenetic background of non-H7 enteropathogenic *Escherichia coli* O157. Front. Microbiol. 7, 1540.

Ferdous, M., Zhou, K., de Boer, R.F., Friedrich, A.W., Kooistra-Smid, A.M., and Rossen, J.W. (2015). Comprehensive characterization of *Escherichia coli* O104:H4 isolated from patients in the Netherlands. Front. Microbio.l *6*, 1348. https://doi.org/10.3389/fmicb.2015.01348

Franz, E., Delaquis, P., Morabito, S., Beutin, L., Gobius, K., Rasko, D.A., Bono, J., French, N., Osek, J., Lindstedt, B.A., Muniesa, M., Manning, S., Lejeune, J., Callaway, T., Beatson, S., Eppinger, M., Dallman, T., Forbes, K.J., Aarts, H., Pearl, D.L., Gannon, V.P.J., Laing, C.R., and Strachan, N.J.C. (2014). Exploiting the explosion of information associated with whole genome sequencing to tackle Shiga toxin-producing *Escherichia coli* (STEC) in global food production systems. Int. J. Food Microbiol. *187*, 57–72.

Franz, E., Gras, L.M., and Dallman, T. (2016). Significance of whole genome sequencing for surveillance, source attribution and microbial risk assessment of food-borne pathogens. Curr. Opin. Food Sci. *8*, 74-79.

Franz, E., van Hoek, A.H., Wuite, M., van der Wal, F.J., de Boer, A.G., Bouw, E.I., and Aarts, H.J. (2015). Molecular hazard identification of non-O157 Shiga toxin-producing *Escherichia coli* (STEC). PLOS ONE *10*, e0120353. https://doi.org/10.1371/journal.pone.0120353

Friesema, I., van der Zwaluw, K., Schuurman, T., Kooistra-Smid, M., Franz, E., van Duynhoven, Y., and van Pelt, W. (2014). Emergence of *Escherichia coli* encoding Shiga toxin 2f in human Shiga toxin-producing *E. coli* (STEC) infections in the Netherlands, January 2008 to December 2011. Euro Surveill. *19*, 26–32.

Friesema, I.H., Keijzer-Veen, M.G., Koppejan, M., Schipper, H.S., van Griethuysen, A.J., Heck, M.E., and van Pelt, W. (2015). Hemolytic uremic syndrome associated with *Escherichia coli* O8:H19 and Shiga toxin 2f gene. Emerging Infect. Dis. *21*, 168–169. https://doi.org/10.3201/eid2101.140515

Gardner, S.N., Slezak, T., and Hall, B.G. (2015). kSNP3.0: SNP detection and phylogenetic analysis of genomes without genome alignment or reference genome. Bioinformatics. *31*, 2877–2878.

Gould, L.H., Mody, R.K., Ong, K.L., Clogher, P., Cronquist, A.B., Garman, K.N., Lathrop, S., Medus, C., Spina, N.L., Webb, T.H., White, P.L., Wymore, K., Gierke, R E., Mahon, B.E., and Griffin, P.M. (2013). Increased recognition of non-O157 Shiga toxin-producing *Escherichia coli* infections in the United States during 2000-2010: epidemiologic features and comparison with *E. coli* O157 infections. Food-borne Pathog. Dis. *10*, 453–460.

Grad, Y.H., Lipsitch, M., Feldgarden, M., Arachchi, H.M., Cerqueira, G.C., Fitzgerald, M., Godfrey, P., Haas, B.J., Murphy, C.I., Russ, C., *et al.* (2012). Genomic epidemiology of the *Escherichia coli* O104:H4 outbreaks in Europe, 2011. Proc. Natl. Acad. Sci. U.S.A. *109*, 3065–3070. https://doi.org/10.1073/pnas.1121491109

Grande, L., Michelacci, V., Bondi, R., Gigliucci, F., Franz, E., Badouei, M.A., Schlager, S., Minelli, F., Tozzoli, R., Caprioli, A., and Morabito, S. (2016). Whole-genome characterization and strain comparison of VT2f-Producing *Escherichia coli* causing hemolytic uremic syndrome. *Emerg Infect Dis*. *22*, 2078–2085.

Grande, L., Michelacci, V., Tozzoli, R., Ranieri, P., Maugliani, A., Caprioli, A., and Morabito, S. (2014). Whole genome sequence comparison of vtx2-converting phages from enteroaggregative haemorrhagic *Escherichia coli* strains. BMC Genomics. *15*, 574.

Haugum, K., Johansen, J., Gabrielsen, C., Brandal, L.T., Bergh, K., Ussery, D.W., Drabløs, F., and Afset, J.E. (2014). Comparative genomics to delineate pathogenic potential in non-O157 Shiga toxin-producing *Escherichia coli* (STEC) from patients with and without haemolytic uremic syndrome (HUS) in Norway. PLOS ONE. 9(10):e111788.

Hazen, T.H., Leonard, S.R., Lampel, K.A., Lacher, D.W., Maurelli, A.T., and Raskoa, D.A. (2016). Investigating the relatedness of enteroinvasive *Escherichia coli* to Other *E. coli* and *Shigella* isolates by using comparative genomics. Infect. Immun. *84*, 2362–2371.

Hazen, T.H., Sahl, J.W., Fraser, C.M., Donnenberg, M.S., Scheutz, F., and Rasko, D.A. (2013). Refining the pathovar paradigm via phylogenomics of the attaching and effacing *Escherichia coli*. Proc. Natl. Acad. Sci. U.S.A. *110*, 12810–12815. https://doi.org/10.1073/pnas.1306836110

Hendrickson, H. (2009). Order and disorder during *Escherichia coli* divergence. PLOS Genet. *5*, e1000335. https://doi.org/10.1371/journal.pgen.1000335

Holmes, A., Allison, L., Ward, M., Dallman, T.J., Clark, R., Fawkes, A., Murphy, L., and Hanson, M. (2015). Utility of whole-genome sequencing of *Escherichia coli* O157 for outbreak detection and epidemiological surveillance. J. Clin. Microbiol. *53*, 3565–3573. https://doi.org/10.1128/JCM.01066-15

Inouye, M., Dashnow, H., Raven, L.A., Schultz, M.B., Pope, B.J., Tomita, T., Zobel, J., and Holt, K.E. (2014). SRST2: Rapid genomic surveillance for public health and hospital microbiology labs. Genome Med. *6*, 90. https://doi.org/10.1186/s13073-014-0090-6

Iyoda, S., Tamura, K., Itoh, K., Izumiya, H., Ueno, N., Nagata, K., Togo, M., Terajima, J., and Watanabe, H. (2000). Inducible stx2 phages are lysogenized in the enteroaggregative and other phenotypic *Escherichia coli* O86:HNM isolated from patients. FEMS Microbiol. Lett. *191*, 7–10.

Jenkins, C., Willshaw, G.A., Evans, J., Cheasty, T., Chart, H., Shaw, D.J., Dougan, G., Frankel, G., and Smith, H.R. (2003). Subtyping of virulence genes in verocytotoxin-producing *Escherichia coli* (VTEC) other than serogroup O157 associated with disease in the United Kingdom. J. Med. Microbiol. 52, 941–947.

Joensen, K.G., Scheutz, F., Lund, O., Hasman, H., Kaas, R.S., Nielsen, E.M., and Aarestrup, F.M. (2014). Real-time whole-genome sequencing for routine typing, surveillance, and outbreak detection of verotoxigenic *Escherichia coli*. J. Clin. Microbiol. *52*, 1501–1510.

Ju, W., Shen, J., Toro, M., Zhao, S., and Menga, J. (2013). Distribution of pathogenicity islands OI-122, OI-43/48, and OI-57 and a high-pathogenicity island in Shiga toxin-producing *Escherichia coli*. Appl. Environ. Microbiol. 79, 3406-3412.

Jünemann, S., Sedlazeck, F.J., Prior, K., Albersmeier, A., John, U., Kalinowski, J., Mellmann, A., Goesmann, A., von Haeseler, A., Stoye, J., *et al.* (2013). Updating benchtop sequencing performance comparison. Nat. Biotechnol. *31*, 294–296. https://doi.org/10.1038/nbt.2522

Karch, H., Denamur, E., Dobrindt, U., Finlay, B.B., Hengge, R., Johannes, L., Ron, E.Z., Tønjum, T., Sansonetti, P.J., and Vicente, M. (2012). The enemy within us: lessons from the 2011 European *Escherichia coli* O104:H4 outbreak. EMBO Mol Med *4*, 841–848. https://doi.org/10.1002/emmm.201201662

Karmali, M. A., Mascarenhas, M., Shen, S., Ziebell, K., Johnson, S., Reid-Smith, R., Isaac-Renton, J., Clark, C., Rahn, K., and Kaper, J.B. 2003. Association of genomic O Island 122 of *Escherichia coli* EDL 933 with verocytotoxin-producing *Escherichia coli* seropathotypes that are linked to epidemic and/or serious disease. J. Clin. Microbiol. *41*, 4930–4940.

Kohl, T.A., Diel, R., Harmsen, D., Rothganger, J., Walter, K.M., Merker, M., Weniger, T., and Niemann, S. (2014). Whole-genome-based *Mycobacterium tuberculosis* surveillance: a standardized, portable, and expandable approach. J. Clin. Microbiol. *52*, 2479-2486.

Lecointre, G., Rachdi, L., Darlu, P., and Denamur, E. (1998). *Escherichia coli* molecular phylogeny using the incongruence length difference test. Mol. Biol. Evol. *15*, 1685–1695.

Leonard, S.R., Mammel, M.K., Rasko, D.A., and Lacher, D.W. (2016). Hybrid Shiga toxin-producing and enterotoxigenic *Escherichia* sp. cryptic lineage 1 strain 7v harbors a hybrid plasmid. Appl. Environ. Microbiol. *82*, 4309–4319.

Losada, L., Debroy, C., Radune, D., Kim, M., Sanka, R., Brinkac, L., Kariyawasam, S., Shelton, D., Fratamico, P.M., Kapur, V., and Feng, P.C. (2016). Whole genome sequencing of diverse Shiga toxin-producing and non-producing *Escherichia coli* strains reveals a variety of virulence and novel antibiotic resistance plasmids. Plasmid. *83*, 8–11.

Maiden, M.C., Bygraves, J.A., Feil, E., Morelli, G., Russell, J.E., Urwin, R., Zhang, Q., Zhou, J., Zurth, K., Caugant, D.A., Feavers, I.M., Achtman, M., and Spratt, B.G. (1998). Multilocus sequence typing: a portable approach to the identification of clones within populations of pathogenic microorganisms. Proc. Natl. Acad. Sci. U.S.A. *95*, 3140–3145.

Makino, K., Ishii, K., Yasunaga, T., Hattori, M., Yokoyama, K., Yutsudo, C.H., Kubota, Y., Yamaichi, Y., Iida, T., Yamamoto, K., Honda, T., Han, C.G., Ohtsubo, E., Kasamatsu, M., Hayashi, T., Kuhara, S., and Shinagawa, H. (1998). Complete nucleotide sequences of 93-kb and 3.3-kb plasmids of an enterohemorrhagic *Escherichia coli* O157:H7 derived from Sakai outbreak. DNA Res. *5*, 1–9.

Mathusa, E.C., Chen, Y., Enache, E., and Hontz, L. (2010). Non-O157 Shiga toxin-producing *Escherichia coli* in foods. J. Food Prot. 73, 1721–1736.

McDaniel, T.K., and Kaper, J.B. (1997). A cloned pathogenicity island from enteropathogenic *Escherichia coli* confers the attaching and effacing phenotype on *E. coli* K-12. Mol. Microbiol. 23, 399–407.

Mellmann, A., Harmsen, D., Cummings, C.A., Zentz, E.B., Leopold, S.R., Rico, A., Prior, K., Szczepanowski, R., Ji, Y., Zhang, W., Mclaughlin, S.F., Henkhaus, J.K., Leopold, B., Bielaszewska, M., Prager, R., Brzoska, P.M., Moore, R.L., Guenther, S., Rothberg, J.M., and Karch, H. (2011). Prospective genomic characterization of the German enterohemorrhagic *Escherichia coli* O104:H4 outbreak by rapid next generation sequencing technology. PLOS ONE. *6*, e22751.

Michelacci, V., Orsini, M., Knijn, A., Delannoy, S., Fach, P., Caprioli, A., and Morabito, S. (2016a). Development of a high resolution virulence allelic profiling (HReVAP) approach based on the accessory genome of *Escherichia coli* to characterize Shiga-toxin producing *E. coli* (STEC). Front Microbiol. 7, 202.

Michelacci, V., Prosseda, G., Maugliani, A., Tozzoli, R., Sanchez, S., Herrera-León, S., Dallman, T., Jenkins, C., Caprioli, A., and Morabito, S. (2016b). Characterization of an emergent clone of enteroinvasive *Escherichia coli* circulating in Europe. Clin. Microbiol. Infect. *22*, 287. e11–9. https://doi.org/10.1016/j.cmi.2015.10.025

Morabito, S., Karch, H., Mariani-Kurkdjian, P., Schmidt, H., Minelli, F., Bingen, E., and Caprioli, A. (1998). Enteroaggregative, Shiga toxin-producing *Escherichia coli* O111:H2 associated with an outbreak of hemolytic-uremic syndrome. J. Clin. Microbiol. *36*, 840–842.

Morabito, S., Tozzoli, R., Oswald, E., and Caprioli, A. (2003). A mosaic pathogenicity island made up of the locus of enterocyte effacement and a pathogenicity island of *Escherichia coli* O157:H7 is frequently present in attaching and effacing *E. coli*. Infect. Immun. *71*, 3343–3348.

Moran-Gilad, J., Prior, K., Yakunin, E., Harrison, T.G., Underwood, A., Lazarovitch, T., Valinsky, L., Luck, C., Krux, F., Agmon, V., Grotto, I., and Harmsen, D. (2015). Design and application of a core genome multilocus sequence typing scheme for investigation of Legionnaires' disease incidents. Euro Surveill. *20*(28):pii=21186.

Newitt, S., MacGregor, V., Robbins, V., Bayliss, L., Chattaway, M.A., Dallman, T., Ready, D., Aird, H., Puleston, R., and Hawker, J. (2016). Two Linked Enteroinvasive *Escherichia coli* Outbreaks, Nottingham, UK, June 2014. Emerging Infect. Dis. *22*, 1178–1184. https://doi.org/10.3201/eid2207.152080

Nyholm, O., Halkilahti, J., Wiklund, G., Okeke, U., Paulin, L., Auvinen, P., Haukka, K., and Siitonen, A. (2015). Comparative Genomics and Characterization of Hybrid Shigatoxigenic and Enterotoxigenic *Escherichia coli* (STEC/ETEC) Strains. PLOS ONE *10*, e0135936. https://doi.org/10.1371/journal.pone.0135936

Perna, N.T., Plunkett, G., Burland, V., Mau, B., Glasner, J.D., Rose, D.J., Mayhew, G.F., Evans, P.S., Gregor, J., Kirkpatrick, H.A., *et al.* (2001). Genome sequence of enterohaemorrhagic *Escherichia coli* O157:H7. Nature *409*, 529–533. https://doi.org/10.1038/35054089

Prager, R., Fruth, A., Siewert, U., Strutz, U., and Tschäpe, H. (2009). *Escherichia coli* encoding Shiga toxin 2f as an emerging human pathogen. Int. J. Med. Microbiol. *299*, 343–353. https://doi.org/10.1016/j.ijmm.2008.10.008

Rasko, D.A., Rosovitz, M.J., Myers, G.S.A., Mongodin, E.F., Fricke, W.F., Gajer, P., Crabtree, J., Sebaihia, M., Thomson, N.R., Chaudhuri, R., Henderson, I.R., Sperandio, V., and Ravel, J. (2008). The pangenome structure of *Escherichia coli*: comparative genomic analysis of *E. coli* commensal and pathogenic isolates. J. Bacteriol. *190*, 6881–6893.

Reid, S.D., Herbelin, C.J., Bumbaugh, A.C., Selander, R.K., and Whittam, T.S. (2000). Parallel evolution of virulence in pathogenic *Escherichia coli*. Nature *406*, 64–67. https://doi.org/10.1038/35017546

Rohde, H., Qin, J., Cui, Y., Li, D., Loman, N.J., Hentschke, M., Chen, W., Pu, F., Peng, Y., Li, J., Xi, F., Li, S., Li, Y., Zhang, Z., Yang, X., Zhao, M., Wang, P., Guan, Y., Cen, Z., Zhao, X., Christner, M., Kobbe, R., Loos, S., Oh, J., Yang, L., Danchin, A., Gao, G.F., Song, Y., Li, Y., Yang, H., Wang, J., Xu, J., Pallen, M.J., Wang, J., Aepfelbacher, M., Yang, R., and *E. coli* O104:H4 Genome Analysis Crowd-Sourcing Consortium. (2011). Open-source genomic analysis of Shiga-toxin-producing *E. coli* O104:H4. N. Engl. J. Med. *365*, 718–724.

Ruppitsch, W., Pietzka, A., Prior, K., Bletz, S., Fernandez, H.L., Allerberger, F., Harmsen, D., and Mellmann, A. (2015). Defining and evaluating a core genome multilocus sequence typing scheme for whole-genome sequence-based typing of *Listeria monocytogenes*. J. Clin. Microbiol. *53*, 2869–2876.

Sahl, J.W., Morris, C.R., Emberger, J., Fraser, C.M., Ochieng, J.B., Juma, J., Fields, B., Breiman, R.F., Gilmour, M., Nataro, J.P., *et al.* (2015). Defining the phylogenomics of *Shigella* species: a pathway to diagnostics. J. Clin. Microbiol. *53*, 951–960. https://doi.org/10.1128/JCM.03527-14

Scavia, G., Morabito, S., Tozzoli, R., Michelacci, V., Marziano, M.L., Minelli, F., Ferreri, C., Paglialonga, F., Edefonti, A., and Caprioli, A. (2011). Similarity of Shiga toxin-producing *Escherichia coli* O104:H4 strains from Italy and Germany. Emerging Infect. Dis. *17*, 1957–1958. https://doi.org/10.3201/eid1710.111072

Scheutz, F., Nielsen, E.M., Frimodt-Moller, J., Boisen, N., Morabito, S., Tozzoli, R., Nataro, J.P., and Caprioli, A. (2011). Characteristics of the enteroaggregative Shiga toxin/verotoxin-producing *Escherichia coli* O104:H4 strain causing the outbreak of haemolytic uraemic syndrome in Germany, May to June 2011. Euro Surveill. *16*(24). pii: 19889.

Selander, R.K., Musser, J.M., Caugant, D.A., Gilmour, M.N., and Whittam, T.S. (1987). Population genetics of pathogenic bacteria. Microb. Pathog. *3*, 1–7.

Smith, D.L., Rooks, D.J., Fogg, P.C., Darby, A.C., Thomson, N.R., McCarthy, A.J., and Allison, H.E. (2012). Comparative genomics of Shiga toxin encoding bacteriophages. BMC Genomics *13*, 311. https://doi.org/10.1186/1471-2164-13-311

Steyert, S.R., Sahl, J.W., Fraser, C.M., Teel, L.D., Scheutz, F., and Rasko, D.A. (2012). Comparative genomics and *stx* phage characterization of LEE-negative Shiga toxin-producing *Escherichia coli*. Front. Cell. Infect. Microbiol. *2*, 133.

Tarr, P.I., Gordon, C.A., and Chandler, W.L. (2005). Shiga-toxin-producing *Escherichia coli* and haemolytic uraemic syndrome. Lancet *365*, 1073–1086.

Touchon, M., Hoede, C., Tenaillon, O., Barbe, V., Baeriswyl, S., Bidet, P., Bingen, E., Bonacorsi, S., Bouchier, C., Bouvet, O., Calteau, A., Chiapello, H., Clermont, O., Cruveiller, S., Danchin, A., Diard, M., Dossat, C., El Karoui, M., Frapy, E., Garry, L., Ghigo, J.M., Gilles, A.M., Johnson, J., Le Bouguénec, C., Lescat, M., Mangenot, S., Martinez-Jéhanne, V., Matic, I., Nassif, X., Oztas, S., Petit, M.A., Pichon, C., Rouy, Z., Ruf, C.S., Schneider, D., Tourret, J., Vacherie, B., Vallenet, D., Médigue, C., Rocha, E.P.C., and Denamur, E. (2009). Organised genome dynamics in the *Escherichia coli* species results in highly diverse adaptive paths. PLOS Genet. *5*(1):e1000344.

Tozzoli, R., Grande, L., Michelacci, V., Ranieri, P., Maugliani, A., Caprioli, A., and Morabito, S. (2014). Shiga toxin-converting phages and the emergence of new pathogenic *Escherichia coli*: a world in motion. Front. Cell. Infect. Microbiol. *4*, 80. https://doi.org/10.3389/fcimb.2014.00080

Trabulsi, L.R., Keller, R., and Tardelli Gomes, T.A. (2002). Typical and atypical enteropathogenic *Escherichia coli*. Emerging Infect. Dis. *8*, 508–513.

Underwood, A.P., Dallman, T., Thomson, N.R., Williams, M., Harker, K., Perry, N., Adak, B., Willshaw, G., Cheasty, T., Green, J., *et al.* (2013). Public health value of next-generation DNA sequencing of enterohemorrhagic *Escherichia coli* isolates from an outbreak. J. Clin. Microbiol. *51*, 232–237. https://doi.org/10.1128/JCM.01696-12

Van Den Beld, M.J.C., and Reubsaet, F.A.G. (2012). Differentiation between *Shigella*, enteroinvasive *Escherichia coli* (EIEC) and noninvasive *Escherichia coli*. Eur. J. Clin. Microbiol. Infect. Dis. *31*, 899-904.

Van Nhieu, G.T., and Sansonetti, P. (2009). *Shigella*. In: Intracellular Niches of Microbes: A Pathogens Guide Through the Host Cell. U.A. Schaible and A. Hass, eds. (Weinheim: Wiley-VCH Verlag GmbH).

Werber, D., Krause, G., Frank, C., Fruth, A., Flieger, A., Mielke, M., Schaade, L., and Stark, K. (2012). Outbreaks of virulent diarrheagenic *Escherichia coli* – are we in control? BMC Med. *10*, 11. https://doi.org/10.1186/1741-7015-10-11

Wirth, T., Falush, D., Lan, R., Colles, F., Mensa, P., Wieler, L.H., Karch, H., Reeves, P.R., Maiden, M.C., Ochman, H., *et al.* (2006). Sex and virulence in *Escherichia coli*: an evolutionary perspective. Mol. Microbiol. *60*, 1136–1151.

Zhou, K., Ferdous, M., de Boer, R.F., Kooistra-Smid, A.M., Grundmann, H., Friedrich, A.W., and Rossen, J.W. (2015). The mosaic genome structure and phylogeny of Shiga toxin-producing *Escherichia coli* O104:H4 is driven by short-term adaptation. Clin. Microbiol. Infect. *21*, 468.e7–18. https://doi.org/10.1016/j.cmi.2014.12.009

Use of Whole-genome Sequencing to Improve Investigations of Outbreaks of *Escherichia coli*

8

Claire Jenkins

Gastrointestinal Bacterial Reference Unit, Public Health England, London, UK.

*Correspondence: claire.jenkins@phe.gov.uk

https://doi.org/10.21775/9781910190777.08

Abstract

The aim of this chapter is to evaluate the use of whole-genome sequencing (WGS) data for detecting and investigating outbreaks of *Escherichia coli.* The methods available for assessing the relatedness between isolates of *E. coli,* and their use in public health microbiology, are summarised. WGS typing methods are suitable for all pathotypes of *E. coli* and provide an unprecedented level of strain discrimination. WGS typing is robust and utilises stable genetic markers that can elucidate the evolutionary context of emerging pathogenic strains. The robustness of the approach ensures confidence in the microbiological identification of linked cases when epidemiological links are obscured. Long-read technology will facilitate the analysis of loss and acquisition of mobile genetic elements in the accessory genome, and improve our understanding of the impact of short-term evolutionary changes during outbreak investigations. There is increasing evidence that the geographical origin of an isolate can be inferred from the phylogeny. Expanding WGS-based typing analysis globally will improve trace-back investigations in the event of a food-borne outbreak, ensuring the rapid implementation of interventions to protect public health. The importance of publicly available WGS data linked to the clinical, epidemiological and environmental context of the sequenced strain cannot be underestimated.

Historical perspective

Typing *Escherichia coli* for outbreak detection and investigation has been performed since the 1940s when John Bray first prepared antiserum to a strain of *E. coli* isolated from an infant with diarrhoea, and used this serum to detect antigenically similar strains in other infants in a north London hospital (Bray, 1945). Throughout the twentieth century serotyping remained the most effective way of detecting outbreaks of *E. coli* in hospitals and in the community (Smith *et al.*, 1996). Currently there are over 180 different O somatic antigen types and over 50 H flagella antigen types, providing a huge number of potential O:H antigen combinations.

In the 1980s, the emergence of Shiga toxin-producing *E. coli* (STEC) O157:H7, and the increasing number of outbreaks associated with this serotype, stimulated the need for a higher level of strain discrimination (Riley *et al.*, 1983). A phage typing scheme was developed by the Canadian Public Health Laboratory Service and was adopted by a number of countries, including the UK where it is still used today (Khakhria *et al.*, 1990). Phage typing is a rapid and cost-effective method for detecting outbreaks and for quickly ruling cases in or out of an outbreak. However, the level of strain discrimination it offers is low.

Pulse field gel electrophoresis (PFGE) was implemented by a number of reference laboratories in the 1990s, and remained the molecular typing method of choice for STEC, notably in North, Central and South America but also in Europe and elsewhere (Gerner-Smidt *et al.*, 2006). PFGE is technically demanding, does not lend itself to high-throughput analysis and, as with most fragment analysis techniques, inter-laboratory data exchange is difficult despite the promotion of standardized protocols. PFGE requires a certain level of expertise and is subject to technological and interpretation challenges (Davis *et al.*, 2003). Generally it is used reactively in outbreaks that have already been identified by epidemiological links.

Subsequently, Lindstedt *et al.* (2004) in Europe and Hyytia-Trees *et al.* (2010) in the USA developed a MultiLocus Variable number tandem repeat Analysis (MLVA) method for typing STEC O157:H7. This method converts the number of short repeat sequences at specific locations in the genome into an eight digit profile. In conjunction with epidemiological evidence linking the cases, isolates with the same profile, or those differing at one locus (a single locus variant), are regarded as likely to be related. This approach lends itself more easily to high through-put and facilitates inter-laboratory data exchange. Difficulties with this approach are related to the interpretation of double locus variants (DLVs) of an outbreak profile and determining the relationship of isolates over an extended time frame. Interpretation of the evolutionary context of a population is not possible using this approach. MLVA schemes for generic *E. coli* do exist but lack the level of discrimination required for outbreak investigation.

Whole-genome sequencing (WGS) has been used to investigate outbreaks of gastrointestinal disease since 2011, when the strain of STEC O104:H4 causing an outbreak of haemolytic uraemic syndrome (HUS) in Germany in 2011, was sequenced in real-time (Mellmann *et al.*, 2011). WGS offers an unprecedented level of strain discrimination and a one step protocol for providing a range of clinically and epidemiologically useful data. In this chapter, the contribution of WGS to the detection and investigation of outbreaks of *E. coli* is discussed, with reference to outbreak scenarios described in the literature (Table 8.1).

Current methods for typing *E. coli* using WGS data

WGS is an ideal approach for typing bacterial pathogens. It is robust, reproducible, provides an unprecedented level of discriminatory power, lends itself to inter-laboratory comparisons, and can be rapid, user-friendly and cost effective. The first step in the WGS process is overnight culture of the bacteria in liquid medium followed by DNA extraction either manually (most commonly using a kit) or using an automated extraction platform, such as the QiaSymphony™ or the MagnaPure™. A library is prepared from the extracted DNA using a kit recommended by the manufacturer of the sequencing platform of choice. Short read sequencers, such as the Ion Torrent™ (Thermo Fisher Scientific USA) or HiSeqTM/MiSeq™ (IlluminaInc USA) are currently the most cost effective platforms, and are most commonly used in clinical and public health settings.

The output from the short read platforms are single-end or paired sequences in Fastq files that contain all the short sequence reads of fragments of DNA ~100bp in length. The short reads are assembled using the most appropriate reference genome, and the assembled genome can then be compared to other genomes to assess similarity. There are a range of methods available to assess relatedness between isolates including analysis of the difference in single nucleotide polymorphisms (SNPs), gene presence or absence throughout the whole genome and gene allele differences, and overall genetic similarity, for example comparing *k*-mers (Parsons *et al.*, 2016) (Table 8.2).

Food-borne outbreaks may be national or international and it is vitally important that WGS typing data can be exchanged across borders (Franz *et al.*, 2014). For *E. coli*, SNP typing and core genome and/or whole-genome MLST (cgMLST/wgMLST) are currently the methods most commonly applied to phylogenetic analysis, and shared databases are in the process of being built. A whole genome MLST (wgMLST) database is in the process ofbeing developed in PulseNet. MLST methods compare most, or all, of the genes in bacterial genomes by creating a barcode by assigning numbers for allelic variants. With respect to the SNP typing approach, each new isolate is compared with a curated regional, national and/or international database of isolates to place it into phylogenetic clusters (Parsons *et al.*, 2016).

Table 8.1 Use of whole-genome sequencing to improve investigations of outbreaks of *Escherichia coli* with reference to published studies

Outbreak Scenario	Description	Reference	Context
1	Outbreak of STEC O157:H7 at an open farm in the south-east of England	Underwood *et al.* (2013)	Robust, high-level strain discrimination compared to traditional molecular typing methods; SNP variation observed during an outbreak involving direct contact with animals
2	Two related outbreaks of STEC O157:H7 PT8 and PT54 associated with the same restaurant	Cowley *et al.*, 2016	Robust, high-level strain discrimination compared to traditional molecular typing methods; Long-read sequencing to explore short-term evolutionary change during an outbreak
3	Outbreak of STEC O157 PT21/28 associated with raw drinking milk	Butcher *et al.* (2016)	Robust, high-level strain discrimination compared to traditional molecular typing methods; Inferring the geographical origin of an outbreak strain from the phylogeny
4	Outbreak of STEC O157 at an open farm in the north-west of England	Rowell *et al.* (2016)	SNP variation observed during outbreaks involving direct contact with animals
5	Two concurrent outbreaks of STEC O157 PT2 associated with contaminated watercress	Jenkins *et al.* (2015)	Inferring the geographical origin of an outbreak strain from the phylogeny
6	Outbreak of STEC O104:H4 associated with fenugreek seeds in 2011	Mellmann *et al.* (2011)	Molecular typing of isolates from outbreak cases; Assessment of pathogenicity profile; Evolutionary history of the outbreak strain
7	Outbreak of STEC O26:H11 linked to nursery	Dallman *et al.* (2015c)	Robust, high-level strain discrimination compared to traditional molecular typing methods
8	Outbreak of STEC O55 in the south of England 2014–16	McFarland *et al.* (2017)	Molecular typing of isolates from outbreak cases; Evolutionary history of the outbreak strain
9	Outbreaks of EIEC O96:H19	Newitt *et al.* (2016)	Molecular typing of isolates from outbreak cases; Evolutionary history of the outbreak strain
10	Outbreak of EAEC O104:H4 associated with curry leaves	Dallman *et al.* (2014)	Molecular typing of isolates from outbreak cases; Assessment of pathogenicity profile; Evolutionary history of the outbreak strain

Outbreak definitions and outbreak detection using SNP typing

Conventionally, outbreaks of STEC are identified by the routine investigation of cases, either through (i) identifying common features between cases in terms of exposures, (ii) indications that the number of cases in a particular location or of a particular serotype, phage type, PFGE or MLVA profile, is higher than expected or (iii) the appearance of microbiological type or profile among cases which are temporally or geographically linked. Surveillance systems often rely on microbiological typing to identify outbreaks that cannot be identified by epidemiological data alone. One of the attractions of the cgMLST/wgMLST approach is that it is possible to assess the gene content of an organism's genome and create a barcode by assigning numbers to allelic variants. The development of a standardized scheme will facilitate inter-laboratory data exchange.

It is also possible to produce a profile, or barcode, using SNP typing data. Hierarchical single linkage clustering can be performed on the pairwise SNP difference between all isolates at various distance thresholds (Δ250, Δ100, Δ50, Δ25, Δ10, Δ5, Δ0). The result of the clustering is a SNP profile, or SNP address, that can be used to describe the population structure based on clonal groups. This is indicated by the number of identical digits in the seven-digit SNP address (Fig. 8.1). Criteria for defining the evolutionary context of WGS clusters where the core genome sequences of the isolates are within 0, 5 or 10 SNPs of each other has been proposed (Table 8.3). Although the current evidence-base indicates that isolates of *E. coli* > 5 SNPs apart are unlikely to be part of the same temporally linked outbreak,

Table 8.2 Commonly used methods to assess relatedness between isolates using whole-genome sequencing data. (T.J. Dallman, personal communication)

Approach	Method	References
Single nucleotide polymorphisms	Map query reads to genome Calculate high-quality SNPs and absent positions Generate sequence alignment of positions that are present in sequences in query set Compute maximum likelihood phylogeny	Neilsen *et al.* (2011) Eppenger *et al.* (2011, 2015) Dallman *et al.*(2015a,b) Joensen *et al.* (2014) Holmes *et al.* (2015)
Gene-by-gene	Define loci that represent 'core' set of genes for scheme of interest Assemble query genome Scan allele set of scheme against assembly to identify alleles present Calculate pairwise allele distance Compute UPMGA tree	Maiden *et al.* (2013) Ferdous *et al.* (2016)
Nucleotide difference – use of *k*-mers mapped to a reference genome	Compute *k*-mers in reference genome Compute *k*-mers in query genome Map query genome *k*-mers to reference *k*-mers Calculate nucleotide difference from reference Create matrix of pairwise nucleotide differences Compute UPMGA tree	Joensen *et al.* (2014)
Alignment free *k*-mer comparisons	Assemble all genomes Generate a set of all *k*-mers Compute *k*-mers in each genome Resolve distance between each genome Generate tree by hierarchical clustering	Parsons *et al.* (2016)

Table 8.3 Criteria used at Public Health England for defining the evolutionary context of WGS clusters, defined as two or more cases, where the core genome sequences of the isolates are within 0, 5 or 10 SNPs (A. Mikhail: Personal communication)

SNP threshold	Epidemiological meaningfulness of the cluster
Zero-SNP cluster	All seven digits of the SNP address are identical, indicating that the core whole-genome sequences of the isolates are identical. Case isolates that fall within a 0–SNP cluster have the strongest genetic evidence for being part of an outbreak with a common exposure or with direct person-to-person transmission. At Public Health England 0-SNP clusters are monitored on a weekly basis and currently two cases falling within a 0-SNP cluster is sufficient to trigger further investigative action.
Five-SNP cluster	The first six digits of the SNP address are identical but the last digit is different. Isolates within a 5-SNP cluster are likely to have been exposed to the same source or linked by person-to-person transmission. Relative to 10-SNP clusters, they are more likely to be temporally clustered and have other similar features (e.g. identical genotypic resistance profile). At Public Health England five-SNP clusters are monitored on a weekly basis and currently five cases falling within a 5-SNP cluster is sufficient to trigger further investigative action.
Ten-SNP cluster	The first five digits of the SNP address are identical but the last two digits are different. Isolates within a 10-SNP cluster may not be directly linked; however clusters at this level may be characterized by person-to-person transmission over an extended period of time, or persistent environmental contamination over an extended period of time. At Public Health England, 10-SNP clusters are not routinely monitored, except where the 10-SNP level has been identified as the best fit for an outbreak characterized by temporally dispersed cases and continuous transmission. An assessment of cases at the 10-SNP level may also be informative for determining the ancestral origin of 5 or 0-SNP clusters.

SNP address

If you imagine all the whole genome sequences that have been identified for STEC O157:H7 laid out with similar sequences next to each other and ones that are very different far away, the SNP address works in a similar way to the American zip code. Imagine the first number represents a certain group of U.S. states, the second and third digits together represent a region in that group and the fourth and fifth digits represent a group of delivery addresses within that region. So with each number you look at a smaller and more detailed or closely related area.

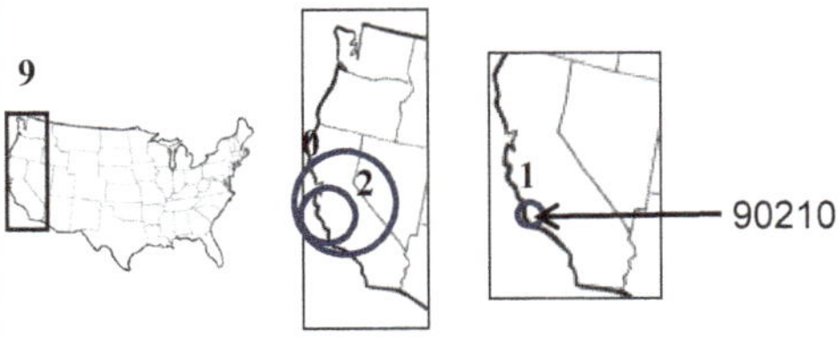

In the diagram below, you can see three SNP levels; blue circles, yellow circles and black circles, which decrease in size as the genetic variation decreases. In this example we have a three digit address for simplicity, the 1st digit corresponds to 10 SNP differences and the 2nd digit corresponds to 5 SNP differences and the final one is 0 SNP differences. If two samples have the same SNP address, they are regarded as identical i.e. two isolates that have the address 1.1.1 have 0 SNP differences between them.

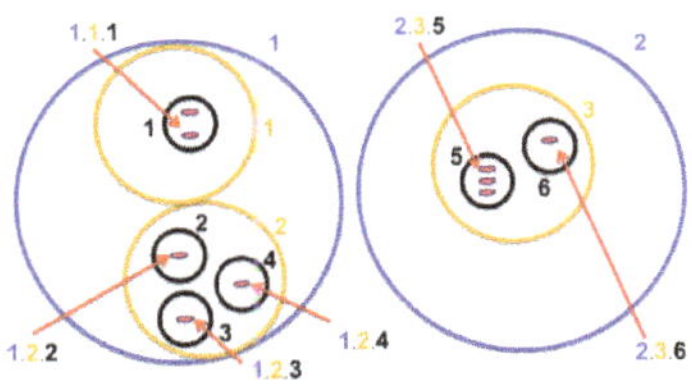

The isolates with the addresses 1.2.2 and 1.2.3 have a different final digit indicating that they are not identical but both have matching 1st and 2nd digits so are within the same 10 SNP cluster and the same 5 SNP cluster.

When we compare the isolate 1.2.2 against 2.3.5, as the first digit is different, they are different by 10 SNPs.

Figure 8.1 Explanation of the principle behind the SNP profile or SNP address. (Courtesy of A. Waldram: Personal communication.)

deeper phylogenetic relationships within the 10- or 25-SNP clusters may provide epidemiologically useful information or associations. Clusters can be categorized by geography (local, national, or international) and type (household, community or travel-associated) to determine further actions.

Initial evaluation studies

New approaches to bacterial typing for use in the clinical and public health settings can be evaluated either by comparisons to the current gold standard technique or by using isolates linked to known outbreaks or sporadic cases. Evaluating a novel approach that offers an unprecedented level of strain discrimination, such as WGS, is problematic as determining the validity of the additional clusters detected by WGS that previously existed under the surveillance radar, can be difficult. Despite the limitations, there have been a number of WGS evaluation studies covering the range of different methods of analysing the data, that attempt to compare traditional and WGS approaches to typing *E. coli*, and these are reviewed below.

The first paper evaluating the use WGS for routine surveillance and outbreak detection of STEC was published by the Danish Technical University in collaboration with the World Health Organization *E. coli* reference laboratory at the Serum Staten Institute in Copenhagen, Denmark. This group sequenced 42 isolates of STEC, including 12 STEC O157:H7 isolates (six of these belonged to an outbreak) received during a 7-week period, analysed the data using their web-based tools NDtree and SNPtree and compared the results to traditional typing methods, including serotyping and PFGE (Joensen *et al.*, 2014). Both NDtree and SNPtree methods were able to distinguish the STEC O157:H7 outbreak isolates from the other six non-outbreak isolates of the same serotype, and the other STEC serotypes. The nucleotide differences within the outbreak strain ranged from 1 to 7, whereas the distance to the other isolates in the STEC O157:H7 group ranged from 607 to 1,617 nucleotide differences. This evaluation also included a comparison of turnaround times of the results and assessed the cost of WGS compared to the equivalent analysis using the battery of traditional tests, and found the WGS approach was cheaper and faster. This study also showcased a variety of WGS analyses ranging from computationally intensive SNP typing to publicly available, user-friendly web-tools for rapid handling of WGS data and extraction of clinically relevant information (Joensen *et al.*, 2014).

In the Netherlands, 132 isolates of STEC were sequenced and the phylogenetic relationship was determined using a gene-by-gene approach (Ferdous *et al.*, 2016). Isolates clustered by sequence type (ST) and subclusters within a single ST cluster generally comprised isolates of the same serotype. They also showed that most of the isolates of the same H type, irrespective of their O serogroup shared a common ancestor, e.g. serogroups O103:H2 and O128:H2. Conversely, in some cases, isolates from the same O serogroups were found to be scattered at distinct positions in the tree, indicating that serogroup is not a robust phylogenetic marker. To explore the genetic relatedness of the isolates based on their virulence genes and other mobile genetic elements, they constructed an additional neighbour-joining tree based on the accessory genome, and this tree was almost identical to the core genome phylogenetic tree.

In a larger study that focused on STEC O157:H7 genomes only, 572 randomly selected isolates from 2012 and 2013 were sequenced (Dallman *et al.*, 2015a). This group attempted to define a common source SNP variation threshold for outbreak detection, based on temporal and epidemiological linkages between isolates. They concluded that the maximum number of SNP differences for isolates to be part of the same household or community outbreak was five SNPs. A number of previously unidentified clusters of isolates that fell within a five-SNP cluster of each other were detected using this approach, although all but two of these clusters were too small to support meaningful epidemiological analysis. Of the two larger clusters, following further investigations, one was ultimately linked to consumption of contaminated salad leaves and the other was associated with visiting a national park. In Scotland, a 5-year retrospective review of 105 isolates of *E. coli* O157:H7 and 11 epidemiologically linked clusters used the same methodology as the study by Dallman *et al.* (2015a). They found that the WGS analysis was concordant with MLVA with respect to detection of outbreaks, and provided further evidence that epidemiologically linked cases exhibited SNP differences of ≤ 4 SNPs, while unrelated cases had SNP differences between 9 and 1632 (Holmes *et al.*, 2015). Two sets of isolates that differed by only one MLVA locus (single locus variant) were 32 and 126 SNPs different from the other isolate in each set, demonstrating the increased discriminatory power of WGS.

In a recent review, Parsons *et al.* (2016) described three outbreaks of STEC in Canada, including a relatively large outbreak of 119 clinical cases linked to the consumption of contaminated pork, and two smaller outbreak associated with a summer fair and a 2012 beef outbreak, respectively. In the outbreak associated with contaminated pork, SNP typing was employed to compare 111 of these clinical cases and six environmental or food isolates, and the data were compared to the current protocol for identifying outbreaks, specifically PFGE and MLVA profiling in combination with epidemiological investigations. Of the 117 isolates sequenced from the pork outbreak, 109 were 0–5 SNPs different from each other, and the intra-outbreak SNP differences in the two other outbreaks were 0–5 SNPs. Overall, clinical, food and environmental isolates from the pork-associated outbreak were found to have ≤ 23 SNPs different from each other and a minimum of 84 SNPs different from isolates not associated with the outbreak. The same isolates were also subjected to *k*-mer analysis and this analysis was able to cluster each outbreak into separate nodes. Furthermore, the *k*-mer method was also able to distinguish isolates from with in the pork-associated outbreak that lacked certain virulence genes present in all other outbreak isolates. The current surveillance methods and SNP analysis could not distinguish the isolates missing virulence genes from other isolates of the same outbreak. The difference between the two WGS methods is likely because SNP analysis compares conserved genomic regions where virulence factors are rarely found, whereas *k*-mer analysis analyses the entire genome. This example highlights the added insight that can be provided when using alternative methods on the same dataset.

The WGS STEC typing evaluation studies described to date all show that the WGS analyses correlate well with traditional methods of outbreak detection and investigation, including serotyping, PFGE, MLVA and epidemiological clustering. Generally, the studies conclude that WGS data provides a higher level of strain discrimination than other molecular methods, including PFGE and MLVA.

Comparison of WGS with current molecular typing techniques

PFGE and MLVA have been used for many years to facilitate outbreak investigations with respect to ruling cases in or out of an outbreak that have been identified within the same time frame and/or geographical location. However, there was often doubt over whether these techniques exhibited the robustness and resolution required to investigate all outbreaks, especially those associated with complex environmental exposures or those that were on-going for an extended time-frame (Eppinger *et al.*, 2011, 2015). The evaluation studies described above demonstrated that WGS data correlates well with the outputs of current typing methods, and provides robust, high-level strain discrimination. These concepts are explored further in the two outbreak scenarios described below.

Outbreak Scenario 1: outbreak of STEC O157:H7 at an open farm in the south-east of England

One of the largest outbreaks of STEC O157:H7 phage type (PT) 21/28 reported in the UK was detected in visitors to an open farm in the south-east of England in 2009 (Underwood *et al.*, 2013). Following the epidemiological investigation and use of MLVA, general contamination of the environment was suspected, and the outbreak was quickly brought under control. Retrospectively, 16 isolates belonging to the outbreak were sequenced in order to compare the resolution gained by this approach with that derived from more traditional epidemiological tools (specifically in this case phage typing and MLVA). 4 SNP differences (five haplotypes) were identified in the 16 sequenced isolates from the Surrey farm. Allele-specific primers were designed to identify the four SNPs in all the isolates associated with this outbreak. No additional haplotypes were identified.

Comparison of the MLVA and SNP analysis showed that a single MLVA profile maybe associated with multiple SNP haplotypes and that multiple MLVA profiles belonged to a single haptotype (Fig. 8.2 and Underwood *et al.*, 2013). This latter observation suggests that the differences in MLVA profiles represent rapid repeat number changes (as well as possible reversions) that occur at a higher rate than single nucleotide mutations responsible for determining phylogenetic lineage. Analysis of these two datasets indicated that MLVA offered a level of strain discrimination comparable to SNP typing. However, the authors argued that SNP typing was a more robust approach with respect to stability of the genetic markers (i.e. SNPs verses tandem repeats). The criteria for interpreting MLVA states that isolates with the same profile or a single locus variant (SLV) of that profile, are

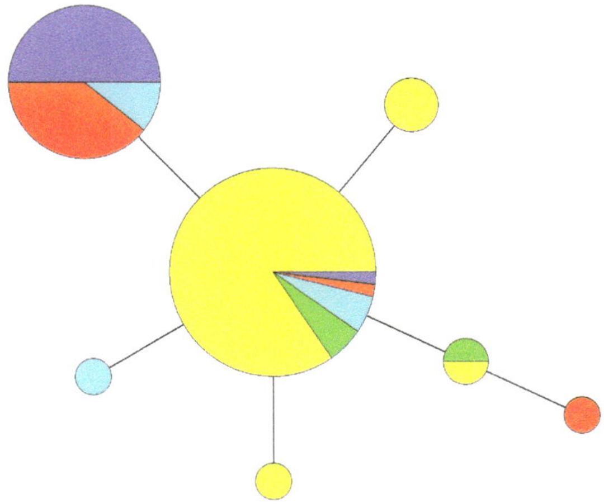

Figure 8.2 Minimal spanning tree of the seven MLVA profiles identified during an outbreak associated with a petting farm (Outbreak Scenario 1). Each MLVA profile is represented by a node and the nodes are coloured according to the haplotypes (H) (SNPs) identified within each node. Red – H1; Green – H2; Blue H3; Yellow – H4; Turquiose – H5.(Data supplied by A. Underwood and M. Day: Personal communication.)

likely to be related, The comparison of MLVA and WGS described in this study provided further evidence that attaching significance to SLVs during outbreaks of STEC O157:H7 is misleading. There is less certainty regarding the meaningfulness of a double locus variant (DLV), and this phenomenon is discussed further in Outbreak Scenario 3.

Outbreak Scenario 2: two related outbreaks of STEC O157:H7 associated with the same food outlet

In August 2012, six cases of STEC O157:H7 PT 8 were epidemiologically linked to the consumption of an unknown food item at a restaurant. Eight weeks later, in October 2012, over 140 confirmed cases of STEC O157:H7 PT54 were linked to the same restaurant. MLVA profiles from both incidents indicated that the August and October outbreaks were caused by the same strain, despite the phage typing difference (Cowley *et al.*, 2016).

WGS was carried out on 59 strains from the August (n=6) and October (n=53) outbreaks alongside 30 other isolates of STEC O157:H7 from sporadic cases residing in same locality. All six strains of PT8 from August fell within the same 5-SNP cluster. The 53 isolates of STEC O157:H7 PT54 from the October outbreak also fell within

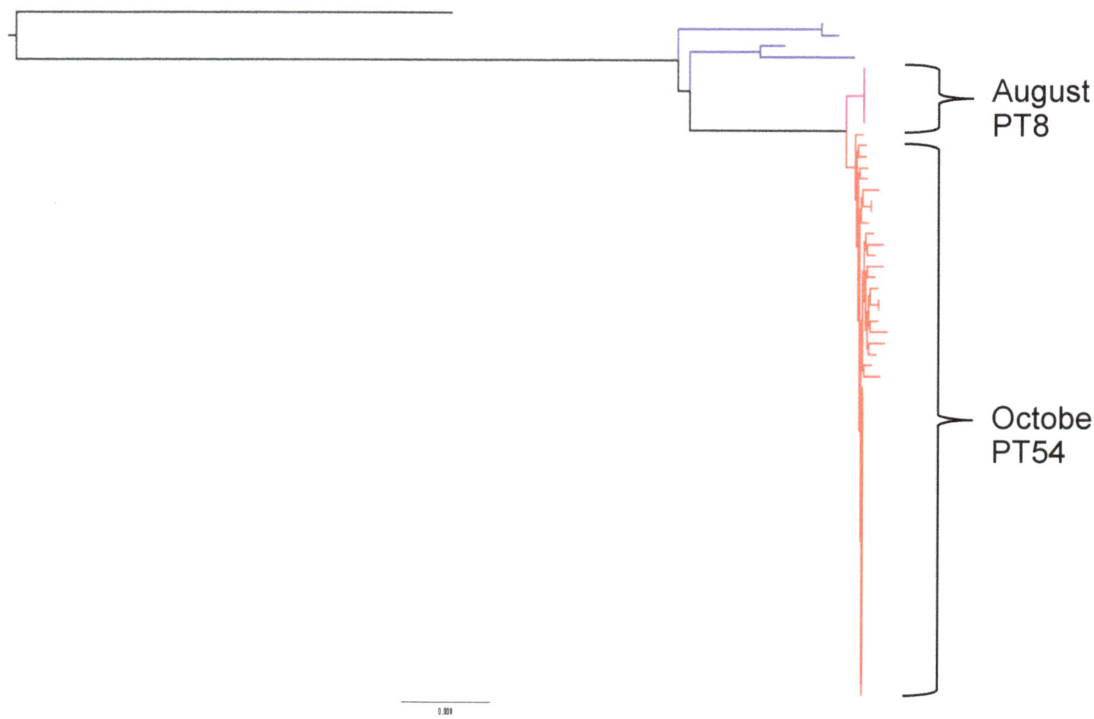

Figure 8.3 Phylogenetic relationship between isolates from two outbreaks at the same restaurant occurring six weeks apart and caused by two different phage types of STEC O157 (Outbreak Scenario 2). The first outbreak (PT8), comprising six isolates within a 5-SNP cluster, is represented in pink. Fifty-three isolates from the second outbreak (PT54) also fell within a 5-SNP cluster, were sequenced and are shown in red. There were three SNPs between the two clusters. The isolates in the database that were most closely related to the outbreak isolates are shown in blue. (Courtesy of T.J. Dallman: Personal communication.)

a 5-SNP cluster. The two clusters were only three SNPs apart (Fig. 8.3). However, the August PT8 strains had a unique SNP not identified in any isolates in the October PT54 cluster indicating that the isolates in the October PT54 did not evolve from the August PT8 outbreak strain. No isolates from the August outbreak, or isolates derived from the August outbreak, were detected during October.

In this example the phage typing data indicated the two strains from the August and October outbreaks were different, and the MLVA analysis showed that they were the same. The WGS data confirmed that the two strains were very closely related and, in addition, shed light on the true phylogenetic relationship between the two strains. The two strains were closely related, indicating that they were from the same original source. The SNP variation observed could be explained by a scenario where by the contaminated vehicle was introduced into the restaurant on two separate, temporarily distinct occasions. Epidemiological analysis implicated imported dried parsley as the contaminated product. This was not confirmed by subsequent microbiological testing of the parsley. However, it is plausible that two separate batches of parsley were the original source of the contamination event in both August and October, and that transmission in October was facilitated by an infected food handler.

This outbreak demonstrated that phage typing does not provide a robust phylogenetic signal. Genetic modifications occurring over a relatively short period of time, either in the core or accessory genome, may result in a change in phage type in closely related strains. Furthermore, comparison of the analysis of WGS and MLVA during this outbreak demonstrated that MLVA did not provide the evolutionary context required to highlight the true relationship between closely related strains. This example neatly illustrates the added value of using WGS data to explore the phylogeny. The relationship between isolates can be assessed more accurately over time and space than the molecular methods used previously. Therefore, typing using WGS data not only offers a robust, high level of phylogenetic resolution, it also provides stable, long-term genetic markers to study the evolutionary context of a strain.

Robust and timely phylogenetic resolution improves case ascertainment during outbreak

The conclusions of published studies comparing PFGE, MLVA and WGS for typing *E. coli* described previously in this chapter, all agree that WGS is the superior technique. However, it is acknowledged that it is difficult to assess this comparison in a quantifiable manner. Dallman *et al.* (2015a) looked at clustering cases based on the WGS defined common source threshold (< 5 SNPs) and whether or not this approach increased sensitivity in identifying linked cases (outbreaks) compared to MLVA. They calculated that there was no significant difference between MLVA vs WGS SNP typing based on clustering a single isolate with another, indicating there was no difference in timeliness of clustering between the two methods. That is to say, MLVA was as sensitive as WGS with respect to detecting an outbreak. However, when the time to cluster completion (all cases of a cluster are clustered) from the initial cluster event (any two cases of a cluster are clustered) was reviewed, there was a significant speed increase in time to completion of clusters with WGS when compared to MLVA. In other words, WGS provided a more robust case definition than MLVA. Other studies have also highlighted the considerable confidence WGS data affords in assigning like versus not-like status to two potentially linked bacteria (Gilchrist *et al.*, 2015). The impact that this level of confidence in the microbiological typing data brings to the outbreak investigation with respect to improved case ascertainment, is illustrated in Outbreak Scenario 3, below.

Outbreak Scenario 3: an outbreak of STEC O15:H7 PT21/28 associated with raw drinking milk

In September 2014, five cases of STEC O157:H7 all reporting the consumption of raw drinking milk (RDM) were identified by the STEC enhanced surveillance system in England (Butcher *et al.*, 2016). An MLVA profile and SNP address were obtained for the outbreak strain. Following the epidemiological identification of the initial five cases, real-time analysis of national MLVA data identified an additional nine cases of STEC O157:H7 PT21/28 with the same MLVA profile or closely related profiles (SLVs and DLVs of the outbreak profile). The WGS data confirmed that the isolates from four of these nine cases fell within the same 5-SNP cluster as the initial five cases.

The case questionnaires of all nine additional cases were reviewed for evidence of the consumption of RDM. None of the cases could be epidemiologically linked to consumption of RDM from the implicated farm following this initial review. The cases were re-interviewed and asked specific questions about their consumption of dairy products, and a search was carried out for their names and postcodes on the distribution list supplied by the operations manager at the implicated farm. Following these in-depth interviews and subsequent follow-up investigations, three of the four cases that fell within the same 5-SNP cluster as the outbreak strain were, ultimately, linked to the consumption of RDM from the implicated farm. The fourth case reported consumption of RDM from an unknown source but a direct link with the implicated farm could not be confirmed.

RDM is a specialist product and it was surprising that four of the cases did not report consumption of RDM when interviewed initially. Sale and consumption of RDM, and other unpasteurized dairy products, is a contentious issue in the UK, and it is possible that certain cases withheld the information deliberately. These cases were identified via the distribution list held by the manager of the implicated farm. Another of the cases was given the RDM by another party and was genuinely unaware that he had consumed the product. There is good evidence in the literature that epidemiological investigations are often confounded by poor patient recall of the food they consumed prior to onset of symptoms, particularly when the product is a side dish (e.g. salad leaves or raw vegetables) or an ingredient of the main dish (e.g. herbs or spices), so-called 'stealth vehicles' (Buchholz *et al.*, 2011; Byrne *et al.*, 2016). Robust microbiological typing, such as that provided by WGS, makes an important contribution to outbreak investigations when the epidemiological links are obscured by poor patient recall or the when the contaminated product is regarded as a stealth vehicle.

Outbreaks associated with direct contact with animals

Outbreaks comprising isolates where the SNP profiles are identical are relatively easy to interpret as they are clearly linked and there is good genetic evidence they are part of an outbreak with a point source, common exposure. Furthermore, for *E. coli* the five SNP threshold has, so far, proven to be a robust marker of relatedness during an outbreak in a number of different evaluation studies, and retrospective and prospective analysis of outbreaks. However, outbreaks associated with different exposures, for example complex food-borne outbreaks (Outbreak Scenario 10 below), direct contact with animals (Outbreak Scenarios 1 and 4 below) or indirect contact with animals via environmental exposures (Outbreak Scenario 8 below) may require different criteria for outbreak detection and investigation. The characteristics of environmental exposures are explored further using the data from the open farm outbreak described in Outbreak Scenario 1 and in Outbreak Scenario 4 below, in order to shed some light on the level of SNP variation observed during outbreaks involving contact with animals.

Outbreak Scenario 1: outbreak of STEC O157:H7 at an open farm in the county of Surrey

As detailed above, five haplotypes, or SNP types, were detected during the retrospective investigation of the outbreak at the petting farm located in the south-east of England (Underwood *et al.*, 2013). The SNP typing data for the outbreak strain was combined with the epidemiological information, such as date of onset of symptoms, associated with each case in an attempt to determine when specific SNP types appeared during the outbreak and what proportion of the outbreak isolates could be attributed to each SNP type. Two of the isolates having the earliest isolation date belonged to a haplotype that represented a terminal node in the minimal spanning tree, and the ancestral SNP type was isolated throughout the outbreak. None of the other SNP types showed any statistically significant relationships with the onset date or animal exposure, indicating that several animals acted as the source of infection and that all SNP types were present at different locations on the farm.

The use of a SNP-based assay to detect the five haplotypes in the outbreak isolates revealed evidence of clonal expansion of STEC O157:H7 on the farm. The clones could be distinguished by their SNP profile into five SNP types, which differed by only a total of four SNPs. All five SNP types were likely to have been present and widely distributed on the farm prior to the first human clinical case from this outbreak. The authors concluded that this diversity was due to the clonal expansion of a dominant strain via widespread distribution of animal faeces or by multiplication of the outbreak strain in the environment prior to the outbreak.

Outbreak Scenario 4: outbreak of STEC O157:H7 at an open farm in the northwest of England

In April 2014 an outbreak of gastrointestinal disease caused by STEC O157:H7 was linked to an animal petting event at a premises located in the northwest of England (Rowell *et al.*, 2016). Both MLVA and WGS analysis provided evidence that the human and animal isolates were phylogentically linked, although WGS data exhibited a higher level of strain discrimination than MLVA. WGS analysis showed that the human cases were infected with isolates belonging to a subset of the diversity of the isolates detected in the lambs (Fig. 8.4). The meaningfulness of this variation is unclear, as it may result from specific lambs (Lambs 1 and 2) being the source of the transmission event or may represent unsampled isolate diversity in either the animals or the human cases. A similar level of diversity was observed in both the human and animal isolates sequenced during this outbreak compared to the open farm outbreak in 2009, described in Outbreak Scenario 1 above. However, in Outbreak Scenario 4, WGS data directly linked specific animals to the human cases, and the exposure associated with handling the lambs was identified as a high risk activity.

Like the outbreak described in Outbreak Scenario 1, despite the higher level of within outbreak strain diversity exhibited by WGS, variation in the MLVA profiles (SLVs and one DLV) was observed between isolates that had identical SNP addresses. This occurs because SNPs (used in WGS analysis)

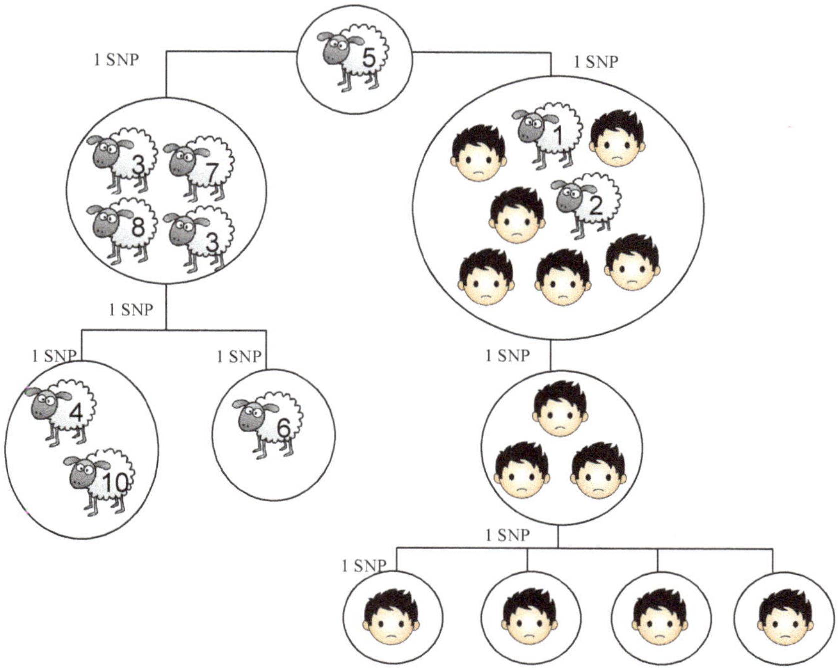

Figure 8.4 WGS analysis of the isolates from human cases and sheep in an outbreak associated with a an animal petting event (Outbreak Scenario 4) showed that the human cases were infected with isolates belonging to a subset of the diversity of the isolates detected in the lambs.

and the polymorphisms of tandemly repeated DNA sequences (used in MLVA) are different biological processes occuring at different rates, possibly under different selection pressures. Previous comparisons of MLVA and SNP analysis during outbreaks have shown that the SNP address is a more robust marker for defining an outbreak than the MLVA profile.

The two outbreaks reviewed in this section highlight the level of diversity we would expect to observe in an outbreak associated with an open farm in the human cases and in the animals sampled on the farm. In both outbreaks, identical SNP profile were detected in human and animal isolates, providing strong evidence the exposure to the animals was the source of the outbreak. However, it is possible, due to unsampled diversity, that identical SNP profiles in animals and humans may not be detected, for example if isolates of STEC O157:H7 had not been successfully recovered from Lambs 1 and 2. In this scenario, it is impoartant to be aware of the level of diversity commonly seen in the farming enviornment to ensure the data are interpreted correctly.

Inferring the geographical origin of an isolate from the WGS phylogeny at the national level

WGS data offers robust, high level of phylogenetic resolution and utilizes stable genetic markers that facilitate the study of the evolutionary context of an outbreak strain. Analysis of the data from the STEC O157:H7 dataset held at Public Health England showed that by exploring the evolutionary context of phylogenetic clusters of isolates, the source of infection could be linked to specific geographical regions of the UK. Dallman *et al.* (2015a) highlighted two phylogenetic clusters comprising isolates < 5 SNP apart where the majority of strains were of Northern Irish provenance. Those patients belonging to the cluster who were not resident in Northern Ireland, reported recent travel prior to onset of symptoms to various parts of Northern Ireland. Similarly, phylogenetic clusters were identified with cases associated with Wales and different regions of England. An example of the latter is described below with reference to Outbreak Scenario 3.

Outbreak Scenario 3: an outbreak of STEC O157:H7 PT21/28 associated with raw drinking milk

The farm implicated in the RDM outbreak described above was located in the south-west of England (Butcher *et al.*, 2016). At the time of the outbreak, five cases had isolates with MLVA profiles that were DLVs of the outbreak MLVA profile, however, no exposure to RDM was identified in these cases. The isolates belonging to these five cases did not fall within the 5-SNP typing cluster, however, they did fall within a 25-SNP cluster of the outbreak cases. Epidemiological analysis of the cases within the 25-SNP cluster, showed that of 70% of the primary cases were resident in the south-west of England or had travelled there within 5 days before the onset of illness (Butcher *et al.*, 2016).

Spatial analysis of the geographical location of the presumed exposure of cases revealed a highly significant cluster in the south-west of England region. Rates of infection with this strain were significantly lower in other parts of England (Fig. 8.5). This analysis provided evidence that the source of infection for some outbreaks and sporadic cases of STEC O157:H7 in the UK may be geographically restricted, and that it may be possible to map the location of the source. The members of the outbreak control team concluded that mapping the potential geographical origin of isolates of STEC O157:H7 may assist in future outbreak investigations as it may be possible to determine the regional source of an implicated food, thus providing an evidence base to direct trace-back investigations to specific locations. Furthermore, this type of geographical analyses may also elucidate the role of the environment as a risk factor for localized transmission of STEC (Butcher *et al.*, 2016).

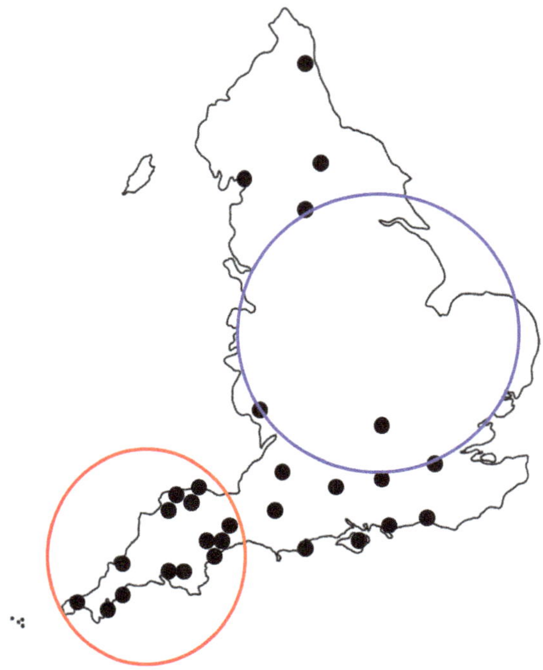

Figure 8.5 Spatial analysis of the geographical location of the presumed exposure of cases in an outbreak associated with raw drinking milk (Outbreak Scenario 3) revealed a highly significant cluster in the south-west of England region (red circle). Rates of infection with this strain were significantly lower in other parts of England (blue circle). (Courtesy of R. Elson: Personal communication.)

Food-borne outbreaks – domestic or non-domestic produce?

Despite high levels of relatedness of the STEC O157:H7 clone, it has long been realized that distinct lineages exist within the population. It is suggested that these arose from the result of geographic spread of an ancestral clone and subsequent regional expansion (Dallman *et al.*, 2015b). Having established that clusters within the UK may be geographically restricted, the possibility that isolates of STEC O157:H7 may also exhibit geographical clustering on a global scale, can be considered (Strachan *et al.*, 2015). Sequencing of STEC O157:H7 bacterial genomes isolated from cases resident in the UK between 1982 and 2016 showed that all three lineages I, I/II and II, were present at the time of the first outbreaks of STEC O157:H7 in 1980s (Dallman *et al.*, 2015b). However, it is clear that certain lineages and sub-lineage were, and still are, common in the UK and others are extremely rare. By tracking the expansion of these lineages and sub-lineages, and by superimposing epidemiological data on the phylogeny, such as known domestic exposures and travel abroad, it is possible to speculate that certain sublineages, clades, or clusters may be domestic in origin and others may be non-domestic.

This approach is currently difficult to evaluate and there are few examples where it has been employed during STEC O157:H7 outbreak investigations to date. However, assumptions can be made regarding the common features of domestic and non-domestic clusters. For example, with reference to the STEC O157:H7 dataset held at Public Health England, approximately 80% of cases have not travelled abroad prior to onset of symptoms, so it is expected that domestic strains would be more common than non-domestic strains in the dataset, and we clearly see a predominance of certain sub-lineages. We see less diversity between the clades within these common sub-lineages because sampling of the pool (i.e. mainly UK cattle and sheep) is more frequent. Included in the domestic clusters, we would expect to see UK animal isolates (e.g. isolates from Outbreak Scenarios 1, 3 and 4) and cases associated with food-borne outbreaks where the food was identified as being of UK origin (e.g. isolates from Outbreak Scenarios 3 and 5). These domestic clusters would also include cases associated with local environmental exposures, such as petting farms and country parks. We would not expect to identify isolates from cases who report recent travel (Table 8.4).

In contrast, non-domestic clusters are more likely to be rare in the UK STEC O157:H7 database and associated with high diversity between isolates within a phylogenetic group. High levels of diversity represent sparse sampling of a large pool of diversity (i.e. a wide variety of zoonotic sources dispersed globally). No UK animal isolates would be present, as the zoonotic source would be located elsewhere. There would be a high proportion of isolates from cases reporting foreign travel. Furthermore, of the cases not reporting travel, we would expect to see links to outbreaks associated with the consumption of imported food (e.g. isolates from Outbreak Scenario 2) or contact with an infected recently returned traveller. We would also anticipate a lack of evidence of domestic acquisition (e.g. no environmental exposures or links to outbreaks associated with domestically produced food) (Table 8.4).

Proof of this principle requires sequencing a large number of isolates from the same region and the data required to accurately identify the regional source for all outbreaks are not currently available. However, evaluating the utility of this concept is important for food-borne outbreak investigations as it may provide data that can guide the trace-back investigations in the right direction. If the outbreak strain can be linked to a likely country of origin,

Table 8.4 Characteristics of domestic and non-domestic clusters of STEC O157:H7

Domestic lineage, clade or cluster	Non-domestic lineage, clade
Examples	
See example 1: Outbreak Scenario 5 outbreak B associated with locally produced watercress	See example 1: Outbreak Scenario 5 outbreak A associated with imported watercress seeds
See example 2: Outbreak Scenario 3 associated with locally produced raw drinking milk	
Characteristic	
Common in domestic dataset	Rare in domestic dataset
Low-level diversity between clusters and clades due to frequent sampling of the pool	High-level diversity between clusters and clades due to infrequent sampling of the pool
High frequency of domestic animal isolates sampled during prevalence studies and sequenced isolates are included in the dataset	Domestic animal isolates not present
Cases do not report recent travel outside the UK prior to onset of symptoms	High frequency of cases reporting recent travel prior to onset of symptoms
Cases from outbreaks known to be associated with domestically produced food	Cases from outbreaks known to be associated with imported food
Cases associated with local environmental exposures, such as petting farms or parks	Cases not associated with local environmental exposures

suspected food items from the same country or region could be targeted for further investigation.

Applying these 'domestic verses non-domestic' criteria to the phylogeny of strains of STEC O157:H7 in the UK indicates that Lineage 1c is a domestic lineage. This lineage mainly comprises PT21/28, currently the most common PT identified in clinical cases and cattle isolates in the UK, and rarely detected in cases reporting recent travel abroad (Byrne *et al.*, 2016, Adams *et al.*, 2016). In other lineages, a higher level of resolution is required to resolve the domestic and non-domestic components. Closer investigation of Lineage I/lla (mainly comprising STEC O157:H7 isolates belonging to PT2) shows that it can be divided in two branches, one branch exhibiting characteristics of a domestic cluster and the other designated a non-domestic cluster (Table 8.4). This concept is explored further with reference to Outbreak Scenario 5 below.

Outbreak Scenario 5: two concurrent outbreaks of STEC O157:H7 PT2 associated with contaminated watercress

In September 2013, an outbreak of STEC O157:H7 PT2 associated with the consumption of contaminated watercress was identified in the UK (Jenkins *et al.*, 2015). Analysis of the WGS data quickly showed that the increase in cases was due to two different outbreaks linked to the same food type but likely to be associated with different sources of contamination. Although the isolates belonged to the same PT, they belonged to different clusters that exhibited different characteristics indicating that one might be from a domestic source and the other might be non-domestic in origin (Table 8.4). Isolates from Outbreak 2 clustered in the same clade (< 39 SNPs) as those from other human cases of domestically acquired infection and domestic cattle. Ultimately, following extensive testing, an isolate linked to the domestic cluster detected in the irrigation water used on one of the implicated farms provided strong evidence that this was a domestic UK strain.

The strain from Outbreak 1 was not isolated from any food, animal or environmental source and therefore, was not conclusively linked to an imported product. However, isolates that fell within the 5-SNP cluster of Outbreak 1 included two cases, identified earlier in the year in February, who had consumed watercress imported from North America and/or mainland Europe. The watercress implicated in both Outbreak 1 and 2 in September was grown in the UK, but the watercress seeds associated with Outbreak 1 were imported, and regarded as a potential source of contamination. The watercress seeds were predominantly supplied by a company based in North America, supplemented by a company based in mainland Europe. Phylogenetic analysis further supported a non-domestic source for the outbreak. Isolates associated with Outbreak 1 were not closely related to any other strains of STEC O157:H7 known to be of UK origin and did not include any UK animal isolates. The cluster was 292 SNPs away from the nearest UK isolate of STEC O157:H7 PT2 (indicating high levels of diversity) and was located on the same branch of the tree as a number of strains of North American origin.

As documented above, currently evidence for the utility of this approach is sparse. However, as more countries implement WGS for routine surveillance of STEC, and the owners of such data ensure open access, cross border exchange of WGS data will have a major impact on facilitating the investigation of national and international outbreaks of food-borne disease.

Outbreaks caused by STEC other than serotype O157:H7

Currently, PFGE and MLVA are the typing methods most commonly employed for typing strains of STEC O157:H7 to inform the epidemiological investigation of cases, and to facilitate outbreak detection and investigation. The MLVA scheme is serotype-specific and, therefore, not applicable to strains of non-O157:H7 STEC. Non-serotype specific MLVA schemes do exist but generally do not exhibit the level of discrimination required for outbreak investigations. Multilocus sequence typing (MLST) is also not discriminatory enough for outbreak investigations of non-O157:H7 STEC and provides approximately the same resolution as serotyping. PFGE is the method of choice in many national reference laboratories for the molecular typing of non-O157:H7 STEC but it is laborious and technically demanding and international laboratory comparisons of the fragment analysis are difficult (Davis *et al.*, 2003).

In contrast, WGS is a versatile approach and the same methods and criteria for outbreak detection and investigation of STEC O157:H7 apply to the other STEC serogroups. Although the majority of publications describing WGS to improve *E. coli* outbreaks investigations focus on STEC O157:H7, WGS has been used to investigate outbreaks of non-O157:H7 STEC. In fact, one of the first STEC outbreaks where WGS was used to facilitate the investigation was the outbreak of HUS caused by STEC O104:H4 in Germany in 2011.

Outbreak Scenario 6: STEC O104:H4 Germany 2011

During the STEC O104:H4 outbreak, WGS was primarily commissioned to determine the characteristics of this highly pathogenic strain, and to map its evolutionary history. Sequencing was carried out on the Ion Torrent PGM™ platform and reads were assembled using the publicly available EAEC 55989 genome sequence (a strain of enteroaggregative *E. coli* (EAEC) O104:H4 isolated in central Africa in the 1990s). Genome comparisons of four outbreak isolates from four different German cities, the reference genome (EAEC 55989) and an historical STEC isolate designated 01–09591 (a strain of STEC O104:H4 isolated from a patient with HUS in 1991), were performed. The analysis demonstrated that all four outbreak isolates were identical, and it was concluded that the outbreak was likely to be clonal and single-sourced (Mellmann *et al.*, 2011). Genome assemblies confirmed that the outbreak strain, designated LB226692, and the historical STEC isolate (01–09591) carried genes typically found in EAEC and STEC and showed a high level of similarity to EAEC 55989. However, unlike typical EAEC strains, both LB226692 and 01-09591 had *stx2*-encoding prophage integrated in *wrbA*, which was also the site of bacteriophage integration for *stx2*-phages in two reference strains of STEC O157:H7, EDL933 and Sakai. In EAEC 55989, *wrbA* was not occupied by a prophage and, unlike the outbreak strain, EAEC 55989 did not harbour the tellurite resistance encoding genes (*ter*). Based on these characteristics, a PCR-based assay targeting *stx2*, the O104 lipopolysaccharide (LPS) encoding gene (*rfb*O104), H4 flagellin-encoding gene (*fli*CH4), and *terD* was developed and implemented for the rapid detection of the outbreak strain (Mellmann *et al.*, 2011).

Plasmid profiling revealed that acquisition and loss of plasmids was a common feature of the evolutionary history of this set of isolates. 01–09591, isolated in 2001, had a 75 kb plasmid encoding AAF/III fimbriae (also present in EAEC 55989) and a 95 kb plasmid encoding type IV pilus system and TEM-1 (not present in EAEC 55989). The outbreak strain (L B226692) had acquired two new plasmids, one encoding AAF/I fimbriae (83 kb) and another encoding TEM-1 and CTX-M-15 beta-lactamases (90 kb), and lost the AAF/III fimbriae encoding plasmid (Mellmann *et al.*, 2011; Rasko *et al.*, 2011).

Mellmann and colleagues (2011) described two evolutionary models for the emergence of the outbreak strain: (i) the 'common ancestor model' proposed a hypothetical STEC O104:H4 progenitor and (ii) the 'linear ancestry model' hypothesised that the STEC O104:H4 strains originated from the prototypical EAEC 55989. Later the same year, after the outbreak was declared over, five sporadic cases of STEC O104:H4 were detected in France (Grad *et al.*, 2013). SNP typing demonstrated that the five post outbreak sporadic isolates detected in France were closely related to each other, and to the outbreak isolate. Importantly, the phylogeny revealed that these five sporadic isolates were not derived from the 2011 outbreak strain, but shared a recent common ancestor. Furthermore, the five sporadic isolates and the 2011 outbreak isolate were more closely related to one another than to the historical *E. coli* O104:H4 isolates (Grad *et al.*, 2013). Grad *et al.* (2013) calculated a linear regression of genetic distance on year of isolation yields estimates of the rate of divergence over time, and concluded that the most recent common ancestor of the STEC clades and EAEC 55989 existed approximately 30 years ago. The most recent common ancestor of the German outbreak strain and sporadic isolates from 2011 existed around 2008 to 2009.

The strain comparison analysis described by Mellmann *et al.* (2011), and subsequently by Grad *et al.* (2013), show that the WGS typing approaches used for investigating outbreaks of STEC O157:H7 are also applicable for non-O157:H7 STEC. In the example above, isolates from cases linked to the outbreak were identical, and were more closely related to contemporary isolates than historical isolates. In Outbreak Scenario 7 below, isolates from cases linked to a nursery outbreak of STEC O26:H11 and other archived isolates of STEC O26:H11 were used to further evaluate the use of genome sequencing as a suitable molecular typing approach for non-O157:H7 STEC outbreak investigations.

Outbreak Scenario 7: STEC O26:H11 associated with a nursery outbreak

The genomes of 19 strains of *E. coli* O26:H11, including the genomes of the four isolates associated with a nursery outbreak were sequenced (Dallman *et al.*, 2015c). All four cases linked to the outbreak were < 3 SNPs apart from each other indicting that the isolates linked to this nursery outbreak had a common source. Moreover, they were clearly differentiated from the sequences of the other 15 strains of *E. coli* O26:H11 from the GBRU archive. Strains from a household outbreak in 2009 were also identified as being closely related to each other with < 3 SNPs between each sequence, and 272 SNPs from the closest isolates in the archive.

In this example, the nursery and household outbreak cases were well-defined, and the outbreaks were relatively simple in that they occurred over a short time frame. In Outbreak Scenario 8 below, SNP typing was used to investigate a more complex outbreak of STEC O55:H7, with cases occurring intermittently over a two year period. The contribution of the WGS data with respect to case identification and in determining the source of the outbreak was assessed.

Outbreak Scenario 8: outbreak of STEC O55:H7 in the south of England

In July 2014, two children were admitted to hospital with HUS in the county of Dorset, in the south of England. Microbiological investigations identified the causative organism as STEC O55:H7 harbouring *stx2a* and *eae*. The combination of these virulence genes are associated with an elevated risk of HUS, and during the course of this outbreak progress to HUS occurred in over 40% of cases. The STEC O55:H7 serotype had not been previously reported in England, although nine cases had occurred in Republic of Ireland (ROI) between 2012 and 2014. Intermittent clusters of cases of STEC O55:H7 continued to occur throughout 2014 and 2015, and by the end of 2015, 30 cases had been identified (McFarland *et al.*, 2017). Extensive investigations were undertaken considering three hypotheses for transmission of STEC O55:H7: (i) contaminated foodstuff or drinking water; (ii) specific recreational/environmental exposure; (iii) epizootic vector and/or general environmental contamination.

WGS was carried out on all 24 of the available isolates from the Dorset cases, two cats owned by two of the cases, and 11 Irish cases. All Dorset isolates from cases and cats were similar to each other forming a distinct clade within five SNP differences of each other (Fig. 8.6). They were similar (1–12 SNPs) to six Irish isolates (five were within five SNPs) from 2013–14 but relatively distant (> 250 SNPs) from earlier Irish isolates from 2012. The isolates from the Dorset cases were all closely genetically related but the diversity between them was greater than might be expected during a simple point source, food-borne outbreak (see Outbreak Scenario 9). Despite very extensive epidemiological investigations no links between the food histories of any of the cases were uncovered, and no common recreational or environmental exposures were revealed. The genetic similarity between isolates taken from Dorset cases and 2013–2014 Irish cases suggests infection by the same or closely related populations of STEC O55:H7. This led to the hypothesis that the exposure was associated with direct contact with an animal carrying the strains, or indirect contact with that animal via their environment. Members of the outbreak control team speculated that the potential link between ROI and Dorset could be explained by animal (e.g. cattle) or bird movements (McFarland *et al.*, 2017). Migratory bird movement corresponded with the seasonality observed in the onset dates of cases, and birds have previously been linked to transmission of STEC O157:H7.

The evidence from WGS of isolates forming a closely related cluster was an important driver behind the extensive investigations undertaken. WGS clearly has a role to play in supporting the investigation of small outbreaks of illness that are not clearly linked epidemiologically.

Outbreaks of DEC other than STEC

So far in this chapter the outbreak scenarios have focused on STEC outbreaks. However, as

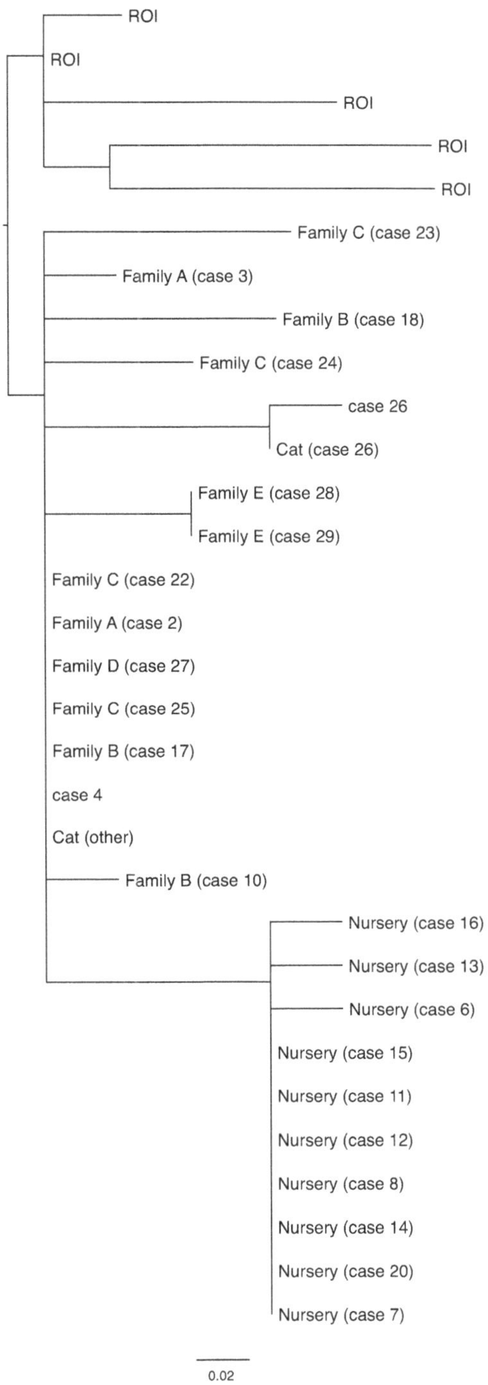

Figure 8.6 WGS was carried out on 24 available isolates from the Dorset human cases, two cats owned by two of the cases, and 11 Irish cases. All Dorset isolates from cases and cats were similar to each other forming a distinct clade within 5 SNP differences of each other. They were similar (1–12 SNPs) to six Irish isolates (five were within 5 SNPs and labelled ROI) from 2013–14 but relatively distant (> 250 SNPs) from earlier Irish isolates from 2012 (not shown). (Courtesy of T.J. Dallman: Personal communication.)

previously stated, WGS is a versatile approach and the same methods and criteria for outbreak detection and investigation of STEC O157:H7 can be applied to other pathotypes of diarrhoeagenic *E. coli* (DEC). The increasing availability and implementation of commercial PCRs for the detection of gastrointestinal pathogens at local hospitals and/or primary health care clinics has resulted in increasing numbers of cases of DEC being identified at the local hospital level. There is, therefore, a need for molecular typing methods that can provide the level of strain discrimination required for outbreak detection and investigation for this diverse group of pathogens.

As discussed above in Outbreak Scenario 6 and indicated in Outbreak Scenario 2 (discussed in more detail in the next section), *E. coli* genomes are dynamic and gene acquisition and loss is frequent and fluid. Novel pathogenic strains of DEC that present a risk to public health are constantly emerging. WGS has a major role to play in the detection and investigation of outbreaks of these emerging pathogens. Not only can WGS data be used to detect outbreaks and link cases where the epidemiological information is unavailable, WGS analysis can also shed light on the pathogenicity profile and evolutionary history of an emerging strain. Outbreak Scenarios 9 and 10 below highlight the utility of WGS during the investigations of outbreaks caused by two emerging strains of DEC, enteroinvasive *E. coli* (EIEC) serotype O96:H19 and EAEC O104:H4.

Outbreak Scenario 9: multiple outbreaks of EIEC O96:H19

In 2012, Escher *et al.* (2014) reported an outbreak of a novel serotype of EIEC, O96:H19, in Italy and this serotype has since been implicated in two outbreaks in the UK in June 2014 associated with two co-located food outlets in the East Midlands (Newitt *et al.*, 2016). In the UK outbreaks, WGS was used to link isolates from human cases, including those from food handlers working at the food outlets, and an isolate from a food item from one of the restaurants.

In the UK outbreak, in July 2014 seven patients were admitted to an emergency department with diarrhoea, vomiting, and fever 24 hours after consuming food purchased at a local takeaway restaurant. A subsequent outbreak was reported linked to a second restaurant located less than 0.1 miles away. Nine isolates from samples in outbreaks A and B, including four patients, one food handler, and the lettuce from outbreak A, and from three patients in outbreak B were sequenced. Phylogenetic analysis showed that all isolates from patients and the food handler were either identical or differed by a single SNP from that sequenced from the lettuce sample. In outbreak A, the combined epidemiological, microbiological and environmental findings implicated lettuce as the vehicle of infection.

The original source of the organism in this outbreak is less clear. Of the 12 food handlers associated with the initial outbreak, nine tested positive for EIEC O96:H19, but most reported they were asymptomatic, so it was not possible to ascertain how or when they acquired their infection. However, one food handler, who tested positive for EIEC, reported becoming ill with gastrointestinal symptoms on return from Pakistan in May 2014. Although the food handler reported not working while symptomatic, there have been reports in the literature of asymptomatic persons shedding EIEC up to 1 year after infection so it is plausible that this food handler may have introduced the organism into the restaurant. Poor food hygiene standards identified at the restaurant may have facilitated cross-contamination among the other food handlers through person-to-person transmission or consumption of contaminated food items.

Other studies investigating the gene content and possible origin of this rare serotype O96:H19 observed that it belonged to sequence type ST99, never reported before in EIEC or *Shigella* lineages. The strain possessed all the virulence genes harboured on the large plasmid conferring the invasive phenotype to EIEC and *Shigella* while showing only some of the known chromosomal virulence genes, and none of the described pathoadaptative mutations, for example those observed in *lacZ* or *CadA* in *Shigella* and other strains of EIEC. At the same time, this serotype displayed motility and biochemical requirements resembling more closely those of the non-pathogenic *E. coli*. Michelacci *et al.* (2015) concluded that strains of *E. coli* O96:H19 strains belong to an emerging EIEC clone, resulting from a recent event of acquisition of the invasion plasmid by commensal *E. coli*.

Outbreak Scenario 10: outbreak of gastrointestinal disease associated contaminated curry leaves

In March 2013, a large outbreak of gastrointestinal disease occurred in the north-east of England and cases were linked to a Street Spice food festival (Dallman *et al.*, 2014). 413 cases reported illness including symptoms of persistent diarrhoea and abdominal pain immediately following the event, and a total of 592 cases were identified following an on-line questionnaire. One hundred and ten specimens were submitted for microbiological testing. Using traditional culture methods, *Salmonella enterica* serotype Agona was isolated from 25 cases and four further patients were found to be infected with other *Salmonella* species. Cohort and trace-back studies implicated a contaminated, fresh curry leaves from Pakistan as the source of the infection. However, the small number of cases testing positive for *Salmonella* species, and the duration of symptoms, which was uncharacteristic of salmonellosis, raised the suspicion that this was a multi-pathogen outbreak and further testing using a pan-pathogen PCR was requested by the Outbreak Control Team. Subsequently, strains of EAEC harbouring *aggR* were isolated from PCR positive faecal specimens.

Phylogeny of EAEC isolated from the cases associated with the outbreak was investigated. Ultimately, ten different serotypes and nine MLSTs were identified among the EAEC isolated from the outbreak cases. The most commonly observed serotypes were O131:H27 ($n=6$), O104:H4 (n = 5) and O20:H19 ($n=2$), and the most frequently identified STs, corresponding with these serotypes, were ST10, ST678 and ST278, respectively. Initially, it was suggested that the variation in serotype in the outbreak strains was masking a closer phylogenetic relationship. However, the phylogenetic tree created by comparing SNPs in the core genome showed that, although strains of the same serotype were relatively closely related, those of different serotypes were diverse. Strains belonging to the same serotype clustered on the same branch of the tree, however, even within the same serotype, isolates were phylogenetically distinct.

A phylogeny based on 3115 core SNPs of 14 strains of *E. coli* O104:H4 illustrated the relationship between the EAEC O104:H4 strains in this study with sporadic strains of *E. coli* O104:H4 and the strain associated with the *E. coli* O104:H4 outbreak in Germany and France in 2011 and with isolates from sporadic cases isolates in France in the same year. Although none of the EAEC O104:H4 isolates in this outbreak had *stx* they shared a common ancestor with the German outbreak strain and the sporadic *stx*-harbouring *E. coli* O104:H4 strains isolated in France in 2011. All five strains of *E. coli* O104:H4 isolated during this outbreak shared the multidrug resistant genomic island conferring resistance to ampicillin, the sulphonamides, streptomycin and tetracycline, and the S83A gyrA mutation, in common with German outbreak strain and the closely related EAEC/STEC sporadic isolates from France (Dallman *et al.*, 2014).

EAEC is known to be a diverse pathotype and this study demonstrates that a high level of diversity in strains recovered from cases associated with a single outbreak. This was a complex outbreak as the diversity observed between the EAEC serotypes and in isolates of the same serotype was high, and atypical for a point source outbreak. Although the EAEC in this study did not harbour *stx*, this outbreak provides further evidence of the pathogenic potential of the EAEC O104:H4 serotype and the WGS analysis contributed to further elucidation of its evolutionary history.

Use of long-read sequencing in outbreak investigations

As discussed previously in this chapter, *E. coli* has a large, dynamic accessory genome comprising multiple mobile genetic elements, such as prophage. The constant gain and loss of mobile genetic elements, such as plasmids and phage is known to be rapid and fluid, but very few studies have explored the consequences of this genetic exchange occurring during an outbreak investigation. In part, this is due to difficulties with the assembly of repetitive and paralogous features of prophages when using short read sequencing caused by multiple assignment in the genomes when reads do not span repeated paralogous prophage genes (Cowley *et al.*, 2016). Long-read sequencing improves *de novo* assemblies and facilitates more accurate characterization of the accessory genome including prophage regions that are a major feature of *E. coli* genomes. In Outbreak Scenario 2 described above and in this section below, WGS resolved the discrepancy between the phage typing and MLVA results (Fig. 8.3), but did not explain how a change in phage type, from PT8 to PT54, could occur in two strains that were so closely related. Cowley *et al.* (2016) used a combination of long-read sequencing technologies to study and document the short-term microevolution events occurring between two closely related outbreak strains.

Outbreak Scenario 2: two related outbreaks of STEC O157:H7 associated with the same food outlet

Fourteen prophage regions were shared between the representative isolates of the PT8 and PT54 clusters. In addition to the 14 shared prophage regions, both PT8 and PT54 isolates each had acquired one unique prophage region. Furthermore, the genome of the PT8 isolate had three repeats of a common prophage region within a 200 kbp inverted repeat. Two prophage regions shared by both PT8 and PT54 isolates, designated P7 and P8, were similar prophages that showed variation between the two representative isolates, indicative of recombination that had contributed to the inverted repeat region. The PT8 isolate had 82 unique genes on the chromosome, 79 of which were prophage associated, and the PT54 isolate had 30 unique genes, 28 of which were prophage associated. Whole genome alignment identified 29 SNPs between the two isolates in fully aligned regions. This was higher than the three SNPs identified between the core genomes based on the short read sequencing but 26 of these SNPs were found in the P7 and P8 shared prophage regions. These prophage regions were not shared with the reference genome, so would not have been identified in the original core genome SNP-calling from the Illumina short read data (Cowley *et al.*, 2016).

Both strains harboured pO157, the O157 virulence plasmid present in nearly all strains of STEC O157:H7. However, the PT54 isolate had acquired an additional ~220 genes introduced on an IncHI2 plasmid not present in the PT8 isolate. While both isolates exhibited tellurite and tetracycline resistance, the PT54 isolate was also resistant to chloramphenicol and streptomycin and this matched with resistance genes located on the IncHI2 plasmid. By conjugating the plasmid into the PT8 isolate and then phage typing the conjugant, it was demonstrated that the acquisition of the IncHI2 plasmid was responsible for the difference in PT observed between the two outbreak clusters. Although, the precise mechanism causing the change in phage type could not be determined, the authors speculated that it may be associated with acquisition of additional tellurite resistance genes or methylase modification genes encoded on the plasmid. They also speculated that acquisition of the plasmid bestowing increased resistance to bacteriophage (PT54 is resistant to more of the T4 typing phages than PT8, which is a phage type susceptible to all the T4 phages in the typing scheme), chloramphenicol, streptomycin, tetracycline, tellurite, and other heavy metal ions enhanced the survival of this strain under certain stressful environmental conditions, thus conferring an evolutionary advantage (Cowley *et al.*, 2016).

Summary

This chapter highlights the ways in which the use of WGS data for typing *E. coli* improves outbreak detection and investigations, and outlines the methods available to assess relatedness between isolates of *E. coli* that have been evaluated in the literature. A review of the use of WGS data in real outbreak scenarios demonstrates a number of advantages of using this approach during outbreak investigations.

- Generic method suitable for all *E. coli*.
- Unprecedented level of strain discrimination.
- Robust, stable genetic markers.
- Case identification when epidemiological links are obscured.
- Tolerates higher levels of diversity seen in environmental exposures.
- Geographical origins of outbreak strains may be inferred from the phylogenetic signal.
- Provides evolutionary context for emerging pathogenic strains.

Long-read technology will facilitate the analysis of loss and acquisition of mobile genetic elements in the accessory genome, and improve our understanding of the impact of short-term evolution during outbreak investigations. Expanding sequenced based WGS globally will improve trace-back investigations in the event of a food-borne outbreak, ensuring the rapid implementation of interventions to protect public health. The importance of publicly available WGS data linked to the clinical, epidemiological and environmental context of the sequenced strain cannot be underestimated.

References

Adams, N.L., Byrne, L., Smith, G.A., Elson, R., Harris, J.P., Salmon, R., Smith, R., O'Brien, S.J., Adak, G.K., and Jenkins, C. (2016). Shiga toxin-producing *Escherichia coli* O157, England and Wales, 1983-2012. Emerging Infect. Dis. *22*, 590–597. https://doi.org/10.3201/eid2204.151485

Bray, J. (1945). Isolation of antigenically homogenous strains of *Bact. Coli neapolitanum* from summer diarrhoea in infants. J. Pathol. Bacteriol. 57, 239-247.

Buchholz, U., Bernard, H., Werber, D., Böhmer, M.M., Remschmidt, C., Wilking, H., Deleré, Y., an der Heiden, M., Adlhoch, C., Dreesman, J., *et al.* (2011). German outbreak of *Escherichia coli* O104:H4 associated with sprouts. N. Engl. J. Med. *365*, 1763–1770. https://doi.org/10.1056/NEJMoa1106482

Butcher, H., Elson, R., Chattaway, M.A., Featherstone, C.A., Willis, C., Jorgensen, F., Dallman, T.J., Jenkins, C., McLauchlin, J., Beck, C.R., and Harrison, S. (2016). Whole genome sequencing improved case ascertainment in an outbreak of Shiga toxin-producing *Escherichia coli* O157 associated with raw drinking milk. Epidemiol. Infect. *144*, 2812-2823. doi: 10.1017/S0950268816000509.

Byrne, L., Adams, N., Glen, K., Dallman, T.J., Kar-Purkayastha, I., Beasley, G., Willis, C., Padfield, S., Adak, G., and Jenkins, C. (2016). Epidemiological and Microbiological Investigation of an Outbreak of Severe Disease from Shiga Toxin-Producing *Escherichia coli* O157 Infection Associated with Consumption of a Slaw Garnish. J. Food Prot. 79, 1161-1168.

Cowley, L.A., Dallman, T.J., Fitzgerald, S., Irvine, N., Rooney, P.J., McAteer, S.P., Day, M., Perry, N.T., Bono, J.L., Jenkins, C., and Gally, D.L. (2016). Short-term evolution of Shiga toxin-producing *Escherichia coli* O157:H7 between two food-borne outbreaks. Microbial. Genomics. 2(9): e000084. doi:10.1099/mgen.0.000084.

Dallman, T.J., Chattaway, M.A., Cowley, L.A., Doumith, M., Tewolde, R., Wooldridge, D.J., Underwood, A., Ready, D., Wain, J., Foster, K., Grant, K.A., and Jenkins, C. (2014). An investigation of the diversity of strains of enteroaggregative *Escherichia coli* isolated from cases associated with a large multi-pathogen food-borne outbreak in the UK. PLOS ONE. *9*, e98103. doi:10.1371/journal.pone.0098103.

Dallman, T.J., Byrne, L., Ashton, P.M., Cowley, L.A., Perry, N.T., Adak, G.K., Petrovska, L., Ellis, R.J., Elson, R., Underwood, A., Green, J., Hanage, W.P., Jenkins, C., Grant, K., and Wain, J. (2015a). Whole-genome sequencing for national surveillance of Shiga toxin-producing *Escherichia coli* O157. Clin. Infect. Dis. *61*, 305-312. doi: 10.1093/cid/civ318.

Dallman, T.J., Ashton, P.M., Byrne, L., Perry, N.T., Petrovska, L., Ellis, R., Allison, L., Hanson, M., Holmes, A., Gunn, G.J., Chase-Topping, M.E., Woolhouse, M.E.J., Grant, K.A., Gally, D.L., Wain, J., and Jenkins, C. (2015b). Applying phylogenomics to understand the emergence of Shiga-toxin-producing *Escherichia coli* O157:H7 strains causing severe human disease in the UK. Microbial. Genomics. *1*(3): e000029. doi: 10.1099/mgen.0.000029.

Dallman, T.J., Byrne, L., Launders, N., Glen, K., Grant, K.A., and Jenkins, C. (2015c). The utility and public health implications of PCR and whole genome sequencing for the detection and investigation of an outbreak of Shiga toxin-producing *Escherichia coli* serogroup O26:H11. Epidemiol. Infect. *143*, 1672–1680. doi: 10.1017/S0950268814002696.

Davis, M.A., Hancock, D.D., Besser, T.E., and Call, D.R. (2003). Evaluation of pulsed-field gel electrophoresis as a tool for determining the degree of genetic relatedness between strains of *Escherichia coli* O157:H7. J. Clin. Microbiol. *41*, 1843–1849.

Escher, M., Scavia, G., Morabito, S., Tozzoli, R., Maugliani, A., Cantoni, S., Fracchia, S., Bettati, A., Casa, R., Gesu, G.P., Torresani, E., and Caprioli, A. (2014). A severe food-borne outbreak of diarrhoea linked to a canteen in Italy caused by enteroinvasive *Escherichia coli*, an uncommon agent. Epidemiol. Infect. *142*, 2559–2566. doi: 10.1017/S0950268814000181.

Eppinger, M., Mammel, M.K., Leclerc, J.E., Ravel, J., and Cebula, T.A. (2011). Genomic anatomy of *Escherichia coli* O157:H7 outbreaks. Proc. Natl. Acad. Sci. U.S.A. *108*, 20142–20147. https://doi.org/10.1073/pnas.1107176108

Eppinger, M., and Cebula, T.A. (2015). Future perspectives, applications and challenges of genomic epidemiology studies for food-borne pathogens: A case study of enterohemorrhagic *Escherichia coli* (EHEC) of the O157:H7 serotype. Gut. Microbes. *6*, 194–201. doi: 10.4161/19490976.2014.969979.

Ferdous, M., Friedrich, A.W., Grundmann, H., de Boer, R.F., Croughs, P.D., Islam, M.A., Kluytmans-van den Bergh, M.F., Kooistra-Smid, A.M., and Rossen, J.W. (2016). Molecular characterization and phylogeny of Shiga toxin-producing *Escherichia coli* isolates obtained from two Dutch regions using whole genome sequencing. Clin. Microbiol. Infect. *22*, 642. e1–9. doi: 10.1016/j.cmi.2016.03.028.

Franz, E., Delaquis, P., Morabito, S., Beutin, L., Gobius, K., Rasko, D.A., Bono, J., French, N., Osek, J., Lindstedt, B.A., Muniesa, M., Manning, S., LeJeune, J., Callaway, T., Beatson, S., Eppinger, M., Dallman, T., Forbes, K.J., Aarts, H., Pearl, D.L,, Gannon, V.P., Laing, C.R., and Strachan, N.J. (2014). Exploiting the explosion of information associated with whole genome sequencing to tackle Shiga toxin-producing *Escherichia coli* (STEC) in global food production systems. Int. J. Food Microbiol. *187*, 57–72. doi: 10.1016/j.ijfoodmicro.2014.07.002.

Grad, Y.H., Godfrey, P., Cerquiera, G.C., Mariani-Kurkdjian, P., Gouali, M., Bingen, E., Shea, T.P., Haas, B.J., Griggs, A., Young, S., Zeng, Q., Lipsitch, M., Waldor, M.K., Weill, F.X., Wortman, J.R., and Hanage, W.P. (2013). Comparative genomics of recent Shiga toxin-producing *Escherichia coli* O104:H4: short-term evolution of an emerging pathogen. MBio. *4*, e00452-12. doi: 10.1128/mBio.00452-12.

Gilchrist, C.A., Turner, S.D., Riley, M.F., Petri, W.A., and Hewlett, E.L. (2015). Whole-genome sequencing in outbreak analysis. Clin. Microbiol. Rev. *28*, 541–563. https://doi.org/10.1128/CMR.00075-13

Gerner-Smidt, P., Hise, K., Kincaid, J., Hunter, S., Rolando, S., Hyytiä-Trees, E., Ribot, E.M., and Swaminathan, B. (2006). PulseNet USA: a five-year update. Foodborne Pathog. Dis. *3*, 9–19. https://doi.org/10.1089/fpd.2006.3.9

Holmes, A., Allison, L., Ward, M., Dallman, T.J., Clark, R., Fawkes, A., Murphy, L., and Hanson, M. (2015). Utility of whole-genome sequencing of *Escherichia coli* O157 for outbreak detection and epidemiological surveillance. J. Clin. Microbiol. 53, 3565–3573. https://doi.org/10.1128/JCM.01066-15

Hyytia-Trees, E., Lafon, P., Vauterin, P., and Ribot, E.M. (2010). Multilaboratory validation study of standardized multiple-locus variable-number tandem repeat analysis protocol for Shiga toxin-producing *Escherichia coli* O157: a novel approach to normalize fragment size data between capillary electrophoresis platforms. Foodborne Pathog. Dis. *7*, 129–136.

Jenkins, C., Dallman, T.J., Launders, N., Willis, C., Byrne, L., Jorgensen, F., Eppinger, M., Adak, G.K., Aird, H., Elviss, N., Grant, K.A., Morgan, D., and McLauchlin, J. (2015). Public health investigation of two outbreaks of Shiga toxin-producing *Escherichia coli* O157 associated with consumption of watercress. Appl. Environ. Microbiol. *81*, 3946–3952. doi: 10.1128/AEM.04188-14.

Joensen, K.G., Scheutz, F., Lund, O., Hasman, H., Kaas, R.S., Nielsen, E.M., and Aarestrup, F.M. (2014). Real-time whole-genome sequencing for routine typing, surveillance, and outbreak detection of verotoxigenic *Escherichia coli*. J. Clin. Microbiol. 52, 1501–1510. doi: 10.1128/JCM.03617-13.

Khakhria, R., Duck, D., and Lior, H. (1990). Extended phage-typing scheme for *Escherichia coli* O157:H7. Epidemiol. Infect. *105*, 511–520.

Lindstedt, B.A., Vardund, T., and Kapperud, G. (2004). Multiple-locus variable-number tandem-repeats analysis of *Escherichia coli* O157 using PCR multiplexing and multi-colored capillary electrophoresis. J. Microbiol. Methods. *58*, 213–222.

McFarland, N., Bundle, N., Jenkins, C., Godbole, G., Mikhail, A., Dallman, T., O'Connor, C., McCarthy, N., O'Connell, E., Treacy, J. *et al.* (2017). Recurrent seasonal outbreak of an emerging serotype of Shiga toxin-producing *Escherichia coli* (STEC O55:H7 Stx2a) in the south west of England, July 2014 to September 2015. Euro Surveill. 22(36), pii: 30610. https://doi.org/10.2807/1560-7917.ES.2017.22.36.30610

Mellmann, A., Harmsen, D., Cummings, C.A., Zentz, E.B., Leopold, S.R., Rico, A., Prior, K., Szczepanowski, R., Ji, Y., Zhang, W., McLaughlin, S.F., Henkhaus, J.K., Leopold, B., Bielaszewska, M., Prager, R., Brzoska, P.M., Moore, R.L., Guenther. S., Rothberg, J.M., and Karch, H. (2011). Prospective genomic characterization of the German enterohemorrhagic *Escherichia coli* O104:H4 outbreak by rapid next generation sequencing technology. PLOS ONE. *6*, e22751.

Michelacci, V., Prosseda, G., Maugliani, A., Tozzoli, R., Sanchez, S., Herrera-León, S., Dallman, T., Jenkins, C., Caprioli, A., and Morabito, S. (2016). Characterization of an emergent clone of enteroinvasive *Escherichia coli* circulating in Europe. Clin. Microbiol. Infect. *22*, 287.e11–9. https://doi.org/10.1016/j.cmi.2015.10.025

Newitt, S., MacGregor, V., Robbins, V., Bayliss, L., Chattaway, M.A., Dallman, T., Ready, D., Aird, H., Puleston, R., and Hawker, J. (2016). Two linked enteroinvasive *Escherichia coli* outbreaks, Nottingham, UK, June 2014. Emerging Infect. Dis. *22*, 1178–1184. https://doi.org/10.3201/eid2207.152080

Parsons, B.D., Zelyas, N., Berenger, B.M., and Chui, L. (2016). Detection, characterization, and typing of Shiga toxin-producing *Escherichia coli*. Front. Microbiol. *7*, 478. https://doi.org/10.3389/fmicb.2016.00478

Rowell, S., King, C., Jenkins, C., Dallman, T.J., Decraene, V., Lamden, K., Howard, A., Featherstone, C.A., Cleary, P. (2016). An outbreak of Shiga toxin-producing *Escherichia coli* serogroup O157 linked to a lamb-feeding event. Epidemiol. Infect. *144*, 2494–2500. doi: 10.1017/S0950268816001229.

Rasko, D.A., Webster, D.R., Sahl, J.W., Bashir, A., Boisen, N., Scheutz, F., Paxinos, E.E., Sebra, R., Chin, C.S., Iliopoulos, D., Klammer, A., Peluso, P., Lee, L., Kislyuk, A.O., Bullard, J., Kasarskis, A., Wang, S., Eid, J., Rank, D., Redman, J.C., Steyert, S.R., Frimodt-Møller, J., Struve, C., Petersen, A.M., Krogfelt, K.A., Nataro, J.P., Schadt, E.E., and Waldor, M.K. (2011). Origins of the *E. coli* strain causing an outbreak of hemolytic-uremic syndrome in Germany. N. Engl. J. Med. *365*, 709-717. doi: 10.1056/NEJMoa1106920.

Riley, L.W., Remis, R.S., Helgerson, S.D., McGee, H.B., Wells, J.G., Davis, B.R., Hebert, R.J., Olcott, E.S., Johnson, L.M., Hargrett, N.T., *et al.* (1983). Hemorrhagic colitis associated with a rare *Escherichia coli* serotype. N. Engl. J. Med. *308*, 681–685. https://doi.org/10.1056/NEJM198303243081203

Smith, H.R., Scotland, S.M., Cheasty, T., Willshaw, G., and Rowe, B. (1996). Enteropathogenic *Escherichia coli* in the United Kingdom. Rev. Microbiol. 27, 45-49.

Strachan, N.J., Rotariu, O., Lopes, B., MacRae, M., Fairley, S., Laing, C., Gannon, V., Allison, L.J., Hanson, M.F., Dallman, T., Ashton, P., Franz, E., van Hoek, A.H., French, N.P., George, T., Biggs, P.J. Forbes, K.J. (2015). Whole genome sequencing demonstrates that geographic variation of *Escherichia coli* O157 genotypes dominates host association. Sci Rep. 5, 14145. doi: 10.1038/srep14145.

Underwood, A.P., Dallman, T., Thomson, N.R., Williams, M., Harker, K., Perry, N., Adak, B., Willshaw, G., Cheasty, T., Green, J., *et al.* (2013). Public health value of next-generation DNA sequencing of enterohemorrhagic *Escherichia coli* isolates from an outbreak. J. Clin. Microbiol. *51*, 232–237. https://doi.org/10.1128/JCM.01696-12

Data Processing of *Escherichia coli* Genome Sequencing, Characterization and Comparison

9

Gian Marco Baranzoni*†, Erin R. Reichenberger† and David S. Needleman*

Eastern Regional Research Center, Agricultural Research Service, United States Department of Agriculture, Wyndmoor, PA, USA.
†These authors have equally contributed to this work

*Correspondence: david.needleman@ars.usda.gov and gmbaranzoni@gmail.com

https://doi.org/10.21775/9781910190777.09

Abstract

Pathogenic *Escherichia coli* continue to raise concerns as one of the major causes of food-borne diseases, bloodstream infections, and urinary tract infections. The remarkable advances in DNA sequencing technologies offer new alternative approaches for detection, characterization, and tracking of pathogenic *E. coli* strains with higher resolution and rapid analysis time. This chapter focuses on the main steps for analysing whole-genome sequencing data and includes tools and databases suitable for *E. coli* genome analysis by scientists with programming and non-programming backgrounds. In particular, different sequencing platforms, typical input/output file formats, and raw data inspection and quality control assessment strategies are reviewed. Also discussed are *de novo* genome assembly, draft assembly improvement, and assembly visualization, as well as *E. coli* predictive genomic tools for serotyping and detection of virulence genes, antimicrobial resistance genes, and mobile genetic elements. Examples of genetic content comparisons and pan-genome characterization and phylogeny are provided.

Introduction

The advent of next-generation sequencing (NGS) and its evolution is significantly changing clinical and food microbiology and may eventually replace traditional cultural, serological and PCR-based methods to detect, type, and characterize bacterial pathogens (Didelot *et al.*, 2012; Ronholm *et al.*, 2016). Due to the increased affordability of NGS technologies and the consequent augmentation of publicly available genomic data, numerous databases and bioinformatics tools[1] have been developed. Additionally, Whole-genome sequencing (WGS) has been shown to significantly reduce analysis time after the prokaryotic colony has been isolated from a food sample. For example, Joensen *et al.* (2014) compared the time needed for typing procedures (serotyping, haemolysin production, β-glucuronidase activity, Shiga toxin (Stx) production, virulence gene detection, pulsed-field gel electrophoresis) using WGS data and routine laboratory procedures and found that the traditional typing techniques took almost twice as long to acquire the results as sequencing-based techniques. Although there is a decrease in analysis time when using NGS data, there is also a new set

1 The words tools and software are used interchangeably.

of skills that are required to fully take advantage of this technology.

Bioinformatics tools can be offered as free General Public License (GPL) tools, web-only interface tools (web tools), and commercially available software packages. GPL tools are typically operated from the command-line and allow the greatest amount of control, specificity, and detailed output, while requiring the steepest learning curve. Web tools are accessible via a browser, are easy to use, and have low user-system requirements. However, they often do not provide the same level of control, specificity, or detailed output that can be obtained by using the same tool from the command-line. Similar to web tools, commercially available software packages incorporate a suite of tools that are accessible using a Graphic User Interface (GUI). While commercial software programs are more user friendly than command-line software, they are proprietary and generally retain their source code, which makes it difficult to understand how the software programs function. Additionally, there are often hidden costs and limitations arriving from licensing and upgrading procedures.

Regardless of which software type is used, the support of a bioinformatician is highly recommended to ensure that the data analysis is performed and interpreted correctly. The authors acknowledge that not all investigators have a bioinformatician on staff and analysing NGS data is not a trivial exercise. We have therefore, ushered forth a chapter that will review NGS sequencing methods, file formats, and bioinformatics tools for prokaryotic genome assembly, annotation, characterization, and multiple genomic comparisons (Fig. 9.1). All the databases and bioinformatics tools mentioned in this chapter are summarized in Table 9.1. For those users with rudimentary or no programming knowledge, a large part of the command-line tools mentioned in this chapter can be used through a Galaxy web-based platform. In particular, the Galaxy versions Orione/CRS4 (Cuccuru *et al.*, 2014) and ARIES (https://www.iss.it/site/aries/) are tailored for bacterial NGS data analysis. Moreover, Biostars (https://www.biostars.org/), SEQanswers (http://seqanswers.com/), and Stack Overflow (https://stackoverflow.com/) are excellent communities that provide helpful information about programming, statistics, software installations, and bioinformatics. As previously stated, in this chapter, we will introduce several tools that could be used to analyse NGS data from prokaryotic organisms and, following the topic of this book, examples will focus on *E. coli*.

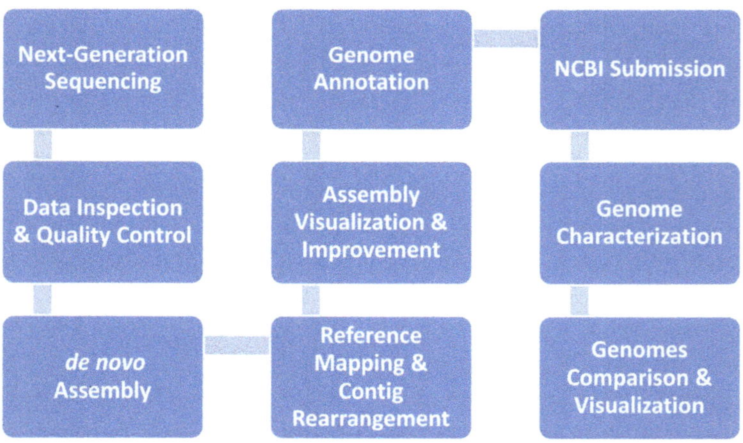

Figure 9.1 General workflow of *E. coli* whole-genome sequencing analysis. Raw data are generated using a high-throughput sequencing platform and quality inspected using FastQC. Low quality reads are filtered or trimmed using Trimmomatic (Bolger *et al.*, 2014). Polished raw data are *de novo* assembled into contigs using e.g. SPAdes (Bankevich *et al.*, 2012). Assemblies can be improved upon using tools such as Pilon (Walker *et al.*, 2014) or PAGIT (Swain *et al.*, 2012), and contigs can be reordered using Mauve (Darling *et al.* 2010) and a reference genome. Draft genomes are annotated and submitted to GenBank for future publication. Serotypes, virulence genes, antimicrobial genes and mobile genetic elements can be identified using predictive genomic tools. If dealing with two or more genomes, they can be compared via visualization of multiple genome alignment, or based on their gene content and genomic variations.

Table 9.1 Selection of databases and bioinformatics tools for *E. coli* next generation sequencing data analysis

Name	Functions	Typology	Website	Reference
Antibiotic Resistance Gene -ANNOTation (ARG-ANNOT)[1]	Antimicrobial resistance gene detection	Database and web tool	http://www.mediterranee-infection.com/article.php?laref=282&titre=arg-annot	Gupta *et al.* (2014b)
Antibiotic Resistance Genes Database (ARDB)	Antimicrobial resistance gene detection	Database and web tool	http://ardb.cbcb.umd.edu/	Liu and Pop (2009)
Bacterial Analysis Pipeline	Genome assembly and assessment, species identification, MLST, virulence and antimicrobial genes identification, and plasmid detection and typing	Web tool	https://cge.cbs.dtu.dk/services/cge/	Thomsen *et al.* (2016)
Bacterial Isolate Genome Sequence Database (BIGSdb)	MLST, whole genome MLST, core genome MLST, and ribosomal MLST	Software for Linux/Unix OS	https://pubmlst.org/software/database/bigsdb/	Jolley and Maiden (2010)
Bandage	Genome assemblies assessment and visualization	Software versions for all OS	https://rrwick.github.io/Bandage/	Wick *et al.* (2015)
BioNumeric	Platform for clustering, identification, statistics, sequence analysis, genome comparison, and metagenomics analysis	Software for Windows OS	http://www.applied-maths.com/bionumerics	Applied Maths
BLAST Ring Image Generator (BRIG)	Multiple genomes comparison and circular visualization	Java software	http://brig.sourceforge.net/	Alikhan *et al.* (2011)
bp_genbank2gff3.pl	Conversion from GenBank to GFF3 format	Perl script	https://github.com/bioperl/bioperl-live/blob/master/scripts/Bio-DB-GFF/bp_genbank2gff3.pl	NP
Burrows-Wheeler Aligner (BWA)	Low-divergent sequence alignment	Software for Linux/Unix OS	http://bio-bwa.sourceforge.net/	Li and Durbin (2009)
Canu	Long-read assembly	Software for Linux/Unix OS	https://github.com/marbl/canu	Koren *et al.* (2017)
CLC Genomics Workbench	Genome assembly, RNA-seq, metagenomic, epigenomic, resequencing, and variant analysis	Software versions for all OS	https://www.qiagenbioinformatics.com/products/clc-genomics-workbench/	QIAgen
Comprehensive Antibiotic Resistance Database (CARD)	Antimicrobial resistance gene detection	Database and web tool	http://arpcard.mcmaster.ca	Jia *et al.* (2017)
Database of Bacterial ExoToxins for Human (DBETH)	Exotoxins gene identification	Database and web tool	http://www.hpppi.iicb.res.in/btox/index.html	Chakraborty *et al.* (2012)
Ensemble algorithm for Genomic Island Detection (EGID)[2]	Genomic island detection	Software for Linux/Unix OS	http://www5.esu.edu/cpsc/bioinfo/software/EGID/	Che *et al.* (2011)
EnteroBase	Genome assembly, MLST, whole genome MLST, core genome MLST, and ribosomal MLST	Database and web tool	http://enterobase.warwick.ac.uk/	NP
Escherichia coli O-groups and H-types (EcOH) database	*Escherichia coli in silico* serotyping	Software for Linux/Unix OS	https://github.com/katholt/srst2	Ingle *et al.* (2016)
FastQC	Quality control of next generation sequencing data	Java software	https://www.bioinformatics.babraham.ac.uk/projects/fastqc/	NP
FASTX-Toolkit	Manage FASTA and FASTQ files	Software for Linux/Unix OS	http://hannonlab.cshl.edu/fastx_toolkit/	NP
Geneious	Genome assembly, mapping, molecular cloning, phylogenetic analysis, annotation and variant analysis	Software versions for all OS	http://www.geneious.com/	Global Headquarters Biomatters Ltd.
ggplot2	Graphics creation and data visualization	R package	https://cran.r-project.org/web/packages/ggplot2/index.html	Wickham (2016)

Table 9.1 Continued

Name	Functions	Typology	Website	Reference
Harvest	Core genome alignment and visualization	Software for Linux/Unix and Mac OS	http://harvest.readthedocs.io/en/v1.1/index.html	Treangen *et al.* (2014)
Horizontal Gene Transfer DataBase (HGT-DB)	Genomic island databases	Database	http://genomes.urv.es/HGT-DB/4	Garcia-Vallve *et al.* (2003)
INTEGRALL	Integrons database	Database	http://integrall.bio.ua.pt/	Moura *et al.* (2009)
Integrative Genomics Viewer (IGV)	Visualization of array-based data, next generation sequence data, and genomic annotation	Java software	http://software.broadinstitute.org/software/igv/	Thorvaldsdóttir *et al.* (2013)
Islander	Genomic Island database	Database	http://bioinformatics.sandia.gov/islander/	Mantri and Williams (2004)
IslandViewer 4	Genomic Island prediction and visualization	Web tool	http://www.pathogenomics.sfu.ca/islandviewer/	Bertelli *et al.* (2017)
ISMapper	Transposable element insertions site detection	Software for Linux/Unix OS	https://github.com/jhawkey/IS_mapper	Hawkey *et al.* (2015)
ISsaga	Annotation of insertion sequences	Web tool	http://issaga.biotoul.fr/issaga_index.php	Varani *et al.* (2011)
iTol	Creation of phylogenetic tree	Web tool	https://itol.embl.de/	Letunic and Bork (2016)
Jalview	Multiple sequence alignment, alignment visualization, phylogenetic tree editing, and clustering	Java software	http://www.jalview.org/	Waterhouse *et al.* (2009)
Lawrence Livermore National Laboratory Virulence Database (MvirDB)	Multi-criteria databases searching tool	Web tool	http://mvirdb.llnl.gov/4	Zhou *et al.* (2007)
Large-Scale BLAST Score Ratio (LS-BSR)	Multiple genome comparison of coding regions	Software for Linux/Unix OS	https://github.com/jasonsahl/LS-BSR	Sahl *et al.* (2010)
Mauve	Multiple genome alignment, alignment visualization, and contigs rearrangement.	Java software	http://darlinglab.org/mauve/mauve.html	Darling *et al.* (2010)
MLST	*In silico* MLST for multiple microorganisms	Web tool	https://cge.cbs.dtu.dk/services/MLST/	Larsen *et al.* (2012)
MP3	Prediction of virulent protein in genomic and metagenomic datasets	Web tool and software for Linux/Unix OS	http://metagenomics.iiserb.ac.in/mp3/index.php	Gupta *et al.* (2014a)
NCBI prokaryotic genome annotation pipeline	Annotation of prokaryotic genomes during the NCBI genome submission	Web tool	https://www.ncbi.nlm.nih.gov/genome/annotation_prok/	Tatusova *et al.* (2016)
Panseq	Determination and analysis of core and accessory genomes	Software for Linux/Unix OS	https://github.com/chadlaing/Panseq	Laing *et al.* (2010)
Pathosystems Resource Integration Center (PATRIC)	Genome assembly and annotation, proteome comparison, RNA-seq analysis, variation analysis and model reconstructions	Web tool	https://www.patricbrc.org/	Wattam *et al.* (2017)
Pathogenicity Island Database v2.0 (PAIDB)	Detection of pathogenicity and antimicrobial resistance islands	Database and web tool	http://www.paidb.re.kr/about_paidb.php	Yoon *et al.* (2015)
PHAge Search Tool Enhanced Release (PHASTER)	Detection, annotation, and visualization of prophages in prokaryotic genomes	Web tool	http://phaster.ca/	Arndt *et al.* (2016)
PhiSpy	Detection and prediction of prophages in prokaryotic genomes	Software for Linux/Unix OS	https://sourceforge.net/projects/phispy/	Akhter *et al.* (2012)
Pilon	Genome assembly improvement and microbial variant detection	Software for Linux/Unix OS	https://github.com/broadinstitute/pilon/	Walker *et al.* (2014)
PlasmidFinder	Plasmid sequences detection in genomes belonging to the Enterobacteriaceae family	Database and web tool	https://cge.cbs.dtu.dk/services/PlasmidFinder/	Carattoli *et al.* (2014)

Table 9.1 Continued

Name	Functions	Typology	Website	Reference
plasmidSPAdes	Detection of plasmid reads in NGS data	Software for Linux/Unix OS	http://spades.bioinf.spbau.ru/plasmidSPAdes/	Antipov *et al.* (2016)
plasmid Multilocus Sequence Typing (pMLST)	Enterobacteriaceae plasmid typing	Web tool	https://cge.cbs.dtu.dk/services/pMLST/	Carattoli *et al.* (2014)
Post-assembly genome-improvement toolkit (PAGIT)	Genome assembly improvement	Software for Linux/Unix OS	http://www.sanger.ac.uk/science/tools/pagit	Swain *et al.* (2012)
Prokka	Annotation of prokaryotic genomes	Software for Linux/Unix OS	https://github.com/tseemann/prokka	Seemann (2014)
PyBamView	Visualization of short read alignments	Software for Linux/Unix OS	https://github.com/mgymrek/pybamview	Gymrek (2014)
Quast	Genome assemblies evaluation	Web tools and software for Linux/Unix OS	http://quast.sourceforge.net/quast	Gurevich *et al.* (2013)
Rapid Annotation using Subsystem Technology (RAST)	Annotation of prokaryotic genomes	Webserver and software versions for all OS	http://rast.nmpdr.org/	Aziz *et al.* (2008)
Repository of antibiotic resistance cassettes (RAC)	Detection and annotation of antibiotic resistance cassettes	Database and web tool	http://rac.aihi.mq.edu.au/rac/	Tsafnat *et al.* (2011)
Resfams	Antibiotic resistance protein families and hidden Markov model profiles database	Database	http://www.dantaslab.org/resfams	Gibson *et al.* (2015)
ResFinder	Acquired antimicrobial resistance gene detection	Database and web tool	https://cge.cbs.dtu.dk/services/ResFinder/	Zankari *et al.* (2013)
Ridom SeqSphere+	Manage NGS and Sanger sequence data, genome assembly, variant analysis, MLST, core genome MLST, and ribosomal MLST	Software for Windows and Linux/Unix OS	http://www.ridom.com/seqsphere/index.shtml	Ridom GmbH
Roary	Prokaryotic pan-genome analysis	Software for Linux/Unix OS	https://sanger-pathogens.github.io/Roary/	Page *et al.* (2015)
Sequence Alignment/Map tools (SAM tools)	Create and manage SAM files	Software for Linux/Unix OS	http://samtools.sourceforge.net/	Li *et al.* (2009)
SerotypeFinder	*Escherichia coli in silico* serotyping	Web tool	https://cge.cbs.dtu.dk/services/SerotypeFinder/	Joensen *et al.* (2015)
Scoary	Determine pan-genome component association to phenotypic traits	Software for Linux/Unix OS	https://github.com/AdmiralenOla/Scoary	Brynildsrud *et al.* (2016)
SOAPdenovo2	Short-read *de novo* assembly	Software for Linux/Unix OS	http://soap.genomics.org.cn/soapdenovo.html	Luo *et al.* (2012)
SPAdes	Short-read *de novo* assembly	Software for Linux/Unix and Mac OS	http://cab.spbu.ru/software/spades/	Bankevich *et al.* (2012)
SuperPhy	*E. coli* predictive genomic tools for virulence and antimicrobial resistance genes detection, genomes comparison, and geospatial distribution	Web tool	https://lfz.corefacility.ca/superphy/home	Whiteside *et al.* (2016)
Trimmomatic	Raw data quality trimming and adapter removal	Java software	http://www.usadellab.org/cms/?page=trimmomatic	Bolger *et al.* (2014)
user-friendly Comprehensive Antibiotic resistance Repository of *Escherichia coli*	Multidrug-resistant *E. coli* database	Database and web tool	http://www.e-bioinformatics.net/ucare/	Saha *et al.* (2015)
Velvet	Short-read *de novo* assembly	Software for Linux/Unix OS	https://www.ebi.ac.uk/~zerbino/velvet/	Zerbino and Birney (2008)

Table 9.1 Continued

Name	Functions	Typology	Website	Reference
VennDiagram	Data visualization using Venn and Euler plots	R package	https://cran.r-project.org/web/packages/VennDiagram/index.html	Chen and Boutros (2011).
Virmugen DB	Detection of virulence factor associated with live attenuated vaccine production	Database and web tool	http://www.violinet.org/virmugendb/index.php	Racz *et al.* (2013)
VirSorter[3]	Phage and prophage detection in single-cell genomes and metagenomic datasets	Software for Linux/Unix OS	https://github.com/simroux/VirSorter/	Roux *et al.* (2015)
Virulence factor database (VFDB)	Virulence factor detection	Database and web tool	http://www.mgc.ac.cn/cgi-bin/VFs/jsif/main.cgi	Chen *et al.* (2016)
VirulenceFinder	Virulence gene detection	Database and web tool	https://cge.cbs.dtu.dk/services/VirulenceFinder/	Joensen *et al.* (2014)
VirulentPred	Prediction of prokaryotic virulent protein	Web tool	http://www.imtech.res.in/raghava/vicmpred/	Garg and Gupta (2008)
VRprofile	Detection of virulence and antimicrobial genes, pathogenicity island and antimicrobial resistance islands	Web tool	http://bioinfo-mml.sjtu.edu.cn/VRprofile	Li *et al.* (2017)

MLST: Multilocus Sequence Typing; NP: Not Published; NGS: Next Generation Sequencing; OS: Operative System. [1] ARG-ANNOT can be run in a local computer using BioEdit (http://www.mbio.ncsu.edu/bioedit/bioedit.html). [2] EGID is also available in the Genomic Island Suite of Tools (http://www5.esu.edu/cpsc/bioinfo/software/GIST/). [3] VirSorter available as a webtool on the CyVerse Discovery Environment (https://de.iplantcollaborative.org/de/) [4] Web page not accessible at the time of writing.

Additionally, we direct the readers to the review by Edwards and Holt (2013) for a step-by-step tutorial that includes example datasets depicting assembly and comparison of *E. coli* genomes.[2]

Escherichia coli: an introduction

Escherichia coli is a Gram-negative bacterium that is found in the intestinal flora of warm-blooded animals. Although *E. coli* is generally harmless, there are pathogenic groups that can provoke intestinal and extraintestinal diseases in humans and other endotherms. Illnesses caused by pathogenic *E. coli* are mainly transmitted by the consumption of contaminated food or water, and thus represent important food safety concerns. Traditionally, *E. coli* strains have been classified by serotype, which is a combination of O-group and H-type that are identified by the presence of somatic and flagellar surface antigens, respectively. There are currently ~186 O-groups and 53 H-types that have been identified, which can yield numerous O-group/H-type combinations (Fratamico *et al.*, 2016). Certain pathogenic strains are strongly associated with specific serotypes. For instance, almost half of the domestically acquired illnesses caused by Shiga toxin-producing *E. coli* (STEC) belong to serotype O157:H7. The remaining O-serogroups that cause a large portion of STEC-related illnesses are O26, O45, O111, O121, and O145 (Smith *et al.*, 2014).

The ability of *E. coli* to provoke human diseases depends on multiple factors such as route of transmission, infective dose, and host resistance abilities. *E. coli* genetics also play a fundamental role in pathogenicity. Certain *E. coli* groups have acquired combinations of virulence factors that allow them to colonize new niches and establish serious infections. Variants of pathogenic *E. coli* are called 'pathotypes' and can be differentiated from one another based on clinical syndromes, infection strategies, and presence of specific virulence factors (Kaper *et al.*, 2004). The diarrhoeagenic *E. coli* include (i) enteropathogenic *E. coli* (EPEC), which provoke profuse watery diarrhoea in children and carry the bundle-forming pilus (*bfp*) and the intimin (*eae*) genes; (ii) STEC, which produce

2 Edwards and Holt have recently released an updated version of the review (https://katholtlab.files.wordpress.com/2017/07/comparativegenomicstutorialv2.pdf).

Shiga toxin (Stx1 and/or Stx2) and are cause diarrhoea (watery and haemorrhagic), haemorrhagic colitis and haemolytic uraemic syndrome (HUS); (iii) enteroinvasive *E. coli* (EIEC), which are intracellular pathogens that carry virulence factors similar to those of *Shigella* and provoke dysentery and potentially HUS; (iv) enteroaggregative *E. coli* (EAEC), which is responsible for traveller's diarrhoea and infection with which is characterized by the presence of the aggregative adherence fimbriae encoded by the pAA plasmid; (v) diffusely adherent *E. coli* (DAEC), which is associated with persistent watery diarrhoea in children and named because of its adherence pattern on specific cell lines *in vitro* and; (vi) enterotoxigenic *E. coli* (ETEC), which use one or more colonization factors and produce toxins known as heat-labile enterotoxin and heat-stable enterotoxin to cause persistent watery diarrhoea; (vii) and adherent invasive *E. coli* (AIEC), which has been associated with Crohn's disease lesions (Croxen *et al.*, 2013; Kaper *et al.*, 2004). Additionally, another important pathotype consists of the extraintestinal pathogenic *E. coli* (ExPEC), which is able to cause infections outside of the gastrointestinal tract (Kaper *et al.*, 2004; Köhler and Dobrindt, 2011).

E. coli has a very flexible genome, and like other bacteria, there are conserved genes that are present in all strains (core genome) and variable sets of genes that are present in some but not all strains (accessory genome). The accessory genome may include genes that confer special abilities to the bacteria that allow them to survive in specific niches and to cause illnesses. As of May 2017, there are 6,142 *E. coli* entries in the National Center for Biotechnology Information (NCBI) nucleotide database.[3] For *E. coli*, the minimum chromosome sequence length is 4,019,802 bp, the maximum sequence length is 6,323,148 bp, and the average length is 5,110,837 bp (NCBI Resource Coordinators, 2017). The *E. coli* genome is also very dynamic as it contains mobile genetic elements such as plasmids, prophages, and transposons that can move within the genome or be transferred to other bacteria. The transfer of mobile genetic elements between one cell and another cell that is not an offspring is called horizontal or lateral gene transfer (Pallen and Wren, 2007). Horizontal gene transfer actively contributes to the distribution of virulence or antimicrobial resistance (AR) genes from strain to strain. For example, the genes encoding Stx are generally transferred by a lambdoid bacteriophage (Asadulghani *et al.*, 2009), and the colistin resistance mechanism MCR-1 was recently found on a plasmid (Liu *et al.*, 2016). The possibility that a highly virulent *E. coli* strain could originate from a less pathogenic strain by horizontal gene transfer was realized in 2011. That year in Germany, an *E. coli* O104:H4 strain caused a large outbreak with 4137 known clinical cases and 54 deaths. The O104:H4 strain, an enteroaggregative *E. coli* with a complete pAA plasmid that harboured virulence associated genes, acquired the Stx2a-encoding phage and the extended-spectrum β-lactamase antibiotic resistance plasmid via horizontal transfer. This unusual combination of virulence factors in this hybrid strain provoked HUS in 22% of the recognized patients (Boisen *et al.*, 2015).

Advances in DNA sequencing technologies

DNA sequencing is the determination of the order of nucleotides within a region of DNA or within an entire genome. The nucleotide sequence contains the genetic instructions for the structure and function of of biological life and the transfer of hereditary information to future generations. In 1977, Sanger and coworkers published a seminal paper on a DNA sequencing method that proved to be less time-consuming and more efficient than previous methodologies (Sanger *et al.*, 1977). Over the next 30 years, refinements of the initial first-generation sequencing method, concomitant with unprecedented advancements in computer processing and genetic analysis software has resulted in the evolution of Sanger sequencing from a manual gel-based technology to a 96-sample well, automated sequencing system, which can yield approximately 1000 bases of DNA sequence with an individual base accuracy as high as 99.999% for a single DNA sample (Shendure and Ji, 2008). Although Sanger sequencing is still used for targeting DNA regions of the genome following PCR amplification of the DNA, newer technologies are now available for sequencing millions of bases in a single run. The first

3 The search parameters were: (*Escherichia coli* [Organism]) AND 4000000:7000000 [Sequence Length]. This approach should exclude plasmids.

E. coli strain sequenced with this technology was the 4,639,221-bp non-pathogenic K-12 (accession number: NC_000913) reference strain (Blattner *et al.*, 1997). This major achievement was followed by sequencing the whole genome of two *E. coli* O157:H7 strains, which were responsible for large outbreaks associated with HUS in Michigan, USA and Sakai, Japan. These early genome comparisons between the *E. coli* K-12 and the two pathogenic *E. coli* O157:H7 strains highlighted the importance of horizontal gene transfer and the accessory genome in bacterial pathogenicity (Hayashi *et al.*, 2001; Perna *et al.*, 2001).

Second-generation sequencing was ushered in by the introduction of the Roche (Basel, Switzerland) 454 sequencing platform (Margulies *et al.*, 2005).[4] A variety of applications and workflow approaches soon followed in conjunction with an explosive increase in bioinformatics tools to decipher and interpret the flow of NGS data. Only a year after the introduction of the Roche sequencing platform, Solexa[5] rolled out their sequencing platform, which could produce over one Gigabase of sequencing data in a single run. At the present time, the newest Illumina (San Diego, California, USA) sequencing platform can generate over one terabase of sequencing data per run. Illumina, the leader in high-throughput sequencing, has manufactured a variety of sequencing platforms from desktop models (MiniSeq and MiSeq) to an integrated cluster of sequencers such as the HiSeq and NovaSeq series that can yield up to 3000 Gb of data. The MiSeq instrument can achieve up to 15 Gb of data with a quality of 99.9% for over 90% of the base sequences. While, the MiniSeq outputs 6.6 Gb to 7.5 Gb of data as 2×150-bp paired reads, and over 80% of the bases have an accuracy of higher than 99.9%. The Ion Torrent (Thermo Fisher Scientific, Waltham, MA, USA) systems are another group of second-generation sequencing platforms with broad appeal. Their latest instrument, the Ion S5 system, can achieve from 0.3 to 15 Gb of sequence data with a quality also comparable to Illumina instruments. Although both the Ion Torrent platforms and the Illumina platforms are capable of doing microbial sequencing and transcriptome (RNA targeted) sequencing, the Ion S5 was developed specifically for targeted sequencing (e.g. sequencing specific regions of a genome) (Bio-IT World, 2015).

Recently third-generation sequencing platforms have become available. Unlike second-generation sequencing that requires PCR-based amplification of the DNA and results in only short-reads that need to be linked together to produce longer fragments, third-generation sequencers comprise single-molecule technologies and generate long-reads of DNA fragments of 10,000s of bases. The ability of longer-read systems that produce more accurate *de novo* bacterial genome assemblies greatly eases the assembly process and enables analysis of genomic variations. In 2011, Pacific Bioscience (Menlo Park, CA, USA) commercially released the first third-generation instrument (PacBio RS), which was used to sequence the *E. coli* O104:H4 strain that was responsible for causing the 2011 HUS-associated outbreak in Germany (Rasko *et al.*, 2011).

A recent, major third-generation achievement is the commercial release of the MinION (Oxford Nanopore Technologies, Oxford, UK). Oxford Nanopore has developed a portable sequencer based on nanopore technology (Jain *et al.*, 2016). The MinION is approximately the size of a midsize flash drive, is controlled by software on a laptop computer, and is capable of single-nucleotide resolution. In 2015, Loman *et al.* (2015) reported using the long-read data from the Oxford Nanopore MinION that performed a *de novo* assembly of the chromosome of *E. coli* strain K-12 with a 99.5% base-level accuracy. Other studies show that nanopore sequencing is a promising technology when sequencing DNA in remote locations (Johnson *et al.*, 2017).

As new developments in NGS technologies continue to evolve, the cost of sequencing an organism has dramatically decreased from millions of dollars to currently around $1000, and it is predicted that the cost will eventually drop even further. Whether using short- or long-read sequencing, the most challenging steps continue to be complete assembly and genome closing, as well as annotation.

4 It should be noted that Roche no longer manufactures or supports the 454 sequencing platform, but is instead vested in developing nanopore sequencing technology.

5 Solexa was later acquired by Illumina.

File formats

There are numerous file formats that are employed to hold biological data. How the data are stored, as well as the type of data contained within those formats have often been the source of much confusion among those who deal in NGS data (Higgins *et al.*, 1992). A major source of the confusion arises from competing file formats that serve the same purpose but are not necessarily compatible across bioinformatics tools or file formats that have several depreciated versions. Many of these formats originated from a strong desire for those without programming experience or knowledge to view, edit, and parse files with NGS data. Additionally, certain applications have specific file types that require conversion into another format for use with a different program or similarly, when shared with a collaborator. This section will focus on some of the most common non-proprietary file formats for whole genome NGS data, their uses, and some of the pitfalls affiliated with these file types. Moreover, file format purposes are summarized in Table 9.2.

Text (.txt, .csv, .tsv)

While most people are probably familiar with text files, this file type is likely to contain more than just NGS data. Additional information in these files may contain metadata that is affiliated with the sequenced organism or the start and stop positions of a particular gene for multiple strains. There are three basic types of text files having the extensions. txt,.csv, and.tsv. A standard text file (.txt) could have data stored in any number of ways, however very common file formats are comma separated values (.csv) and tab separated values (.tsv). These formats can store very large tables where the columns are distinguished by special characters such as a comma (.csv) or a tab (.tsv).

FASTA (.fasta, .fa, .fsa, .faa, .fna, .frn)

One of the first NGS files that many people will encounter is the FASTA file. This text-based format tool developed by Pearson and Lipman (1988) compares protein and DNA sequences. These files contain either nucleic acid or amino acid sequence data, and they are commonly used for sequence alignment and similarity searches of databases. General FASTA files will have .fa, .fsa, or .fasta file extensions. The extension .fna is specific for a FASTA file containing nucleic acids whereas a .faa FASTA file is specific for amino acid sequences. Lastly, the .frn FASTA file contains RNA data. The typical format of a FASTA file consists of a header line that is followed by lines of the sequence. The header line immediately precedes the line of sequence and must start with the '>' character (Fig. 9.2). In the event of a multi-sequence FASTA file, all headers must be uniquely named. Additionally, although it is not required, headers should be informative and provide some information about the sequence.

FASTQ (.fastq)

FASTQ files are similar to FASTA files, but they contain additional information such as sequencing instrument, the sequencing run, and the quality score for each base within the sequence read (Cock *et al.*, 2010). The FASTQ format is widely used for short-read data obtained from sequencing instruments. An example of this file format can be seen in

Table 9.2 File Formats and their corresponding utility

Utility	FASTA	FASTQ	SAM, BAM	GenBank	GFF, GTF	Tabular	Sequin
Sequencing		X					
Quality Control		X					
Assembly	X	X	X				
Annotation				X	X	X	X
Transcriptomics		X	X		X		
NCBI Genome Submission	X			X		X	X
Visualization	X		X	X	X		
Comparative Genomics	X			X	X		

```
>NC_000913.3 Escherichia coli str. K-12 substr. MG1655, complete genome
AGCTTTTCATTCTGACTGCAACGGGCAATATGTCTCTGTGTGGATTAAAAAAAGAGTGTCTGATAGCAGCTTCTGAACTG
GTTACCTGCCGTGAGTAAATTAAAATTTTATTGACTTAGGTCACTAAATACTTTAACCAATATAGGCATAGCGCACAGAC
AGATAAAAATTACAGAGTACACAACATCCATGAAACGCATTAGCACCACCATTACCACCACCATCACCATTACCACAGGT
AACGGTGCGGGCTGACGCGTACAGGAAACACAGAAAAAAGCCCGCACCTGACAGTGCGGGCTTTTTTTTTCGACCAAAGG
TAACGAGGTAACAACCATGCGAGTGTTGAAGTTCGGCGGTACATCAGTGGCAAATGCAGAACGTTTTCTGCGTGTTGCCG
ATATTCTGGAAAGCAATGCCAGGCAGGGGCAGGTGGCCACCGTCCTCTCTGCCCCCGCCAAAATCACCAACCACCTGGTG
GCGATGATTGAAAAAAACCATTAGCGGCCAGGATGCTTTACCCAATATCAGCGATGCCGAACGTATTTTTGCCGAACTTTT
```

Figure 9.2 FASTA file format. The first line (header) starts with a '>' and the subsequent lines contain the sequence. A FASTA file can contain a single sequence or multiple sequences.

```
39 @M03838:16:000000000-ANNB2:1:1101:14554:2187 1:N:0:6
40 CTAACAAGCATTGAACGTTTAGGTGGGCTTCGGGATAAATTAGCGTCGAACTTTACTATTGATGATTCTTATTATACCGA
41 +
42 AAAAAFFFBFFFGGGGGFGGGGGHFGGDHGHGGGGFCGGHHCFHGEEEEFEGFGHHHBGGHHFHFFGFGFFFHHFFHFGD
```

Figure 9.3 FASTQ file format. FASTQ files generally contain multiple genomic reads generated during the sequencing process. Each read is composed by four elements: (i) title and optional description sequence line marked by the initial character '@' (line 39), (ii) nucleotide sequence (line 40), (iii) '+' which signals the end of the sequence and optionally the title line could be repeated here (line 41), (iv) PHRED quality scores shown as ASCII codes (line 42).

Fig. 9.3. The first line of a FASTQ file commences with the character '@'. This initial line is identical in purpose to the header line of the FASTA file and uniquely identifies the sequence (second line). The FASTQ headers generally contain information about the name of the sequencing instrument and possibly the flow cell lane and positional information. The second line contains the nucleotide sequence, which is also identical in format and purpose to the sequences found in a FASTA file. The third line begins with the '+' character and is meant to signal the end of the sequence line. The '+' character may be followed by the sequence identifier found in the header line or other descriptive information. The final line shows the quality score (Q-score) for each base in the sequence (note that the length of the quality scores must be the same length as the sequence). The Q-score is organized from lowest to highest quality and consists of ASCII codes ranging from the base number 33 ('!') to the 75 ('K'), which correspond to the PHRED Q-score 0 and 42, respectively. The complete scoring scheme can be found at http://www.drive5.com/usearch/manual/quality_score.html. The Q-score for a particular base is related to the probability that the sequencer has incorrectly assigned the base (Cock *et al.*, 2010). As an example of base calling accuracy, the first nucleotide in Table 9.3 is 'G', and the Q-score for that nucleotide is '3'. The corresponding Q-score for the ASCII character '3' is 18 with a probability error (P) of 0.0158[6] (Table 9.3). In other words, P is the probability of an incorrect assignment by the sequencer; in the case of the first base 'G', it is approximately 1/63. Cock *et al.* (2010) provide the historical context of the

6 The PHRED Q-score (Q) is related to the probability (P) by the following equation: $Q = -10\log_{10}P$.

Table 9.3 Base calling accuracy per nucleotide in a FASTQ file

Nucleotide Sequence	G	C	A	G	C	G	C	G	G	A
Quality Score[1]	3	:	<	8	B	=	<	A	@	C
ASCII base number	51	58	60	56	66	61	60	65	64	67
Q-score[2]	18	25	27	23	33	28	27	32	31	34
P[3]	0.0158	0.0032	0.0020	0.0050	0.0005	0.0016	0.0020	0.0006	0.0008	0.0004

[1]ASCII codes ranging from '!' to 'K'.
[2]PHRED Q-score can be derived by subtracting 33 from the base number corresponding to the ASCII character.
[3]The probability error (P) formula is $P = 10^{-Q/10}$.

FASTQ format in a manner that is both succinct and descriptive while also including information on how to convert PHRED scores between alternative sequencing methods.

SAM/BAM Files (.sam, .bam, .bai)

SAM stands for Sequence Alignment Map and is a format that contains positional information about how a sequence/read is mapped to the reads of a reference genome (Li *et al.*, 2009). These files are largely used for reference mapping, sorting, and indexing purposes. SAM files may be the output from certain programs, or they can be created from tools such as Burrows–Wheeler Aligner (BWA) (Li and Durbin, 2009) or The Sequence Alignment/Map tools (SAM tools) (Li *et al.*, 2009). A SAM file contains two sections, a header section and an alignment section. Headers commence with a '@' character, and alignments do not. Although an extensive header is not required, the header may contain generic information or more specific information such as the state of ordering that was performed on the file. Lines originating from the alignment portion of the SAM file contain 11 mandatory tab-delimited fields with the option for additional fields (Li *et al.*, 2009). If a mandatory field is missing a value, it can be represented as a zero or a '*', depending on the field. A SAM file has the following 11 fields: (i) QNAME – name of template query; (ii) FLAG – bitwise descriptor; (iii) RNAME – reference sequence name; (iv) POS – leftmost position of 1st matching base; (v) MAPQ[6] – mapping quality; (vi) CIGAR – 'concise idiosyncratic gapped alignment report' string; (vii) RNEXT – reference sequence name of the next read in the alignment; (viii) PNEXT – alignment position of next read; (ix) TLEN – length of template sequence; (x) SEQ – sequence; and (xi) QUAL – ASCII code of base quality. Specifications for some fields are extensive (see https://samtools.github.io/hts-specs/SAMv1.pdf for additional information).

SAM files are large, though easy to parse. The compressed version of a SAM file is a Binary Alignment Map (BAM) file (Li *et al.*, 2009). Although, the information in the BAM file is, at its essence, identical to a SAM file, it is compressed and only readable by a computer. There is a special file that accompanies a BAM file. It has a .bai extension and is an index to what is in the BAM file. This allows programs that employ SAM/BAM files to quickly jump to specific areas of the BAM file without having to read through the entire BAM file. A.bai file can be thought of a hyperlink that takes you from the top of the webpage directly to the bottom of the webpage without having to scroll through the entire page.

GenBank (.gb, .gbk, .gbf)

GenBank formatted files contain the sequence of the organism's genome, information about all features (e.g. CDS, tRNA, etc…), and the protein sequence of the translated feature. A GenBank file is generally the output of genome annotation and is split into three main portions: header, features, and origin. The header of a GenBank file contains information about the size of the genome, whether the chromosome is circular or linear, the organism in question, and other information such as published references (Fig. 9.4). The genome may have been used in one publication or several publications. Therefore, the reference section may have multiple listings of journals and titles. The feature portion of the GenBank file displays regions within the genome that perform a biological function (Fig. 9.5). CDS, mRNA, tRNA, and gene are some of the typical GenBank features. One of the most commonly observed features is the CDS or coding sequence feature (note: the gene feature is always paired with a CDS feature). Each feature will contain information about the location of the feature on the genome (e.g. 190…273), the name of gene (if known), the name of the product, a translation of the features nucleotide sequence into a protein sequence, and additional attributes that provide further information about the feature. Lastly, there is the origin, which contains the nucleotide sequence of the entire genome, as shown in Fig. 9.6.

GFF/GTF

GFF and GTF stand for General Feature Format and General Transfer Format, respectively. These file formats are two possible means to store genomic features and can be used for many purposes such as genome visualization, RNA-Seq analysis, and pan-genome analysis. There are several different versions of these feature files: GFF2, GFF3, and GTF. Each file type is essentially a.tsv file that contains genomic features in a column-based format. Each line represents a single annotated feature and each

```
LOCUS       BA000007             5498450 bp    DNA     circular BCT 22-DEC-2016
DEFINITION  Escherichia coli O157:H7 str. Sakai DNA, complete genome.
ACCESSION   BA000007 AP002550-AP002569
VERSION     BA000007.2
DBLINK      BioProject: PRJNA226
            BioSample: SAMN01911278
KEYWORDS    .
SOURCE      Escherichia coli O157:H7 str. Sakai
  ORGANISM  Escherichia coli O157:H7 str. Sakai
            Bacteria; Proteobacteria; Gammaproteobacteria; Enterobacterales;
            Enterobacteriaceae; Escherichia.
REFERENCE   1
  AUTHORS   Makino,K., Yokoyama,K., Kubota,Y., Yutsudo,C.H., Kimura,S.,
            Kurokawa,K., Ishii,K., Hattori,M., Tatsuno,I., Abe,H., Iida,T.,
            Yamamoto,K., Onishi,M., Hayashi,T., Yasunaga,T., Honda,T.,
            Sasakawa,C. and Shinagawa,H.
  TITLE     Complete nucleotide sequence of the prophage VT2-Sakai carrying the
            verotoxin 2 genes of the enterohemorrhagic Escherichia coli O157:H7
            derived from the Sakai outbreak
  JOURNAL   Genes Genet. Syst. 74 (5), 227-239 (1999)
   PUBMED   10734605
```

Figure 9.4 GenBank header for *E. coli* O157:H7 strain Sakai (accession number: BA000007). The header includes general information such as accession, BioProject, and BioSample numbers, organism name, genome size, and reference information (when available).

```
FEATURES             Location/Qualifiers
     source          1..5498450
                     /organism="Escherichia coli O157:H7 str. Sakai"
                     /mol_type="genomic DNA"
                     /strain="Sakai"
                     /sub_strain="RIMD 0509952"
                     /serovar="O157:H7"
                     /db_xref="taxon:386585"
     gene            complement(25..711)
                     /gene="ECs5361"
     CDS             complement(25..711)
                     /gene="ECs5361"
                     /inference="non-experimental evidence, no additional
                     details recorded"
                     /note="similar to LAST_ECOLI gi|1790865 percent identity
                     97 in 228 aa (Conserved in E.coli K-12)"
                     /codon_start=1
                     /transl_table=11
                     /product="hypothetical protein"
                     /protein_id="BAB38784.1"
                     /translation="MRITIILVAPARAENIGAAARAMKTMGFSELRIVDSQAHLEPAT
                     RWVAHGSGDIIDNIKVFPTLAESLHDVDFTVATTARSRAKYHYYATPVELVPLLEEKS
                     SWMSHAALVFGREDSGLTNEELALADVLTGVPMVADYPSLNLGQAVMVYCYQLAILIQ
                     QPTKSDTTADQHQLQALRERAMALLTTLAVADDIKLVDWLQQRLGLLEQRDTAMLHRL
                     LHDIEKNITK"
```

Figure 9.5 GenBank features for *E. coli* O157:H7 strain Sakai (accession number: BA000007). Image shows the source feature (genome size, organism name, data type, strain name, etc.), one gene feature, and one CDS feature. Note that the gene name for the CDS feature (ECs001) is the same as the gene feature that precedes it since they are paired together.

```
ORIGIN
        1 gaaaatact tactaaggcg tttttattt ggtgatattt ttttcaatat catgcagcaa
       61 acggtgcaac attgccgtgt ctcgttgctc taaaagcccc aggcgttgtt gtaaccagtc
      121 gaccagtttt atgtcatctg ccactgccag agtcgtcagc aatgccatgg ctcgttcgcg
      181 taaagcttgc agttgatgtt ggtctgccgt tgtgtcactt ttcgtcggtt gttgtattaa
      241 tattgctaat tgatagcaat agaccatcac cgcctgcccc agattgagcg aaggataatc
      301 cgccaccatc ggcacaccag taagaacgtc agccaacgct aactcttcgt tagtcagccc
      361 ggaatcttcg cgaccaaaca ccagcgcggc atggctcatc catgaagatt tttcctctaa
      421 cagcggcacc agttctactg gcgtagcgta gtaatgatat ttcgcccgac tgcgcgcagt
      481 ggtggcgaca gtgaaatcga catcgtgtaa cgattcagcc agtgtcggga aaactttaat
      541 attatcaata atatcaccag atccatgtgc gacccagcgg gtggctggct ccaggtgtgc
      601 ctgactatcg acaatccgca gttcgctaaa ccccatcgtt ttcattgccc gcgccgctgc
      661 cccaatattt tctgctctgg cgggtgcgac cagaataatc gttatacgca tattgccact
      721 cttcttgacc aaatgaccgc gaacagggtg atcactgtca acttattacg cggtgcgaat
      781 ttacaaattc ttaacgtaag tcgcagaaaa agccctttac ttagcttaaa aaaggctaaa
      841 ctatttcctg actgtactaa cggtttaatt gttaaaaaat gctacatatc cttctgttta
```

Figure 9.6 GenBank nucleotide sequence for *E. coli* O157:H7 strain Sakai (accession number: BA000007). This portion of the GenBank file contains the genomic sequence of the organism.

column must be separated from other columns by a singular tab, not white spaces. A GFF file contains the following nine fields: (i) seqname – name of chromosome, scaffold, or contig; (ii) source – generally this is the name of the program that annotated the feature; (iii) feature – type of feature (e.g. CDS, gene, source); (iv) start – start position of the feature; (v) end – end position of the feature (note that this number must be higher than start number); (vi) score – a floating point value that may represent similarity or some other measurement; (vii) strand – '+' for forward, '-' for reverse; (viii) frame – either '0', '1' or '2'. A '0' indicates the first base of the feature is the first base of a codon, '1' the second base is the first base of a codon, and '2' means the third base is the first base of the codon; and (ix) attribute – a semicolon-separated list of tag–value pairs, providing additional information about each feature [e.g. *'ID=PROKKA_00013;inference=ab initio prediction:Prodigal:2.6,protein motif:Pfam:PF06894.5;locus_tag=PROKKA_00013;product=Bacteriophage lambda minor tail protein (GpG)'*]. GFF2 files are generally depreciated, mainly because of their inability to effectively handle alternatively spliced transcripts.

The main difference between GTF and GFF is the final (ninth) column in the file. In a GTF file, the last column (attributes) must commence with these mandatory items: (i) gene id value – a globally unique identifier for the genomic source of the sequence; and (ii) transcript id value – a globally unique identifier for the predicted transcript. An example of this field would be *gene id 'Em:U62317.C22.6.mRNA'; transcript id 'Em:U62317.C22.6.mRNA'; exon number 1*. In a GFF3 file, the last column is a series of feature attributes that are in a feature tab = value format. Features include items such as ID, Name, Alias, Parent, Target, and Gap (e.g. *'ID=contig00005;Parent=mrna8743;Name=tobE...'*). One of the ways that a user could determine which version they are dealing with is to look at the first line of the file. The program making these files will often include the version at the top of the file (e.g. *'##gff-version 3'*). GFF3 annotation files can be generated using Prokka and a FASTA file or from a GenBank file using the BioPerl script bp_genbank2gff3.pl.[7] For more detailed information about these file formats, see the Ensemble writings (http://www.ensembl.org/info/website/upload/gff.html) and the writing of Lincoln Stein (https://github.com/The-Sequence-Ontology/Specifications/blob/master/gff3.md).

File pitfalls

It is important to avoid placing spaces in file names, though this usually does not cause an error when working with web tools, it may cause some confusion when working in a Linux/Unix environment (note: many web tools operate from a Linux/Unix OS). As an example, the file entitled 'discarded contigs.fasta' should be changed to 'discarded_contigs.fasta'.

Another important pitfall relates to the end of line (EOL) characters. Microsoft Windows and UNIX employ different end of line (EOL) characters. Most users are unaware of these characters as they cannot be seen without specifically setting the display parameters of the editing program. The EOL for Windows is '^M' whereas it is '$' for UNIX. If a DOS/Windows EOL character is encountered during the running of a program, it will throw an error. Copying and pasting the necessary information into a basic text editor like Notepad rather than Microsoft Word can avoid many of these errors. As mentioned previously, programs that require tab-delimited files require a tab to distinguish between columns. It is important to verify that the file contains tabs and not a series of white spaces.

Text files (FASTA and FASTQ files included) can be generated using different line width settings, resulting in a different line division along the text. An example is represented in Fig. 9.7 where the line width in the same FASTA file is set to 70 characters (Fig. 9.7A) or unlimited (Fig. 9.7B). It can be noted from the line numbers (left-hand side panel) that in Fig. 9.7A the nucleotide sequence is wrapped after 70 characters generating nine (and more) lines of text. Conversely, the other FASTA file is only composed of two lines (Fig. 9.7B). In the latter case, the text editor applied a 'soft' wrap in order to fit the entire sequence on a computer screen; without this setting, line 2 would have continued outside the screen. It could come down to personal preference, but there are software and scripts that would only recognize the first line of the sequence when this is spread across multiple lines. When dealing with FASTA files, many of the current bioinformatics

7 The bp_genbank2gff3.pl script can be downloaded at: https://github.com/bioperl/bioperl-live/blob/master/scripts/Bio-DB-GFF/bp_genbank2gff3.pl

Figure 9.7 Text file differences in line width are shown using the FASTA file containing the *E. coli* O157:H7 strain EDL933 plasmid sequence (accession number: CP008958). (A) Line width set to 70 characters (NCBI FASTA files often have a line width of up to 80 characters). (B) Line width was changed to unlimited using the FASTAX-Toolkit (FASTA Formatter). From the line numbers on the left side of the panels, it can be noted that the sequence in A is distributed from line 2 to line 10 (and more), whereas the sequence in B is placed entirely in line 2.

tools will search through each line as part of the same sequence until they encounter a line that starts with '>', identify another header line and the following sequence. However, there are tools that work differently, and they are likely to issue an error. Additionally, unlimited line width can facilitate the functions of certain file parsers. For example, in FASTQ files, the header starts with the '@'. However, this character is also the 64th ASCII code and can be used in the Q-score line. If the line width of the FASTQ file is limited, it may occur that the Q-score is broken into multiple lines and one of these could start with the '@' character. The program is forced go back and count all the ASCII codes (one for each nucleotide) and determine if the line is a header or a Q-score line. This will inevitably extend the time of analysis. In short, it is wise to check the manual to see how the software reads the input files and, when in doubt or after obtaining unexpected results, the re-formatting of files with an unlimited line width using the FASTX-Toolkit (http://hannonlab.cshl.edu/fastx_toolkit/) should be considered.

Genome assembly and annotation

Genome assembly is the procedure by which randomly fragmented reads (output from the sequencing process) are reconstituted into a sequence (or sequences when plasmids are present) that accurately represents the organism's genome in both size and nucleotide order. Advances in sequencing technology have increased the read lengths and the accuracy of base pair identification while rendering a cheaper process with quicker turnaround times (Kisand and Lettieri, 2013; Magoc *et al.*, 2013; Powers *et al.*, 2013; Walker *et al.*, 2014; Liao *et al.*, 2015). Additionally, competing tools with refined algorithms have improved the quality of the assembly and decreased the computational demands necessary to complete the assembly process. Before the assembly process and before any sequencing occurs, the motivation behind the sequencing should be understood. For example, is the goal to assembly a draft genome or to close the genome? Different sequencer platforms have distinctive read lengths and error rates, and sequencers producing short-read data will not be sufficient to close the genome. Although there are approaches available to complete the genome, such as the Illumina mate pair libraries and gap closure using Sanger sequencing, third-generation sequencing is the first choice when the goal is to generate a complete genome sequence.

The importance of an accurate assembly cannot be understated, as it can affect the order and/or identification of genes. Incorrectly assembled genomes can introduce misleading results, especially when comparing genomes. In this section we will briefly cover the assembly process, quality control and how it can affect the final assembly and conclude by listing some popular assembly and post-assembly tools. Although Sanger sequencing is still used, this discussion will be restricted to WGS from second- and third-generation sequencing platforms.

Assembly process and guiding principles

E. coli genomes contain a significant number of repeats, as well as mobile elements such as plasmids and transposons, which complicate the assembly process. Because of mobile elements and their proclivity to act as their name implies, simply taking the reads and mapping them to a reference can yield an incorrectly assembled genome unless the sequenced organism is known to align well to the reference genome. This makes *E. coli* a good candidate for *de novo* assembly. A general approach, which incorporates the *de novo* process is to inspect the data and run a data cleaning tool to remove adapters, reads with ambiguous characters (e.g. N), uncharacteristically long reads (generally an error), and reads with poor quality scores. The next step is to build consensus sequence, referred to as a contig, from the overlapping sequenced reads as shown in Fig. 9.8. If a good reference genome sequence is available, it may be used to map and re-order the obtained contigs (Fig. 9.9). The draft genome can be also improved by employing post-assembly tools that are used to correct assembly errors.

Items that can drastically affect the assembly are quality control, coverage, insert size, and *k*-mer size. Insufficient coverage will reduce the likelihood that the aligned reads are able to make it into a consensus sequence, poor quality scores may create false alignments, and as mentioned previously, clean data can greatly improve the quality of an assembly. Quality control is very likely the least engaging aspect of any type of production; however, it is arguably the most important step. Two popular tools for inspecting and cleaning the data, respectively are FastQC (https://www.bioinformatics.babraham.ac.uk/projects/fastqc/) and Trimmomatic (Bolger *et al.*, 2014); however, some assembly tools such as Canu (Koren *et al.*, 2017) incorporate an error correction and cleansing process into their pipeline, while other software requires the user to accomplish this by other means. Regardless of how the data are cleaned, it is highly advisable to review the sequenced reads both before and after the tidying process (Fig. 9.10). Special care should be used when trimming data. Too much trimming can remove valuable reads leaving insufficient bases to provide the overlaps required to form consensus sequences (Fig. 9.10B). For examples of Trimmomatic settings, refer to the caption of Fig. 9.10.

```
TGTCGGTGGTGGG
    GGTGGTGGGGATTATCAACTTT
TGTCGGTGGTGGGGATTAT
TGTCGGTGGTGGGGATTATCAAC                             Aligned
            GGGATTATCAACTTTGGCTGCTCGGG              Reads
        GTGGGGATTATCAACTTTGG
                      CTTTGGCTGCTCGGGCTGATG
                        TTGGCTGCTCGGGCTGATGGTG
                     CAACTTTGGCTGCTCGGGCTGATGGT

TGTCGGTGGTGGGGATTATCAACTTTGGCTGCTCGGGCTGATGGTG  — Consensus
                                                  Sequence
```

Figure 9.8 Overlapping aligned reads are built into a consensus contig during the assembling process.

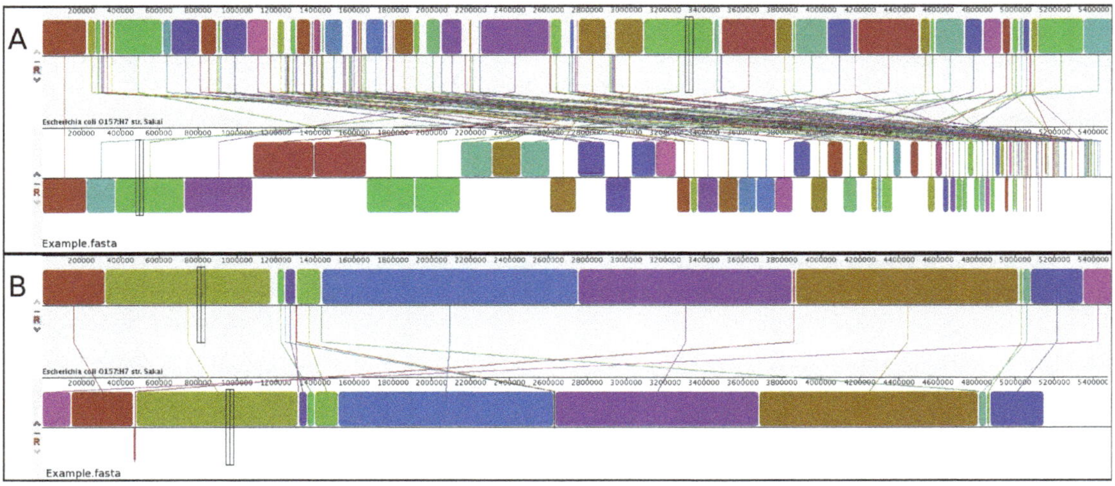

Figure 9.9 Contigs mapping against a reference genome and re-ordering. This figure is divided into two panels: (A) draft genome contigs aligned against a reference and (B) the contigs re-ordered according to the reference genome. Additionally, each panel contains the reference genome (top) and draft assembly contigs (bottom). These processes were performed using Mauve (build 2015–02–13) (Darling *et al.*, 2010), an *E. coli* O157:H7 draft genome (unpublished data), and *E. coli* O157:H7 strain Sakai genome (accession number: BA000007) as a reference.

In order to create a quality assembly, the number of overlaps between the reads for each base must be plentiful, implying a sufficient amount of coverage. Coverage or sequencing depth refers to the number of times each particular nucleotide base has been sequenced. A good rule of thumb is to perform the sequencing with the goal of 50X coverage (Desai *et al.*, 2013).

Assembly tools have different performance standards when analysing data generated from different platforms; a good comparison of short-data assemblers is presented by Jünemann and coworkers (Jünemann *et al.*, 2014). Popular assembler programs for short-read data are SOAPdenovo2 (Luo *et al.*, 2012), SPAdes (Bankevich *et al.*, 2012) and Velvet (Zerbino and Birney, 2008), while Canu (Koren *et al.*, 2017) can be used to assemble long-read data. Tools such as SPAdes (Bankevich *et al.*, 2012) can create a hybrid assembly that incorporate both short and long-reads from different sequencing machines. Once the assembly is complete, the draft genome can be corrected using Pilon (Walker *et al.*, 2014) or the Post-Assembly Genome-Improvement Toolkit (PAGIT) (Swain *et al.*, 2012), which was developed to improve the quality of the assembly by using a reference genome to close gaps and to correct base and assembly errors. There are commercially available GUI software such as CLC Genomics Workbench (QIAgen, Hilden, Germany), Geneious (Global Headquarters Biomatters Ltd, Auckland, New Zealand), and free webserver tools such as Pathosystems Resource Integration Center (PATRIC) (Wattam *et al.*, 2017), EnteroBase (http://enterobase.warwick.ac.uk/), and the Bacterial Analysis Pipeline (Thomsen *et al.*, 2016) that can also create draft assemblies.

Assembly visualization and assessment

There are multiple GUI programs that can be run with the goal of visualizing the alignments within a web browser and/or run locally (Fig. 9.11). Programs like Integrative Genomics Viewer (IGV) (Robinson *et al.*, 2011; Thorvaldsdóttir *et al.*, 2013), Jalview (Waterhouse *et al.*, 2009)[8], and PyBamView (Gymrek, 2014) allow the user to see how the overlapping reads (found in SAM/BAM or indexed BAM files) map to the reference genome. There are also programs like Bandage that allow you to visualize how the contigs are connected as shown in Fig. 9.12 (Wick *et al.*, 2015). This user-friendly

8 Jalview does more than simple visualization. It also performs multiple sequence alignment, alignment editing, phylogenetic tree creation, and clustering.

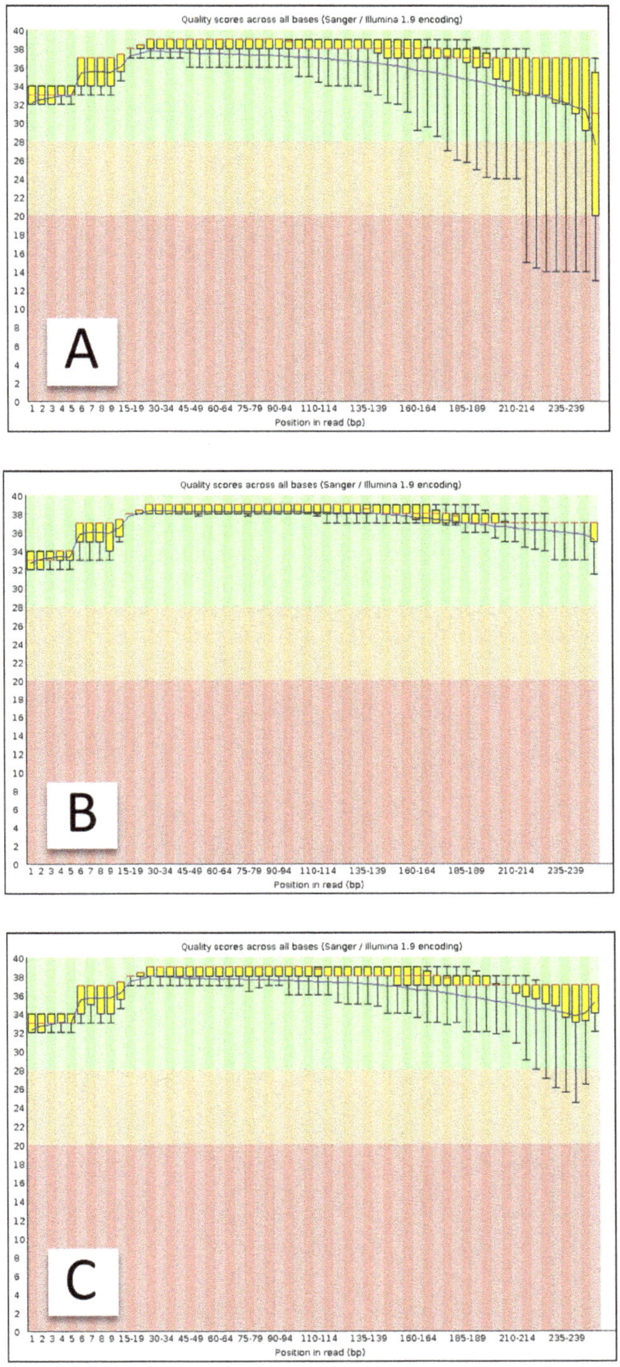

Figure 9.10 Quality assessment of short-read data in FASTQ format, where three levels of data filtering are shown: (A) unfiltered raw data, (B) stringent filtering, and (C) moderate filtering which is generally preferred. Quality per nucleotide of raw short-reads data (unpublished data) are plotted as yellow box-plots using FastQC (v0.11.4). The X-axis and Y-axis pertain to the read position (bp) and PHRED quality score (Q-score), respectively, while the background colours indicate areas of low (red), reasonable (orange), high (green) quality. Quality filtering was performed with Trimmomatic (v0.36) (Bolger *et al.*, 2014) using the functions: (i) LEADING to remove nucleotides under a specific Q-score from the beginning of the read, (ii) TRAILING to remove nucleotides under a specific Q-score from the end of the read, (iii) SLIDINGWINDOW to screen the read from beginning to the end in groups of a selected amount of nucleotides (first argument) which is trimmed when the average Q-score of the nucleotide group is below a selected Q-score (second argument), (iv) MINLEN to remove shorter than a selected length. The parameters set for stringent filtering are *'LEADING:3 TRAILING:3 SLIDINGWINDOW:4:28 MINLEN:40'* and for moderate filtering *'LEADING:3 TRAILING:3 SLIDINGWINDOW:4:15 MINLEN:36'*.

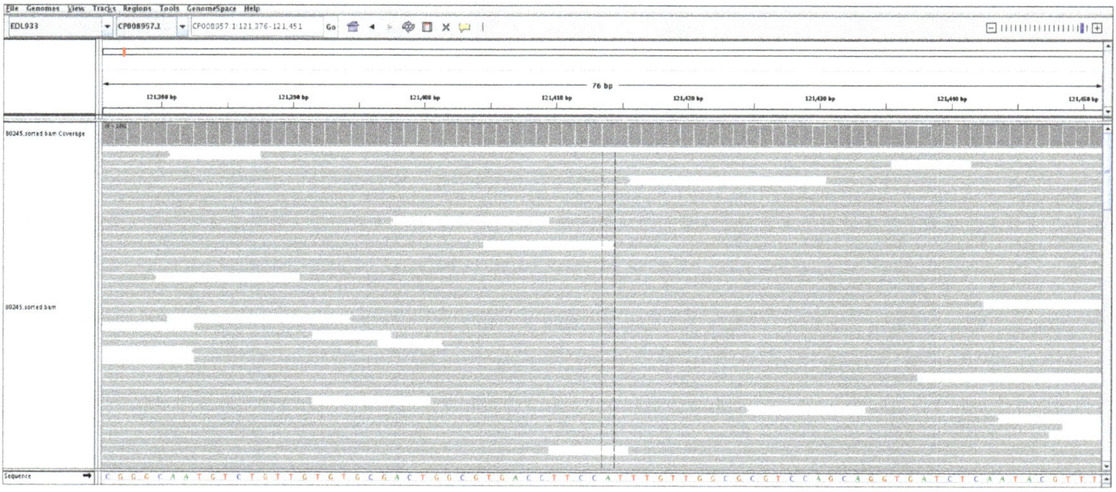

Figure 9.11 BAM file visualization. After assembling, the aligned reads in BAM format can be visualized using the Integrative Genomics Viewer (Thorvaldsdóttir *et al.*, 2013). Data were generated using a MiSeq instrument by sequencing an *E. coli* O157:H7 strain (unpublished data) and aligned to the *E. coli* O157:H7 strain EDL933 genome (accession number: CP008957). Individual sequence reads are represented as grey lines, and the scheme depicts the overlapping aligned reads that will later produce consensus contigs as shown in the bottom of the figure.

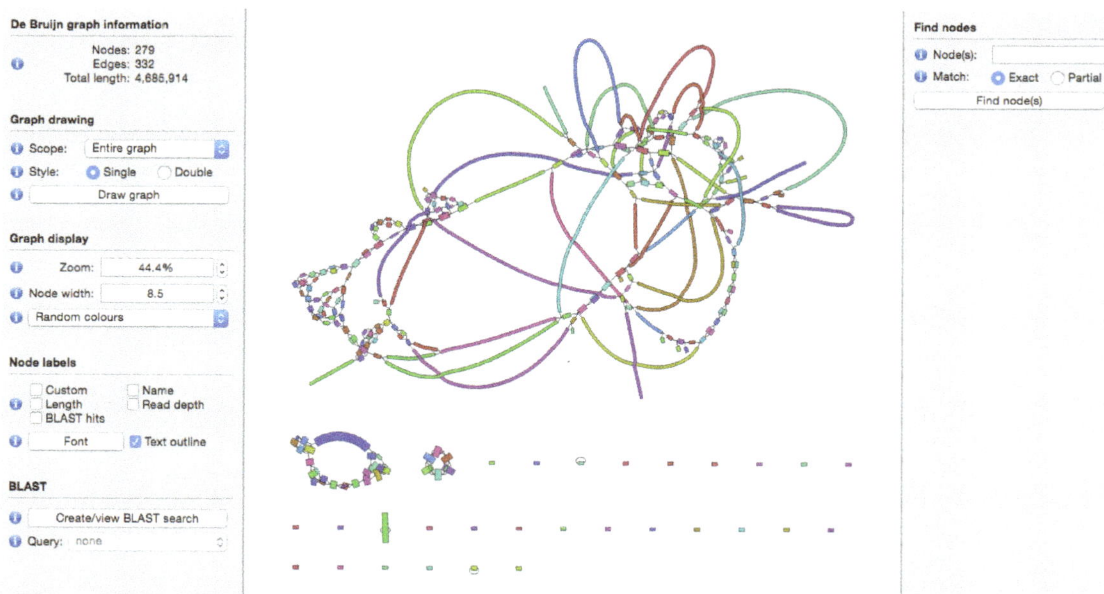

Figure 9.12 Draft assembly visualization. Graphical representations generated by Bandage (Wick *et al.* 2015) are helpful in showing how the contigs are assembled together into the draft genome. Each coloured line represents a contig. Re-printed with permission from https://rrwick.github.io/Bandage/. Copyright 2015, Ryan R. Wick.

and interactive program works on all computer platforms and uses output files from Velvet, SPAdes, and other assembly programs.

Assessment programs like QUAST can provide some basic information that will aid in determining the quality of the assembly (Gurevich *et al.*, 2013). Such general information should include the number of contigs, the size of largest contig, length of assembly, and the N50 size.[9] For example, two assemblies of the same *E. coli* strain that were created with different programs or different quality control measures can be compared to each other by looking at the number of contigs. If one assembly has 270 contigs and the other has 180, barring assembly errors, the 180-contig assembly will be better than the 270-contig assembly because there were more overlapping reads that created fewer consensus contigs sequences. QUAST will provide these numbers, and they can do additional analysis if a reference genome is provided. Furthermore, assessment tools can be used to compare a draft assembly of the same genome assembled by different software.

Genome annotation

Once there is a draft assembly, it can be used with a genome annotation tool to identify the location of the genes and coding regions in the genome. A typical approach to predict gene function is to compare the gene in question to a curated database, and if a 'hit' is found, the function of the matched gene will be assigned to the queried gene. There are multiple prokaryotic gene annotation programs that vary in their platform, annotation methods, employed databases, and output files (Delcher *et al.*, 1999; Besemer *et al.*, 2001; Aziz *et al.*, 2008; Seemann, 2014; Tatusova *et al.*, 2016). One of the most thorough annotation programs to use is Prokka (Seemann, 2014). This program works from the command-line, and it generates numerous output files that include: (i) a GenBank file (.gbk); (ii) annotation files in tabular (.tbl) and GFF3 (.gff) formats; (iii) protein (.faa) and nucleotide sequences (.ffn) of CDS features; (iv) sequences of all predicted transcripts; and (v) a Sequin file (.sqn) that can be used for GenBank submissions. Alternatively, Prokka can be used as a web tool in the Galaxy/CRS4 platform (Cuccuru *et al.*, 2014). The Rapid Annotation using Subsystem Technology (RAST) (Aziz *et al.*, 2008), the PATRIC genome annotation service (Wattam *et al.*, 2017), and the NCBI prokaryotic genome annotation pipeline (Tatusova *et al.*, 2016) are additional prokaryotic genome annotation services that can also be utilized.

9 To calculate the N50 size, contigs are first ordered from largest to smallest. Then, the largest contigs are grouped together until the total length is larger than half size of the genome. The N50 value corresponds to the length of the last contig needed to reach half of the genome size.

NCBI genome submission

One of the purposes of NCBI is to act as a repository for sequenced data. In order to submit a new genome sequence to NCBI, it is necessary to first register for a NCBI account that will enable the user to both submit data and track the submission status. NCBI has recently updated their portal for submission of a complete, draft, or incomplete prokaryotic genome. More detailed information can be found on their website (https://www.ncbi.nlm.nih.gov/genbank/genomesubmit/). To summarize, once a user account has been created, the user will be prompted to enter or apply for a BioProject number. The BioProject database contains information on the sequences or related sequences being submitted from a single or collective group of scientists. The BioProject number is a unique identifier that links the sequence information to a specific place within the BioProject database. Information about BioProjects can be found at www.ncbi.nlm.nih.gov/bioproject. Once the BioProject number has been received, the next step is to provide the BioSample number or request a number from the BioSample database (www.ncbi.nlm.nih.gov/books/NBK169436). This database is used to store metadata on the project's organism and is reciprocally linked to the BioProject. The BioProject number and BioSample number procurement are similar and require the filling in of textboxes and selecting options (e.g. radio buttons). After receiving the BioSample number, you will have the option to create a Sequence Read Archive (SRA) database. This step is not required, but this SRA database will store the raw sequence data for access by the public. Information on the SRA can be obtained at www.ncbi.nlm.nih.gov/sra/docs.

In silico characterization of *Escherichia coli*

Gold standard methods to type and characterize *E. coli* strains are based on well-established cultural and molecular techniques that are laborious and time-consuming (Fratamico *et al.*, 2016). A growing alternative to these approaches is based on predictive genomics, which aims to infer phenotypic traits from the genome sequence. In recent years, genomic, transcriptomic, and proteomic studies have strongly contributed to the identification of new markers, virulence factors, and AR factors. As WGS becomes more common, even small laboratories will be able to sequence a newly isolated *E. coli* strain, extract its gene content, predict the serotype, and detect the carried virulence and AR genes.

Most of the tools reviewed in this section are accessible on-line, and they often require an assembled genome. Nevertheless, there are some exceptions, such as PATRIC (Wattam *et al.*, 2017) and the Center for Genomic Epidemiology (CGE) server (http://www.genomicepidemiology.org/) that work directly with the raw data using their own assembly pipelines. It is important to remind the users that the quality of the assembly may affect the results of the downstream bioinformatics analysis (e.g. sequence target may be included in a mobile element that is missed due to gaps in the draft genome). Additionally, different results may be obtained analysing the same data assembled with different algorithms (Ingle *et al.*, 2016). One drawback of web-based tools is that they usually do not support the upload of a large number of genomes. This problem can be overcome by downloading the database of interest onto a local computer. However, programming and/or UNIX skills are necessary to set up and perform the analysis.

Serotyping

Traditionally, *E. coli* serotypes are determined by agglutination using antisera generated from rabbits immunized with *E. coli* reference strains possessing the different O-groups and H-types. Despite the simplicity of the method, reagent preparation is laborious and equivocal results can arise from cross-reactions, autoagglutinating strains, and non-typeable strains (Wang *et al.*, 2003; DebRoy *et al.*, 2011). The O-groups are distinguished by the O-antigen, which is located in the distal part of the lipopolysaccharide on the bacterial outer membrane. The lipopolysaccharide hydrophilic domain can be divided into a core unit consisting of a non-repeating oligosaccharide and the O-antigen polysaccharide, which typically consists of a unit of two to seven sugar residues repeated 10 to 25 times, and this is generally what is recognized by the antibodies. The O-antigen location and diversity make these biopolymers highly immunogenic and good candidates for *E. coli* classification. The genes necessary to synthesize and transfer the sugars, and then process the O-antigen are grouped in the O-antigen gene cluster (O-AGC), which is located between the 39 bp-element JUMPstart sequence and *galF* at the 5′-end and *gnd* at the 3′-end (DebRoy *et al.*, 2011). However, genes such as *gne* (UDP-GalNAc-4-epimerase) and *wzz* (O-antigen chain length determination protein) can be positioned outside this region (Iguchi *et al.*, 2015). The sequences of all of the known O-AGC in *E. coli* were determined, including the OX-antigens[10] (Iguchi *et al.*, 2015; DebRoy *et al.*, 2016). No O-AGC was found in strains belonging to O14 and O57 (Iguchi *et al.*, 2015; DebRoy *et al.*, 2016), which could explain the absence of O-polysaccharide (O-rough) in the O14 reference strain (Joensen *et al.*, 2015). The nucleotide sequence of 21 O-AGCs from the reference strains were almost identical (> 98% sequence identity) to one or more O-AGCs belonging to different O-group reference strains. Additionally, in some cases serological cross-reactions have been observed among strains, raising the question of whether they could be considered as belonging to the same O-group (Iguchi *et al.*, 2015; DebRoy *et al.*, 2016).[11] The majority of the O-antigens are generated by the Wzx/Wzy-dependent pathway, in which the sugar residue units are transported to the periplasm through the O-antigen flippase (Wzx) where they are assembled into a polysaccharide by the O-antigen polymerase (Wzy). A smaller number of O-antigens, including O9, O52, O60, O89, O92, O95, O97, O99, O101, and O162 are processed by the Wzm/Wzt ATP-binding cassette transporter-dependent pathway, in which the polysaccharide is assembled on the internal surface of the inner membrane and subsequently transported to the outer membrane.

The H-antigens are determined by the different variants of the flagellin protein, which is the subunit that polymerizes to form the *E. coli* flagella. Flagellin contains fairly conserved regions at the N- and C-terminal regions, with the central region conferring

10 OX-antigens are used by some laboratories, but they are not recognized by the Statens Serum Institute in Denmark.

11 It should be mentioned that there are still some disagreements regarding the similarity between of O153 and O178 and between O20 and O137, care should be taken when analysing these O-groups.

serological specificity. Flagellin is encoded by the *fliC* gene in 44 out of 53 H-types whereas the remaining nine H-antigens are encoded by *flkA*, *fllA*, *flmA*, or *flnA* (Wang *et al.*, 2003). With the use of molecular methods for H-type determination, it has been reported that non-*fliC* H-type *E. coli* may also carry a non-expressed *fliC* gene, which causes the strains to test positive for more than one H-type (Tominaga, 2004; Tominaga and Kutsukake, 2007).

The expanding use of NGS technologies is opening the possibility of predicting the *E. coli* serotype directly from whole-genome sequence data. At present, the only web tool designed for this purpose is SerotypeFinder (Joensen *et al.*, 2015), which is supported by the CGE. SerotypeFinder predicts the serotype by blasting a genome or contigs against a custom-made database containing a large collection of O-antigen processing genes and flagellin-encoding genes. To date, O-group and H-type determination is made based on 202 *wzx*, 209 *wzy*, 14 *wzm*, 14 *wzt*, 93 *fliC*, and 9 non-*fliC* variants for a total of 169 O-groups and 53 H-types. The user can customize the default parameters that are set at 85% and 60% for identity and coverage, respectively. The SerotypeFinder database can be downloaded, or SerotypeFinder can be used directly on pre-assembled genomes or raw data. More than 500 *E. coli* strains were used to benchmark SerotypeFinder against traditional serotyping methods. Researchers found that there was 98% and 99% agreement of Serotypefinder with conventional serotyping results for the O-groups and H-types, respectively (Joensen *et al.*, 2015).

Recently, Ingle *et al.* (2016) created the *Escherichia coli* O-groups and H-types (EcOH) database that includes the SerotypeFinder database and other newly discovered O-AGCs. Although there is no accompanying web tool, the EcOH database is available to download with software that uses raw sequencing reads as input (Ingle *et al.*, 2016). Ingle and coworkers tested the EcOH database against several large datasets of *E. coli* genomes. The O-antigen was properly assigned by the presence of at least one O-antigen processing gene in 91% of 300 genomes associated with food-borne outbreaks (Ingle *et al.*, 2016). They also showed that the choice of software for assembly influences the ability to detect the O-antigen processing genes using BLAST. Almost two hundred EPEC strains were assembled with two different software, and the O-groups were predicted in 91% of the draft genomes using Velvet and Velvet optimizer (Zerbino and Birney, 2008) compared to 64% of the sequences assembled with SPAdes (v3.0) (Bankevich *et al.*, 2012). It should be noted that this study did not utilize the most updated version of SPAdes available at the time (Ingle *et al.*, 2016).

Although molecular methods for O-AGC detection are already in use, they are PCR-based (usually targeting only one O-antigen processing gene), and many target O-groups frequently associated with human diseases. The validation of SerotypeFinder and the EcOH database represent comprehensive studies that investigated all known serotypes. Results revealed that the relationship between genotype (O-AGC) and phenotype (O-antigen) was not as straightforward as was expected. In some cases, the same O-antigen processing gene was shared with multiple O-groups. Identical *wzx* genes were found in O2/O50, O17/O77, O118/O151, O169/O183, and O141ab/O141ac. Additionally, combinations of O-antigen processing genes from unrelated O-groups were detected in some strains. For instance, one strain carried both O93 *wzx*/*wzy* and O8 *wzm*/*wzt*, and another strain tested positive for O104 *wzx*/*wzy* and O9 *wzm*/*wzt* (Joensen et al., 2015). Moreover, novel O-AGCs were discovered by sequencing not only un-typeable *E. coli* strains, but also strains with known serotype, which suggests that the same O-antigen can be biosynthesized by different O-AGCs (Ingle *et al.*, 2016; Iguchi *et al.*, 2016; Delannoy *et al.*, 2017). The H-type results are more consistent since, generally, a single gene encodes the flagellar H-antigens. However, some strains may test positive for multiple H-types due to the presence of a cryptic *fliC* allele together with a non-*fliC* gene. Joensen *et al.* (2015), and subsequently Ingle *et al.* (2016), have demonstrated that *in silico* serotyping is more discriminatory than traditional methods and can be applied to un-typeable or non-motile strains.

Virulence factors

The characterization of the virulence factor profile carried by an *E. coli* strain can give valuable information about its potential pathogenicity. The presence of the *eae* gene in conjunction with the Shiga toxin subtype genes (*stx*2a, *stx*2c, or *stx*2d) increases the probability of the host developing haemorrhagic colitis or HUS (Ethelberg *et al.*, 2004;

Melton-Celsa, 2014). Moreover, food authorities, such as the Food Safety and Inspection Service of the US Department of Agriculture, use serogroup-associated genes and the *stx* and *eae* virulence genes as markers to screen food and environmental samples as safety measures (Smith *et al.*, 2014). Virulence factors have to be assessed both *in vitro* and *in vivo*, and are defined as molecules produced by the bacteria that are essential in provoking host diseases (Cross *et al.*, 1984; Wassenaar and Gaastra, 2001; Pallen and Wren, 2007). Numerous genes involved in pathogenesis have been identified and their functions can be divided into the following categories: (i) adhesion to the host endothelial cell membrane; (ii) aggregation with other bacterial cells; (iii) translocation of molecules into the host cell; (iv) toxin production; (v) host cell invasion; (vi) suppression of host immune response; and (vii) iron intake (Wu *et al.*, 2008; Ribet and Cossart, 2015). Various databases were developed to collect sequences from virulence factors, including also virulence-associated and virulence life-style factors that increase the likelihood of triggering diseases. Some of these databases are discussed below.

Using sequence homology, it is possible for investigators to quickly detect known virulence genes in a newly sequenced *E. coli* genome. The Virulence Factor Database (VFDB) was created in 2005 and has been upgraded over the past 11 years (Chen *et al.*, 2005, 2012, 2016; Yang *et al.*, 2008;). In its latest version, VFDB is organized in two main groups; 'Set A', a collection of non-redundant, well curated, and experimentally verified virulence factors, and 'Set B' includes homologues and predicted virulence factors.[12] *E. coli* is an important part of the VFDB, which includes specific virulence factors found in the major *E. coli* pathotypes. The database also describes the function of the virulence factors, how these factors cluster together, and the general characterizations of the pathogenic *E. coli* strains that carry these virulence genes. VFDB allows the user to browse virulence factors by name or organism, to query a sequence using BLAST, and to directly download 'Set A' and 'Set B' (Chen *et al.*, 2016). VirulenceFinder is a user-friendly web tool hosted by the CGE to quickly detect virulence genes from raw reads or assembled genomes of *Listeria*, *Staphylococcus aureus*, *E. coli*, and *Enterococcus*. When it was created, the VirulenceFinder database contained the sequences of numerous variants of 76 *E. coli* virulence genes (Joensen *et al.*, 2014).

Virulence factors can also be screened from the predicted proteome that is generated post-annotation, which is an effective approach for the identification of new putative virulence genes. Two tools can be used for this purpose: VirulentPred (Garg and Gupta, 2008) and MP3 (Gupta *et al.*, 2014a). VirulentPred is based on a machine learning technique called two-layer support vector machines (SVM). This approach can be used for such purposes as clustering and classification and is less computationally expensive than performing the same tasks via pairwise alignments. This web tool receives FASTA (.faa) files as input, where the number of protein sequences is limited to 500 when a multi-sequence FASTA file is used (Garg and Gupta, 2008). MP3 uses a combination of SVM and hidden Markov models (HMM) to identify virulence genes and performs better than VirulentPred because of its blended approach. Although MP3 was originally designed for metagenomic studies, it also provides a 'Genomic' option to analyse all the translated proteins obtained from a single *E. coli* genome. The MP3 web tool accepts large.faa files (up to 10 MB) as input or it can be locally installed onto a Unix/Linux operating system (Gupta *et al.*, 2014a).

Lastly, some databases such as the Database of Bacterial Exotoxins for Human (DBETH) and Virmugen DB are specialized for particular categories of virulence factors (Chakraborty *et al.*, 2012; Racz *et al.*, 2013). DBETH contains sequence, structure, interactions network information, and analytical results of 33 exotoxins produced by *E. coli* and offers web tools to query domains, sequence, and structures for the identification of potential toxins (Chakraborty *et al.*, 2012). Virmugen DB is a database of virulence factors that could be mutated to reduce virulence of the pathogen for use as a live attenuated vaccine (Racz *et al.*, 2013).

It has been previously stated that many factors play a role in *E. coli* pathogenesis and that the distinction between pathogenic and commensal strains is not well defined. Regardless of whether the virulence factors are detected by sequence homology or by other methods, care should

12 Non-characterized genes in pathogenicity islands are not included.

be taken when interpreting the results since (i) virulence factors are also found in non-pathogenic bacteria; (ii) incomplete virulence loci are not expected to produce functional protein complexes, and (iii) the presence of anti-virulence genes could suppress the expression of virulence factors (Wassenaar and Gunzer, 2015). These difficulties have been described throughout the literature. Wassenaar *et al.* (2015) analysed the genomes of *E. coli* strain K-12 and other *E. coli* strains that have been used as probiotics for several years using VFDB and the Lawrence Livermore National Laboratory Virulence Database (MvirDB). Based on these databases, genes that encode for a suite of adhesion factors and the cytolysin *hly* were identified. Adhesion factors promote intestinal colonization but are insufficient to provoke diseases. The *hly* gene encodes for the cytotoxic *E. coli* α-haemolysin, which is thought to enhance *E. coli* virulence in extra-intestinal infections but is not involved in intestinal infections (Mainil, 2013). Usually virulence factors are the result of a complex interaction of proteins encoded by many genes located in one or more virulence locus, as is sometimes the case with loci encoding fimbriae or involved in protein secretion. The detection of a single gene may be insufficient to confirm the production a functional protein complex (Wassenaar *et al.*, 2015; Vazquez-Rosas-Landa *et al.*, 2017). Using WGS, it is possible to determine the gene content of a bacterial strain and verify the completeness of the known virulence loci. Unfortunately, at present, there is no tool that provides a clear output facilitating this kind of analysis. Additionally, even if the entire virulence loci are complete, other genes that are capable of suppressing the functions of virulence genes (anti-virulence genes) could be present. For example, in EAEC, if the anti-virulence gene *aar* is present, the expression of the regulator AggR is down-regulated. As a consequence, the production of the aggregative adherence fimbriae and other virulence-associated factors are reduced, decreasing the virulence of the strain (Santiago *et al.*, 2014).

Antimicrobial resistance

Although the use of antibiotics remains controversial in STEC infections, they are the first line of defence for bloodstream infections, urinary tract infections, and systemic diseases (Agger *et al.*, 2015). Bacteria can develop antimicrobial resistance very quickly, and resistance has been observed less than one year after a product is made available for clinical use. Therefore, it is very important to use antibiotics only when it is necessary to limit the selection and diffusion of resistant strains (Adu-Oppong *et al.*, 2017). Additionally, a growing number of antimicrobial-resistant ExPEC cases have been the source of great concern (Bryce *et al.*, 2016). The 2015 annual report of the European Antimicrobial Resistance Surveillance Network (EARS-Net) showed that more than half of the 72,123 *E. coli* strains tested positively for antibiotic resistance to at least one antimicrobial group. Furthermore, the number of strains resistant to third generation cephalosporins, fluoroquinolones, and aminoglycosides rose in the period between 2012 and 2015 (ECDC EARS-Net, 2017).

The increased availability of sequenced bacterial genomes stimulated the development of databases and tools to detect AR genes and identify new genes directly from the genomes. Identification methods are mainly based on aligning the gene content of an *E. coli* strain against a well-curated database of AR genes (Adu-Oppong *et al.*, 2017). The Antibiotic Resistance Genes Database (ARDB) was one of the first AR databases, and despite a lack of updates since 2009, it continues to be very popular (Liu and Pop, 2009). It includes 23,137 (3912 from *E. coli*) resistance genes for 249 antibiotics, as well as a BLAST tool, which compares strains to the genes in the database. ResFinder is an up-to-date database that specializes in AR genes that are acquired by horizontal gene transfer, and therefore, it does not include information on AR caused by mutations on chromosomal genes (Zankari *et al.*, 2012). It includes data from ARDB (Liu and Pop, 2009), the Tetracycline and Macrolide-Lincosamide-Streptogramin Nomenclature database (http://faculty.washington.edu/marilynr/) and the Lahey Clinic website (http://www.lahey.org/Studies/). ResFinder includes a BLAST tool that accepts complete genomes, draft genomes, or raw data from the main sequencing platforms as input. The tool is set by default at 90% identity and 60% minimum length, but the user can change these parameters. ResFinder is frequently updated (latest update was performed in February 2017), and the database can also be downloaded for a local use. ResFinder predictions were also compared to phenotypic tests, and the results were in agreement in 99.74% of the

3051 tests (Zankari *et al.*, 2013). Antibiotic Resistance Gene-ANNOTation (ARG-ANNOT) was developed in 2014 to detect antibiotic resistance genes and identify novel putative ones (Gupta *et al.*, 2014b). It collects AR gene sequences along with their mutations and includes data from previously created databases such as Lahey Clinic website (http://www.lahey.org/Studies/), the Antibiotic Resistance Genes Online (ARGO) (Scaria *et al.*, 2005), the Lawrence Livermore National Laboratory Virulence Database (MvirDB) (Zhou *et al.*, 2007), ARDB (Liu and Pop, 2009), and ResFinder (Zankari *et al.*, 2012). ARG-ANNOT offers a BLAST tool on its web page and a tutorial to help the download of the database and perform pairwise-alignment using the graphic user interface software BioEdit (http://www.mbio.ncsu.edu/bioedit/bioedit.html) onto a local computer. The Comprehensive Antibiotic Resistance Database (CARD) was developed in 2013 and recently expanded (McArthur *et al.*, 2013; Jia *et al.*, 2017). CARD is a complete database that not only includes AR gene and protein sequences but also mutations, publications, and links to external public databases. Additionally, CARD follows a very controlled vocabulary that is based on several ontologies. In particular, it contains the Antibiotic Resistance Ontology and the Model Ontology. The first classifies AR gene data according to antimicrobial molecules, drug targets, resistance mechanisms, mutations, and relationship with other molecules. While the second defines reference sequences, search parameters to detect AR genes, or mutations. CARD presents an easy-to-use interface to browse the database, the antibiotic resistance ontology, and the model ontology. Analysis can be performed using BLAST and the resistance gene identifier tools, which detects protein homologues and their variants. The results can be visualized using the resistance gene identifier visualization tool, and they are reported in three categories: 'perfect' specifies an identical match to the reference, 'strict' indicates a new variant of a known gene, and 'loose' signifies a distant homologue to a protein encoding gene. Data, ontologies and tools are available to download and run locally using command-line software (Jia *et al.*, 2017). The database is periodically updated by the CARD team and the latest update occurred in May 2017. The user-friendly Comprehensive Antibiotic resistance Repository of *Escherichia coli* (u-CARE) consists of all the information regarding multidrug-resistant *E. coli* including publications, resistance genes sequences, Single Nucleotide Polymorphisms (SNPs), transcription factors, and links to other public databases. Analysis can be performed using a BLAST tool directly on the u-CARE web page, or data can be downloaded and used locally (Saha *et al.*, 2015). Unlike previously discussed BLAST-based tools, Resfams is a curated database of protein families with antimicrobial function that uses HMM to identify AR genes. In particular, it is especially effective in metagenomic studies as demonstrated by Gibson *et al.* (2015), who showed that 64% of AR genes identified by Resfams were not detected by BLAST-based methods (e.g. ARDB and CARD) in soil and human-gut samples. Because Resfams is not a web-based tool, the user is required to download the database and set up the necessary software in a command-line environment (Gibson *et al.* 2015).

Nevertheless, the detection of AR genes in a genome does not always correspond to an antimicrobial resistance phenotype. For example, the research group of Gordon *et al.* (2014) compared the AR results generated by genotypic and phenotypic techniques in 501 strains of *Staphylococcus aureus* and found that not all strains carrying the *blaZ* gene were resistant to methicillin. Susceptible *blaZ*-positive strains were further characterized, and it was discovered that they also carried a frameshift mutation that inactivated the gene (Gordon *et al.* 2014). Moreover, resistance to antimicrobials may result from a combination of systems and cannot be determined by a single gene (Martinez-Martinez *et al.*, 1999).

Mobile genetic elements

Plasmids, bacteriophages, transposable elements, and integrons are examples of DNA sequences that can move around the genome and are referred to as mobile elements. Mobile elements are acquired from horizontal gene transfer and have played a fundamental role in *E. coli* evolution as they can rapidly confer new phenotypes with more adaptive capabilities. In general, mobile elements are of interest and have been identified and characterized because they carried virulence or antimicrobial resistance genes (Asadulghani *et al.*, 2009; Johnson and Nolan, 2009; Liu *et al.*, 2016).

A plasmid is a double-stranded DNA molecule

that can replicate independently from the chromosome and can be transferred to other bacteria by conjugation. When interpreting WGS data, distinguishing between chromosomal sequences and those that arise from a plasmid can be challenging, especially when employing short-read data. SPAdes and plasmidSPAdes are two command-line tools that can be used to identify reads belonging to plasmids (Bankevich *et al.*, 2012; Antipov *et al.*, 2016). Additionally, the PlasmidFinder web tool was developed to identify plasmids associated with bacterial species belonging to the Enterobacteriaceae family. The detected plasmids can also be typed using the plasmid Multilocus Sequence Typing (pMLST) tool. Both tools are hosted by the CGE, and thus can be used from raw sequence data, contigs, or completely assembled and closed plasmids (Carattoli *et al.*, 2014).

Bacteriophages are another group of mobile elements that can integrate their genetic material into a bacterial host and survive indefinitely. An integrated bacteriophage is classified as a prophage. When the *E. coli* strain Sakai was sequenced, it was reported to contain 18 prophages that covered 16% of its genome (Canchaya *et al.*, 2003). PHAge Search Tool Enhanced Release (PHASTER) is a user-friendly web-based resource for the identification and annotation of prophage sequences found within bacterial genomes and plasmids by evaluating the DNA sequence homology against a bacteriophage and prophage database. It is based on a previous web server that has recently been updated to a significantly improved version (Zhou *et al.*, 2011; Arndt *et al.*, 2016). Other tools have been developed for prophage detection in a microbial genome including PhiSpy, which incorporates an algorithm to predict novel prophage sequences (Akhter *et al.*, 2012) and VirSorter, which works with both single-cell genomes and metagenomic datasets (Roux *et al.*, 2015).

Transposable elements are DNA sequences that can change their position within the genome or can be transferred to other bacteria. They can alter a genome by inactivating genes located at the insertions sites, or they extract genetic material from one region and transfer it in a new location. The simplest transposable elements (also called insertion sequence elements) are comprised of inverted repeats, genes involved in the transposition. ISsaga is a web tool that can be used to detect and annotate insertion sequence elements in assembled genomes (Varani *et al.*, 2011), while ISMapper is a tool used to identify insertion sites by mapping the short reads data to the insertion elements (Hawkey *et al.*, 2015).

Integrons are site-specific mobile genetic elements that can be incorporated into bacterial chromosomes and plasmids. They can carry one or more genes (called gene cassettes) and are known to diffuse antimicrobial resistance proprieties (Gillings *et al.*, 2008). Databases for integrons (Moura *et al.*, 2009) and antibiotic resistance gene cassettes (Tsafnat *et al.*, 2011) are publicly available.

After horizontal gene transfer has occurred, the portion of alien DNA present in the bacterial genome is known as a genomic island. If the genomic island includes virulence or AR genes they are also referred to as pathogenicity islands (PAI) or resistance islands (REI), respectively. The study of transposable elements is important to understand rapid changes of pathogenicity among bacteria of the same species and detect virulence or AR gene clusters. Several genomic island databases have been developed such the Horizontal Gene Transfer database (Garcia-Vallve *et al.*, 2003), Islander (Mantri and Williams, 2004), and the Pathogenicity Island Database v2.0 (PAIDB) (Yoon *et al.*, 2015). Methods to predict genomic islands are vast and can be based on genome comparison, sequence homology, or differences in genomic signatures (e.g. GC content, GC-skew, dinucleotide frequencies, codon usage) between the host and the microorganisms from which the genomic island originated. PAIDB has a built-in BLAST and text search tool that can query the database for PAIs or REIs (Yoon *et al.*, 2015). IslandViewer 4 is a popular web tool that integrates four different prediction programs (IslandPick, IslandPath-DIMOB, SIGI-HMM, and Islander) into a pipeline that is used to identify generic genomic islands in bacterial genomes (Bertelli *et al.*, 2017). To identify PAIs and REIs, IslandViewer 4 also compares the gene included in the identified genomic islands to the following databases: CARD (McArthur *et al.*, 2013), VFDB (Chen *et al.*, 2005), VICTORS (Xiang *et al.*, 2007), and PATRIC virulence factors (Mao *et al.*, 2015). IslandViewer 4 stores pre-computed analysis results of all the microbial genomes included in RefSeq. A single or group of complete or incomplete genomes may be uploaded for analysis so long as the genomes are EMBL or GenBank formatted.[13] Additionally, IslandViewer 4 supports both linear and circular visualization and permits a comparison of two genomes (Bertelli *et al.*, 2017). If stand-alone software are preferred, the free Ensemble algorithm for Genomic Island Detection (EGID)[14] can be used. EGID utilizes previously developed software (AlienHunter, COLOMBO SIGI-HMM, INDeGenIUS, IslandPath, and PAI-IDA) and incorporates the results of the six individual programs into one, achieving better performance than each individual tool (Che *et al.*, 2011; Hasan *et al.*, 2012). For more information regarding bioinformatics approaches for genomic island identification, the review written by Che *et al.* (2014) is highly recommended.

Multi-criteria predictive genomic tools

The increasing number of web tools and databases to analyse NGS data can be overwhelming for researchers. Luckily, some webservers provide a suite of tools that integrate multiple publicly available databases. The Lawrence Livermore National Laboratory Virulence Database (MvirDB) is a centralized system that collects information about toxins, virulence factors, and AR genes from Tox-Prot (Jungo and Bairoch, 2005), SCORPION (Srinivasan *et al.*, 2002), PRINTS (Attwood *et al.*, 2003), VFDB (Chen *et al.*, 2005), TVFac toxin and virulence factor database (Los Alamos National Laboratory, New Mexico, USA), Islander (Mantri and Williams, 2004), the Antibiotic Resistance Genes Online ARGO (Scaria *et al.*, 2005), and the Virus Database system VIDA (Albà *et al.*, 2001). Although redundant information is preserved, all the sequences are annotated with an internal tool and taxonomically clustered in order to facilitate quick searches and simplify the analysis. The user can utilize the MvirDB web interface to perform Boolean searches or query a sequence using the BLAST tool. MvirDB is automatically updated every week and may be downloaded for more intensive computational investigations (Zhou *et al.*, 2007).

PATRIC is a large database of pre-computed genomic, transcriptomic, and proteomic data of numerous bacterial pathogens (including *Escherichia coli*) that also offers numerous cloud services such as genome assembly, genome annotation, proteome comparison, RNA-Seq analysis, variation analysis, and model reconstructions (Wattam *et al.*, 2017). The PATRIC genome annotation service is also able to detect virulence and AR genes by performing BLAST searches against the following databases: VICTORS (Xiang *et al.*, 2007), VFDB (Chen *et al.*, 2005), PATRIC virulence factors (Mao *et al.*, 2015), ARDB (Liu and Pop, 2009), CARD (McArthur *et al.*, 2013), the Therapeutic Drug Targets database (Yang *et al.*, 2016), DrugBank (Law *et al.*, 2014), and the human genome for human homologues (Genome Reference Consortium Human Build 38). Moreover, when a user submits a genome, its antibiotic susceptibility is predicted by a machine learning approach using a large number of genomes carrying AR genes and associated metadata deposited in PATRIC (Wattam *et al.*, 2017).

SuperPhy is an *E. coli* predictive genomic platform that offers tools and pre-computed analysis results of virulence and AR gene detection, genome comparisons, and geospatial distribution (Whiteside *et al.*, 2016). The SuperPhy virulence and AR gene database includes sequences from CARD (McArthur *et al.*, 2013), VFDB (Chen *et al.*, 2005), and other sources (Whiteside *et al.*, 2016).

VRprofile is a newly developed web tool for the analysis of virulence and AR genes and their associated pathogenicity islands and antimicrobial resistance islands (Li *et al.*, 2017). MobilomeDB is the database built in VRprofile that includes data from (i) VFDB (Chen *et al.*, 2005) and VICTORS (Xiang *et al.*, 2007) for virulence factors; (ii) ARDB (Liu and Pop, 2009), CARD (McArthur *et al.*, 2013), and BacMet (Pal *et al.*, 2014) for AR factors; (iii) T3DB (Wang *et al.*, 2012), SecReT4 (Bi *et al.*, 2013), and SecReT6 (Li *et al.*, 2015) for bacterial secretion systems; (iv) ICEberg (Bi *et al.*, 2012), Phage_Finder (Fouts, 2006), and PHAST (Zhou *et al.*, 2011) for conjugative elements and prophages; (v) PAIDB (Yoon *et al.*, 2015), INTEGRALL (Moura *et al.*, 2009), and ISfinder (Siguier *et al.*, 2006) for PAIs, REIs, integrons, and insertion elements; and other sequences from the literature (Li *et al.*, 2017). VRprofile uses both sequence homology and HMM approaches to detect single

13 Batch uploading is possible.

14 The Ensemble algorithm for Genomic Island Detection software is also available in the Genomic Island Suite of Tools (http://www5.esu.edu/cpsc/bioinfo/software/GIST/)

gene sequences and gene clusters or to predict homologous genes. VRprofile output also includes the results of the composition-based module, which predicts island-like regions. Furthermore, the VRprofile server provides three additional tools, CDSeasy, CGCfinder, and COGviewer for annotation, customized BLAST searches, and species-specific arrangement analysis, respectively.

Comparison and visualization of genetic content

Once the characterization of a genome has been completed, it can be compared across multiple strains. Determining which genomes carry particular genes, how similar two sequences are to each other, or whether a collection of isolates shares genes not found in isolates collected from differing environments are but a few tasks that can be performed. If an investigator has a gene or genes of interest and wants to observe whether these are found in multiple genomes and how their sequence(s) differ from the reference sequence(s), the Large Scale – BLAST Score Ratio (LS-BSR) tool is an excellent starting point (Sahl *et al.*, 2014). LS-BSR requires a reference for comparative purposes, the file of which may contain a single gene sequence, or it can be a multi-sequence and can be run against single or numerous query genomes.[15] To calculate the similarity of a gene, the reference bit score is initially calculated by taking the protein reference sequence and aligning it to its nucleotide sequence, and then invoking TBLASTN to obtain the value. The BLAST score ratio is calculated by dividing the bit score of the queried sequence by the reference bit score. The output of LS-BSR is a matrix (Table 9.4) where the gene names were extracted from the headers of the reference FASTA file. The matrix can be visualized as a heatmap (Fig. 9.13) using the R package ggplot2 (Wickham, 2016; R Core Team, 2017) in order to rapidly locate missing genes or genomes that differ significantly from other genomes (e.g. Strain D in Fig. 9.13). Some tools, such as LS-BSR and Scoary,[16] allow the user to look for genomic differences between sets of isolates (Sahl *et al.*, 2014; Page *et al.*, 2015; Brynildsrud *et al.*, 2016). For example, if the studied genomes can be subcategorized into groups (e.g. based on isolation source or level of pathogenicity), the software can be used to determine the presence of genes that are unique to those groups.

Graphical representation of genetic content comparisons is often a necessary component to understanding differences between multiple organisms. One way to visualize the comparison of two or more genomes is using BLAST Ring Image Generator (BRIG) (Alikhan *et al.*, 2011) or Mauve (Darling *et al.*, 2010). Both programs require a reference, are Java-based, and can be executed from any operating system.[17] Additionally, they can be

15 All files should be in FASTA format.

16 Scoary works with the output from Roary and LS-BSR.

17 Mauve can also be used from the command-line.

Table 9.4 Large-Scale BLAST Score Ratio (LS-BSR build version 12.13.2016) output matrix

Gene_Products	Strain A	Strain B	Strain C	Strain D	Strain E	Strain F	Strain G
hchA	0.99	0.99	0.99	0.86	0.99	0.99	0.99
gnd	0.82	0.82	0.82	0.89	0.82	0.82	0.82
fbaB	0.9	0.9	0.9	0.94	0.9	0.9	0.9
mlrA	0.93	0.93	0.93	1	0.93	0.99	0.93
ompC	0.88	0.88	0.88	0.8	0.88	0.88	0.88
gcvR	1	1	1	1	1	1	1
glyA	0.96	0.96	0.96	0.98	0.96	0.96	0.96
clpB	0.97	0.97	0.97	1	0.97	0.97	0.97
grpE	0.96	0.95	0.96	1	0.96	0.96	0.95

The first column contains the gene name and the first row contains the *E. coli* O157:H7 strains (unpublished). BSR values are shown for each gene-strain set, which represents the level of similarity between the genes of each strain and the reference genes of *E. coli* strain Sakai (accession number: BA000007).

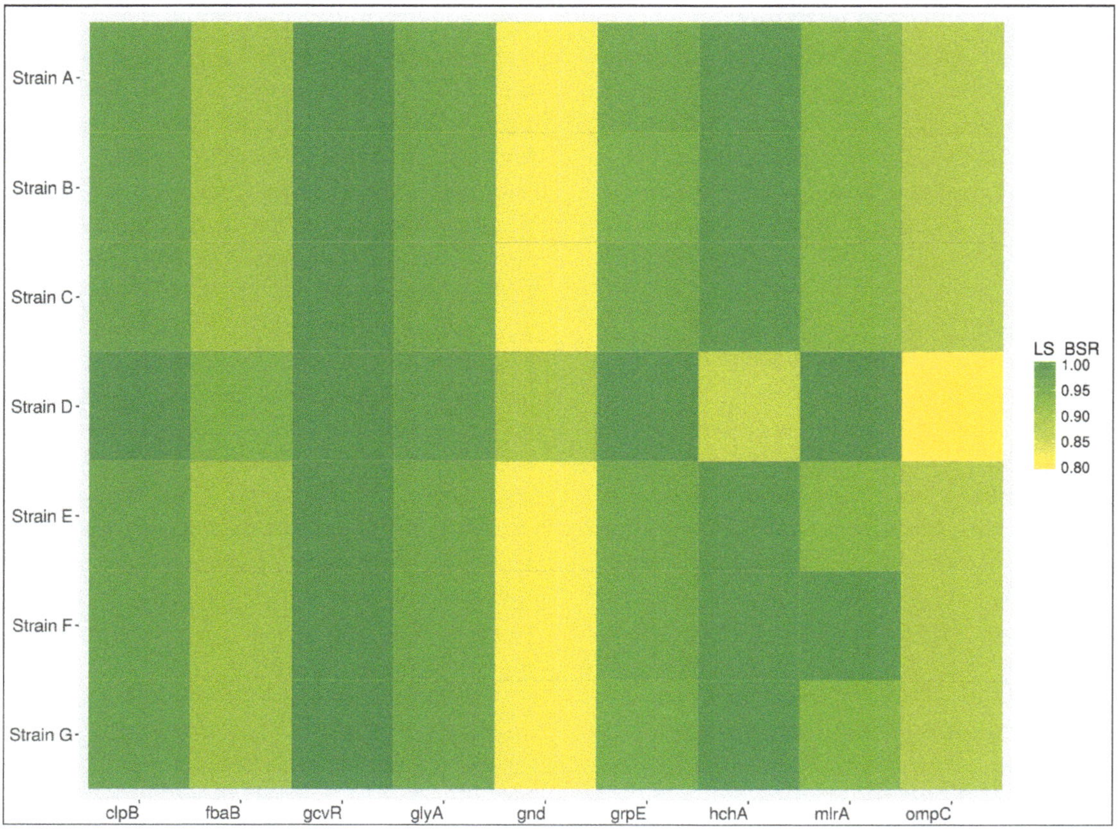

Figure 9.13 Visualization of gene similarity. The heatmap was created from the Large-Scale BLAST Score Ratio matrix in Table 9.4, using the R package ggplot2 (Wickham, 2016; R Core Team, 2017). The level of similarity between the genes in each *E. coli* strain and the reference strain EDL933 is represented by a scale of colours from yellow (low) to green (high). This heatmap quickly highlights differences between strains, e.g. strain D appears to have more variation than other strains.

used with draft or finished assemblies in FASTA or GenBank file formats. These tools can incorporate numerous genomes into the analysis; however, the user should be aware that these reference-based tools will miss regions that are present in a query genome (non-reference) but not in the reference genome. BRIG uses BLAST to compare the similarity of query genomes to the reference genome and will depict the genomes in a circular fashion.

The three strains that were compared by Welch *et al.* (2002): STEC strain EDL933 (CP008957), the ExPEC strain CFT073 (AE014075), and the non-pathogenic *E. coli* strain K-12 (NC000913) will be used as an example to further describe these visualization tools. A typical BRIG output is shown in Fig. 9.14. Genomes are represented by concentric rings, the outer ring corresponds to the reference genome *E. coli* strain EDL933, while the subsequent rings shown in different colours correspond to *E. coli* strain K-12 and *E. coli* strain CFT073. The gaps observed in the inner concentric rings represent the lack of genetic material of the query genome when compared to the reference, and the degree of sequence similarity between the query and the reference genome is illustrated by the colour scale (see Fig. 9.14).[18] BRIG output images can also include annotations found in GenBank files and/or custom annotations. In order to use custom annotations, the start and stop position of the feature in question along with a label must be provided in a tab-delimited file. In our example, the annotations of 20 prophage regions in *E. coli* strain EDL933 (in red) were identified using the PHASTER web server, and the start and stop positions of each region and their label were recorded

18 The identity levels can be set by the user.

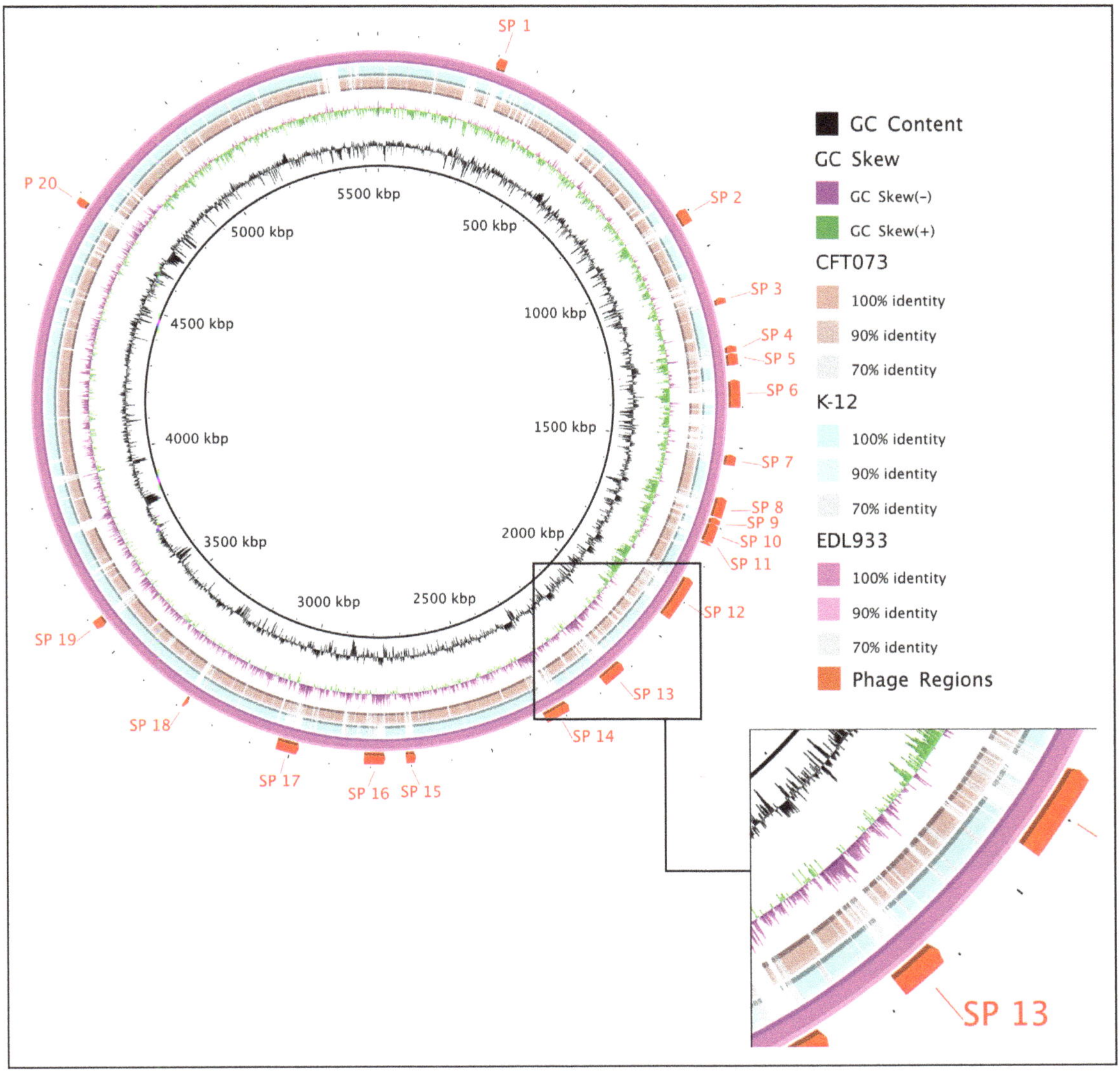

Figure 9.14 Circular visualization of multiple genomes alignment. This figure displays the information in a concentric fashion. From the centre: (i) sequence length, (ii) GC content, (iii) GC skew, (iv) *E. coli* strain CFT073 (accession number: AE014075), (v) *E. coli* strain K-12 (accession number: NC000913), (vi) *E. coli* strain EDL933 (accession number: CP008957), (vii) custom annotation of the 20 prophage regions in the EDL933 genome (SP 1–20). Similarity level between the queries and the reference genome is represented by different colour scales, as shown in the accompanying legend. As expected 100% of similarity in all its length is shown for the genome of EDL933 since it was used as both query and reference. The gaps corresponding to prophage regions SP 12 and SP 13 were also magnified and shown in an inset. This figure and custom annotations were generated using BLAST Ring Image Generator (Alikhan *et al.*, 2011) and PHASTER (Arndt *et al.*, 2016), respectively.

into a text file and supplied to BRIG (Fig. 9.14). Although numerous gaps are seen in the concentric rings of *E. coli* strains K-12 and CFT073, many of these breaks correspond to the prophage regions of the EDL933 strain. This result supports the low number of PHASTER-identified prophage regions found in the K-12 and CFT073 genomes (six and eight, respectively). The BRIG manual is an excellent resource that contains step-by-step instructions describing the upload of sequence files, the creation of custom annotations, and the use of SAM files to show assembly coverage. As of this writing, the BRIG software, manual, and tutorial example files can be downloaded from https://sourceforge.net/projects/brig/files/.

Mauve aligns the query genomes to the

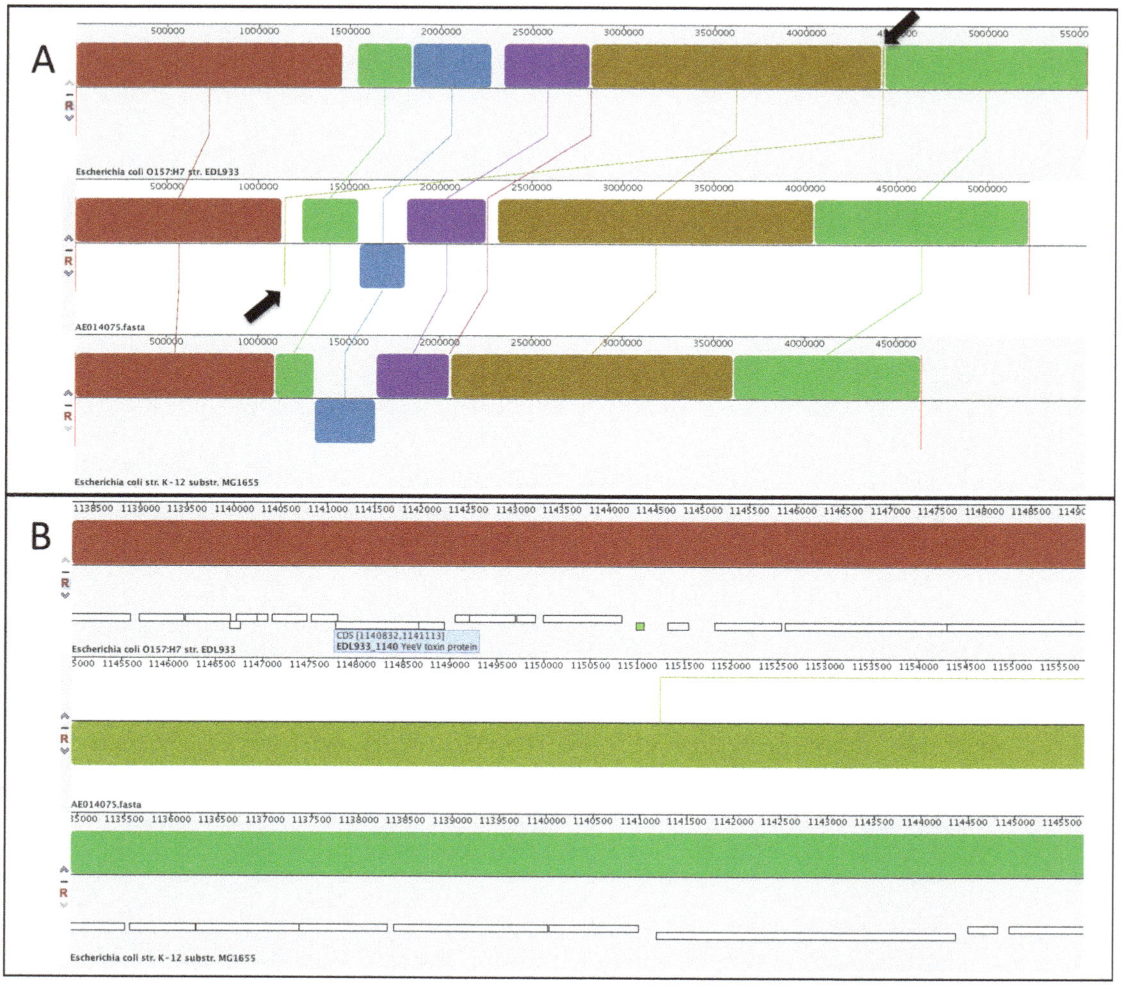

Figure 9.15 Linear visualization of multiple genomes alignment, where *E. coli* strain K-12 (accession number: NC000913) and *E. coli* strain CFT073 (accession number: AE014075) were aligned to the reference *E. coli* strain EDL933 (accession number: CP008957) using with Mauve (build 2015–02–13) (Darling *et al.*, 2010). Panel (A) shows the reference genome at the top and the two aligned *E. coli* strains below. Collinear and homologous regions are represented as blocks of the same colour connected between the genomes. Inversions are illustrated by blocks that rest on the opposite side of the centre line (blue block) and rearrangements are identified by the different order of the blocks (arrows). The rearranged region was magnified at a level where the annotation is visible and shown in panel (B). Genomic features are colour-coded based on their type. For example, white and green represents CDSs and tRNAs, respectively. The strain CFT073 lacks of annotation because a FASTA file was used as input, whereas GenBank files were utilized for the other strains.

reference genome, visualizes them in a linear fashion, and provides an output containing the genome alignments in eXtended Multi-FastA (XMFA) file format (Fig. 9.15A). One of the benefits of using Mauve is that the user can easily see rearrangements and inversions, and as this alignment is interactive, the user may zoom-in and -out of interest areas. Fig. 9.15A shows the alignment of *E. coli* strains K-12 and CFT073 to the EDL933 reference strain. Similarly coloured blocks across multiple strains represent collinear and homologous regions. Inversions are represented by coloured blocks that rest below the genome's centre line, and rearrangements are identified by the coloured lines that connect a block from one genome to another. As an example, a noticeable inversion between the reference and strains CFT073 and K-12 is represented by the blue blocks in Fig. 9.15A. While, one slight rearrangement and inversion between the reference EDL933 and strain CFT073 is located at ~1200 kb, and it is represented by a slender green line and inverted block (see arrows in Fig. 9.15A and the block is also magnified in Fig. 9.15B). Additionally, Fig. 9.15B shows that the genome annotations of *E. coli* strains EDL933 (top) and K-12 (bottom) are represented by consecutive thin blocks. This information is lacking in strain CFT073 (middle), as a FASTA file containing no annotated information was used as input instead of GenBank files used for the other two strains. Those blocks correspond to features (e.g. CDS, tRNA) found in GenBank files; the user can zoom-in on a block to the point where the actual nucleotides can be seen, and they can hover the cursor over a block to see the specifics of each feature.

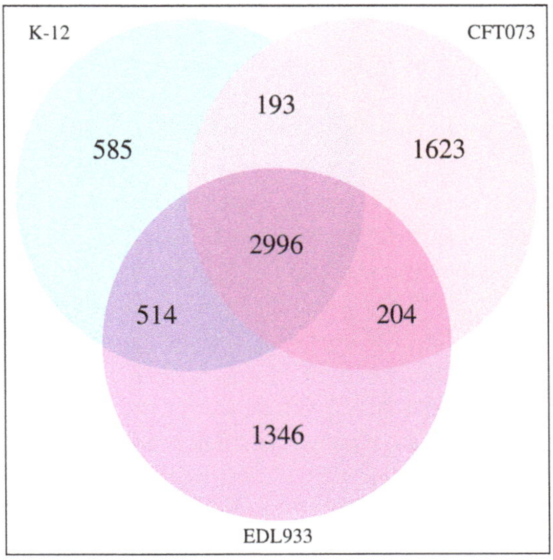

Figure 9.16 Pan-genome Venn diagram based on the study of three *E. coli* strains performed by Welch *et al*. (2002). This diagram offers a rapid assessment of the distribution of genes among the three strains analysed. In this study, 2996 genes are shared among the strains, termed 'core genes'. Each strain has a number of unique genes of 585, 1623 and 1346 for the strain K-12, CFT073, and EDL933, respectively. The total number of accessory genes is 911 (193+204+514) when all the strains are considered. Looking at pairs of strains, the accessory genes present in the strains K-12 and CFT073 are 193, strains K-12 and EDL933 are 514, and strains CFT073 and EDL933 are 204. The Venn diagram was created using the R 'VennDiagram' package (Chen and Boutros, 2011; R Core Team, 2017) with data from the study published by Welch *et al*. (2002). Copyright (2002) National Academy of Sciences, U.S.A.

Pan-genome characterization and phylogenetic tools

The pan-genome[19] is a comprehensive and non-redundant collection of all the genes that are present in a group of closely related strains[20] (Medini *et al*., 2005; Tettelin *et al*., 2005; Tettelin *et al*., 2008; Lapierre and Gogarten, 2009; Guimarães *et al*., 2015). Information gleaned from this approach can be used to observe and/or identify putative virulence genes, evolutionary trends, niche adaptation schemes, and genes present in the genomic cohort or environment. The pan-genome can be divided into three gene groups: (i) core genes that are found in all analysed strains, (ii) unique genes belonging to a specific strain, and (iii) accessory genes that are found within subsets of the studied dataset (Feil, 2004; Medini *et al*., 2005; Tettelin *et al*., 2005; Tettelin *et al*., 2008; Lapierre and Gogarten, 2009; Guimarães *et al*., 2015). As an example, Fig. 9.16 shows a pan-genome Venn diagram created with data from a 2002 study on *E. coli* (Welch *et al*., 2002) using the R package VennDiagram (Chen and Boutros, 2011; R Core Team, 2017). The number of core genes is 2,996. The number of unique genes is 585, 1623, and 1346 for *E. coli* strains K-12, CFT073, and EDL933, respectively. The number of accessory genes can be assessed in a number of ways: (i) if all the strains are considered, the total number of accessory genes is 911 (193+204+514); (ii) if two strains are considered, the number of accessory genes is 193 for the strains K-12 and CFT073, 514 for the strains K-12 and EDL933, and 204 for the strains CFT073 and EDL933.

Unique and accessory genes are often specific to an environmental niche and confer some level of fitness advantage to the organisms that possess them. These genes are often hypothetical and/or connected to bacteriophages, transposons, and plasmids that deal with ancillary biochemical pathways or are specific to dealings with the

19 'Pan' in pan-genome comes from the Greek word '*παν*', which means '*whole*'.

20 It is not necessary to focus exclusively on closely related strains; however, introducing diversity into the analysis will likely reduce the number of core genes and make it more challenging to elucidate fine distinctions or similarities among more homogenous strains.

external environment (Feil, 2004; Medini *et al.*, 2005; Rasko *et al.*, 2008; Lapierre and Gogarten, 2009; Guimarães *et al.*, 2015; Barth *et al.*, 2016). Contrary to core genes, unique and accessory genes are not subjected to strong selective pressures, and therefore display higher mutation rates (Lapierre and Gogarten, 2009; Guimarães *et al.*, 2015). These mutations can lead to pseudogenes or to the alteration of the gene content, which affects the adaptability of an organism. The core genes are present in all genomes within a studied dataset and are generally responsible for cellular functions such as replication, translation, and housekeeping (Medini *et al.*, 2005; Guimarães *et al.*, 2015; Barth *et al.*, 2016). In contrast to accessory and unique genes, the highly conserved core genes are under a great deal of selective pressure and exhibit low mutation rates (Lapierre and Gogarten, 2009; Guimarães *et al.*, 2015).

Numerous pan-genomic studies have been performed and have helped elucidate the concept of open and closed pan-genomes, the size and characterization of the core genes, as well as evolutionary characteristics, and they have reported a broad range of pan-genome sizes. Rasko *et al.* (2008) reported the size of the pan-genome and core genome of 17 *E. coli* strains to be ~2344 and ~2,00 genes, respectively. Lukjancenko *et al.* (2010) analysed 53 pathogenic and commensal *E. coli* strains and estimated a pan-genome size of 13,296 gene families and a core size of 1472 gene families. Moreover, other researchers have estimated the *E. coli* core genome to comprise 2356 and 2865 gene families (Chen *et al.*, 2006; Chaudhuri *et al.*, 2010). Although these discrepancies depend in large part on which strains are included in the analysis, some could arise from differences in annotation. Therefore, tools based on sequence identity (BLAST or similar) have also been used to perform pan-genome analysis (Vieira *et al.*, 2011). Several researchers have developed mathematical equations to estimate pan-genome and core genome sizes, but they have found that some species of bacteria are so diverse that unique genes will continue to be detected even after the addition of hundreds or thousands of genomes (Medini *et al.*, 2005; Tettelin *et al.*, 2005; Chen *et al.*, 2006; Chaudhuri *et al.*, 2010; Laing *et al.*, 2010; Guimarães *et al.*, 2015). As *E. coli* is a facultative anaerobe that can be found within the gastrointestinal tract of animals, as well as in the environment, there are numerous opportunities for environmental niche adaptations (Chen *et al.*, 2006; Rasko *et al.*, 2008; Chaudhuri *et al.*, 2010; Lukjancenko *et al.*, 2010; Vieira *et al.*, 2011). Therefore, given a pool of studied genomes, the greater the species diversity, the lower the number of core genes. Species with this trait are described as having an open pan-genome. By contrast, bacteria with a stable genome will not see an increase in pan-genome size after sequencing few strains; in this case, their pan-genome is referred to as closed (Tettelin *et al.*, 2005; Guimarães *et al.*, 2015).

There are three excellent tools that can be used to perform pan-genome analysis; LS-BSR (Sahl *et al.*, 2014), Panseq (Laing *et al.*, 2010), and Roary (Page *et al.*, 2015). Each tool operates from the command-line and requires the installation of numerous dependencies. The LS-BSR toolkit contains a series of scripts to analyse the pan-genome and compare genetic content across multiple genomes, and as mentioned previously, uses FASTA files as input (Sahl *et al.*, 2014). Panseq performs pan-genome analysis from input FASTA files, and has the added benefit of identifying SNPs within the core genome. Roary functions similarly to LS-BSR and Panseq, though it uses GFF3 annotation files that can be obtained using Prokka during the annotation process (Page *et al.*, 2015).

Once the pan-genome structure has been examined, the strains can be clustered together into a tree to investigate the evolutionary relationship among the analysed strains. From the tools mentioned above, only Panseq output files can be further used to create phylogenetic trees based on the core genes, SNPs, or presence/absence of accessory genes (Laing *et al.*, 2010). In an earlier section, Mauve was used to create a draft assembly by reordering the query genome contigs based on the reference genome. Mauve has additional capabilities such as exporting SNPs, or creating a Newick tree file from multiple genome alignments that can be used to make phylogenetic trees. Once the user has a tree file, it can be imported into the Interactive Tree Of Life (iTOL) web tool for visualization and exported for publication purposes (Letunic and Bork, 2016). The Harvest suite offers a set of tools that are used for core-genome alignment (Parsnp) and dynamic visualization of alignments and dendrograms (Gingr) (Treangen *et al.*, 2014). Harvest is optimized to operate on highly similar genomes, accepts a directory of FASTA or GenBank files as input, can be used to align thousands of genomes, and is reference-based. Gingr can also accept standard alignment files (e.g. multi-Fasta, XMFA, Newick and VCF) and act as a multiple genome alignment viewer.

Another WGS-based approach for evolutionary studies, SNP typing, is based on variant detection. This detection method is achieved by read-mapping against a reference genome, and it is becoming popular in food-borne pathogen surveillance since it is effective in differentiating closely related strains (Franz *et al.*, 2016). Applications of this technique as utilized in retrospective investigations of *E. coli* outbreaks are extensively reviewed by Claire Jenkins in Chapter 8. Among all its services, PATRIC includes a 'Variation Analysis' pipeline that allows the user to upload multiple short-read data files and align them against a reference of choice, selecting among five alignment tools and two SNP callers (Wattam *et al.*, 2017).

Furthermore, NGS data can be used to reproduce previous phylogenetic analyses that were developed using Sanger sequencing such as 16S rRNA gene sequence analysis and multilocus sequence typing (MLST). Analysis based on 16S rRNA gene sequencing has been used in evolutionary studies and became the standard for prokaryote classification (Fox *et al.*, 1977). The 16S rRNA gene is present in all bacteria, consists of conserved and variable segments, and it has been proposed to undergo minimal amounts of horizontal gene transfer, though this hypothesis remains controversial (Kitahara *et al.*, 2012; Lan *et al.*, 2016; Větrovský and Baldrian, 2013). Multiple copies of the 16S rRNA gene have been found within a single organism, and variation can exist between the gene copies (within one genome), and the gene can be highly similar among different bacterial species (Kitahara *et al.*, 2012; Lan *et al.*, 2016; Lukjancenko *et al.*, 2010; Větrovský and Baldrian, 2013). Hence, evolutionary trees based on the 16S rRNA gene may not sufficiently distinguish *E. coli* from other species and may also not distinguish bacteria at the genus level (Lan *et al.*, 2016; Lukjancenko *et al.*, 2010). Nevertheless, the 16S rRNA gene is still commonly used in metagenomic studies for identification of bacteria (Petrosino *et al.*, 2009; Turnbaugh *et al.*, 2009; Reichenberger *et al.*, 2013; Nguyen *et al.*, 2016).

MLST is based on the variability found across a set of housekeeping genes and, depending on the combination of identified alleles, a sequence type is assigned (Maiden *et al.*, 1998). MLST was developed to establish a standardized method, easy to reproduce, and able to distinguish bacterial strains, particularly those that are pathogenic in nature (Maiden *et al.*, 1998; Feil, 2004). Web tools used for *in silico* MLST typing that accept NGS data are available at CGE (Larsen *et al.*, 2012) and EnteroBase (http://enterobase.warwick.ac.uk/). Unfortunately, it has been consistently demonstrated that MLST has insufficient discriminatory power to distinguish closely related strains or to completely segregate pathogenic strains from commensal strains when clustered or viewed on a phylogenic tree (Laing *et al.*, 2010; Lukjancenko *et al.*, 2010; Medini *et al.*, 2005). Researchers have found better agreement in pathotype/serotype strain clustering when using a pan-genome approach, highlighting the importance of extending the analysis to include core and accessory genomes (Laing *et al.*, 2010; Lukjancenko *et al.*, 2010).

Recognizing these challenges, the MLST concept was recently extended to use a gene-by-gene approach that includes core genes (cgMLST) and genes that are part of the whole genome (wgMLST) (Sheppard *et al.*, 2012; Maiden *et al.*, 2013). These methods do not require the analysis of closely related strains, and as they use a curated allele database, a reference sequence is not required (Maiden *et al.*, 2013). Although standardized schemes and databases are currently under development, extended MLST (gene-by-gene approach) has been successfully applied to several pathogens, and PulseNet International will implement it for routine food-borne pathogen surveillance analysis (Taboada *et al.*, 2017). Schemes for *E. coli* cgMLST and wgMLST are available in EnteroBase (http://enterobase.warwick.ac.uk/) where a user can upload and analyse newly sequenced *E. coli* strains together with the numerous genomes already present in the database. Alternatively, the Bacterial Isolate Genome Sequence Database (BIGSdb) software (Jolley and Maiden, 2010) or modules for the commercially available software BioNumerics (Applied Maths, Sint-Martens-Latem, Belgium), CLC Genomics Workbench (QIAgen), and Ridom SeqSphere+ (Ridom GmbH, Münster, Germany) can be used.

Future perspectives

The high level of resolution and evolutionary information provided by WGS is moving clinical and food microbiology towards the implementation of culture-free methods for routine analysis for epidemiological and molecular risk assessment studies as NGS technologies continue to evolve. Companies are investing a large amount of resources to further reduce sequencing costs, improve the accuracy, and extend the read length. If this trend continues, it could be expected that complete genome sequences and draft genomes constituted by few contigs will be more common. However, significant obstacles are still hindering the routine use of WGS. Currently, some of the main limitations can be ascribed to NGS heterogeneous platforms that have different library preparation protocols and instrument technologies. Each platform has intrinsic characteristics such as library type, read length, or error rate, and these differences require varied workflows for data processing. The vast number of data processing software available requires competent bioinformatics skills to perform analyses and interpret results. Moreover, while sequencing costs per base are generally dropping across all available platforms, the cost of instruments, reagents, and expert personnel for sample preparation and data analysis remain inaccessible to small laboratories.

Bacterial predictive genomic tools have notably improved. They can accurately identify serotypes, virulence genes, and antimicrobial resistance genes. Compared to previous methods, these tools are fast, can detect multiple targets, and can be optimized for high throughput analysis. However, their ability to predict the emergence of pathogenic bacteria is still questionable due to the current lack of knowledge surrounding the interplay of virulence and AR genes in the actual expression of a hazardous phenotype. Hence, more *in vitro* and *in vivo* studies are necessary to associate gene targets to bacterial phenotypic properties.

The unprecedented discriminatory power of WGS-based typing techniques has been used successfully to source and geographically track retrospective outbreak investigations. If implemented, WGS-based approaches could potentially detect outbreaks that were previously unnoticed. However, defined standard protocols, including quality parameters, pipelines, and ontology-based data integration must be established for *E. coli* and other bacterial genomes.

References

Adu-Oppong, B., Gasparrini, A.J., and Dantas, G. (2017). Genomic and functional techniques to mine the microbiome for novel antimicrobials and antimicrobial resistance genes. Ann. N. Y. Acad. Sci. *1388*, 42–58. https://doi.org/10.1111/nyas.13257

Agger, M., Scheutz, F., Villumsen, S., Molbak, K., and Petersen, A.M. (2015). Antibiotic treatment of verocytotoxin-producing *Escherichia coli* (VTEC) infection: a systematic review and a proposal. J. Antimicrob. Chemother. *70*, 2440-2446. doi: 10.1093/jac/dkv162

Akhter, S., Aziz, R.K., and Edwards, R.A. (2012). PhiSpy: a novel algorithm for finding prophages in bacterial genomes that combines similarity- and composition-based strategies. Nucleic Acids Res. *40*, e126. doi: 10.1093/nar/gks406

Albà, M.M., Lee, D., Pearl, F.M., Shepherd, A.J., Martin, N., Orengo, C.A., and Kellam, P. (2001). VIDA: a virus database system for the organization of animal virus genome open reading frames. Nucleic Acids Res. *29*, 133–136.

Alikhan, N.F., Petty, N.K., Ben Zakour, N.L., and Beatson, S.A. (2011). BLAST Ring Image Generator (BRIG): simple prokaryote genome comparisons. BMC Genomics *12*, 402. https://doi.org/10.1186/1471-2164-12-402

Antipov, D., Hartwick, N., Shen, M., Raiko, M., Lapidus, A., and Pevzner, P.A. (2016). plasmidSPAdes: assembling plasmids from whole genome sequencing data. Bioinformatics 32, 3380–3387.

Arndt, D., Grant, J.R., Marcu, A., Sajed, T., Pon, A., Liang, Y., and Wishart, D.S. (2016). PHASTER: a better, faster version of the PHAST phage search tool. Nucleic Acids Res. *44*, W16–21. https://doi.org/10.1093/nar/gkw387

Asadulghani, M., Ogura, Y., Ooka, T., Itoh, T., Sawaguchi, A., Iguchi, A., Nakayama, K., and Hayashi, T. (2009). The defective prophage pool of *Escherichia coli* O157: prophage-prophage interactions potentiate horizontal transfer of virulence determinants. PLOS Pathog. *5*, e1000408. doi: 10.1371/journal.ppat.1000408

Attwood, T.K., Bradley, P., Flower, D.R., Gaulton, A., Maudling, N., Mitchell, A.L., Moulton, G., Nordle, A., Paine, K., Taylor, P., *et al.* (2003). PRINTS and its automatic supplement, prePRINTS. Nucleic Acids Res. *31*, 400–402.

Aziz, R.K., Bartels, D., Best, A.A., DeJongh, M., Disz, T., Edwards, R.A., Formsma, K., Gerdes, S., Glass, E.M., Kubal, M., *et al.* (2008). The RAST Server: rapid annotations using subsystems technology. BMC Genomics *9*, 75. https://doi.org/10.1186/1471-2164-9-75

Bankevich, A., Nurk, S., Antipov, D., Gurevich, A.A., Dvorkin, M., Kulikov, A.S., Lesin, V.M., Nikolenko, S.I., Pham, S., Prjibelski, A.D., *et al.* (2012). SPAdes: a new genome assembly algorithm and its applications to single-cell sequencing. J. Comput. Biol. *19*, 455–477. https://doi.org/10.1089/cmb.2012.0021

Barth, S.A., Menge, C., Eichhorn, I., Semmler, T., Wieler, L.H., Pickard, D., Belka, A., Berens, C., and Geue, L. (2016). The accessory genome of Shiga toxin-producing *Escherichia coli* defines a persistent colonization type in cattle. Appl. Environ. Microbiol. *82*, 5455–5464. doi: 10.1128/AEM.00909-16

Bertelli, C., Laird, M.R., Williams, K.P., Simon Fraser University Research Computing, G., Lau, B.Y., Hoad, G., Winsor, G.L., and Brinkman, F.S. (2017). IslandViewer 4: expanded prediction of genomic islands for larger-scale datasets. Nucleic Acids Res. *17*, 132. doi: 10.1093/nar/gkx343

Besemer, J., Lomsadze, A., and Borodovsky, M. (2001). GeneMarkS: a self-training method for prediction of gene starts in microbial genomes. Implications for finding sequence motifs in regulatory regions. Nucleic Acids Res. 29, 2607-2618.

Bi, D., Liu, L., Tai, C., Deng, Z., Rajakumar, K., and Ou, H.Y. (2013). SecReT4: a web-based bacterial type IV secretion system resource. Nucleic Acids Res *41*, D660–665. doi: 10.1093/nar/gks1248

Bi, D., Xu, Z., Harrison, E.M., Tai, C., Wei, Y., He, X., Jia, S., Deng, Z., Rajakumar, K., and Ou, H.Y. (2012). ICEberg: a web-based resource for integrative and conjugative elements found in Bacteria. Nucleic Acids Res. *40*, D621–626. doi: 10.1093/nar/gkr846

Bio-IT World (2015). Thermo Fisher Clarifies Its Vision for Sequencing with Release of Ion S5 Instruments. [online] Available at: http://www.bio-itworld.com/2015/9/1/thermo-fisher-clarifies-vision-sequencing-release-ion-s5-instruments.html [accessed 15 May 2017].

Blattner, F.R., Plunkett, G., Bloch, C.A., Perna, N.T., Burland, V., Riley, M., Collado-Vides, J., Glasner, J.D., Rode, C.K., Mayhew, G.F., *et al.* (1997). The complete genome sequence of *Escherichia coli* K-12. Science *277*, 1453–1462.

Boisen, N., Melton-Celsa, A.R., Scheutz, F., O'Brien, A.D., and Nataro, J.P. (2015). Shiga toxin 2a and Enteroaggregative *Escherichia coli* – a deadly combination. Gut Microbes 6, 272–278. https://doi.org/10.1080/19490976.2015.1054591

Bolger, A.M., Lohse, M., and Usadel, B. (2014). Trimmomatic: a flexible trimmer for Illumina sequence data. Bioinformatics *30*, 2114–2120. https://doi.org/10.1093/bioinformatics/btu170

Bryce, A., Hay, A.D., Lane, I.F., Thornton, H.V., Wootton, M., and Costelloe, C. (2016). Global prevalence of antibiotic resistance in paediatric urinary tract infections caused by *Escherichia coli* and association with routine use of antibiotics in primary care: systematic review and meta-analysis. BMJ. *15*, 352. doi: i939.10.1136/bmj.i939

Brynildsrud, O., Bohlin, J., Scheffer, L., and Eldholm, V. (2016). Rapid scoring of genes in microbial pan-genome-wide association studies with Scoary (vol *17*, 238, 2016). Genome Biol. *17*, 238. doi: 10.1186/s13059-016-1132-8

Canchaya, C., Proux, C., Fournous, G., Bruttin, A., and Brüssow, H. (2003). Prophage genomics. Microbiol. Mol. Biol. Rev. *67*, 238–276.

Carattoli, A., Zankari, E., Garcia-Fernandez, A., Voldby Larsen, M., Lund, O., Villa, L., Moller Aarestrup, F., and Hasman, H. (2014). *In silico* detection and typing of plasmids using PlasmidFinder and plasmid multilocus sequence typing. Antimicrob. Agents Chemother. *58*, 3895-3903. doi: 10.1128/AAC.02412-14

Chakraborty, A., Ghosh, S., Chowdhary, G., Maulik, U., and Chakrabarti, S. (2012). DBETH: a database of bacterial exotoxins for human. Nucleic Acids Res. *40*, D615-620. doi: 10.1093/nar/gkr942

Chaudhuri, R.R., Sebaihia, M., Hobman, J.L., Webber, M.A., Leyton, D.L., Goldberg, M.D., Cunningham, A.F., Scott-Tucker, A., Ferguson, P.R., Thomas, C.M., and Frankel, G. (2010). Complete genome sequence and comparative metabolic profiling of the prototypical enteroaggregative *Escherichia coli* strain 042. PLOS ONE. *5*, e8801. doi: 10.1371/journal.pone.0008801.

Che, D., Hasan, M.S., and Chen, B. (2014). Identifying pathogenicity islands in bacterial pathogenomics using computational approaches. Pathogens. *3*, 36–56. doi: 10.3390/pathogens3010036.

Che, D., Hasan, M.S., Wang, H., Fazekas, J., Huang, J., and Liu, Q. (2011). EGID: an ensemble algorithm for improved genomic island detection in genomic sequences. Bioinformation *7*, 311–314.

Chen, H., and Boutros, P.C. (2011). VennDiagram: a package for the generation of highly-customizable Venn and Euler diagrams in R. BMC Bioinformatics *12*, 35. https://doi.org/10.1186/1471-2105-12-35

Chen, L., Xiong, Z., Sun, L., Yang, J., and Jin, Q. (2012). VFDB 2012 update: toward the genetic diversity and molecular evolution of bacterial virulence factors. Nucleic Acids Res. *40*, D641-645. doi: 10.1093/nar/gkr989

Chen, L., Yang, J., Yu, J., Yao, Z., Sun, L., Shen, Y., and Jin, Q. (2005). VFDB: a reference database for bacterial virulence factors. Nucleic Acids Res. 33, D325–328. doi: 10.1093/nar/gki008

Chen, L., Zheng, D., Liu, B., Yang, J., and Jin, Q. (2016). VFDB 2016: hierarchical and refined dataset for big data analysis – 10 years on. Nucleic Acids Res. *44*, D694–D697. https://doi.org/10.1093/nar/gkv1239

Chen, S.L., Hung, C.S., Xu, J.A., Reigstad, C.S., Magrini, V., Sabo, A., Blasiar, D., Bieri, T., Meyer, R.R., Ozersky, P., and Armstrong, J.R. (2006). Identification of genes subject to positive selection in uropathogenic strains of *Escherichia coli*: A comparative genomics approach. Proc. Natl. Acad. of Sci. USA. *103*, 5977–5982. doi: 10.1073/pnas.0600938103

Cock, P.J.A., Fields, C.J., Goto, N., Heuer, M.L., and Rice, P.M. (2010). The Sanger FASTQ file format for sequences with quality scores, and the Solexa/Illumina FASTQ variants. Nucleic Acids Res. *38*, 1767–1771. https://doi.org/10.1093/nar/gkp1137

Cross, A.S., Gemski, P., Sadoff, J.C., Orskov, F., and Orskov, I. (1984). The importance of the K1 capsule in invasive infections caused by *Escherichia coli*. J. Infect. Dis. *149*, 184–193.

Croxen, M.A., Law, R.J., Scholz, R., Keeney, K.M., Wlodarska, M., and Finlay, B.B. (2013). Recent advances in understanding enteric pathogenic *Escherichia coli*. Clin. Microbiol. Rev. *26*, 822–880. https://doi.org/10.1128/CMR.00022-13

Cuccuru, G., Orsini, M., Pinna, A., Sbardellati, A., Soranzo, N., Travaglione, A., Uva, P., Zanetti, G., and Fotia, G. (2014). Orione, a web-based framework for NGS

analysis in microbiology. Bioinformatics *30*, 1928–1929. https://doi.org/10.1093/bioinformatics/btu135
Darling, A.E., Mau, B., and Perna, N.T. (2010). progressiveMauve: multiple genome alignment with gene gain, loss and rearrangement. PLOS ONE *5*, e11147. https://doi.org/10.1371/journal.pone.0011147
DebRoy, C., Fratamico, P.M., Yan, X., Baranzoni, G., Liu, Y., Needleman, D.S., Tebbs, R., O'Connell, C.D., Allred, A., Swimley, M., and Mwangi, M. (2016). Comparison of O-antigen gene clusters of all O-serogroups of *Escherichia coli* and proposal for adopting a new nomenclature for O-typing. PLOS ONE. *11*, e0147434. doi: 10.1371/journal.pone.0147434
DebRoy, C., Roberts, E., and Fratamico, P.M. (2011). Detection of O antigens in *Escherichia coli*. Anim. Health Res. Rev. *12*, 169–185. https://doi.org/10.1017/S1466252311000193
Delannoy, S., Beutin, L., Mariani-Kurkdjian, P., Fleiss, A., Bonacorsi, S., and Fach, P. (2017). The *Escherichia coli* serogroup O1 and O2 lipopolysaccharides are encoded by multiple O-antigen gene clusters. Front. Cell. Infect. Microbiol. *7*, 30. https://doi.org/10.3389/fcimb.2017.00030
Delcher, A.L., Harmon, D., Kasif, S., White, O., and Salzberg, S.L. (1999). Improved microbial gene identification with GLIMMER. Nucleic Acids Res. *27*, 4636–4641.
Desai, A., Marwah, V.S., Yadav, A., Jha, V., Dhaygude, K., Bangar, U., Kulkarni, V., and Jere, A. (2013). Identification of optimum sequencing depth especially for *de novo* genome assembly of small genomes using next generation sequencing data. PLOS ONE. *8*, e60204. doi: 10.1371/journal.pone.0060204
Didelot, X., Bowden, R., Wilson, D.J., Peto, T.E.A., and Crook, D.W. (2012). Transforming clinical microbiology with bacterial genome sequencing. Nat. Rev. Genet. *13*, 601–612. https://doi.org/10.1038/nrg3226
ECDC EARS-Net (2017). European Centre for Disease Prevention and Control (ECDC). Antimicrobial resistance surveillance in Europe 2015. Annual report of the European Antimicrobial Resistance Surveillance Network (EARS-Net). [online] Available at: https://ecdc.europa.eu/sites/portal/files/media/en/publications/Publications/antimicrobial-resistance-europe-2015.pdf (accessed 15 May 2017).
Edwards, D.J., and Holt, K.E. (2013). Beginner's guide to comparative bacterial genome analysis using next-generation sequence data. Microb. Inform. Exp. *3*, 2. https://doi.org/10.1186/2042-5783-3-2
Ethelberg, S., Olsen, K.E., Scheutz, F., Jensen, C., Schiellerup, P., Enberg, J., Petersen, A.M., Olesen, B., Gerner-Smidt, P., and Mølbak, K. (2004). Virulence factors for hemolytic uremic syndrome, Denmark. Emerging Infect. Dis. *10*, 842–847.
Feil, E.J. (2004). Small change: keeping pace with microevolution. Nat. Rev. Microbiol. *2*, 483–495. https://doi.org/10.1038/nrmicro904
Fouts, D.E. (2006). Phage_Finder: automated identification and classification of prophage regions in complete bacterial genome sequences. Nucleic Acids Res. *34*, 5839–5851. doi: 10.1093/nar/gkl732
Fox, G.E., Pechman, K.R., and Woese, C.R. (1977). Comparative cataloging of 16S ribosomal ribonucleic acid: molecular approach to procaryotic systematics. Int. J. Syst. Evol. Microbiol. *27*, 44–57. doi: 10.1099/00207713-27-1-44
Franz, E., Gras, L.M., and Dallman, T. (2016). Significance of whole genome sequencing for surveillance, source attribution and microbial risk assessment of food-borne pathogens. Curr. Opin. Food Sci. *8*, 74–79. doi: 10.1016/j.cofs.2016.04.004
Fratamico, P.M., DebRoy, C., Liu, Y., Needleman, D.S., Baranzoni, G.M., and Feng, P. (2016). Advances in molecular serotyping and subtyping of *Escherichia coli*. Front. Microbiol. *7*, 644. https://doi.org/10.3389/fmicb.2016.00644
Garcia-Vallve, S., Guzman, E., Montero, M.A., and Romeu, A. (2003). HGT-DB: A database of putative horizontally transferred genes in prokaryotic complete genomes. Nucleic Acids Res. *31*, 187–189. doi: 10.1093/nar/gkg004
Garg, A., and Gupta, D. (2008). VirulentPred: a SVM based prediction method for virulent proteins in bacterial pathogens. BMC Bioinformatics *9*, 62. https://doi.org/10.1186/1471-2105-9-62
Gibson, M.K., Forsberg, K.J., and Dantas, G. (2015). Improved annotation of antibiotic resistance determinants reveals microbial resistomes cluster by ecology. ISME J. *9*, 207-216. doi: 10.1038/ismej.2014.106
Gillings, M., Boucher, Y., Labbate, M., Holmes, A., Krishnan, S., Holley, M., and Stokes, H.W. (2008). The evolution of class 1 integrons and the rise of antibiotic resistance. J. Bacteriol. *190*, 5095–5100. https://doi.org/10.1128/JB.00152-08
Gordon, N.C., Price, J.R., Cole, K., Everitt, R., Morgan, M., Finney, J., Kearns, A.M., Pichon, B., Young, B., Wilson, D.J., and Llewelyn, M.J. (2014). Prediction of *Staphylococcus aureus* antimicrobial resistance by whole-genome sequencing. J. Clin. Microbiol. *52*, 1182–1191. doi: 10.1128/JCM.03117-13
Guimarães, L.C., Florczak-Wyspianska, J., de Jesus, L.B., Viana, M.V., Silva, A., Ramos, R.T., Soares, S.d.e.C., and Soares, S.d.e.C. (2015). Inside the pan-genome – methods and software overview. Curr. Genomics *16*, 245–252. https://doi.org/10.2174/1389202916666150423002311
Gupta, A., Kapil, R., Dhakan, D.B., and Sharma, V.K. (2014a). MP3: a software tool for the prediction of pathogenic proteins in genomic and metagenomic data. PLOS ONE *9*, e93907. https://doi.org/10.1371/journal.pone.0093907
Gupta, S.K., Padmanabhan, B.R., Diene, S.M., Lopez-Rojas, R., Kempf, M., Landraud, L., and Rolain, J.M. (2014b). ARG-ANNOT, a new bioinformatic tool to discover antibiotic resistance genes in bacterial genomes. Antimicrob. Agents Chemother. *58*, 212–220. doi: 10.1128/AAC.01310-13
Gurevich, A., Saveliev, V., Vyahhi, N., and Tesler, G. (2013). QUAST: quality assessment tool for genome assemblies. Bioinformatics *29*, 1072–1075. https://doi.org/10.1093/bioinformatics/btt086
Gymrek, M. (2014). PyBamView: a browser-based application for viewing short read alignments. Bioinformatics *30*, 3405–3407. https://doi.org/10.1093/bioinformatics/btu565

Hasan, M.S., Liu, Q., Wang, H., Fazekas, J., Chen, B., and Che, D. (2012). GIST: Genomic island suite of tools for predicting genomic islands in genomic sequences. Bioinformation *8*, 203–205. https://doi.org/10.6026/97320630008203

Hawkey, J., Hamidian, M., Wick, R.R., Edwards, D.J., Billman-Jacobe, H., Hall, R.M., and Holt, K.E. (2015). ISMapper: identifying transposase insertion sites in bacterial genomes from short read sequence data. BMC Genomics. *16*, 667. doi: 10.1186/s12864-015-1860-2

Hayashi, T., Makino, K., Ohnishi, M., Kurokawa, K., Ishii, K., Yokoyama, K., Han, C.G., Ohtsubo, E., Nakayama, K., Murata, T., and Tanaka, M. (2001). Complete genome sequence of enterohemorrhagic *Escherichia coli* O157:H7 and genomic comparison with a laboratory strain K-12. DNA Res. *8*, 11–22. doi: 10.1093/dnares/8.1.11

Higgins, D.G., Bleasby, A.J., and Fuchs, R. (1992). CLUSTAL V: improved software for multiple sequence alignment. Comput. Appl. Biosci. *8*, 189–191.

Iguchi, A., Iyoda, S., Kikuchi, T., Ogura, Y., Katsura, K., Ohnishi, M., Hayashi, T., and Thomson, N.R. (2015). A complete view of the genetic diversity of the *Escherichia coli* O-antigen biosynthesis gene cluster. DNA Res. *22*, 101–107. https://doi.org/10.1093/dnares/dsu043

Iguchi, A., Iyoda, S., Seto, K., Nishii, H., Ohnishi, M., Mekata, H., Ogura, Y., and Hayashi, T. (2016). Six novel O genotypes from Shiga toxin-producing *Escherichia coli*. Front. Microbiol. *7*, 765. https://doi.org/10.3389/fmicb.2016.00765

Ingle, D.J., Valcanis, M., Kuzevski, A., Tauschek, M., Inouye, M., Stinear, T., Levine, M.M., Robins-Browne, R.M., and Holt, K.E. (2016). *In silico* serotyping of *E. coli* from short read data identifies limited novel O-loci but extensive diversity of O:H serotype combinations within and between pathogenic lineages. Microb. Genom. *2*, e000064. doi: 10.1099/mgen.0.000064

Jain, M., Olsen, H.E., Paten, B., and Akeson, M. (2016). The Oxford Nanopore MinION: delivery of nanopore sequencing to the genomics community. Genome Biol. *17*, 239. doi: 10.1186/s13059-016-1103-0

Jia, B., Raphenya, A.R., Alcock, B., Waglechner, N., Guo, P., Tsang, K.K., Lago, B.A., Dave, B.M., Pereira, S., Sharma, A.N., and Doshi, S. (2017). CARD 2017: expansion and model-centric curation of the comprehensive antibiotic resistance database. Nucleic Acids Res. *45*, D566–D573. doi: 10.1093/nar/gkw1004

Joensen, K.G., Scheutz, F., Lund, O., Hasman, H., Kaas, R.S., Nielsen, E.M., and Aarestrup, F.M. (2014). Real-time whole-genome sequencing for routine typing, surveillance, and outbreak detection of verotoxigenic *Escherichia coli*. J. Clin. Microbiol. *52*, 1501–1510. doi: 10.1128/JCM.03617-13

Joensen, K.G., Tetzschner, A.M., Iguchi, A., Aarestrup, F.M., and Scheutz, F. (2015). Rapid and easy In Silico serotyping of *Escherichia coli* isolates by use of whole-genome sequencing data. J. Clin. Microbiol. *53*, 2410–2426. https://doi.org/10.1128/JCM.00008-15

Johnson, S.S., Zaikova, E., Goerlitz, D.S., Bai, Y., and Tighe, S.W. (2017). Real-time DNA sequencing in the Antarctic dry valleys Using the Oxford nanopore sequencer. J. Biomol. Tech. *28*, 2–7. https://doi.org/10.7171/jbt.17-2801-009

Johnson, T.J., and Nolan, L.K. (2009). Pathogenomics of the virulence plasmids of *Escherichia coli*. Microbiol. Mol. Biol. Rev. *73*, 750–774. https://doi.org/10.1128/MMBR.00015-09

Jolley, K.A., and Maiden, M.C. (2010). BIGSdb: Scalable analysis of bacterial genome variation at the population level. BMC Bioinformatics *11*, 595. https://doi.org/10.1186/1471-2105-11-595

Jünemann, S., Prior, K., Albersmeier, A., Albaum, S., Kalinowski, J., Goesmann, A., Stoye, J., and Harmsen, D. (2014). GABenchToB: a genome assembly benchmark tuned on bacteria and benchtop sequencers. PLOS ONE *9*, e107014. https://doi.org/10.1371/journal.pone.0107014

Jungo, F., and Bairoch, A. (2005). Tox-Prot, the toxin protein annotation program of the Swiss-Prot protein knowledgebase. Toxicon *45*, 293–301.

Kaper, J.B., Nataro, J.P., and Mobley, H.L. (2004). Pathogenic *Escherichia coli*. Nat. Rev. Microbiol. *2*, 123–140. https://doi.org/10.1038/nrmicro818

Kisand, V., and Lettieri, T. (2013). Genome sequencing of bacteria: sequencing, *de novo* assembly and rapid analysis using open source tools. BMC Genomics *14*, 211. https://doi.org/10.1186/1471-2164-14-211

Kitahara, K., Yasutake, Y., and Miyazaki, K. (2012). Mutational robustness of 16S ribosomal RNA, shown by experimental horizontal gene transfer in *Escherichia coli*. Proc. Natl. Acad. Sci. U.S.A. *109*, 19220–19225. https://doi.org/10.1073/pnas.1213609109

Köhler, C.D., and Dobrindt, U. (2011). What defines extraintestinal pathogenic *Escherichia coli*? Int. J. Med. Microbiol. *301*, 642–647. https://doi.org/10.1016/j.ijmm.2011.09.006

Koren, S., Walenz, B.P., Berlin, K., Miller, J.R., Bergman, N.H., and Phillippy, A.M. (2017). Canu: scalable and accurate long-read assembly via adaptive k-mer weighting and repeat separation. Genome Res. *27*, 722–736. https://doi.org/10.1101/gr.215087.116

Laing, C., Buchanan, C., Taboada, E.N., Zhang, Y., Kropinski, A., Villegas, A., Thomas, J.E., and Gannon, V.P. (2010). Pan-genome sequence analysis using Panseq: an online tool for the rapid analysis of core and accessory genomic regions. BMC Bioinformatics. *11*, 461. doi: 10.1186/1471-2105-11-461

Lan, Y.M., Rosen, G., and Hershberg, R. (2016). Marker genes that are less conserved in their sequences are useful for predicting genome-wide similarity levels between closely related prokaryotic strains. Microbiome. *4*,18. doi: 10.1186/s40168-016-0162-5

Lapierre, P., and Gogarten, J.P. (2009). Estimating the size of the bacterial pan-genome. Trends Genet. *25*, 107–110. https://doi.org/10.1016/j.tig.2008.12.004

Larsen, M.V., Cosentino, S., Rasmussen, S., Friis, C., Hasman, H., Marvig, R.L., Jelsbak, L., Sicheritz-Pontén, T., Ussery, D.W., Aarestrup, F.M., *et al.* (2012). Multilocus sequence typing of total-genome sequenced bacteria. J. Clin. Microbiol. *50*, 1355–1361. https://doi.org/10.1128/JCM.06094-11

Law, V., Knox, C., Djoumbou, Y., Jewison, T., Guo, A.C., Liu, Y., Maciejewski, A., Arndt, D., Wilson, M., Neveu, V.,and Tang, A. (2014). DrugBank 4.0: shedding new light on drug metabolism. Nucleic Acids Res. *42*, D1091–1097. doi: 10.1093/nar/gkt1068

Letunic, I., and Bork, P. (2016). Interactive tree of life (iTOL) v3: an online tool for the display and annotation of phylogenetic and other trees. Nucleic Acids Res. *44*, W242–W245. https://doi.org/10.1093/nar/gkw290

Li, H., and Durbin, R. (2009). Fast and accurate short read alignment with Burrows-Wheeler transform. Bioinformatics *25*, 1754–1760. https://doi.org/10.1093/bioinformatics/btp324

Li, H., Handsaker, B., Wysoker, A., Fennell, T., Ruan, J., Homer, N., Marth, G., Abecasis, G., and Durbin, R. (2009). The Sequence Alignment/Map format and SAMtools. Bioinformatics *25*, 2078–2079. https://doi.org/10.1093/bioinformatics/btp352

Li, J., Tai, C., Deng, Z., Zhong, W., He, Y., and Ou, H.Y. (2017). VRprofile: gene-cluster-detection-based profiling of virulence and antibiotic resistance traits encoded within genome sequences of pathogenic bacteria. Brief. Bioinform. *bbw141*. doi: 10.1093/bib/bbw141

Li, J., Yao, Y., Xu, H.H., Hao, L., Deng, Z., Rajakumar, K., and Ou, H.Y. (2015). SecReT6: a web-based resource for type VI secretion systems found in bacteria. Environ. Microbiol. *17*, 2196–2202. https://doi.org/10.1111/1462-2920.12794

Liao, Y.C., Lin, S.H., and Lin, H.H. (2015). Completing bacterial genome assemblies: strategy and performance comparisons. Sci. Rep. *5*, 8747. https://doi.org/10.1038/srep08747

Liu, B., and Pop, M. (2009). ARDB – Antibiotic Resistance Genes Database. Nucleic Acids Res. *37*, D443-447. doi: 10.1093/nar/gkn656

Liu, Y.Y., Wang, Y., Walsh, T.R., Yi, L.X., Zhang, R., Spencer, J., Doi, Y., Tian, G., Dong, B., Huang, X., and Yu, L.F. (2016). Emergence of plasmid-mediated colistin resistance mechanism MCR-1 in animals and human beings in China: a microbiological and molecular biological study. Lancet Infect. Dis. *16*, 161-168. doi: 10.1016/S1473-3099(15)00424-7

Loman, N.J., Quick, J., and Simpson, J.T. (2015). A complete bacterial genome assembled *de novo* using only nanopore sequencing data. Nat. Methods *12*, 733–735. https://doi.org/10.1038/nmeth.3444

Lukjancenko, O., Wassenaar, T.M., and Ussery, D.W. (2010). Comparison of 61 sequenced *Escherichia coli* genomes. Microb. Ecol. *60*, 708–720. https://doi.org/10.1007/s00248-010-9717-3

Luo, R., Liu, B., Xie, Y., Li, Z., Huang, W., Yuan, J., He, G., Chen, Y., Pan, Q., Liu, Y., and Tang, J. (2012). SOAPdenovo2: an empirically improved memory-efficient short-read *de novo* assembler. Gigascience. *1*, 18. doi: 10.1186/2047-217X-1-18

Magoc, T., Pabinger, S., Canzar, S., Liu, X., Su, Q., Puiu, D., Tallon, L.J., and Salzberg, S.L. (2013). GAGE-B: an evaluation of genome assemblers for bacterial organisms. Bioinformatics *29*, 1718–1725. https://doi.org/10.1093/bioinformatics/btt273

Maiden, M.C.J., Bygraves, J.A., Feil, E., Morelli, G., Russell, J.E., Urwin, R., Zhang, Q., Zhou, J.J., Zurth, K., Caugant, D.A., and Feavers, I.M. (1998). Multilocus sequence typing: A portable approach to the identification of clones within populations of pathogenic microorganisms. Proc. Natl. Acad. Sci. U.S.A. *95*, 3140–3145. doi: 10.1073/pnas.95.6.3140

Maiden, M.C., Jansen van Rensburg, M.J., Bray, J.E., Earle, S.G., Ford, S.A., Jolley, K.A., and McCarthy, N.D. (2013). MLST revisited: the gene-by-gene approach to bacterial genomics. Nat. Rev. Microbiol. *11*, 728–736. https://doi.org/10.1038/nrmicro3093

Mainil, J. (2013). *Escherichia coli* virulence factors. Vet. Immunol. Immunopathol. *152*, 2–12. doi: 10.1016/j.vetimm.2012.09.032

Mantri, Y., and Williams, K.P. (2004). Islander: a database of integrative islands in prokaryotic genomes, the associated integrases and their DNA site specificities. Nucleic Acids Res. *32*, D55-58. doi: 10.1093/nar/gkh059

Mao, C., Abraham, D., Wattam, A.R., Wilson, M.J., Shukla, M., Yoo, H.S., and Sobral, B.W. (2015). Curation, integration and visualization of bacterial virulence factors in PATRIC. Bioinformatics *31*, 252–258. https://doi.org/10.1093/bioinformatics/btu631

Margulies, M., Egholm, M., Altman, W.E., Attiya, S., Bader, J.S., Bemben, L.A., Berka, J., Braverman, M.S., Chen, Y.J., Chen, Z., *et al.* (2005). Genome sequencing in microfabricated high-density picolitre reactors. Nature *437*, 376–380.

Martinez-Martinez, L., Pascual, A., Hernandez-Alles, S., Alvarez-Diaz, D., Suarez, A.I., Tran, J., Benedi, V.J., and Jacoby, G.A. (1999). Roles of beta-lactamases and porins in activities of carbapenems and cephalosporins against *Klebsiella pneumoniae*. Antimicrob. Agents Chemother. *43*, 1669–1673.

McArthur, A.G., Waglechner, N., Nizam, F., Yan, A., Azad, M.A., Baylay, A.J., Bhullar, K., Canova, M.J., De Pascale, G., Ejim, L., and Kalan, L. (2013). The comprehensive antibiotic resistance database. Antimicrob. Agents Chemother. *57*, 3348–3357. doi: 10.1128/AAC.00419-13

Medini, D., Donati, C., Tettelin, H., Masignani, V., and Rappuoli, R. (2005). The microbial pan-genome. Curr. Opin. Genet. Dev. *15*, 589–594.

Melton-Celsa, A.R. (2014). Shiga toxin (Stx) classification, structure, and function. Microbiol. Spectr. *2*, EHEC–0024–2013. https://doi.org/10.1128/microbiolspec.EHEC-0024-2013

Moura, A., Soares, M., Pereira, C., Leitão, N., Henriques, I., and Correia, A. (2009). INTEGRALL: a database and search engine for integrons, integrases and gene cassettes. Bioinformatics *25*, 1096–1098. https://doi.org/10.1093/bioinformatics/btp105

NCBI Resource Coordinators (2017). Database Resources of the National Center for Biotechnology Information. Nucleic Acids Res. *45*, D12–D17. doi: 10.1093/nar/gkw1071

Nguyen, N.P., Warnow, T., Pop, M., and White, B. (2016). A perspective on 16S rRNA operational taxonomic unit clustering using sequence similarity. NPJ Biofilms Microbiomes *2*, 16004. https://doi.org/10.1038/npjbiofilms.2016.4

Page, A.J., Cummins, C.A., Hunt, M., Wong, V.K., Reuter, S., Holden, M.T., Fookes, M., Falush, D., Keane, J.A., and Parkhill, J. (2015). Roary: rapid large-scale prokaryote pan genome analysis. Bioinformatics *31*, 3691–3693. https://doi.org/10.1093/bioinformatics/btv421

Pal, C., Bengtsson-Palme, J., Rensing, C., Kristiansson, E., and Larsson, D.G.J. (2014). BacMet: Antibacterial biocide and metal resistance genes database. Nucleic Acids Res. *42*, D737-D743. doi: 10.1093/nar/gkt1252

Pallen, M.J., and Wren, B.W. (2007). Bacterial pathogenomics. Nature *449*, 835–842.

Pearson, W.R., and Lipman, D.J. (1988). Improved tools for biological sequence comparison. Proc. Natl. Acad. Sci. U.S.A. *85*, 2444–2448.

Perna, N.T., Plunkett, G., Burland, V., Mau, B., Glasner, J.D., Rose, D.J., Mayhew, G.F., Evans, P.S., Gregor, J., Kirkpatrick, H.A., *et al.* (2001). Genome sequence of enterohaemorrhagic *Escherichia coli* O157:H7. Nature *409*, 529–533. https://doi.org/10.1038/35054089

Petrosino, J.F., Highlander, S., Luna, R.A., Gibbs, R.A., and Versalovic, J. (2009). Metagenomic pyrosequencing and microbial identification. Clin. Chem. *55*, 856–866. https://doi.org/10.1373/clinchem.2008.107565

Powers, J.G., Weigman, V.J., Shu, J., Pufky, J.M., Cox, D., and Hurban, P. (2013). Efficient and accurate whole genome assembly and methylome profiling of *E. coli*. BMC Genomics *14*, 675. https://doi.org/10.1186/1471-2164-14-675

R Core Team (2017). The' R Project for Statistical Computing. [online] Available at: https://www.r-project.org/ [Accessed 5 September 2017].

Racz, R., Chung, M., Xiang, Z., and He, Y. (2013). Systematic annotation and analysis of 'virmugens'-virulence factors whose mutants can be used as live attenuated vaccines. Vaccine *31*, 797-805. doi: 10.1016/j.vaccine.2012.11.066

Rasko, D.A., Rosovitz, M.J., Myers, G.S., Mongodin, E.F., Fricke, W.F., Gajer, P., Crabtree, J., Sebaihia, M., Thomson, N.R., Chaudhuri, R., *et al.* (2008). The pangenome structure of *Escherichia coli*: comparative genomic analysis of *E. coli* commensal and pathogenic isolates. J. Bacteriol. *190*, 6881–6893. https://doi.org/10.1128/JB.00619-08

Rasko, D.A., Webster, D.R., Sahl, J.W., Bashir, A., Boisen, N., Scheutz, F., Paxinos, E.E., Sebra, R., Chin, C.S., Iliopoulos, D., *et al.* (2011). Origins of the *E. coli* strain causing an outbreak of hemolytic-uremic syndrome in Germany. N. Engl. J. Med. *365*, 709–717. https://doi.org/10.1056/NEJMoa1106920

Reichenberger, E.R., Alexander, G.M., Perreault, M.J., Russell, J.A., Schwartzman, R.J., Hershberg, U., and Rosen, G. (2013). Establishing a relationship between bacteria in the human gut and complex regional pain syndrome. Brain, Behav. Immun. *29*, 62--69. doi: 10.1016/j.bbi.2012.12.005

Ribet, D., and Cossart, P. (2015). How bacterial pathogens colonize their hosts and invade deeper tissues. Microbes Infect. *17*, 173–183. https://doi.org/10.1016/j.micinf.2015.01.004

Robinson, J.T., Thorvaldsdóttir, H., Winckler, W., Guttman, M., Lander, E.S., Getz, G., and Mesirov, J.P. (2011). Integrative genomics viewer. Nat. Biotechnol. *29*, 24–26. https://doi.org/10.1038/nbt.1754

Ronholm, J., Nasheri, N., Petronella, N., and Pagotto, F. (2016). Navigating microbiological food safety in the era of whole-genome sequencing. Clin. Microbiol. Rev. *29*, 837-857. doi: 10.1128/CMR.00056-16

Roux, S., Enault, F., Hurwitz, B.L., and Sullivan, M.B. (2015). VirSorter: mining viral signal from microbial genomic data. PeerJ *3*, e985. https://doi.org/10.7717/peerj.985

Saha, S.B., Uttam, V., and Verma, V. (2015). u-CARE: user-friendly comprehensive antibiotic resistance repository of *Escherichia coli*. J. Clin. Pathol. *68*, 648–651. doi: 10.1136/jclinpath-2015-202927

Sahl, J.W., Caporaso, J.G., Rasko, D.A., and Keim, P. (2014). The large-scale blast score ratio (LS-BSR) pipeline: a method to rapidly compare genetic content between bacterial genomes. Peer J. 2,e332. doi: 10.7717/peerj.332

Sanger, F., Nicklen, S., and Coulson, A.R. (1977). DNA sequencing with chain-terminating inhibitors. Proc. Natl. Acad. Sci. U.S.A. *74*, 5463–5467.

Santiago, A.E., Ruiz-Perez, F., Jo, N.Y., Vijayakumar, V., Gong, M.Q., and Nataro, J.P. (2014). A large family of antivirulence regulators modulates the effects of transcriptional activators in Gram-negative pathogenic bacteria. PLOS Pathog. *10*, e1004153. doi: 10.1371/journal.ppat.1004153

Scaria, J., Chandramouli, U., and Verma, S.K. (2005). Antibiotic Resistance Genes Online (ARGO): a Database on vancomycin and beta-lactam resistance genes. Bioinformation *1*, 5–7.

Seemann, T. (2014). Prokka: rapid prokaryotic genome annotation. Bioinformatics *30*, 2068–2069. https://doi.org/10.1093/bioinformatics/btu153

Shendure, J., and Ji, H. (2008). Next-generation DNA sequencing. Nat. Biotechnol. *26*, 1135–1145. https://doi.org/10.1038/nbt1486

Sheppard, S.K., Jolley, K.A., and Maiden, M.C. (2012). A gene-by-gene approach to bacterial population genomics: whole genome MLST of *Campylobacter*. Genes 3, 261–277. https://doi.org/10.3390/genes3020261

Siguier, P., Perochon, J., Lestrade, L., Mahillon, J., and Chandler, M. (2006). ISfinder: the reference centre for bacterial insertion sequences. Nucleic Acids Res. *34*, D32-36. doi: 10.1093/nar/gkj014

Smith, J.L., Fratamico, P.M., and Gunther, N.W. (2014). Shiga toxin-producing *Escherichia coli*. Adv. Appl. Microbiol. *86*, 145–197. https://doi.org/10.1016/B978-0-12-800262-9.00003-2

Srinivasan, K.N., Gopalakrishnakone, P., Tan, P.T., Chew, K.C., Cheng, B., Kini, R.M., Koh, J.L., Seah, S.H., and Brusic, V. (2002). SCORPION, a molecular database of scorpion toxins. Toxicon *40*, 23–31.

Swain, M.T., Tsai, I.J., Assefa, S.A., Newbold, C., Berriman, M., and Otto, T.D. (2012). A post-assembly genome-improvement toolkit (PAGIT) to obtain annotated genomes from contigs. Nat Protoc 7, 1260–1284. https://doi.org/10.1038/nprot.2012.068

Taboada, E.N., Graham, M.R., Carriço, J.A., and Van Domselaar, G. (2017). Food Safety in the age of next generation sequencing, bioinformatics, and open data access. Front. Microbiol. *8*, 909. https://doi.org/10.3389/fmicb.2017.00909

Tatusova, T., DiCuccio, M., Badretdin, A., Chetvernin, V., Nawrocki, E.P., Zaslavsky, L., Lomsadze, A., Pruitt, K.D., Borodovsky, M., and Ostell, J. (2016). NCBI prokaryotic genome annotation pipeline. Nucleic Acids Res. *44*, 6614–6624. https://doi.org/10.1093/nar/gkw569

Tettelin, H., Masignani, V., Cieslewicz, M.J., Donati, C., Medini, D., Ward, N.L., Angiuoli, S.V., Crabtree, J., Jones, A.L., Durkin, A.S., and DeBoy, R.T. (2005). Genome analysis of multiple pathogenic isolates of *Streptococcus agalactiae*: implications for the microbial 'pan-genome'. Proc. Natl. Acad. of Sci. USA. *102*, 13950-13955. doi: 10.1073/pnas.0506758102

Tettelin, H., Riley, D., Cattuto, C., and Medini, D. (2008). Comparative genomics: the bacterial pan-genome. Curr. Opin. Microbiol. *11*, 472–477.

Thomsen, M.C., Ahrenfeldt, J., Cisneros, J.L., Jurtz, V., Larsen, M.V., Hasman, H., Aarestrup, F.M., and Lund, O. (2016). A bacterial analysis platform: an integrated system for analysing bacterial whole genome sequencing data for clinical diagnostics and surveillance. PLOS ONE. *11*, e0157718. doi: 10.1371/journal.pone.0157718

Thorvaldsdóttir, H., Robinson, J.T., and Mesirov, J.P. (2013). Integrative Genomics Viewer (IGV): high-performance genomics data visualization and exploration. Brief. Bioinformatics *14*, 178–192. https://doi.org/10.1093/bib/bbs017

Tominaga, A. (2004). Characterization of six flagellin genes in the H3, H53 and H54 standard strains of *Escherichia coli*. Genes Genet. Syst. *79*, 1–8.

Tominaga, A., and Kutsukake, K. (2007). Expressed and cryptic flagellin genes in the H44 and H55 type strains of *Escherichia coli*. Genes Genet. Syst. *82*, 1–8.

Treangen, T.J., Ondov, B.D., Koren, S., and Phillippy, A.M. (2014). The Harvest suite for rapid core-genome alignment and visualization of thousands of intraspecific microbial genomes. Genome Biol. *15*, 524.

Tsafnat, G., Copty, J., and Partridge, S.R. (2011). RAC: Repository of antibiotic resistance cassettes. Database *2011*, bar054. https://doi.org/10.1093/database/bar054

Turnbaugh, P.J., Hamady, M., Yatsunenko, T., Cantarel, B.L., Duncan, A., Ley, R.E., Sogin, M.L., Jones, W.J., Roe, B.A., Affourtit, J.P., *et al.* (2009). A core gut microbiome in obese and lean twins. Nature *457*, 480–484. https://doi.org/10.1038/nature07540

Varani, A.M., Siguier, P., Gourbeyre, E., Charneau, V., and Chandler, M. (2011). ISsaga is an ensemble of web-based methods for high throughput identification and semi-automatic annotation of insertion sequences in prokaryotic genomes. Genome Biol. *12*. R30. doi: 10.1186/gb-2011-12-3-r30

Vazquez-Rosas-Landa, M., Ponce-Soto, G.Y., Eguiarte, L.E., and Souza, V. (2017). Comparative genomics of free-living Gamma-Proteobacteria. Pathogenesis-related genes or interaction-related genes? Pathog. Dis. *75*, ftx059. doi: 10.1093/femspd/ftx059

Větrovský, T., and Baldrian, P. (2013). The variability of the 16S rRNA gene in bacterial genomes and its consequences for bacterial community analyses. PLOS ONE *8*, e57923. https://doi.org/10.1371/journal.pone.0057923

Vieira, G., Sabarly, V., Bourguignon, P.Y., Durot, M., Le Fèvre, F., Mornico, D., Vallenet, D., Bouvet, O., Denamur, E., Schachter, V., *et al.* (2011). Core and panmetabolism in *Escherichia coli*. J. Bacteriol. *193*, 1461–1472. https://doi.org/10.1128/JB.01192-10

Walker, B.J., Abeel, T., Shea, T., Priest, M., Abouelliel, A., Sakthikumar, S., Cuomo, C.A., Zeng, Q., Wortman, J., Young, S.K., and Earl, A.M. (2014). Pilon: an integrated tool for comprehensive microbial variant detection and genome assembly improvement. PLOS ONE. *9*, e112963. doi: 10.1371/journal.pone.0112963

Wang, L., Rothemund, D., Curd, H., and Reeves, P.R. (2003). Species-wide variation in the *Escherichia coli* flagellin (H-antigen) gene. J. Bacteriol. *185*, 2936–2943.

Wang, Y., Huang, H., Sun, M., Zhang, Q., and Guo, D. (2012). T3DB: an integrated database for bacterial type III secretion system. BMC Bioinformatics *13*, 66. https://doi.org/10.1186/1471-2105-13-66

Wassenaar, T.M., and Gaastra, W. (2001). Bacterial virulence: can we draw the line? FEMS Microbiol. Lett. *201*, 1–7. doi: 10.1111/j.1574-6968.2001.tb10724.x

Wassenaar, T.M., and Gunzer, F. (2015). The prediction of virulence based on presence of virulence genes in *E. coli* may not always be accurate. Gut Pathog. *7*, 15. https://doi.org/10.1186/s13099-015-0062-4

Wassenaar, T.M., Zschüttig, A., Beimfohr, C., Geske, T., Auerbach, C., Cook, H., Zimmermann, K., and Gunzer, F. (2015). Virulence genes in a probiotic *E. coli* product with a recorded long history of safe use. Eur. J. Microbiol. Immunol. *5*, 81–93. https://doi.org/10.1556/EUJMI-D-14-00039

Waterhouse, A.M., Procter, J.B., Martin, D.M.A., Clamp, M., and Barton, G.J. (2009). Jalview Version 2—a multiple sequence alignment editor and analysis workbench. Bioinformatics. *25*, 1189-1191. doi: 10.1093/bioinformatics/btp033

Wattam, A.R., Davis, J.J., Assaf, R., Boisvert, S., Brettin, T., Bun, C., Conrad, N., Dietrich, E.M., Disz, T., Gabbard, J.L., and Gerdes, S. (2017). Improvements to PATRIC, the all-bacterial Bioinformatics Database and Analysis Resource Center. Nucleic Acids Res. *45*, D535–D542. doi: 10.1093/nar/gkw1017

Welch, R.A., Burland, V., Plunkett, G.I.I.I., Redford, P., Roesch, P., Rasko, D., Buckles, E.L., Liou, S.R., Boutin, A., Hackett, J., and Stroud, D. (2002). Extensive mosaic structure revealed by the complete genome sequence of uropathogenic *Escherichia coli*. Proc. Natl. Acad. Sci. *99*, 17020-17024. doi: 10.1073/pnas.252529799

Whiteside, M.D., Laing, C.R., Manji, A., Kruczkiewicz, P., Taboada, E.N., and Gannon, V.P. (2016). SuperPhy: predictive genomics for the bacterial pathogen *Escherichia coli*. BMC Microbiol. *16*, 65. https://doi.org/10.1186/s12866-016-0680-0

Wick, R.R., Schultz, M.B., Zobel, J., and Holt, K.E. (2015). Bandage: interactive visualization of *de novo* genome assemblies. Bioinformatics *31*, 3350-3352. doi: 10.1093/bioinformatics/btv383

Wickham, H. (2016). ggplot2: elegant graphics for data analysis. (Cham, Switzerland: Springer International Publishing. doi: 10.1007/978-3-319-24277-4

Wu, H.J., Wang, A.H., and Jennings, M.P. (2008). Discovery of virulence factors of pathogenic bacteria. Curr. Opin. Chem. Biol. *12*, 93–101. https://doi.org/10.1016/j.cbpa.2008.01.023

Xiang, Z., Tian, Y., and He, Y. (2007). PHIDIAS: a pathogen-host interaction data integration and analysis system. Genome Biol. *8*, R150.

Yang, H., Qin, C., Li, Y.H., Tao, L., Zhou, J., Yu, C.Y., Xu, F., Chen, Z., Zhu, F., and Chen, Y.Z. (2016). Therapeutic target database update 2016: enriched resource for bench to clinical drug target and targeted pathway information. Nucleic Acids Res. *44*, D1069–1074. doi: 10.1093/nar/gkv1230

Yang, J., Chen, L., Sun, L., Yu, J., and Jin, Q. (2008). VFDB 2008 release: an enhanced web-based resource for comparative pathogenomics. Nucleic Acids Res. *36*, D539-542. doi: 10.1093/nar/gkm951

Yoon, S.H., Park, Y.K., and Kim, J.F. (2015). PAIDB v2.0: exploration and analysis of pathogenicity and resistance islands. Nucleic Acids Res. *43*, D624-630. doi: 10.1093/nar/gku985

Zankari, E., Hasman, H., Cosentino, S., Vestergaard, M., Rasmussen, S., Lund, O., Aarestrup, F.M., and Larsen, M.V. (2012). Identification of acquired antimicrobial resistance genes. J. Antimicrob. Chemother. *67*, 2640-2644. doi: 10.1093/jac/dks261

Zankari, E., Hasman, H., Kaas, R.S., Seyfarth, A.M., Agerso, Y., Lund, O., Larsen, M.V., and Aarestrup, F.M. (2013). Genotyping using whole-genome sequencing is a realistic alternative to surveillance based on phenotypic antimicrobial susceptibility testing. J. Antimicrob. Chemother. *68*, 771-777. doi: 10.1093/jac/dks496

Zerbino, D.R., and Birney, E. (2008). Velvet: algorithms for *de novo* short read assembly using de Bruijn graphs. Genome Res. *18*, 821–829. https://doi.org/10.1101/gr.074492.107

Zhou, C.E., Smith, J., Lam, M., Zemla, A., Dyer, M.D., and Slezak, T. (2007). MvirDB – a microbial database of protein toxins, virulence factors and antibiotic resistance genes for bio-defence applications. Nucleic Acids Res. *35*, D391-394. doi: 10.1093/nar/gkl791

Zhou, Y., Liang, Y., Lynch, K.H., Dennis, J.J., and Wishart, D.S. (2011). PHAST: a fast phage search tool. Nucleic Acids Res. *39*, W347-352. doi: 10.1093/nar/gkr485

Culture-independent Sequence-based Approaches for Diagnostics and Food Safety Testing

10

Susan R. Leonard* and Christopher A. Elkins

Division of Molecular Biology, Office of Applied Research and Safety Assessment, Center for Food Safety and Applied Nutrition, United States Food and Drug Administration, Laurel, MD, USA.

Correspondence: susan.leonard@fda.hhs.gov

https://doi.org/10.21775/9781910190777.10

Abstract

Technological advances in genome sequencing have been leveraged to determine microbial compositions of sample matrices without culturing for individual bacterial species. Such metagenomics approaches have only recently been effectively utilized in pathogenic *Escherichia coli* research. This chapter provides an overview of current metagenomics research and applications, bioinformatic analysis tools, and considerations for detection sensitivity as it relates to *E. coli* pathogens in diagnostics and food safety testing. By bioinformatically mining the metagenomic dataset, *E. coli* serotype, virulence determinants, and antimicrobial susceptibility can be determined. Strain-level discrimination is complex and may be complicated by indigenous *E. coli* strains in the sample microbiota. Nevertheless, the application of metagenomics has the potential benefit of rapid diagnosis or characterization of a contaminating *E. coli* pathogen in food products. This would be particularly valuable for epidemiological efforts and source attribution during outbreak investigations. Furthermore, as the entire microbial community and relative abundances are defined, the effect of perturbations in the microbial community on the growth or survival of *E. coli* pathogens can be elucidated. Bioinformatic analysis tools yielding reliably accurate results presented with a straightforward interpretation will enhance the practicality of metagenomics as a surveillance method in the future.

Introduction

Advances in DNA sequencing technology over the past decade have resulted in the ability to quickly obtain massive quantities of genomic sequence data. The advent of next-generation sequencing (NGS) technology has had a profound impact on clinical diagnostics and testing of food products and food-associated environments. Typically, bacterial pathogens are cultured for isolation followed by whole-genome sequencing (WGS) and characterization. Characterization often includes molecular subtyping and determination of virulence and antimicrobial susceptibility profiles. Furthermore, the isolated strain can be compared genomically to other strains for source attribution and epidemiological studies. However, it would be ideal to bypass the isolation step and perform pathogen detection and characterization directly from clinical samples, food matrixes, or food processing environments. Thus, metagenomic analysis, in which the genetic composition of culture-independent samples is determined, is an attractive rapid technique. The increasing throughput of NGS technologies coupled with decreasing cost has significantly advanced metagenomic studies involving human or environmental microbiomes as it permits the necessary sequencing depth to detect and characterize low abundance microbes. Microbial community analysis has been applied to clinical samples and food safety research, however, these applications are in their infancy and currently there are only a limited number of studies involving pathogenic *E. coli*. When employing culture-independent sequence-based methods, a continuum exists in the comprehensiveness of characterization of an *E. coli* pathogen with a corresponding increase in complexity of the bioinformatic analysis required. If the objective is to merely detect the presence of a pathogen through identification of specific virulence genes, the metagenomic analysis is reasonably straightforward. Conversely, if a *de novo* assembly of the *E. coli* pathogen or strain-level attribution is the goal, the data analysis is very challenging particularly if the per cent abundance of *E. coli* in the sample matrix is low. This chapter describes approaches to generating and analysing metagenomic data, including describing microbial communities to strain-level phylogenetic resolution. Limits of detection and sensitivity as it is applied to food safety and clinical diagnostics is also discussed.

Approaches to sequencing and analysis of metagenomic samples

Targeted amplicon sequencing

By far the most common method of determining bacterial members of a microbial community by targeted amplicon sequencing involves classifying bacteria using 16S rRNA gene sequences. A PCR amplification step is used to selectively amplify a variable region in the 16S rRNA gene, followed by amplicon sequencing utilizing NGS technology. This approach to microbial community analysis has been referred to as metaprofiling, deep amplicon sequencing, and/or metataxonomics. Relying solely on the sequence of the 16S rRNA gene, lower phylogenetic resolution can be obtained than when using a shotgun metagenomics approach. Additionally, the inability to distinguish pathogenic and non-pathogenic *E. coli* renders sole use of this method unsuitable for clinical or food safety applications involving *E. coli*.

Other targeted amplicon sequencing approaches utilize multiplex PCR prior to sequencing in which highly discriminating regions of the genomes of known targets are amplified. These regions may include phylogenetically discriminating single nucleotide polymorphisms (SNPs) and/or specific genes such as virulence determinants, multilocus sequence typing (MLST) loci, or genes with functional importance. This approach has been used in a study to assess the efficacy of detecting biothreat agents present in low copy numbers in a human genomic background (Gardner *et al.*, 2015). That work makes use of a panel of 467 primer pairs that result in amplicons designed to distinguish organisms at the species level. Another study made use of a set of 44 molecular inversion probes rather than using conventional multiplex PCR to selectively capture target genes for detection of ciprofloxacin resistance markers (Stefan *et al.*, 2016).

Although no studies utilizing discriminative targeted amplicon sequencing for detection of pathogenic *E. coli* have been reported, it remains an option as a culture-independent sequencing approach for *E. coli* pathogens. Virulence determinants of all *E. coli* pathotypes, along with phylogenetically discriminative loci can be targeted in a highly parallel multiplexed PCR strategy. Classification of an *E. coli* pathogen at the subspecies level is possible by including a set of conserved genes used for MLST in the amplicon set. Sequencing specific amplified targets results in a pronounced increase in the number of *E. coli*-relevant reads compared to shotgun metagenomic sequencing where the percentage of background reads associated with the sample matrix can be very high. Preliminary results obtained with this approach using STEC-spiked spinach samples demonstrate the increased sensitivity of this technique compared to shotgun metagenomic sequencing (unpublished results). However, the improved sensitivity does require additional time for the target amplification step. A significant disadvantage of targeted sequencing is the inability to detect novel genes, and gene variants may even be undetected. For example, universal primers failed to amplify the *stx2f* gene even though they were designed in conserved regions of other *stx2* variants carried by *E. coli* (Feng *et al.*, 2011). This limitation can be overcome by periodic updates to capture newly discovered targets, but would potentially require redesigning the PCR primer multiplex panel.

Shotgun metagenomic sequencing

In contrast to amplicon sequencing, shotgun metagenomic sequencing requires no initial amplification step; therefore PCR bias is avoided. The entire microbial DNA in the sample is sequenced, thus discovery of novel organisms and genes is possible with adequate sequencing depth. In addition, sequence data can be queried for an unlimited number of known genes of interest. This is particularly important for detection of pathogenic *E. coli* as partial characterization is required to determine pathogenicity potential and pathotype. A detection method that incorporates the ability to simultaneously determine the presence/absence of multiple *E. coli* virulence genes is ideal since hybrid pathotype strains have been reported to cause severe disease (Bielaszewska *et al.*, 2014; Leonard *et al.*, 2016b; Mariani-Kurkdjian *et al.*, 2014; Rasko *et al.*, 2011). Addition of genes or discriminative genomic regions to the analysis is straightforward, requiring only that sequence regions of interest are added to a database.

Taxonomic classification

Ideally, a 16S rRNA amplicon dataset or shotgun metagenomic dataset contains sequence arising from all microbes present in a clinical or food sample. The data can be used to identify the microbes present in the sample community as well as their relative abundances. Performing community analysis, also referred to as taxonomic profiling, provides biological context that cannot be achieved by traditional culture-based detection methods. Bioinformatic tools available for taxonomic profiling utilize several different analysis regimes and differ in the level of taxonomic resolution attained (Table 10.1). Resolving the community members at the family and genus level is fairly straightforward, but resolution at the species or subspecies level can be difficult, especially if relative abundances are of importance. Furthermore, strain level resolution is particularly challenging given the limited availability of unique sequences that can be used for differentiation. If 16S rRNA amplicon sequencing is used for identifying bacterial members of the community, results are generally reported anywhere from the phylum to family taxonomic level, and at times the genus level. QIIME is a commonly used bioinformatic analysis pipeline for community profiling based on targeted 16S rRNA sequencing (Caporaso *et al.*, 2010). Finer phylogenetic resolution is possible when shotgun metagenomic sequencing is utilized, and taxonomic results often include the genus or species level. In some cases, community profiles determined from metagenomic data are reported at the subspecies level, but most bioinformatic tools are not yet capable of reliable subspecies identification.

Bioinformatic tools used for phylogenetic classification of shotgun metagenomic sequence reads rely on binning, or separating reads into taxonomic categories to determine the composition (i.e. organisms present and their relative per cent abundances) of the sampled community. They may take the less sophisticated but time-consuming approach of BLASTn algorithms in which reads are aligned to many genome sequences constituting a database. In the lowest common ancestor approach, the read is categorized into the lowest taxonomic clade that includes all significant hits based on a set cut-off value (e.g. MG-RAST (Meyer *et al.*, 2008) and MEGAN (Huson and Weber, 2013)). Alternatively, some algorithms make use of clade-specific marker genes (e.g. MetaPhlAn (Segata *et al.*, 2012; Truong *et al.*, 2015)), amino acid sequences of conserved proteins (e.g. PhyloPhlAn (Segata *et al.*, 2013)) or discriminative *k*-mer sequences of varying length (e.g. Kraken (Wood and Salzberg, 2014) and CLARK (Ounit *et al.*, 2015)). These approaches increase the speed at which data can be processed. Another advantage of this method is that if unique *k*-mers can be identified at the subspecies level, superior taxonomic resolution can be reached. Utilizing an in-house custom bioinformatic tool based on matching *k*-mers (25 bp), subspecies resolution at the *E. coli* phylogroup level was achieved for STEC-spiked leafy produce experiments (Leonard *et al.*, 2016a). This proves useful in determining relative abundances of different *E. coli* strains in the sample, e.g. spiked STEC and indigenous strain, provided they belong to different *E. coli* phylogroups. A different algorithm relying on SNP patterns in a set of universal bacterial and archaeal genes, ConStrains, was developed to obtain strain level resolution (Luo *et al.*, 2015), however it has not been utilized on metagenomic datasets involving pathogenic *E. coli* at this time.

Taxonomic classification tools rely on reference databases, namely genomes or discriminative sequences, that must be updated as additional microbial sequences become available. Computing the proportions with which a taxonomic clade populates a sample is challenging as not every read in the metagenomic dataset will contain

Table 10.1 Examples of bioinformatic tools for analysis of metagenomic sequencing data

Software	Application	Website	Reference
MetaVelvet	Metagenome assembly	http://metavelvet.dna.bio.keio.ac.jp/	Namiki *et al.* (2012)
IDBA-UD	Metagenome assembly	http://i.cs.hku.hk/~alse/hkubrg/projects/idba_ud/	Peng *et al.* (2012)
Ray Meta	Metagenome assembly	http://denovoassembler.sourceforge.net/	Boisvert *et al.* (2012)
Bowtie 2	Align sequencing reads to reference (mapping)	http://bowtie-bio.sourceforge.net/bowtie2/	Langmead and Salzberg (2012)
MG-RAST	Microbial community taxonomic profiling	http://metagenomics.anl.gov/	Meyer *et al.* (2008)
MEGAN5	Microbial community taxonomic profiling and functional analysis of metatranscriptomic data	http://ab.inf.uni-tuebingen.de/software/megan5/	Huson and Weber (2013)
MetaPhlAn2	Microbial community taxonomic profiling	http://segatalab.cibio.unitn.it/tools/metaphlan2/	Truong *et al.* (2015)
Kraken	Microbial community taxonomic profiling	http://ccb.jhu.edu/software/kraken/	Wood and Salzberg (2014)
CLARK	Microbial community taxonomic profiling	http://clark.cs.ucr.edu/	Ounit *et al.* (2015)
PhyloPhlAn	Taxonomy and phylogeny of microbes	http://segatalab.cibio.unitn.it/tools/metaphlan2/	Segata *et al.* (2013)
PanPhlAn	Microbial community taxonomic profiling (subspecies resolution)	http://segatalab.cibio.unitn.it/tools/panphlan/	Scholz *et al.* (2016)
ConStrains	Microbial community taxonomic profiling (subspecies resolution)	http://bitbucket.org/luo-chengwei/constrains/	Luo *et al.* (2015)
PathoScope 2	Microbial community taxonomic profiling (subspecies resolution)	http://sourceforge.net/projects/pathoscope/	Hong *et al.* (2014)
WG-FAST	Phylogenetic placement based on SNPs from metagenomic data	https://github.com/jasonsahl/wgfast/	Sahl *et al.* (2015)
CONCOCT	Binning metagenome contigs	https://github.com/BinPro/CONCOCT/	Alneberg *et al.* (2014)
MaxBin	Binning metagenome contigs	https://sourceforge.net/projects/maxbin/	Wu *et al.* (2014)
MyCC	Binning metagenome contigs	https://sourceforge.net/projects/sb2nhri/files/MyCC/	Lin and Liao (2016)
LSA	Binning metagenomic reads prior to *de novo* assembly	https://github.com/brian-cleary/LatentStrainAnalysis/	Cleary *et al.* (2015)

sequence that is discriminative to the same level of taxonomic resolution. The use of more than one bioinformatic pipeline for corroboration of the community analysis results provides greater confidence. Another consideration is that as the reference databases associated with bioinformatic tools change over time, the resulting community profile obtained may be altered thus making analysis a dynamic process especially when comparing across past studies.

Microbial community analysis applications

Several recent examples demonstrate the use of taxonomic profiling for culture-independent identification of *E. coli* within sample microbiota. *E. coli* was identified using metagenomic datasets obtained from urine samples of patients exhibiting urinary tract infections (Hasman *et al.*, 2014). Community analysis identified *E. coli* mixed with *Enterococcus faecalis* that was not detected using conventional culturing methods. The ability to identify more than one possible pathogen simultaneously is a considerable advantage of culture-independent detection methods. Using metagenomic sequencing, *Clostridium difficile* was discovered along with STEC O104:H4 in several patient faecal samples used in the investigation of the STEC O104:H4 outbreak in Germany (Loman *et al.*, 2013). Furthermore, several patients exhibiting diarrhoea had faecal samples negative for STEC, but positive by metagenomic data analysis for *Campylobacter jejuni, Campylobacter concisus, Salmonella enterica,* or *C. difficile.* Herein, the demonstrated ability to quickly differentiate patients presenting similar clinical symptoms but with different causative pathogens would significantly aid epidemiological efforts during an STEC outbreak. In this regard, real-time sequencing in which sequence data can be retrieved and analysed as it is generated is possible with the Oxford Nanopore MinION platform. An example of the power of this method can be seen in the identification of viral pathogens in human blood (Greninger *et al.*, 2015). A low complexity sample, a mixture of *E. coli* and *Staphylococcus aureus,* was used to evaluate a streaming bioinformatic algorithm developed for bacterial pathogen detection in clinical samples in real time sequenced on the MinION platform (Cao *et al.*, 2016). *E. coli* was detected as the predominant species in the sample after only 35 minutes of sequencing by aligning reads to a reference genome database and periodically computing relative abundances. This has likely implications for rapid clinical diagnosis of *E. coli* pathogens if it can be applied to the relevant sample matrixes.

Taxonomic profiling for *E. coli* identification has also been applied to food safety studies. Bhatt *et al.* (2012) report community profiling results from shotgun metagenomic sequencing of milk samples collected from cattle with subclinical mastitis. *E. coli* was found to be the most abundant bacteria in the milk microbiota of two of the three bovine breeds tested. No further *E. coli* characterization was attempted; thus, while the study is informative, it exemplifies the limited conclusions that can be drawn from community analysis at the species level involving *E. coli* without further genome characterization. To address this issue, Blagden *et al.* (2016) developed a BLAST-based algorithm that includes datasets of specific nucleotide sequences 40 to 140 nt in length, referred to as e-probes. One e-probe set was designed to specifically detect *E. coli* O157:H7 and a different set indicates the presence of pathogenic *E. coli* but cannot identify the serotype. Alfalfa sprouts were inoculated with STEC O157:H7 Sakai followed by metagenomic sequencing, and the resultant dataset revealed *E. coli* in the sample. The metagenomic data were then analysed using the e-probe algorithm and the presence of *E. coli* O157:H7 was successfully detected in the sprout microbiota. Furthermore, pathogenic *E. coli* was detected in samples prepared with either an *E. coli* O26:H11 or O145:H28 strain (Blagden *et al.*, 2016). In an example of subspecies taxonomic profiling, the correct STEC genome was identified in the metagenomic data of all 40 STEC-positive faecal samples (positive by conventional culture-based methods) associated with the STEC O104:H4 outbreak in Germany (Hong *et al.*, 2014; Loman *et al.*, 2013). However, the correct STEC strain genome was the highest ranked STEC strain in only 27 of the samples, and STEC with relative abundance 1.4% was identified in an STEC-negative sample.

In addition to pathogen identification, taxonomic profiling is a valuable tool in deciphering how microbial communities shift with environmental perturbations, and this information may provide clues for controlling pathogen growth. 16S rRNA amplicon sequencing was used to investigate composition and diversity of the phyllosphere on packaged spinach leaves after storage for 15 days at either 4°C or 10°C (Lopez-Velasco *et al.*, 2011). After storage, the relative abundance of *Escherichia* on spinach stored at 4°C was only 0.5%, while the relative abundance was determined to be 23.07% on spinach stored at 10°C. The diversity of the spinach microbiome was observed to decrease upon storage, with an increasing relative abundance of psychrotrophic bacteria such at *Pseudomonas* spp. (Lopez-Velasco *et al.*, 2011). The authors suggest using the results of community analyses of the microbiota on ready-to-eat vegetables to improve food safety by identifying potential bio-control strains/species that persist during post-harvest processing and compete with pathogens such as STEC for growth during storage. Yang *et al* used shotgun metagenomic sequencing to study the compositional changes in environmental samples collected from following groups of cattle through the processing steps in the beef production chain (Yang *et al.*, 2016). The results suggest that interventions during the production chain decrease microbial diversity and the overall abundance of *E. coli*. However, the relative abundance of *E. coli* was higher in the market-ready samples than in the arrival samples, suggesting *E. coli* may be better able to survive the interventions than some other bacterial species. Future similar longitudinal studies including food processing from farm to ready-to-eat product can potentially be used to examine the effect of steps taken to reduce growth or eliminate pathogens.

Profiling sample microbiota may elucidate how the presence of STEC alters the composition of the community or, conversely, how the microbial community affects the growth or survival of STEC. For example, 16S rRNA amplicon profiling of pre-harvest cattle hides demonstrated that low bacterial community diversity is correlated with STEC presence (Chopyk *et al.*, 2016). Interestingly, the community composition associated with hides harbouring O157:H7 compared to non-O157 STEC was different. Two genera exhibiting higher relative abundance in the absence of STEC, *Soilbacillus* and *Streptococcus*, may be degrading *N*-acylhomoserine lactones and producing lactic acid, respectively, which may be inhibiting colonization of STEC on the hide (Chopyk *et al.*, 2016). As was the case for storage of ready-to-eat produce mentioned above, the use of biotherapeutic agents, individually or as mixtures, was also suggested. A correlation between low microbial diversity and STEC prevalence was also observed in another study with cattle but from a colonization perspective. In that work, 16S rRNA amplicon profiling of beef calf microflora was combined with STEC detection using selective plating and colony PCR for Shiga toxin genes (Mir *et al.*, 2016). The dynamics of STEC colonization and composition of gut microbiota were followed during the first year of life using faecal samples at four time-points. The prevalence of STEC colonization decreased with age, while the diversity of the gut microbiota increased. Furthermore, the relative abundances of some bacterial phylum were significantly changed as the calves aged. The results demonstrated that the Firmicutes:Bacteroidetes ratio was higher in older calves and in STEC-negative faecal samples.

In a food-related example, utilizing 16S rRNA amplicon profiling of bacterial communities inhabiting romaine lettuce in field trials, Williams *et al.* noted that the microbiota associated with plants inoculated with *E. coli* O157:H7 differed from that of control plants at 10 of 12 time points examined (Williams *et al.*, 2013). Furthermore, there was a distinct difference in the microbial community on lettuce inoculated with *E. coli* O157:H7 at most time points when comparing plants that remained positive for *E. coli* O157:H7 with those that were negative for *E. coli* O157:H7. However, the survival or loss of *E. coli* O157:H7 could not be attributed to a specific genus as no consistent differences in significantly enriched or depleted genera were observed between all samples for either positive or negative samples (Williams *et al.*, 2013). In clinical diagnostic research, differences in gut microbiomes have been observed between healthy and ill patients. For example, using a metagenomic approach for profiling the faecal microbiome of infants in a necrotizing enterocolitis (NEC) cohort, Ward *et al.* observed a decrease in relative abundance of *Veillonella* and increase in relative abundance of *E. coli* associated with development of NEC (Ward *et al.*, 2016). Additionally, the presence of *E. coli* was correlated with NEC-associated death. Further metagenomic data analysis to obtain greater phylogenetic resolution of the colonizing *E. coli* strains was performed which is discussed in the respective section below.

In the food safety industry, food matrixes are incubated in specific media with the purpose of enriching for the pathogen of interest. Increasing the proportion of *E. coli* relative to other bacteria in the food sample/enrichment media milieu allows for detection of *E. coli* contaminating a food sample at very low levels pre-enrichment. Increasingly, studies determining shifts in the microbial community during the enrichment process are performed with the aim of tailoring the protocol (media, temperature, additives, etc.) to preferentially enhance the growth of the pathogen of interest. While shifts in microbial composition are the focus here, enrichment as it relates more specifically to detection sensitivity is discussed in greater detail in a later chapter section. Several examples of these types of studies are metagenomic sequencing to determine the shifts in the tomato phyllosphere during enrichment for *Salmonella* (Ottesen *et al.*, 2013), 16S rRNA and metagenomic sequencing to determine the changes in the cilantro microbiome during enrichment for *Salmonella* (Jarvis *et al.*, 2015), and the effect on the mung bean sprout microbiome using two different enrichment media and temperatures for detection of STEC using 16S rRNA gene sequencing (Margot *et al.*, 2016). Margot *et al.* discovered that the bacterial communities were strikingly different after enriching unspiked sprouts in either buffered peptone water (BPW) or Enterobacteriaceae enrichment broth (EE-broth) for 16 hours. Using plate counts, it was determined that STEC O157:H7 grew to 1.5 logs higher when spiked sprouts were enriched 16 hours in EE-broth compared to BPW (Margot *et al.*, 2016). In another study, metagenomic sequencing was performed on bagged spinach samples spiked with STEC O157:H7 Sakai and enriched according to the FDA BAM protocol (Leonard *et al.*, 2015). According to the protocol, samples are incubated statically for 5 hours at 37°C, antimicrobials are added, followed by static enrichment for 18 hours at 42°C (Feng *et al.*, 2013). Samples taken pre-enrichment, and after 5 hours and 23 hours enrichment demonstrate the dramatic alteration in the bacterial community during the enrichment process as the relative abundance of *Pseudomonas* spp. decreases markedly and a variety of genus in the *Enterobacteriaceae* family become the most abundant organisms (Leonard *et al.*, 2015). Furthermore, the difference in the relative abundances of the bacterial species between spiked samples at 8 hours enrichment in which the initial 5 hours included a static versus shaking incubation is revealed. When the samples are shaken, the per cent abundance of *Aeromonas* spp. is increased, while that of the spiked STEC Sakai (reported as *E. coli* phylogroup E) is lower. In another protocol change, antimicrobial supplements were not added and the change in bacterial composition of the enriched samples is demonstrated primarily in the growth of the Gram-positive bacteria *Bacillus* and *Paenibacillus*, with a decrease in relative per cent abundance of STEC Sakai (Leonard *et al.*, 2015). Studies such as these provide a useful guide for future work directed at optimizing enrichment protocols for detecting STEC as early as possible within the enrichment time-frame.

Strain characterization using shotgun metagenomics

Molecular serotyping

While serotype of an *E. coli* strain in a clinical or food sample is valuable for gaining medical perspective, it is particularly important when assessing the pathogenicity of STEC strains. Some STEC serotypes are strongly correlated with outbreaks and severe illness in humans (Feng and Reddy, 2013; Gould *et al.*, 2013; Rasko *et al.*, 2011; Wang *et al.*, 2013). Currently, there are over 185 different O-groups which are defined largely by the sequence of the *wzx*, *wzy*, *wzm*, and *wzt* genes. The 53 H-types are defined by the sequence of the flagellin genes *fliC*, *flkA*, *fllA*, *flmA*, and *flnA* (Fratamico *et al.*, 2016; Joensen *et al.*, 2015). By associating sequences of these loci with antigenic specificity, molecular serotyping of *E. coli* isolates can be performed directly from whole-genome sequence data, and in fact, this method has been shown to be superior to conventional serotyping methods (Fratamico *et al.*, 2016; Joensen *et al.*, 2015). Applying metagenomic sequencing to detection of STEC Sakai spiked in bagged spinach, the serotype of the spiked strain, O157:H7, was successfully determined by mapping reads to serotyping loci sequences (Leonard *et al.*, 2015). This investigation was extended to include 12 STEC isolates representing eight serotypes which confirmed the feasibility of performing molecular serotyping directly from metagenomic data provided the relative abundance of the STEC strain is high enough (Leonard *et al.*, 2016a).

Loman *et al* performed metagenomic sequencing on clinical faecal samples that had been stored from individuals exhibiting diarrhoea during the STEC O104:H4 outbreak in Germany (Loman *et al.*, 2013). Sequence for the O-antigen determining cluster from a draft genome of the outbreak strain obtained from the metagenomic data were used to confirm the O104 serogroup. Sequence reads from 40 samples positive for STEC O104:H4 by conventional detection methods were mapped to the draft genome and the flagellar H antigen type, H4, was confirmed in 21 of the samples. The presence of H-antigens other than, or in addition to, H4 was noted in seven of the faecal samples (Loman *et al.*, 2013). The presence of indigenous *E. coli* in bagged spinach was also noted by molecular serotyping using metagenomic data (Leonard *et al.*, 2015, 2016a). The presence of more than one *E. coli* strain in a sample presents one of the major challenges for detection and characterization of *E. coli* pathogens in culture-independent clinical diagnostics and food safety testing. When more than one *E. coli* strain is present in a metagenomic sample and the read counts mapping to the respective serotyping loci are disparate, the associated O- and H-types making up the individual serotypes of the *E. coli* strains can be deduced. Molecular serotyping of *E. coli* in metagenomic samples can be performed using the same database utilized for molecular serotyping using whole-genome sequencing (WGS) data.

Virulence gene detection

Similar to molecular serotyping, detection and subtyping of *E. coli* virulence genes can be performed assembly-free by mapping sequence reads to gene sequences of interest. Molecular markers defining the *E. coli* pathotypes as well as other genes of interest can be included in a bioinformatic analysis pipeline and would aid in diagnosis of clinical samples or assessing food samples for public health safety. WGS data can be used to predict antimicrobial resistance profiles in *E. coli* with a sensitivity and specificity comparable to phenotypic methods (Koser *et al.*, 2014; Stoesser *et al.*, 2013; Tyson *et al.*, 2015). Inclusion of antimicrobial resistance gene discovery in analysis of metagenomic datasets may expedite patient care decisions as high-throughput sequencing becomes more rapid and cost effective. As with molecular serotyping, well curated virulence gene and antibiotic resistance gene databases are important for improving accuracy especially with accession of novel determinants and resistance alleles. Interrogating metagenomic data for specific virulence factors does not allow for novel gene discovery, but does complement the detection of *E. coli* using microbial community analysis by providing essential risk assessment information.

STEC strain pathogenicity is linked, in part, to the Shiga toxin gene subtypes considering some alleles have been associated with more serious human illness, namely, *stx1a*, *stx2a*, *stx2c*, and *stx2d* (Melton-Celsa, 2014). The genes encoding intimin, *eae*, and enterohaemolysin, *ehxA*, are also considered indicators of pathogenicity (Feng, 2014; Feng and Reddy, 2013; Wang *et al.*, 2013) and are included in compliance decisions by the US FDA. Several examples of *E. coli* virulence gene detection using shotgun metagenomic data have been reported. Using STEC O157:H7 Sakai for spiking bagged spinach, *stx1*, *stx2*, *ehxA*, and *eae* were included in the detection scheme (Leonard *et al.*, 2015). Furthermore, subtyping was successfully performed on both the Shiga toxin genes and the intimin gene to confirm the presence of *stx1a*, *stx2a*, and *γ-eae* that are carried in the STEC Sakai genome. Subsequent work including detection of STEC strains spiked in bagged spinach carrying a variety of Shiga toxin alleles, namely, *stx1a*, *stx2a*, *stx2c*, and *stx2d* demonstrated the ability to detect and correctly subtype the Shiga toxin genes from metagenomic data given adequate sequencing depth (Leonard *et al.*, 2016a). In addition, two isolates included in that work are considered hybrid pathotype strains. One is a STEC/enteroaggregative strain carrying *aggR* and the other is a STEC/enterotoxigenic strain carrying the heat-stable enterotoxin gene *stb*. These genes were also detected directly from the metagenomic data. Importantly, no false positives for virulence genes were observed in that study for samples in which the spiked STEC strain was the only serogroup detected in the metagenomic sample (Leonard *et al.*, 2016a).

In another example, Bhatt *et al.* constructed metabolic sequence profiles of the microbial community (which included *E. coli* as discussed above) associated with milk from cattle with subclinical mastitis using shotgun metagenomic data (Bhatt *et al.*, 2012). Widespread use of antibiotics in cattle may be a contributing factor to the finding that approximately half the identified sequences in the metagenomes were associated with antimicrobial resistance including multidrug resistance and resistance to metals. Interrogating metagenomic samples associated with host animals colonized with pathogenic *E. coli* and their environment for antimicrobial resistance genes may provide a useful way to trace the spread of resistance to particular antibiotics. In fact, several studies have been conducted in which antimicrobial resistance genes within the microbiota of cattle faeces have been identified using metagenomic sequencing (Chambers *et al.*, 2015; Durso *et al.*, 2011; Zhou *et al.*, 2016). Using shotgun metagenomic sequencing, Yang *et al.* (2016) detected food-borne pathogens including *E. coli* within the microbiome of the beef production chain including the arrival and exit from the feedlot, the truck used to transport the cattle to the abattoir, the holding pen at the abattoir, and the market-ready beef products. Non-host (*Bos taurus*) reads were aligned to the Virulence Factor Database (Chen *et al.*, 2012) but are reported in per cent samples in each sample collection category containing at least one virulence factor from superfamilies (adhesion and invasion, iron acquisition, secretion system, toxin) rather than virulence factors belonging to a particular food-borne pathogen. While this study was clearly aimed at determining the abundance of specific bacterial pathogens throughout the beef production system, the lack of specificity in determining/reporting the *E. coli* virulence genes present does not allow for tracing the fate of STEC throughout the process. However, studies like this that make use of metagenomic sequencing for longitudinal sampling could prove very useful in the food production industry if careful attention is paid to specific virulence genes considered molecular markers for *E. coli* pathogenicity/strain identification.

Virulence gene detection directly from metagenomic data may also prove valuable in the clinical setting as the cost of deep sequencing decreases. Ward *et al.* utilized shotgun metagenomic sequencing of faecal samples from 144 preterm and 22 term infants including both NEC and control cases (Ward *et al.*, 2016). The *E. coli* specific gene content of the samples was retrieved using an assembly-free analysis tool, PanPhlAn (pangenome-based phylogenomic analysis) that can be used for genomic profiling of a strain within a complex community (Scholz *et al.*, 2016). Genes were extracted from a set of 116 *E. coli* reference genomes chosen for maximum diversity and used to build a pangenome consisting of gene family clusters. Sequence reads from the samples were mapped to each of the 116 *E. coli* genomes using Bowtie2 (Langmead and Salzberg, 2012) and coverages for gene families were determined. In this way, presence/absence of each *E. coli* gene family was obtained for each metagenomic sample. The study demonstrates that detection of UPEC strains correlates with a risk of NEC in preterm infants.

In the above mentioned report describing the metagenomic sequencing of clinical samples associated with the STEC O104:H4 outbreak in Germany, sequences from the *stx2AB* genes were detected in the metagenomes of 27 of the 40 STEC-positive samples (Loman *et al.*, 2013). Importantly, this included six samples that were negative for Shiga toxin by ELISA. In addition, antibiotic resistance genes were discovered in the draft genome obtained from bioinformatic analysis of a combination of data from the metagenomic samples. In developing and validating PanPhlAn (Scholz *et al.*, 2016), Scholz *et al.* made use of the sequence data from 12 faecal samples associated with the *E. coli* O104:H4 German outbreak (Loman *et al.*, 2013). Gene family profiles of 110 reference *E. coli* strains and the 12 metagenomic samples obtained using PanPhlAn were used in a simple cluster analysis. The results demonstrate that for the metagenomic samples in which the outbreak strain is the only *E. coli* present, the outbreak samples cluster together (Scholz *et al.*, 2016). However, when other *E. coli* strains are present in the faecal sample, clustering based on gene family absence/presence does not produce accurate results. Within the genes identified as present for those samples clustering closely, the same virulence genes and resistance factors were identified as those identified by WGS of the outbreak isolate.

In a different clinically related study, Hasman *et al.* performed metagenomic sequencing directly on 23 urine samples, 19 of which were *E. coli* culture-positive samples (Hasman *et al.*, 2014). Sequence data were mapped to a database of known antibiotic resistance gene variants, and all resistance genes detected by WGS of cultured *E. coli* single isolates were also detected directly from the metagenomic data. In fact, a greater number of resistance genes were detected in the metagenomic data than for pure culture sequence data for some samples. These additional resistance genes had a lower coverage in most cases and were most likely due to genomic content from bacteria other than the *E. coli* of interest. Filtering out the low coverage resistance genes resulted in antimicrobial resistance profiles similar to that obtained from pure cultures (Hasman *et al.*, 2014). In the clinical setting, rapid antibiotic resistance profiles may be possible in the future using the Nanopore MinION for metagenomic sequencing as the challenges of high error rate and low throughput are addressed. Proof of principle was performed by direct sequencing of clinical urine samples utilizing the MinION (Schmidt *et al.*, 2016). *E. coli* was correctly detected and, among the samples, 51 antimicrobial resistance genes were identified of the 55 genes detected by Illumina sequencing of cultured *E. coli* isolates. However, reliable identification of resistance-conferring mutations and allelic variants was problematic (Schmidt *et al.*, 2016). That study was performed on samples of very low complexity as the urine contained only a single pathogen.

Determining strain relatedness using metagenomic data

WGS provides superior resolution for strain-level discrimination compared to other subtyping methods currently used, and thus is a valuable and effective tool in characterizing pathogens during an outbreak investigation (Jackson *et al.*, 2016; Kvistholm Jensen *et al.*, 2016; Rasko *et al.*, 2011). These real-time outbreak investigations and other retrospective studies (Jenkins *et al.*, 2015; Newitt *et al.*, 2016; Rowell *et al.*, 2016; Underwood *et al.*, 2013; Wilson *et al.*, 2016) have demonstrated the utility of WGS in linking epidemiologically relevant clinical and food-borne isolates including pathogenic *E. coli*. The use of shotgun metagenomic data to infer strain relatedness would decrease outbreak response time as it does not rely on pathogen isolation before sequencing. Methods for subtyping *E. coli* have advanced with a corresponding increase in phylogenetic resolution. Ultimately, to fully characterize a pathogenic *E. coli* strain, recovery of the genome of the strain from the clinical or food matrix is required. Regardless, longitudinal studies and outbreaks are two areas within the clinical or food safety fields where significant advancements can be made with metagenomic sampling.

Phylogenetic analysis

Both MLST and SNP analysis can be applied to shotgun metagenomic data. As with SNP analysis performed on WGS data, the genome coverage must be sufficient for accurate results, and this can present a greater problem with metagenomic data since the relative abundance of the pathogenic *E. coli* strain is very low in some samples. The fact that culture-independent clinical and food samples may contain indigenous *E. coli* along with the pathogenic *E. coli* of interest further complicates the bioinformatic analysis. However, in the absence of indigenous *E. coli*, or when the abundance of the *E. coli* pathogen is much greater than that of the indigenous *E. coli*, strain-level phylogenetic resolution can be achieved.

MLST analysis to determine *E. coli* sequence type was performed on the metagenomic data in the study mentioned above conducted by Ward *et al.* in which DNA extracted from faecal samples from infants exhibiting NEC was subjected to metagenomic sequencing (Ward *et al.*, 2016). Sequence data were mapped against a database of *E. coli* MLST sequences to retrieve the sequences of the *E. coli* genes *adk*, *fumC*, *gyrB*, *icd*, *mdh*, *purA*, and *recA* from the metagenomic samples. The investigators were able to assign a MLST sequence type to most samples in which the relative abundance of *E. coli* was at least 1% and MLST types associated with UPEC were identified. Furthermore, it was confirmed from faecal samples collected over a period of days postpartum that with the exception of one infant, the dominant *E. coli* sequence type was constant in individual infants (Ward *et al.*, 2016). MLST typing was also implemented on sequence data obtained from *E. coli* isolates from pure culture and directly on urine samples (Hasman *et al.*, 2014). The MLST types obtained from metagenomic data and the WGS data from the isolate corresponding to the metagenomic sample were in agreement for all 17 cases included in the study (Hasman *et al.*, 2014). In addition to MLST analysis, Hasman *et al.* (2014) performed SNP analysis on both the metagenomic data and sequence data from pure *E. coli* isolates cultured from the urine samples. SNP differences were observed between the metagenomic data and WGS data; however, the metagenomic sample and the corresponding *E. coli* isolated from the sample cluster more closely in a phylogenetic tree than to other samples or pure isolates included in the study.

Loman *et al.* were able to definitively confirm the MLST sequence type of the STEC O104:H4 German outbreak strain from the metagenomic data for 9 of the 40 STEC-positive samples and obtain partial MLST information for the correct sequence type for an additional six samples (Loman *et al.*, 2013). Although Loman *et al.* (2013) did not perform finer scale phylogenetic resolution than can be obtained by MLST, Sahl *et al.* (2015) used the metagenomic data from the faecal samples to test the Whole Genome Focused Array SNP Typing (*WG-FAST*) pipeline developed for phylogenetic placement of bacterial strains from low coverage genome datasets. The results demonstrated that *WG-FAST* correctly genotyped the 12 samples reported to contain only O104:H4 sequence and no other *E. coli* strains (unmixed samples) (Loman *et al.*, 2013). Furthermore, two of the mixed *E. coli* samples were correctly placed in a phylogenetic tree and another five samples with mixtures of *E. coli* were classified as near-misses (phylogenetic placement not accurate but close to the outbreak reference genome). The remaining samples were scattered throughout the *E. coli* phylogroups rather than clustering with the STEC O104:H4 reference genome (Sahl *et al.*, 2015). This example illustrates the difficulty of obtaining strain-level phylogenetic resolution for *E. coli* pathogens from faecal samples as the level of indigenous *E. coli* strains in the faecal sample may be equal to or higher than that of the pathogen in some cases. This same obstacle to SNP typing *E. coli* in metagenomic samples containing indigenous *E. coli* was encountered when performing *E. coli* core gene SNP analysis on metagenomic data from enriched STEC-spiked spinach samples (Leonard *et al.*, 2016a). That study demonstrates, however, that in the absence of indigenous *E. coli* on the spinach, SNP analysis based on metagenomic sequence data are effective in clearly differentiating the spiked STEC strain from other closely related STEC strains and places the *E. coli* in the metagenomic sample as nearest neighbour in an *E. coli* phylogeny to the genome of the STEC strain used for spiking. Future research on the application of phylogenetic analysis of a contaminating pathogenic *E. coli* strain directly from food samples is anticipated as the number and complexity of studies utilizing metagenomics in the area of food safety grows.

Genome assembly

De novo assembly of the genome of an *E. coli* pathogen from a metagenomic dataset arising from a complex clinical or food matrix is a formidable process. Performing a reference-based assembly is far less daunting, but assumes *a priori* knowledge of the pathogenic *E. coli* strain of interest. Reference-based assemblies were successfully performed for STEC O157:H7 Sakai resulting in a draft genome containing only six gaps totalling 200 bp from metagenomic samples of STEC-spiked spinach (0.1 CFU/g), but only after sample enrichment to increase the relative abundance of STEC in the sample (Leonard *et al.*, 2015). By employing a bioinformatic pipeline in which both human DNA sequence and sequence matching reads from faecal samples taken from healthy individuals were screened out, Loman *et al* obtained a *de novo* assembled draft genome of the STEC O104:H4 German outbreak strain (Loman *et al.*, 2013). This was only possible, however, by using metagenomic data from multiple faecal samples in the analysis.

Shotgun metagenomic reads can be assembled *de novo* into a metagenome using existing assemblers created for individual genome assemblies. However, assemblers specifically developed for metagenomic data are reported to demonstrate superior performance (Table 10.1). *De novo* assembly of metagenomic data does not result in contigs belonging only to the strain of interest, thus contigs must be binned into species or strain level clusters. Binning contigs from metagenomes is challenging, particularly if similar species or strains in the sample have similar relative abundances, however, there have been bioinformatic tools developed to assist in this endeavour (Table 10.1). Binning strategies make use of both contig coverage and contig composition (typically tetranucleotide frequency) across multiple samples in which the relative abundance of identical target community members varies (Alneberg *et al.*, 2014; Imelfort *et al.*, 2014; Kang *et al.*, 2015; Wu *et al.*, 2014). Alneberg *et al.* (2014) developed an algorithm, CONCOCT, that was validated with synthetic mock metagenome datasets and was also tested for pathogen discovery using metagenomic data arising from sequencing faecal samples associated with the STEC O104:H4 outbreak strain (Loman *et al.*, 2013). The fact that faecal samples taken at different days after onset of diarrhoea would be expected to contain different relative abundances of the STEC O104:H4 pathogen was exploited. Two bins containing *E. coli* sequence were obtained. A higher percentage of contigs in one bin were outbreak-specific (determined by mapping to the reference outbreak genome), while the other bin contained many more contigs that map to the *E. coli* core genome (Alneberg *et al.*, 2014). The outbreak genome, therefore, was found in a combination of two bins, most likely partly due to the complication of indigenous *E. coli* strains in the faecal samples. A more recent binning algorithm, MyCC, was developed to recover individual genomes from metagenomes utilizing 40 universal phylogenic marker genes in addition to tetranucleotide frequency (Lin and Liao, 2016). It is reported to demonstrate superior performance compared to other binning tools, but binned some *E. coli* and *Salmonella* contigs together.

In a different binning strategy in which metagenomic reads are binned prior to *de novo* assembly, latent strain analysis, Cleary *et al.* (2015) validate the method using multiple mock datasets containing sequences from a mixture of *Salmonella* genomes spiked into gut metagenome sequence. Although the method was able to create partitions (bins) each of which was relatively specific to one of the *Salmonella* genomes, the core of the *Salmonella* genomes were found in a different partition. Gkanogiannis *et al.* (2016) report another bioinformatic pipeline in which longer *k*-mers (≥ 20-mers) are used to capture coverage and composition information directly from reads for binning prior to assembly. The binning algorithm was applied to metagenomic data from the STEC O104:H4 German outbreak-associated faecal samples (Loman *et al.*, 2013). Impressively, almost 70% of the genome was contained in just one cluster; however, that cluster also contained sequence arising from other *E. coli* strains in the faecal samples (Gkanogiannis *et al.*, 2016). In general, the results from current binning algorithms (Alneberg *et al.*, 2014; Cleary *et al.*, 2015; Gkanogiannis *et al.*, 2016) highlight the difficulties of separating the genomes of strains that possess a common core along with variable regions as is the case for *E. coli* pathogens. Hence, a single bin does not contain a single genome.

Currently, *de novo* assembly of pathogenic *E. coli* genomes from metagenomic data has been restricted to testing of *in silico* datasets or datasets from retrospective studies such as the STEC O104:H4 outbreak (Loman *et al.*, 2013). This may change in the near future as additional bioinformatic programs are developed and become more easily accessible to researchers without training in bioinformatics. The availability of new datasets as research in the field accelerates may aid in the development of reliably accurate analysis tools. Additionally, while throughput is lower with Pacific Biosciences Single Molecule Real-time sequencing compared to short-read sequencing methods (e.g. Illumina), the longer reads allow superior assemblies and this may prove useful for metagenome assembly of samples containing pathogenic *E. coli.* A practical alternate strategy preferable to a reference genome assembly is available that may be used on individual samples that does not assume knowledge of a specific *E. coli* strain in a sample. In this strategy, reads are binned as *E. coli*-specific by BLASTn matches to a database of *E. coli* genomes with subsequent *de novo* assembly of the *E. coli*-specific reads. A large database containing many disparate genomes of all phylogroups and all *E. coli* pathotypes would be imperative to obtaining the most accurate and complete genome of interest. An obvious drawback to this method is the inability to discover novel genes or phages in the *E. coli* pathogen, and the presence of two or more *E. coli* strains of similar per cent abundance in the sample would remain problematic. A completely different approach that has been suggested would be to apply Hi-C technology wherein DNA in close physical proximity is chemically cross-linked before DNA extraction, ultimately resulting in long-range contiguity information provided by the sequenced reads (Beitel *et al.*, 2014). However, this technique requires more complicated and time-consuming laboratory preparation methods prior to sequencing thus would be less suitable for routine clinical or food safety work.

Detection sensitivity

The ability to detect low levels of pathogenic *E. coli* in a food or clinical sample using metagenomic sequencing depends on both the depth of sequencing and the proportion of reads belonging to *E. coli* compared with all other reads arising from the sample. Clinical samples contain human DNA and can be quite complex; for example, faecal samples contain gut microbiota and may also contain human intestinal epithelial cells. Urine samples are much less complex, thus detection/strain characterization of *E. coli* in urine is less complicated. Similar to clinical samples, metagenomic sequencing libraries derived from *E. coli* hosts such as cattle also contain DNA from both host cells and indigenous microbiota. Food samples have indigenous microbiota that remains even if a wash step is implemented before sampling, and DNA from the food product itself may constitute a portion of the total sample DNA. If possible, enrichment of the *E. coli*-associated DNA can be performed in the laboratory as part of sample processing before constructing the sequencing library. In order to develop a robust metagenomic sequencing-based detection method for *E. coli*, it is important to determine the protocol sensitivity including sample preparation, depth of sequencing, and sequence read length.

Sequencing depth

A combination of the indigenous microbial load on a food sample and the level of pathogenic *E. coli* contamination in large part determine the depth of sequencing required to detect the pathogen if no enrichment step is utilized. Fresh leafy vegetables have been implicated in many food-borne outbreaks associated with STEC (Herman *et al.*, 2015). The bacterial load on packaged ready-to-eat spinach has been reported to be 10^8 to 10^9 CFU/g (Jackson *et al.*, 2013) or, in another report, 10^6 to 10^7 CFU/g but increasing to 10^{10} CFU/g after 8 days of storage at 10°C (Babic *et al.*, 1996). This high microbial load creates difficulty in attempting to detect very low contamination levels of STEC on spinach using shotgun metagenomic sequencing without an enrichment step. In fact, the contamination level necessary to detect spiked STEC O157:H7 Sakai on bagged spinach pre-enrichment using a total of ten million paired-end 250 bp reads per sample was determined (Leonard *et al.*, 2015). Although the presence of an STEC strain could be detected at a spike level of 10,000 CFU/g, the complete serotyping and virulence gene profile could not be confidently determined unless the spike level is increased to 100,000 CFU/g. Mapping the reads to the STEC Sakai genome demonstrated that 0.84% of the reads derive from the spiked STEC Sakai strain and the genome coverage was estimated to be 3× (Leonard *et al.*, 2015). Sequencing to this depth on a routine basis is cost and time prohibitive as only two samples could be multiplexed on an Illumina MiSeq instrument.

Both the Illumina MiSeq and HiSeq instruments were used during the retrospective study in which the faecal microbiota of the patients exhibiting diarrhoea during the STEC O104:H4 outbreak in Germany was sequenced, (Loman *et al.*, 2013). The average total reads per sample were 14.4 million (150 bp paired-end) and 17.31 million (151 bp paired-end) for samples sequenced using the MiSeq and HiSeq, respectively. The STEC O104:H4 chromosomal coverage was determined to be less than 5× for 26 of the STEC-positive samples. The lowest coverage obtained for which the MLST type could be fully determined was 4.03×, and the lowest chromosomal coverage for which the H-type could be determined was 1.91×. However, the H-type and full MLST type could not be determined for several samples with coverages greater than these. Thus greater sequencing depth would be required to have confidence in characterization of the contaminating *E. coli* at these very low coverages since coverage is expected to be uneven across the genome. Reads mapped to the complete full length *stx2AB* genes for some samples, but below a chromosomal coverage of 3.44× only partial length fragments of the genes could be found. Despite partial coverage of the Shiga toxin genes, it was possible to subtype the Shiga toxin genes in 13 samples. Overall, sequences from the Shiga toxin genes were detected in 67% (27/40) of the STEC-positive samples (Loman *et al.*, 2013). In order to avoid false negatives, either longer read lengths and/or additional reads would be necessary utilizing the sample processing method in that study. Illumina sequencing technology has progressed since that work was performed to achieve read lengths of 300 bp utilizing the MiSeq instrument. Utilizing this more recent technology would have resulted in a two-fold increase in genome coverage with the same number of total reads leading to the possible outcome of a higher Shiga toxin gene detection rate.

Ultimately, the choice of sequencing depth and read length, the combination of which determines genome coverage, will depend on the intended aim(s). While a coverage of 5× would be sufficient to determine the molecular serotype and virulence gene profile of an *E. coli* pathogen, performing accurate SNP analysis for strain-level phylogenetic resolution necessitates greater coverage. This was demonstrated by spiking a variety of STEC strains in bagged spinach (Leonard *et al.*, 2016a). When determining read length to use for a given total number of reads, there is a tradeoff between level of characterization of the *E. coli* pathogen and time. Sequencing a metagenomic sample to great depth (greater number of total reads) with shorter read lengths is more rapid than sequencing with less depth using longer reads even though both methods may result in similar *E. coli* genome coverages. In the first case, time to detection is faster, but in the second case more samples can be sequenced simultaneously in a single sequencing run. If the goal of the study involves not only detection of the *E. coli* pathogen, but also includes a determination of the composition of the microbial community in the sample, then sequencing depth becomes important so that organisms present in very low abundances will not be undetected. Comparing the microbial diversity at different sequencing depths indicates whether sufficient depth has been attained.

Food sample enrichment

Given the low proportion of pathogenic *E. coli* to the remainder of the sample microbiota in many food samples, enrichment has been the standard practice before attempting to culture an *E. coli* or detect it utilizing qPCR. Ideally, a culture-independent sequenced-based method of detecting STEC in food matrixes would not include an enrichment step, however as exemplified above, detecting STEC pre-enrichment on fresh leafy green produce is not practical. Although enriching a sample to increase the relative abundance of the pathogenic *E. coli* adds to the detection time, the time-consuming step of discriminative culturing is avoided, thus decreasing the overall detection time compared to conventional methods. Acquiring the requisite sequencing depth, while retaining practicality, is possible using sample enrichment. This is demonstrated by the fact that STEC O157:H7 Sakai spiked at a level of 100 CFU/g in bagged spinach could be detected and characterized after 5 hours of enrichment following the FDA BAM protocol (Feng *et al.*, 2013; Leonard *et al.*, 2015). This spike level is three logs lower than could be detected pre-enrichment. After an additional 18 hours of enrichment (23 hours total), a spike level of 0.1 CFU/g could be detected. In fact, the relative abundance of STEC was high enough that 12 samples could be multiplexed in a single sequencing run on the Illumina MiSeq after 23 hours of enrichment. However, the relative abundance of STEC after enrichment varies between samples even when using the same STEC strain and spike level suggesting that even small changes in the composition of the indigenous microbial community pre-enrichment can be magnified during enrichment as the bacterial species compete for growth (Leonard *et al.*, 2016a). Packaged ready-to-eat spinach and romaine lettuce were spiked with STEC O157:H7 Sakai at 0.1 CFU/g and enriched according to the FDA BAM protocol for 23 hours. Variation was observed in the read counts mapping to the respective serotyping and virulence gene loci even after normalizing to 1.5 million total reads per sample as evidenced by the high standard deviations (Fig. 10.1). However, molecular serotyping and virulence characterization was achieved for all samples, demonstrating the utility of shotgun metagenomic sequencing for detection of *E. coli* pathogens on food provided that an enrichment step is included.

Expanding the above detection sensitivity investigation, 12 different STEC strains were spiked in bagged spinach at 0.1 CFU/g. Samples were enriched 23 hours and subjected to shotgun metagenomic sequencing. The serotype and virulence genes of the spiked strain, including correct Shiga toxin subtype, were detected in 94% (34/36) of the samples with an average of 2.4 million total reads (250 bp paired-end) per sample (Leonard *et al.*, 2016a). During the enrichment process, all *E.*

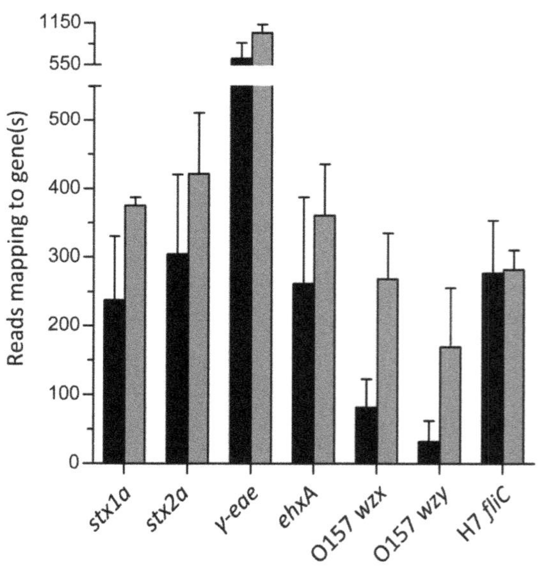

Figure 10.1 Detection of STEC O157:H7 Sakai in fresh ready-to-eat leafy produce using metagenomic sequencing. Six brands of packaged spinach (black) and three brands of packaged romaine lettuce (grey) were spiked with STEC O157:H7 Sakai at a level of 0.1 CFU/g and enriched according to the FDA BAM protocol for 23 hours. Shotgun metagenomic sequencing was performed on aliquots of the enriched media utilizing the Illumina MiSeq platform, and reads (250 bp paired-end) were mapped to the respective serotyping and virulence loci. Read counts were normalized to 1.5 million total reads per sample for comparison. Error bars indicate standard deviation.

coli strains present in the sample grow in competition with each other and with other members of the sample microbiota. In some cases, non-pathogenic indigenous *E. coli* grow to a much higher relative abundance than the contaminating *E. coli* pathogen and prevent characterization of the pathogen. This was the case for two of the samples (Leonard *et al.*, 2016a). This result illustrates the caution that must be used to avoid false negatives when designing food safety testing protocols for metagenomic sequencing. A sufficient number of samples of a particular food product along with a variety of STEC strains must be tested so that natural variation in indigenous microbial load and STEC strain-specific growth is taken into account when determining an adequate total number of sequencing reads of a given length to use with a particular enrichment protocol.

Modifications can be made to the enrichment protocol in an attempt to enhance the sensitivity of detection or shorten the enrichment time necessary for detection. For example, shortening the initial 37°C incubation to 2 hours rather than 5 hours before increasing the temperature to 42°C increases the relative abundance of the spiked STEC strain in bagged spinach after a total of 8 hours enrichment (Fig. 10.2). However, further testing utilizing a variety of STEC strains and stressed STEC cells is important before an alteration to the protocol such as this can be implemented with confidence. A significant amount of experimental work aimed at modifying *E. coli* enrichment protocols for different food products is expected in the future as the food safety industry incorporates metagenomic sequencing into applied research and method validation studies.

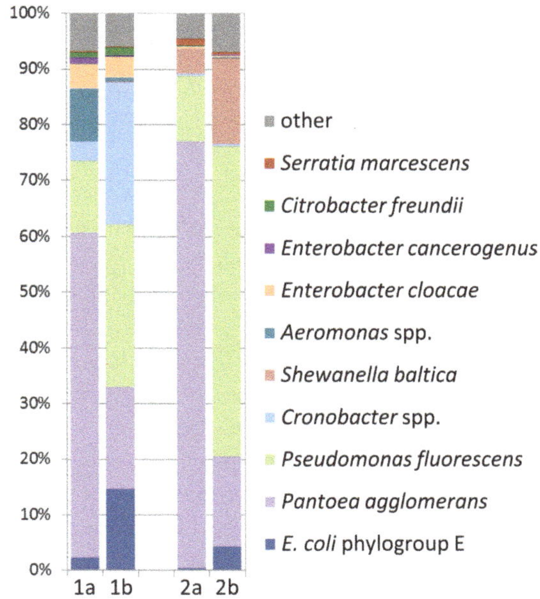

Figure 10.2 Relative abundance of bacterial species associated with spinach spiked with STEC O157:H7 Sakai and enriched using different protocols. Two different packaged spinach samples were each divided into 100 g samples that were spiked with STEC Sakai at a level of 0.1 CFU/g and enriched 8 hours. For each spinach sample, one portion was enriched for 5 hours at 37°C followed by 3 hours at 42°C (1a, 2a) while the other portion was enriched for 2 hours at 37°C followed by 6 hours at 42°C (1b, 2b). Community analysis was performed on metagenomic sequence data using a custom *k*-mer signature database. STEC Sakai is detected as *E. coli* phylogroup E. Bacterial genera or species contributing ≥1% relative abundance are shown, and the sum of all other species identified in lesser amounts is included in 'other'.

Host DNA depletion/DNA extraction

Samples for which host DNA comprises a large proportion of the total DNA may require orders of magnitude greater sequencing depth to detect and characterize an *E. coli* pathogen with a concomitant increase in cost and decrease in throughput. Preprocessing samples to reduce human or host DNA is one strategy that can be employed to increase sensitivity of pathogen detection. Both physical and chemical methods have been utilized. In the detection of *E. coli* in the milk microbiome of cattle with subclinical mastitis, Bhatt *et al.* filtered diluted milk samples using a 3 micron nitrocellulose filter to remove host cells prior to DNA extraction (Bhatt *et al.*, 2012). In utilizing metagenomic sequencing to detect *E. coli* in urine samples, Hasman *et al.* used differential centrifugation to separate human and bacterial cells (Hasman *et al.*, 2014). In another study, several methods were investigated to selectively lyse human cells followed by DNase treatment before collecting bacterial cells from cerebrospinal fluid and nasopharyngeal aspirate (Hasan *et al.*, 2016). A preprocessing method utilizing saponin, a non-ionic surfactant, to lyse human cells was found to significantly improve the sensitivity of pathogen detection. However, the authors do caution that this strategy may not work well for frozen samples as bacterial cell walls may be disrupted, thus selective lysing of human cells will not be effective. A combination of physical and chemical methods was used to deplete human DNA from the metagenomic sample prepared from urine for sequencing on the MinION Nanopore to detect *E. coli* and antibiotic resistance genes (Schmidt *et al.*, 2016).

An alternative strategy to depleting host DNA involves removing host-derived reads from the metagenomic dataset after sequencing but prior to further bioinformatic processing. While this strategy serves to increase the proportion of pathogenic *E. coli* reads in the dataset, it does not decrease the depth of sequencing necessary for pathogen detection as the host DNA is also sequenced. Despite the fact that detection sensitivity is not increased, downstream bioinformatic analysis is facilitated and this approach has been used in several studies involving *E. coli*. Human DNA sequences from faecal samples were screened out by mapping reads to a human reference genome in both the study investigating the outbreak in Germany caused by STEC O104:H4 (Loman *et al.*, 2013) and the study implicating uropathogenic *E. coli* in NEC in preterm infants (Ward *et al.*, 2016). In the work involving detection of pathogens including *E. coli* in the beef production chain, Yang *et al.* removed reads arising from host DNA by aligning reads to the host (*Bos taurus*) reference genome prior to further analysis (Yang *et al.*, 2016).

While differential lysis of eukaryotic and prokaryotic cells is a valuable strategy to deplete host DNA, differential lysis efficiency among various bacterial members comprising the sample microbiota can distort the relative abundances resulting from community analyses. Many evaluations of DNA extraction methods have demonstrated how different DNA extraction methods result in variations in relative abundances of bacteria in microbial communities, for example, the human microbiome (Yuan *et al.*, 2012) and poultry faeces and litter (Rothrock *et al.*, 2014). DNA extraction methods utilize either a mechanical step such as bead beating or an enzymatic lysis step such as application of lysozyme. Hybrid protocols are also commonly used for environmental samples in which bead beating is followed by enzymatic lysis. Including a mechanical lysis step has been found to increase the relative abundances of Gram-positive bacteria in the community profile results as the cell wall is more difficult to disrupt than for Gram-negative bacteria.

If community analysis is important to the study, care must be taken to extract DNA from the Gram-positive bacteria, although for some applications such as screening food samples for pathogenic *E. coli* it is not necessary. In fact, when using the FDA BAM protocol for enrichment prior to collecting the metagenomic sample, antimicrobials are added to suppress the growth of Gram-positive bacteria (Feng *et al.*, 2013). Without the addition of antimicrobials, however, even the addition of lysozyme to the lysis buffer during DNA extraction without the use of any mechanical lysis step affects the composition of the microbial community. This effect is exemplified with cilantro spiked with STEC O157:H7 Sakai and enriched according to the FDA BAM protocol without the addition of antimicrobials (Fig. 10.3). Two identical media aliquots were taken from each of two enriched spiked cilantro samples and subjected to metagenomic sequencing.

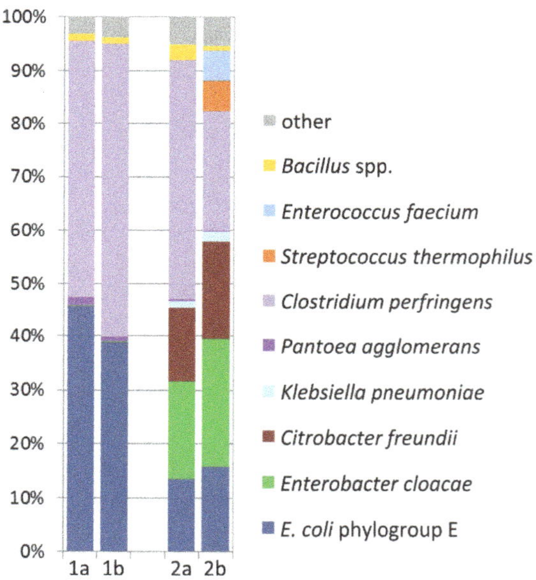

Figure 10.3 Effect on relative abundance of bacterial species associated with cilantro spiked with STEC O157:H7 Sakai using different DNA extraction buffers. Two different samples of 25 g cilantro were spiked with STEC Sakai at a level of 0.4 CFU/g and enriched 23 hours. DNA from each of the two metagenomic samples was extracted using lysis buffer without (1a, 2a) and with lysozyme (1b, 2b). Community analysis was performed on metagenomic sequence data using a custom *k*-mer signature database. STEC Sakai is detected as *E. coli* phylogroup E. Bacterial genera or species contributing ≥1% relative abundance are shown, and the sum of all other species identified in lesser amounts is included in 'other'.

For each cilantro sample, the DNA from one aliquot was extracted using the Qiagen DNeasy kit with the lysis buffer included in the kit, while DNA from the other aliquot was extracted using the protocol included with the kit for Gram-positive bacteria in which lysis buffer is prepared with lysozyme. The relative abundance of *Clostridium perfringens* is increased by approximately 6% when utilizing the lysis buffer with lysozyme for one of the cilantro samples, with a concomitant decrease in per cent abundance of the spiked STEC strain (Fig. 10.3). The difference in the composition of the bacterial communities associated with a second cilantro sample utilizing the two different DNA extraction lysis buffers is particularly striking with *Streptococcus thermophilus* and *Enterococcus faecium* revealed only in the community profile utilizing lysis buffer with lysozyme (Fig. 10.3). However, the detection sensitivity for STEC Sakai is not significantly impacted for that sample. Nevertheless, it is imperative that consistency be used in the DNA extraction procedure in order to compare community composition results across all samples. Additionally, even when the sensitivity of a metagenomics protocol for detection/virulence characterization of *E. coli* pathogens is assessed in the absence of community profiling, a single DNA extraction method must be adhered to for sample comparison unless the study is specifically aimed at evaluating DNA extraction methods.

Challenges and future directions

Significant challenges exist which must be considered when utilizing a metagenomics approach for diagnostics and food safety testing involving pathogenic *E. coli*. The discovery of more than one *E. coli* strain in some samples, e.g. human faecal samples (Loman *et al.*, 2013) and spinach samples (Leonard *et al.*, 2016a), exemplifies one inherent complexity of detecting *E. coli* pathogens using metagenomic data. Specifically, while attributes such as species, serotype, or virulence genes can be easily detected, deconvoluting the data to definitively associate these with the same genome cannot be done without further more complicated analysis. In fact, currently, in some cases bioinformatic analysis may not be sufficient thus pathogen isolation for WGS is required. Nevertheless, metagenomic sequencing provides a rapid screening method for determining presence of virulence genes that can be followed by culturing an *E. coli* strain. Another drawback to metagenomic sequencing is the inability to definitively determine whether an *E. coli* pathogen is viable in the sample. The use of metatranscriptomics, i.e. sequencing total messenger RNA rather than DNA in a sample, has been suggested to overcome this difficulty as RNA decays more rapidly than DNA. While DNA extraction, sequencing library construction, and use of sequencers are straightforward and in some cases streamlined by automation, bioinformatic analysis can prove relatively complex if characterization beyond serotyping and virulence gene detection of an *E. coli* pathogen in the sample matrix is the intended aim. As bioinformatic tools incorporate algorithms to address increasingly complex analyses of metagenomic samples with highly accurate results, research in the metagenomics field will be accelerated. Additionally, while metagenomics is a powerful method to rapidly detect *E. coli* pathogens, characterize their genomes, and determine antibiotic susceptibility, results from analyses must be displayed in an easily interpreted manner before metagenomics can be used routinely in clinical diagnostics and food safety surveillance.

References

Alneberg, J., Bjarnason, B.S., de Bruijn, I., Schirmer, M., Quick, J., Ijaz, U.Z., Lahti, L., Loman, N.J., Andersson, A.F., and Quince, C. (2014). Binning metagenomic contigs by coverage and composition. Nat. Methods *11*, 1144–1146. https://doi.org/10.1038/nmeth.3103

Babic, I., Roy, S., Watada, A.E., and Wergin, W.P. (1996). Changes in microbial populations on fresh cut spinach. Int. J. Food Microbiol. *31*, 107–119.

Beitel, C.W., Froenicke, L., Lang, J.M., Korf, I.F., Michelmore, R.W., Eisen, J.A., and Darling, A.E. (2014). Strain- and plasmid-level deconvolution of a synthetic metagenome by sequencing proximity ligation products. Peer J. *2*, e415. https://doi.org/10.7717/peerj.415

Bhatt, V.D., Ahir, V.B., Koringa, P.G., Jakhesara, S.J., Rank, D.N., Nauriyal, D.S., Kunjadia, A.P., and Joshi, C.G. (2012). Milk microbiome signatures of subclinical mastitis-affected cattle analysed by shotgun sequencing. J. Appl. Microbiol. *112*, 639–650. https://doi.org/10.1111/j.1365-2672.2012.05244.x

Bielaszewska, M., Schiller, R., Lammers, L., Bauwens, A., Fruth, A., Middendorf, B., Schmidt, M.A., Tarr, P.I., Dobrindt, U., Karch, H., *et al.* (2014). Heteropathogenic virulence and phylogeny reveal phased pathogenic metamorphosis in *Escherichia coli* O2:H6. EMBO Mol. Med. *6*, 347–357. https://doi.org/10.1002/emmm.201303133

Blagden, T., Schneider, W., Melcher, U., Daniels, J., and Fletcher, J. (2016). Adaptation and validation of E-Probe diagnostic nucleic acid analysis for detection of *Escherichia coli* O157:H7 in metagenomic data from complex food matrices. J. Food Prot. *79*, 574–581. https://doi.org/10.4315/0362-028X.JFP-15-440

Boisvert, S., Raymond, F., Godzaridis, E., Laviolette, F., and Corbeil, J. (2012). Ray Meta: scalable de novo metagenome assembly and profiling. Genome Biol. *13*, R122. https://doi.org/10.1186/gb-2012-13-12-r122

Cao, M.D., Ganesamoorthy, D., Elliott, A.G., Zhang, H., Cooper, M.A., and Coin, L.J. (2016). Streaming algorithms for identification of pathogens and antibiotic resistance potential from real-time MinION(TM) sequencing. Gigascience *5*, 32. https://doi.org/10.1186/s13742-016-0137-2

Caporaso, J.G., Kuczynski, J., Stombaugh, J., Bittinger, K., Bushman, F.D., Costello, E.K., Fierer, N., Pena, A.G., Goodrich, J.K., Gordon, J.I., *et al.* (2010). QIIME allows analysis of high-throughput community sequencing data. Nat. Methods 7, 335-336. https://doi.org/10.1038/nmeth.f.303

Chambers, L., Yang, Y., Littier, H., Ray, P., Zhang, T., Pruden, A., Strickland, M., and Knowlton, K. (2015). Metagenomic analysis of antibiotic resistance genes in dairy cow feces following therapeutic administration of third generation cephalosporin. PLoS One *10*, e0133764. https://doi.org/10.1371/journal.pone.0133764

Chen, L., Xiong, Z., Sun, L., Yang, J., and Jin, Q. (2012). VFDB 2012 update: toward the genetic diversity and molecular evolution of bacterial virulence factors. Nucleic Acids Res. *40*, D641–645. https://doi.org/10.1093/nar/gkr989

Chopyk, J., Moore, R.M., DiSpirito, Z., Stromberg, Z.R., Lewis, G.L., Renter, D.G., Cernicchiaro, N., Moxley, R.A., and Wommack, K.E. (2016). Presence of pathogenic *Escherichia coli* is correlated with bacterial community diversity and composition on pre-harvest cattle hides. Microbiome *4*, 9. https://doi.org/10.1186/s40168-016-0155-4

Cleary, B., Brito, I.L., Huang, K., Gevers, D., Shea, T., Young, S., and Alm, E.J. (2015). Detection of low-abundance bacterial strains in metagenomic datasets by eigengenome partitioning. Nat. Biotechnol *33*, 1053-1060. https://doi.org/10.1038/nbt.3329

Durso, L.M., Harhay, G.P., Bono, J.L., and Smith, T.P. (2011). Virulence-associated and antibiotic resistance genes of microbial populations in cattle feces analyzed using a metagenomic approach. J. Microbiol. Methods *84*, 278–282. https://doi.org/10.1016/j.mimet.2010.12.008

Feng, P. (2014). Shiga toxin-producing *Escherichia coli* (STEC) in Fresh Produce – a food safety dilemma. Microbiol. Spectr. *2*, EHEC-0010-2013. https://doi.org/10.1128/microbiolspec.EHEC-0010-2013

Feng, P.C., and Reddy, S. (2013). Prevalences of Shiga toxin subtypes and selected other virulence factors among Shiga-toxigenic *Escherichia coli* strains isolated from fresh produce. Appl. Environ. Microbiol. *79*, 6917–6923. https://doi.org/10.1128/AEM.02455-13

Feng, P.C., Jinneman, K., Scheutz, F., and Monday, S.R. (2011). Specificity of PCR and serological assays in the detection of *Escherichia c*oli Shiga toxin subtypes. Appl. Environ. Microbiol. *77*, 6699-6702. https://doi.org/10.1128/AEM.00370-11

Feng, P., Weagant, S.D., and Jinneman, K. (2013). Diarrheagenic *Escherichia coli*. In Bacteriological Analytical Manual, U.S. Food and Drug Administration, ed. (Silver Spring, MD: USFDA).

Fratamico, P.M., DebRoy, C., Liu, Y., Needleman, D.S., Baranzoni, G.M., and Feng, P. (2016). Advances in Molecular Serotyping and Subtyping of *Escherichia coli*. Front. Microbiol. *7*, 644. https://doi.org/10.3389/fmicb.2016.00644

Gardner, S.N., Frey, K.G., Redden, C.L., Thissen, J.B., Allen, J.E., Allred, A.F., Dyer, M.D., Mokashi, V.P., and Slezak, T.R. (2015). Targeted amplification for enhanced detection of biothreat agents by next-generation sequencing. BMC Res. Notes *8*, 682. https://doi.org/10.1186/s13104-015-1530-0

Gkanogiannis, A., Gazut, S., Salanoubat, M., Kanj, S., and Bruls, T. (2016). A scalable assembly-free variable selection algorithm for biomarker discovery from metagenomes. BMC Bioinformatics *17*, 311. https://doi.org/10.1186/s12859-016-1186-3

Gould, L.H., Mody, R.K., Ong, K.L., Clogher, P., Cronquist, A.B., Garman, K.N., Lathrop, S., Medus, C., Spina, N.L., Webb, T.H., *et al.* (2013). Increased recognition of non-O157 Shiga toxin-producing *Escherichia coli* infections in the United States during 2000-2010: epidemiologic features and comparison with *E. coli* O157 infections. Food-borne Pathog. Dis. *10*, 453–460. https://doi.org/10.1089/fpd.2012.1401

Greninger, A.L., Naccache, S.N., Federman, S., Yu, G., Mbala, P., Bres, V., Stryke, D., Bouquet, J., Somasekar, S., Linnen, J.M., *et al.* (2015). Rapid metagenomic identification of viral pathogens in clinical samples by real-time nanopore sequencing analysis. Genome Med. *7*, 99. https://doi.org/10.1186/s13073-015-0220-9

Hasan, M.R., Rawat, A., Tang, P., Jithesh, P.V., Thomas, E., Tan, R., and Tilley, P. (2016). Depletion of human DNA in spiked clinical specimens for improvement of sensitivity of pathogen detection by next-generation sequencing. J. Clin. Microbiol. *54*, 919–927. https://doi.org/10.1128/JCM.03050-15

Hasman, H., Saputra, D., Sicheritz-Ponten, T., Lund, O., Svendsen, C.A., Frimodt-Moller, N., and Aarestrup, F.M. (2014). Rapid whole-genome sequencing for detection and characterization of microorganisms directly from clinical samples. J. Clin. Microbiol. *52*, 139–146. https://doi.org/10.1128/JCM.02452-13

Herman, K.M., Hall, A.J., and Gould, L.H. (2015). Outbreaks attributed to fresh leafy vegetables, United States, 1973-2012. Epidemiol. Infect. *143*, 3011–3021. https://doi.org/10.1017/S0950268815000047

Hong, C., Manimaran, S., Shen, Y., Perez-Rogers, J.F., Byrd, A.L., Castro-Nallar, E., Crandall, K.A., and Johnson, W.E. (2014). PathoScope 2.0: a complete computational framework for strain identification in environmental or clinical sequencing samples. Microbiome *2*, 33. https://doi.org/10.1186/2049-2618-2-33

Huson, D.H., and Weber, N. (2013). Microbial community analysis using MEGAN. Methods Enzymol. *531*, 465-485. https://doi.org/10.1016/B978-0-12-407863-5.00021-6

Imelfort, M., Parks, D., Woodcroft, B.J., Dennis, P., Hugenholtz, P., and Tyson, G.W. (2014). GroopM: an automated tool for the recovery of population genomes from related metagenomes. PeerJ *2*, e603. https://doi.org/10.7717/peerj.603

Jackson, B.R., Tarr, C., Strain, E., Jackson, K.A., Conrad, A., Carleton, H., Katz, L.S., Stroika, S., Gould, L.H., Mody, R.K., *et al.* (2016). Implementation of nationwide real-time whole-genome sequencing to enhance listeriosis outbreak detection and investigation. Clin. Infect. Dis. *63*, 380-386. https://doi.org/10.1093/cid/ciw242

Jackson, C.R., Randolph, K.C., Osborn, S.L., and Tyler, H.L. (2013). Culture dependent and independent analysis of bacterial communities associated with commercial salad leaf vegetables. BMC Microbiol. *13*, 274. https://doi.org/10.1186/1471-2180-13-274

Jarvis, K.G., White, J.R., Grim, C.J., Ewing, L., Ottesen, A.R., Beaubrun, J.J., Pettengill, J.B., Brown, E., and Hanes, D.E. (2015). Cilantro microbiome before and after nonselective pre-enrichment for *Salmonella* using 16S rRNA and metagenomic sequencing. BMC Microbiol. *15*, 160. https://doi.org/10.1186/s12866-015-0497-2

Jenkins, C., Dallman, T.J., Launders, N., Willis, C., Byrne, L., Jorgensen, F., Eppinger, M., Adak, G.K., Aird, H., Elviss, N., *et al.* (2015). Public health investigation of two outbreaks of Shiga toxin-producing *Escherichia coli* O157 associated with consumption of watercress. Appl. Environ. Microbiol. *81*, 3946–3952. https://doi.org/10.1128/AEM.04188-14

Joensen, K.G., Tetzschner, A.M., Iguchi, A., Aarestrup, F.M., and Scheutz, F. (2015). Rapid and easy in silico serotyping of *Escherichia coli* isolates by use of whole-genome sequencing data. J. Clin. Microbiol. *53*, 2410–2426. https://doi.org/10.1128/JCM.00008-15

Kang, D.D., Froula, J., Egan, R., and Wang, Z. (2015). MetaBAT, an efficient tool for accurately reconstructing single genomes from complex microbial communities. PeerJ. 3, e1165. https://doi.org/10.7717/peerj.1165

Koser, C.U., Ellington, M.J., and Peacock, S.J. (2014). Whole-genome sequencing to control antimicrobial resistance. Trends Genet. *30*, 401–407. https://doi.org/10.1016/j.tig.2014.07.003

Kvistholm Jensen, A., Nielsen, E.M., Bjorkman, J.T., Jensen, T., Muller, L., Persson, S., Bjerager, G., Perge, A., Krause, T.G., Kiil, K., *et al.* (2016). Whole-genome sequencing used to investigate a nationwide outbreak of listeriosis caused by ready-to-eat delicatessen meat, Denmark, 2014. Clin. Infect. Dis. *63*, 64–70. https://doi.org/10.1093/cid/ciw192

Langmead, B., and Salzberg, S.L. (2012). Fast gapped-read alignment with Bowtie 2. Nat. Methods *9*, 357–359. https://doi.org/10.1038/nmeth.1923

Leonard, S.R., Mammel, M.K., Lacher, D.W., and Elkins, C.A. (2015). Application of metagenomic sequencing to food safety: detection of Shiga toxin-producing *Escherichia coli* on fresh bagged spinach. Appl. Environ. Microbiol. *81*, 8183–8191. https://doi.org/10.1128/AEM.02601-15

Leonard, S.R., Mammel, M.K., Lacher, D.W., and Elkins, C.A. (2016a). Strain-level discrimination of Shiga toxin-producing *Escherichia coli* in spinach using metagenomic sequencing. PLoS One *11*, e0167870. https://doi.org/10.1371/journal.pone.0167870

Leonard, S.R., Mammel, M.K., Rasko, D.A., and Lacher, D.W. (2016b). Hybrid Shiga toxin-producing and enterotoxigenic *Escherichia* sp. cryptic lineage 1 strain 7v harbors a hybrid plasmid. Appl. Environ. Microbiol. *82*, 4309–4319. https://doi.org/10.1128/AEM.01129-16

Lin, H.H., and Liao, Y.C. (2016). Accurate binning of metagenomic contigs via automated clustering sequences using information of genomic signatures and marker genes. Sci. Rep. *6*, 24175. https://doi.org/10.1038/srep24175

Loman, N.J., Constantinidou, C., Christner, M., Rohde, H., Chan, J.Z., Quick, J., Weir, J.C., Quince, C., Smith, G.P., Betley, J.R., *et al.* (2013). A culture-independent sequence-based metagenomics approach to the investigation of an outbreak of Shiga-toxigenic *Escherichia coli* O104:H4. JAMA *309*, 1502–1510. https://doi.org/10.1001/jama.2013.3231

Lopez-Velasco, G., Welbaum, G.E., Boyer, R.R., Mane, S.P., and Ponder, M.A. (2011). Changes in spinach phylloepiphytic bacteria communities following minimal processing and refrigerated storage described using pyrosequencing of 16S rRNA amplicons. J. Appl. Microbiol. *110*, 1203–1214. https://doi.org/10.1111/j.1365-2672.2011.04969.x

Luo, C., Knight, R., Siljander, H., Knip, M., Xavier, R.J., and Gevers, D. (2015). ConStrains identifies microbial strains in metagenomic datasets. Nat. Biotechnol. *33*, 1045–1052. https://doi.org/10.1038/nbt.3319

Margot, H., Tasara, T., Zwietering, M.H., Joosten, H., and Stephan, R. (2016). Effects of different media on the enrichment of low numbers of Shiga toxin-producing *Escherichia coli* in mung bean sprouts and on the development of the sprout microbiome. Int. J. Food Microbiol. *232*, 26–34. https://doi.org/10.1016/j.ijfoodmicro.2016.05.005

Mariani-Kurkdjian, P., Lemaitre, C., Bidet, P., Perez, D., Boggini, L., Kwon, T., and Bonacorsi, S. (2014). Haemolytic-uraemic syndrome with bacteraemia caused by a new hybrid *Escherichia coli* pathotype. New Microbes New Infect. *2*, 127–131. https://doi.org/10.1002/nmi2.49

Melton-Celsa, A.R. (2014). Shiga Toxin (Stx) classification, structure, and function. Microbiol. Spectr. *2*, EHEC-0024-2013. https://doi.org/10.1128/microbiolspec.EHEC-0024-2013

Meyer, F., Paarmann, D., D'Souza, M., Olson, R., Glass, E.M., Kubal, M., Paczian, T., Rodriguez, A., Stevens, R., Wilke, A., *et al.* (2008). The metagenomics RAST server – a public resource for the automatic phylogenetic and functional analysis of metagenomes. BMC Bioinformatics *9*, 386. https://doi.org/10.1186/1471-2105-9-386

Mir, R.A., Weppelmann, T.A., Elzo, M., Ahn, S., Driver, J.D., and Jeong, K.C. (2016). Colonization of beef cattle by Shiga toxin-producing *Escherichia coli* during the first year of life: A cohort study. PLoS One *11*, e0148518. https://doi.org/10.1371/journal.pone.0148518

Namiki, T., Hachiya, T., Tanaka, H., and Sakakibara, Y. (2012). MetaVelvet: an extension of Velvet assembler to de novo metagenome assembly from short sequence reads. Nucleic Acids Res. *40*, e155. https://doi.org/10.1093/nar/gks678

Newitt, S., MacGregor, V., Robbins, V., Bayliss, L., Chattaway, M.A., Dallman, T., Ready, D., Aird, H., Puleston, R., and Hawker, J. (2016). Two linked enteroinvasive *Escherichia coli* outbreaks, Nottingham, UK, June 2014. Emerg. Infect. Dis. *22*, 1178-1184. https://doi.org/10.3201/eid2207.152080

Ottesen, A.R., Gonzalez, A., Bell, R., Arce, C., Rideout, S., Allard, M., Evans, P., Strain, E., Musser, S., Knight, R., *et al.* (2013). Co-enriching microflora associated with culture based methods to detect *Salmonella* from tomato phyllosphere. PLoS One *8*, e73079. https://doi.org/10.1371/journal.pone.0073079

Ounit, R., Wanamaker, S., Close, T.J., and Lonardi, S. (2015). CLARK: fast and accurate classification of metagenomic and genomic sequences using discriminative *k*-mers. BMC Genomics *16*, 236. https://doi.org/10.1186/s12864-015-1419-2

Peng, Y., Leung, H.C., Yiu, S.M., and Chin, F.Y. (2012). IDBA-UD: a de novo assembler for single-cell and metagenomic sequencing data with highly uneven depth. Bioinformatics *28*, 1420–1428. https://doi.org/10.1093/bioinformatics/bts174

Rasko, D.A., Webster, D.R., Sahl, J.W., Bashir, A., Boisen, N., Scheutz, F., Paxinos, E.E., Sebra, R., Chin, C.S., Iliopoulos, D., *et al.* (2011). Origins of the *E. coli* strain causing an outbreak of hemolytic-uremic syndrome in Germany. N. Engl. J. Med. *365*, 709-717. https://doi.org/10.1056/NEJMoa1106920

Rothrock, M.J., Jr., Hiett, K.L., Gamble, J., Caudill, A.C., Cicconi-Hogan, K.M., and Caporaso, J.G. (2014). A hybrid DNA extraction method for the qualitative and quantitative assessment of bacterial communities from poultry production samples. J. Vis. Exp. *10*. https://doi.org/10.3791/52161

Rowell, S., King, C., Jenkins, C., Dallman, T.J., Decraene, V., Lamden, K., Howard, A., Featherstone, C.A., and Cleary, P. (2016). An outbreak of Shiga toxin-producing *Escherichia coli* serogroup O157 linked to a lamb-feeding event. Epidemiol. Infect. *144*, 2494–2500. https://doi.org/10.1017/S0950268816001229

Sahl, J.W., Schupp, J.M., Rasko, D.A., Colman, R.E., Foster, J.T., and Keim, P. (2015). Phylogenetically typing bacterial strains from partial SNP genotypes observed from direct sequencing of clinical specimen metagenomic data. Genome Med. *7*, 52. https://doi.org/10.1186/s13073-015-0176-9

Schmidt, K., Mwaigwisya, S., Crossman, L.C., Doumith, M., Munroe, D., Pires, C., Khan, A.M., Woodford, N., Saunders, N.J., Wain, J., *et al.* (2016). Identification of bacterial pathogens and antimicrobial resistance directly from clinical urines by nanopore-based metagenomic sequencing. J. Antimicrob. Chemother. 72, 104–114. https://doi.org/10.1093/jac/dkw397

Scholz, M., Ward, D.V., Pasolli, E., Tolio, T., Zolfo, M., Asnicar, F., Truong, D.T., Tett, A., Morrow, A.L., and Segata, N. (2016). Strain-level microbial epidemiology and population genomics from shotgun metagenomics. Nat. Methods *13*, 435–438. https://doi.org/10.1038/nmeth.3802

Segata, N., Bornigen, D., Morgan, X.C., and Huttenhower, C. (2013). PhyloPhlAn is a new method for improved phylogenetic and taxonomic placement of microbes. Nat. Commun. *4*, 2304. https://doi.org/10.1038/ncomms3304

Segata, N., Waldron, L., Ballarini, A., Narasimhan, V., Jousson, O., and Huttenhower, C. (2012). Metagenomic microbial community profiling using unique clade-specific marker genes. Nat. Methods *9*, 811–814. https://doi.org/10.1038/nmeth.2066

Stefan, C.P., Koehler, J.W., and Minogue, T.D. (2016). Targeted next-generation sequencing for the detection of ciprofloxacin resistance markers using molecular inversion probes. Sci. Rep *6*, 25904. https://doi.org/10.1038/srep25904

Stoesser, N., Batty, E.M., Eyre, D.W., Morgan, M., Wyllie, D.H., Del Ojo Elias, C., Johnson, J.R., Walker, A.S., Peto, T.E., and Crook, D.W. (2013). Predicting antimicrobial susceptibilities for *Escherichia coli* and *Klebsiella pneumoniae* isolates using whole genomic sequence data. J. Antimicrob. Chemother. *68*, 2234–2244. https://doi.org/10.1093/jac/dkt180

Truong, D.T., Franzosa, E.A., Tickle, T.L., Scholz, M., Weingart, G., Pasolli, E., Tett, A., Huttenhower, C., and Segata, N. (2015). MetaPhlAn2 for enhanced metagenomic taxonomic profiling. Nat. Methods *12*, 902–903. https://doi.org/10.1038/nmeth.3589

Tyson, G.H., McDermott, P.F., Li, C., Chen, Y., Tadesse, D.A., Mukherjee, S., Bodeis-Jones, S., Kabera, C., Gaines, S.A., Loneragan, G.H., *et al.* (2015). WGS accurately predicts antimicrobial resistance in *Escherichia coli*. J. Antimicrob. Chemother. *70*, 2763–2769. https://doi.org/10.1093/jac/dkv186

Underwood, A.P., Dallman, T., Thomson, N.R., Williams, M., Harker, K., Perry, N., Adak, B., Willshaw, G., Cheasty, T., Green, J., *et al.* (2013). Public health value of next-generation DNA sequencing of enterohemorrhagic *Escherichia coli* isolates from an outbreak. J. Clin. Microbiol. *51*, 232–237. https://doi.org/10.1128/JCM.01696-12

Wang, F., Yang, Q., Kase, J.A., Meng, J., Clotilde, L.M., Lin, A., and Ge, B. (2013). Current trends in detecting non-O157 Shiga toxin-producing *Escherichia coli* in food. Food-borne Pathog. Dis. *10*, 665–677. https://doi.org/10.1089/fpd.2012.1448

Ward, D.V., Scholz, M., Zolfo, M., Taft, D.H., Schibler, K.R., Tett, A., Segata, N., and Morrow, A.L. (2016). Metagenomic sequencing with strain-level resolution implicates uropathogenic *E. coli* in necrotizing enterocolitis and mortality in preterm infants. Cell Rep. *14*, 2912–2924. https://doi.org/10.1016/j.celrep.2016.03.015

Williams, T.R., Moyne, A.L., Harris, L.J., and Marco, M.L. (2013). Season, irrigation, leaf age, and *Escherichia coli* inoculation influence the bacterial diversity in the lettuce phyllosphere. PLoS One *8*, e68642. https://doi.org/10.1371/journal.pone.0068642

Wilson, M.R., Brown, E., Keys, C., Strain, E., Luo, Y., Muruvanda, T., Grim, C., Jean-Gilles Beaubrun, J., Jarvis, K., Ewing, L., *et al.* (2016). Whole genome DNA sequence analysis of *Salmonella* subspecies *enterica* serotype Tennessee obtained from related peanut butter food-borne outbreaks. PLoS One *11*, e0146929. https://doi.org/10.1371/journal.pone.0146929

Wood, D.E., and Salzberg, S.L. (2014). Kraken: ultrafast metagenomic sequence classification using exact alignments. Genome Biol. *15*, R46. https://doi.org/10.1186/gb-2014-15-3-r46

Wu, Y.W., Tang, Y.H., Tringe, S.G., Simmons, B.A., and Singer, S.W. (2014). MaxBin: an automated binning method to recover individual genomes from metagenomes using an expectation-maximization algorithm. Microbiome 2, 26. https://doi.org/10.1186/2049-2618-2-26

Yang, X., Noyes, N.R., Doster, E., Martin, J.N., Linke, L.M., Magnuson, R.J., Yang, H., Geornaras, I., Woerner, D.R., Jones, K.L., *et al.* (2016). Use of metagenomic shotgun sequencing technology to detect food-borne pathogens within the microbiome of the beef production chain. Appl. Environ. Microbiol. *82*, 2433-2443. https://doi.org/10.1128/AEM.00078-16

Yuan, S., Cohen, D.B., Ravel, J., Abdo, Z., and Forney, L.J. (2012). Evaluation of methods for the extraction and purification of DNA from the human microbiome. PLoS One *7*, e33865. https://doi.org/10.1371/journal.pone.0033865

Zhou, B., Wang, C., Zhao, Q., Wang, Y., Huo, M., Wang, J., and Wang, S. (2016). Prevalence and dissemination of antibiotic resistance genes and coselection of heavy metals in Chinese dairy farms. J. Hazard. Mater. *320*, 10-17. https://doi.org/10.1016/j.jhazmat.2016.08.007

Use of Omics Technologies to Develop Strategies to Control *Escherichia coli* from Farm to Table

11

Teresa M. Bergholz* and Manoj K. Shah

Department of Microbiological Sciences, North Dakota State University, Fargo, ND, USA.

*Correspondence: teresa.bergholz@ndsu.edu

https://doi.org/10.21775/9781910190777.11

Abstract

The food supply is a series of complex environments which can pose a number of stresses to microbes. These stresses include intrinsic stresses inherent to a given food, such as pH and water activity. These stresses also include extrinsic factors that can vary during each step of food harvesting and processing, including temperature and relative humidity, as well as stresses specific to food processing. Bacteria possess a wide range of strategies to adapt to changing environmental conditions in order to survive under suboptimal conditions. These stress responses can be complex, involving activation of regulatory networks and cascades of gene and protein expression. Omics technologies allowing for the measurement of global changes in gene or protein expression, or changes in the metabolome, can provide insights into the molecular mechanisms that a bacterium utilizes to adapt to different environmental stresses. The aim of this chapter is to provide an overview of the available data assessing global changes in Shiga-toxin-producing *Escherichia coli* (STEC) that facilitate our understanding of how this pathogen adapts to stresses presented at different stages of food production, with the goal of using these insights to develop control measures that will effectively reduce the presence of this pathogen in foods.

Shiga toxin-producing *E. coli* (STEC) as a public health concern

STEC are human pathogens of zoonotic origin, which have been linked to serious disease such as haemorrhagic colitis and haemolytic uraemic syndrome (Smith *et al.*, 2014). Over 400 serotypes of STEC are identified to be pathogenic due to the presence of one or more genes encoding Shiga toxin (Blanco *et al.*, 2004; Brooks *et al.*, 2005; Hussein, 2007; Tozzoli and Scheutz, 2014). The most studied serotype of STEC to date has been O157:H7, due to large outbreaks in the 1980s and 1990s associated with strains of this serotype (Goldwater and Bettelheim, 1998). The importance of non-O157 STEC serotypes has been increasingly recognized, and the most recent estimates from the USA Centers for Disease Control and Prevention indicate that non-O157 STEC are responsible for the majority of human illness caused by STEC (Scallan *et al.*, 2011). Surveillance data of food-borne outbreaks in the USA from 2010 to 2014 reported 34 outbreaks due to STEC, resulting in 636 illnesses, 178 hospitalizations, and one death (Crowe *et al.*, 2015). These outbreaks were attributed to a variety of sources, including vegetable row crops, beef, sprouts, dairy, nuts/seeds, fish, and wild game (Crowe *et al.*, 2015). In 2015, two STEC outbreaks were reported due to O26 at Chipotle restaurants (55 illnesses and 21 hospitalizations) (FDA, 2016), and O157 in rotisserie chicken salad (19 illnesses, five hospitalizations with two HUS patients) (CDC, 2015).

STEC infection can result in haemorrhagic colitis and serious complications such as haemolytic uraemic syndrome (HUS), which can result in kidney failure and death (Griffin and Tauxe, 1991). Collectively, the virulence factors, low infectious dose, and stress resistance properties make STEC a significant public health concern. While the presence of Shiga toxin is what defines STEC, the presence of Shiga toxins along with other virulence factors such as the locus of enterocyte effacement (LEE), fimbrial adhesins, and outer membrane proteins, as well as plasmids such as pO157 contribute to pathogenesis (Law, 2000; Lim *et al.*, 2010). Variation in virulence among STEC O157:H7 has been described, with a hypervirulent clade where strains containing Stx2a and Stx2c are associated with significantly higher levels of severe disease in the form of HUS (Manning *et al.*, 2008). Typically, if a STEC strain possesses LEE and Stx2, particularly the Stx2a and Stx2c subtypes, the strain is likely to cause severe illness such as haemorrhagic colitis and HUS (Ethelberg *et al.*, 2004), though there are examples of STEC serotypes capable of causing severe disease that do not possess LEE (Karmali *et al.*, 1985). One of the largest STEC outbreaks occurred in 2011, resulting in 845 patients with HUS, and was linked to *E. coli* O104:H4 that possessed both the typical set of virulence factors associated with enteroaggregative *E. coli* and Stx2a (Bielaszewska *et al.*, 2011; Grad *et al.*, 2012). The low infectious dose of STEC O157:H7, O26:H11, and O111:H- has been determined from assessment of foods implicated in outbreaks (Paton *et al.*, 1996; Michino *et al.*, 1999; Tuttle *et al.*, 1999; Buvens *et al.*, 2011), as well as environmental sources (Strachan *et al.*, 2001). While similar data supporting the low infectious dose of many other important non-O157 serotypes are not available, the similar stress resistance properties of these non-O157 serotypes, in particular the ability to withstand extremely low pH (Large *et al.*, 2005; Kim *et al.*, 2015), is thought to contribute to a low infectious dose of these serotypes as well.

Cattle are considered to be one of the major reservoirs of STEC (Hussein, 2007; Ferens and Hovde, 2011) and contribute either directly or indirectly to the presence of this pathogen in the food supply (Fig. 11.1). Other animals, such as birds, horses, sheep, and pigs, can also carry STEC, though are considered to be secondary reservoirs (Mathusa *et al.*, 2010). On farms, cattle shedding STEC contribute to circulation of the pathogen in the farm environment (Shere *et al.*, 2002) and acquisition by other cattle in the herd (Matthews *et al.*, 2006) and on nearby farms (Herbert *et al.*, 2014). Cattle and other farm animals can also acquire STEC through consumption of contaminated drinking water (Shere *et al.*, 1998) and feed (Davis *et al.*, 2003). Cattle shedding STEC can then lead to contamination of water and soil, resulting in the contamination of surface water (Johnson *et al.*, 2003; Nadya *et al.*, 2016) and produce (Cooley *et al.*, 2013). Use of irrigation water containing STEC can lead to contamination of leafy greens and other vegetables in the field, as well as potential for long-term contamination of soil (Wachtel *et al.*, 2002; Islam *et al.*, 2004). Wild animals, such as deer and pigs, can carry STEC, and can also be a source of contamination of produce (Jay *et al.*, 2007; Laidler *et al.*, 2013) and water (Cooley *et al.*, 2013).

During harvest, STEC present on hides or in fecal material of cattle can lead to contamination of carcasses (Elder *et al.*, 2000). STEC present in soil or water has the potential to contaminate produce such as lettuce during harvest via transfer by harvesting tools (Taormina *et al.*, 2009) and by workers (Monaghan and Hutchison, 2016). Processing procedures for both meats and produce have the potential to spread STEC from contaminated product to non-contaminated product. For example, wash water recirculation during processing can magnify the spread of contamination at large, centralized leafy green processing facilities (Luo, 2007; Herdt and Feng, 2009). Food contact surfaces, utensils, and workers all have the potential to introduce STEC into a product during processing (Garcia *et al.*, 2006). Additional STEC contamination from outside sources is unlikely for foods that have been packaged during storage and transport. Foods that are stored and transported in bulk have additional potential for STEC contamination via handling and cross-contamination during storage and transport (Abban *et al.*, 2012). There are also concerns for spread of food-borne pathogens such as STEC at home or in restaurants during food preparation. Pathogens can spread from raw foods (chicken, beef or produce) to other foods or utensils such as knives or cutting boards, which can further contaminate more foods and surfaces (Cogan *et al.*, 1999; Erickson *et al.*, 2015).

STEC stress responses characterized by omics technologies

STEC general stress response

The alternate sigma factor RpoS governs gene expression during stationary phase and in response to stressful environments (Lange and Hengge-Aronis, 1991). RpoS is induced under many different stresses, including acid shock (Small *et al.*, 1994), heat shock (Hengge-Aronis *et al.*, 1991), osmotic stress (Hengge-Aronis *et al.*, 1993), oxidative stress (Sammartano *et al.*, 1986), and low temperature (White-Ziegler *et al.*, 2008). A core set of RpoS-regulated genes has been identified, the dependence of the genes within the regulon on RpoS can vary depending on the specific stress conditions (Weber *et al.*, 2005), as well as the strain background (Dong and Schellhorn, 2009a,b). While the majority of information on the RpoS regulon has been described in *E. coli* K-12, the genes regulated by RpoS during stationary phase growth in LB has been defined for *E. coli* O157:H7 strain EDL933 (Dong and Schellhorn, 2009b). These include the sets of genes previously identified in K-12, such as *osmY*, *otsBA*, and *treF*, involved in responding to osmotic stress, *gadAXW* involved in responding to acid stress, and *katE* involved in responding to oxidative stress. The RpoS regulon in O157:H7 EDL933 also includes a number of strain-specific genes, many hypothetical or unknown function, as well as known virulence genes, such as *ler*, the regulator of the LEE island, and a number of LEE-encoded genes (Dong and Schellhorn, 2009b).

STEC acid stress response

The response to acid stress has been well-studied in *E. coli*, with most data generated from study of non-pathogenic strains. After a number of notable outbreaks of haemorrhagic colitis linked to STEC O157:H7 present in apple cider and apple juice (Cody *et al.*, 1999), a number of studies were focused on characterizing the acid-resistance capabilities of STEC, yet omics data related to characterizing the genome-wide response to acidic conditions is limited for STEC. King and colleagues assessed how the STEC O157:H7 from the Sakai, Japan, outbreak responded to three different acidulants, lactic acid, acetic acid, and hydrochloric acid (King *et al.*, 2010). They identified 136 genes that were significantly induced in stationary phase cells exposed to pH 5.5 with all three of the acidulants, representing the universal response to acid stress. This universal response includes genes involved in the oxidative stress response, cell envelope stress responses controlled by RpoE, iron and manganese uptake, and cold shock proteins. They also observed that expression of *stx1A* and *hha* was significantly higher when STEC O157:H7 was exposed to pH 5.5 with HCl or lactic acid (King *et al.*, 2010). In a transcriptome comparison across multiple different environments, Landstorfer *et al.*, measured STEC O157:H7 EDL933 transcript levels in LB at pH 4 relative to those in LB at pH 7, and found that the LEE genes were significantly down-regulated at pH 4 (Landstorfer *et al.*, 2014).

Omics techniques have also been used to understand more about the relationships between specific cellular features linked to survival and/or virulence and acid stress. Gunther *et al.* (2014) compared protein expression between STEC O157 isolates that could and could not produce curli, coiled surface structures composed of curlin protein subunits that are involved in binding fibronectin, laminin, and various serum proteins (Uhlich *et al.*, 2001). Proteins that contribute to acid resistance, including GadA and HdeAB, were expressed at higher levels in the curli negative strain, and subsequent acid-resistance assays demonstrated that the curli negative strain was significantly more resistant to pH 2.5 compared to the curli positive strain, indicating a relationship between the regulation of curli expression and expression of acid-resistance proteins (Gunther *et al.*, 2014). Similarly, Veses-Garcia *et al.* compared transcriptomes of *E. coli* with and without Stx phage, and identified a link between the presence of the Stx phage $\Phi 24_B$ and increased expression of genes encoding the glutamate dependent acid-resistance system (Veses-Garcia *et al.*, 2015).

STEC oxidative stress response

There are three major regulators of the oxidative stress response in *E. coli*, which can be activated under different types of oxidative stress. OxyR controls a regulon of ~ 40 genes, and is activated primarily by exposure of the cell to hydrogen peroxide (Chiang and Schellhorn, 2012). SoxR, when activated by oxidation, induces expression of *soxS*, which in turn regulates expression of ~ 100 genes. The SoxRS system is activated by exposure of the cell to nitric oxide, paraquat, and hydrogen peroxide (Chiang and Schellhorn, 2012). A number of RpoS regulated genes play important roles in responding to oxidative stress, such as *katE*, *xthA*, and *sodC*, and *rpoS* transcript levels are regulated by the MqsR–MqsA toxin–antitoxin system under oxidative stress conditions (Wang *et al.*, 2011).

The response of STEC to oxidative stress has been evaluated under different types of oxidative stressors: hydrogen peroxide, paraquat, and sodium hypochlorite. Allen and Griffiths (2012) examined changes in O157:H7 gene expression during exposure of exponentially growing cells to two levels (1mM and 2.5mM) of hydrogen peroxide, and found evidence for upgrelation of the RpoS and OxyR regulons, as well as increased transcript levels of virulence genes *stx1AB*, *stx2AB*, *ler*, and *lpfA*. They also assessed differentially expressed genes from exponential and stationary phase cells exposed to 300μM paraquat and identified that in exponential phase, the RpoS, SoxRS, and OxyR regulons were activated, along with increased expression of the Shiga toxin genes, while in stationary phase, only the SoxRS regulon was activated (Allen and Griffiths, 2012). Wang and colleagues (2009) investigated the effects of chlorine and hydrogen peroxide stress on two different strains of STEC O157:H7; the strain involved in the radish sprout outbreak in Japan in 1998 and the 2006 US spinach outbreak strain. Across both strains, they identified 133 genes that were up-regulated under oxidative stress, with chlorine exposure leading to a greater magnitude of change in gene expression compared to peroxide stress. In addition to the SoxRS, OxyR, and RpoS regulons, they identified significant up-regulation of genes involved in the heat shock response, osmotic stress, and cell envelope stress responses (Wang *et al.*, 2009), indicating some overlaps between these stress responses.

While the majority of data are available for STEC O157:H7 strains, Mei and colleagues evaluated changes in expression of a selected subset of known stress response genes in STEC strains representing the big six serotypes (O26, O45, O103, O111, O121, and O145) of concern in the USA (Mei *et al.*, 2015). They found that while some genes, such as *sodA*, were consistently up-regulated by all the strains tested when exposed to 2.5mM H_2O_2, others had significant variation in expression amongst the strains, including *dps*, which was up-regulated in O145 and O111, though not the other serotypes, and *stx1A*, which was up-regulated in O157:H7, but none of the other strains (Mei *et al.*, 2015).

STEC temperature response

E. coli response to temperature stress is fairly well described based on data from K-12 and other non-pathogenic strains. There are two main regulators controlling the heat shock response, RpoH and RpoE (Schumann, 2016). RpoH controls ~100 genes, including the heat shock chaperones encoded by *groES/groEL* and *dnaK*, *dnaJ*, and *grpE* (Nonaka *et al.*, 2006). RpoE controls 50 genes, including the chaperones encoded by *degP* and *skp* (Rhodius *et al.*, 2006). Carruthers and Minion (2009) investigated the heat shock response in STEC O157:H7 strain EDL933 by measuring global changes in gene expression upon shifting from 37°C to 50°C. In addition to the previously described RpoH and RpoE regulon members, they identified 15 O157-specific genes that were also up-regulated by heat shock, including 12 genes that are found on the Stx2-encoding phage BP-933W.

The *E. coli* response to low temperatures and cold shock has also been described mainly in K-12, and includes activation of RNA chaperones and ribosomal factors that ensure translation can occur at lower temperatures (Wick and Egli, 2004). A combination of transcriptome and proteome data demonstrated that STEC O157:H7 Sakai activated the RpoS regulon, as well as the cell envelope stress response regulons controlled by Rcs and CpxRA during adaptation and steady-state exposure to 14°C (Kocharunchitt *et al.*, 2012, King *et al.*, 2014). Another proteome study examined differences between STEC O157:H7 growth at 37°C and 15°C, and identified significant up-regulation of Cfa, cyclopropane fatty acid synthesis, and FabD, involved in fatty acid synthesis, highlighting the importance of membrane fluidity modifications during adaptation to low temperature (Vidovic *et al.*, 2011).

STEC osmotic stress response

The majority of omics data detailing the response of STEC to osmotic stress is from the STEC O157:H7 Sakai strain, and is described using both transcriptomic and proteomic data sets. The dynamic response to osmotic stress, during a shift from a water activity of 0.993–0.967, as well as the steady state response to a water activity of 0.967 have been described (Kocharunchitt *et al.*, 2012; Kocharunchitt *et al.*, 2014). STEC O157:H7 Sakai adapts to osmotic stress by activation of the RpoS general stress response regulon, the RpoH heat shock regulon, as well as a number of regulons involved in the cell envelope stress response, including CpxRA, Rcs, and RpoE (Kocharunchitt *et al.*, 2014). During this adaptation to osmotic stress, Ler, the regulator for the LEE genes, is induced, as well as many genes encoding components of the type three secretion system (Kocharunchitt *et al.*, 2014), demonstrating that osmotic stress at 35°C is one of the signals that activates this virulence factor in STEC. During steady state growth at a_w = 0.967 and 25°C, O157:H7 Sakai had significantly lower expression of *stx1AB*, *ler*, and the LEE genes compared to at a_w = 0.993, while the RpoS and Rcs regulons were significantly up-regulated (Kocharunchitt *et al.*, 2012). Together, these two studies show that STEC activates a number of stress response regulons to manage the adaptation to an osmotic shock, and utilizes two of those stress response regulons to maintain growth under high osmolarity once adaptation occurs. The two studies were conducted at two different growth temperatures, so it is harder to draw conclusions about the overall effects of osmotic stress on expression of virulence factors, as optimal expression of these factors occurs at 37°C.

STEC response to stresses in the primary production environment

The majority of information on specific stress responses, the regulators and regulons activated or repressed by a given environmental stress, and the molecular mechanisms *E. coli* uses to adapt to the stress are based on carefully controlled studies in laboratory medium, typically under conditions that are optimal with the exception of the stress examined. These studies have provided a wealth of information on stress response mechanisms, and are a key resource when assessing how STEC adapts to complex, multi-stress environments that it is likely to encounter during transmission through the food supply (Fig. 11.1). In many cases, the complete view of how STEC adapts to different stresses in the food supply is lacking, though considerable effort has recently been focused on this issue. Table 11.1 highlights the STEC responses to different foods and food-related environments based on available omics data.

STEC associated with cattle and cattle manure

STEC, in particular STEC O157:H7, have been well-studied in cattle, with the majority of research conducted on understanding factors that influence carriage and shedding of O157:H7 by cattle. A number of DNA-based subtyping methods have identified specific genotypes of O157:H7 that are commonly isolated from cattle, which are distinct from genotypes that are commonly isolated from human disease (Kim *et al.*, 1999; Besser *et al.*, 2007). O157 strains with bovine-associated genotypes were found to express genes involved in acid resistance (*gadEWA*, *gadBC*, *hdeABD*), RpoS regulated stress response genes (*dps*, *cspD*, *osmC*, *osmY*), and genes regulated by RpoN and Nac that are involved in nitrogen assimilation and metabolism at significantly higher levels than O157 strains with human associated genotypes (Vanaja *et al.*, 2010). While STEC do not cause disease in healthy adult cattle (Russell *et al.*, 2000), STEC that colonize the GI tract of cattle do so through the faecal–oral route. The first stressful environment that STEC would encounter upon ingestion by cattle is the rumen, and changes in the proteome of STEC O157:H7 strain 86–24 was assessed to determine the bacterial factors that may contribute to the ability of STEC to adapt to the rumen (Kudva *et al.*, 2014). Rumen fluid contains a number of short chain fatty acids, including acetic acid, propionic acid, and butyric acid, and O157:H7 proteins that were significantly up-regulated in rumen fluid included those involved in osmotic stress response (OsmE), oxidative stress (Ygge, MoaB, DmsB, FdoH, KatG), heat shock response (DnaK, GroEL), nitrogen metabolism, and the glycolytic pathway (Kudva *et al.*, 2014). STEC O157:H7 strain EDL 933 was

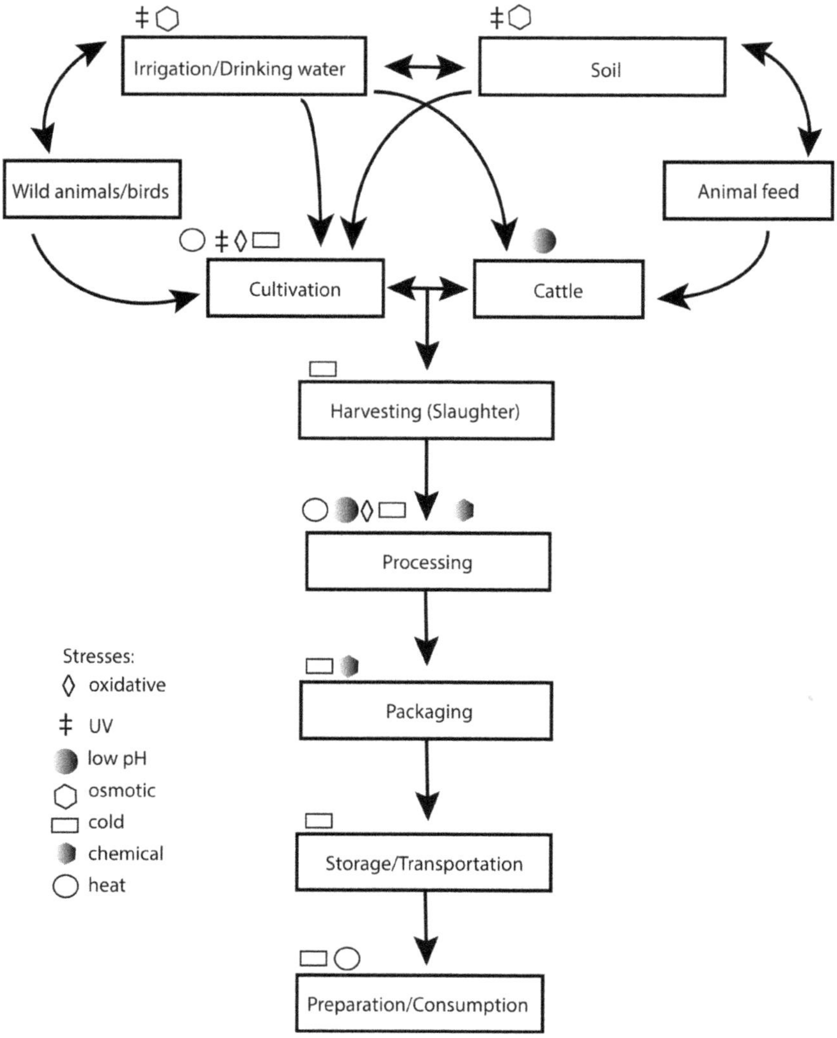

Figure 11.1 Schematic diagram of food flow with potential stresses encountered by STEC during the 'farm to fork' cycle. Arrows denote potential for spread of STEC from reservoirs into and through the food supply.

demonstrated to have a competitive growth advantage when intestinal mucus-derived carbohydrates were the main source of available energy, indicating that this feature may allow persistence of this pathogen in the cattle GI tract (Bertin *et al.*, 2013).

The primary colonization site for STEC O157 in cattle is the terminal recto-anal junction (RAJ) (Naylor *et al.*, 2003), while other STEC serotypes can be found throughout the large intestine (Lim *et al.*, 2007). Some colonized cattle are known to shed O157 in quite high numbers, greater than 10^4 CFU/g faeces, and are referred to as 'super shedders' (Omisakin *et al.*, 2003). The exact mechanism used by O157 to attach to the RAJ is not currently known, though it does occur in a LEE-independent manner. A comparative genomics analysis was conducted comparing an O157 strain isolated from a super shedder and other O157 isolates from human outbreaks of illness, and was able to rule out type I fimbriae, intimin, Tir, EspA, and EspB as contributing to the super shedder phenotype, though the mechanism still remains elusive (Cote *et al.*, 2015). A proteomics-based expression library screening experiment identified OmpA (outer membrane protein A) as a potential modulator for increased adherence of O157 to RAJ cells (Kudva *et al.*, 2015).

After shedding from colonized cattle, the faecal environment, as well as manure composting and addition of manure to soil, pose a different set of

Table 11.1 Summary of STEC stress response and virulence genes regulated at different stages in the food supply chain. ↓ indicates genes that are down-regulated, ↑ indicates genes that are up-regulated under a given condition

Primary production	Stage	Genes/proteins regulated	Functional response	References
Cattle and cattle manure				
	Cattle	*gadEWA*, *gadBC*, *hdeABD* ↑	Acid resistance	Vanaja *et al*. (2010)
		RpoS (*dps*, *cspD*, *osmC*, *osmY*) ↑	General stress response	
		RpoN and Nac ↑	Nitrogen assimilation and metabolism	
		OsmE ↑	Osmotic stress response	Kudva *et al*. (2014)
		YggE, MoaB, DmsB, EdoH, KatG ↑	Oxidative stress response	
		DnaK, GroEL ↑	Heat shock response	
	Manure	*dnaKJ*, *htpX*, *ibpAB* ↑	Heat shock response	Landstorfer *et al*. (2014)
		cpxP, *pspABC* ↑	Cell envelope stress response	
	Dairy compost	RpoH ↑	Heat shock response	Singh and Jiang (2015)
		chpS, *dinJ* ↑	Toxin–antitoxin gene	
	Manure-amended soil	*rpoS* ↑	General stress response	van Hoek *et al*. (2013)
Fresh produce				
	Lettuce root	*ybiM* and *ycfR* ↑	Biofilm formation and colanic acid production	Hou *et al*. (2013)
		rpoS ↑	General stress response	
		RpoE ↑	Cell envelope stress response	
	Lettuce/radish sprouts	*xspA*, *asr*, *htpX*, *lpxP* ↑	General stress response	Van der Linden *et al*.(2016), Landstorfer *et al*. (2014)
		marRAB ↑	Multidrug resistance	
		azoR, *acpD* ↑	Degradation of inhibitory phenolic compounds	
		ybiM, *bssS* ↑	Biofilm formation	Fink *et al*. (2012)
		csgAB ↑	Attachment	
		rpoS, *uspA*, *hdeABD*, *gadBC*, *gadEWA* ↑	General stress response	
		rpoE ↑	Cell envelope stress response	
	Lettuce leaf lysate	OxyR and SoxS ↑	Oxidative stress response	Kyle *et al*. (2010)
		betAB, *caiAT* ↑	Osmotic stress response	
		MarR ↑	Multidrug resistance	
	Lettuce leaf extract	*ycfR* (BhsA) ↑	Multiple stress resistance and biofilm formation	Hou *et al*. (2013), Kyle *et al*. (2010)
		soxS ↑	Oxidative stress response	
Ground beef				
	Ground beef extract	*sodC* ↑	Oxidative stress response	Fratamico *et al*. (2011)
		osmB ↓	Osmotic stress response	
		gadB ↓	Acid resistance	

Table 11.1 Summary

Primary production	Stage	Genes/proteins regulated	Functional response	References
	Ground beef at 48°C	*eae, toxB* ↑	Type three secretion system, toxin	Slanec and Schmidt (2011)
		groEL, hcpA, ipfA ↑	Heat shock response	
Fruit juice				
	Apple juice pH 3.5	*cpxRA* ↑	Cell envelope stress response	Bergholz *et al.* (2009)
	Pineapple juice pH 3.8	*gadA and adiA* ↑	Acid resistance	Kim *et al.* (2016); De Biase and Lund (2015)
Food processing				
	High pressure processing	RpoS and RpoE ↑	General and cell envelope stress response	Malone *et al.* (2006)
		Dps and OtsA (RpoS) ↑	General stress response	Robey *et al.* (2001); Malone *et al.* (2006)
Antimicrobial ingredients				
	Carvacrol	*ler* and *stx2B, fliC, and luxS* ↓	Type three secretion system, quorum sensing	Mith *et al.* (2015)
	Probiotic bacteria	*stx2A* ↓	Shiga toxin	Carey *et al.* (2008)
	L. acidophilus supernatants	LEE ↓	Type three secretion system	Medellin-Pena *et al.* (2007)

environmental conditions that STEC must adapt to in order to survive. STEC O157:H7 EDL 933 inoculated into cattle faeces was found to down-regulate genes on the LEE pathogenicity island and *stx2A*, and up-regulate genes involved in the heat shock response (*dnaKJ, htpX, ibpAB*) and the cell envelope stress response (*cpxP, pspABC*) indicating that the cells are experiencing stresses that negatively impact protein folding and the cell membrane (Landstorfer *et al.*, 2014). Cattle manure is typically composted prior to use, and during the composting process, STEC that are present is likely experiencing heat stress. STEC O157:H7 was exposed to 47.5°C for 10 minutes in dairy compost, and was found to up-regulate heat shock genes regulated by RpoH, and two toxin–antitoxin genes (*chpS, dinJ*) (Singh and Jiang, 2015) which contribute to a number of stress resistance features, such as the generation of persister cells (Magnuson, 2007). The cells also responded to the heat shock by down-regulating expression of key virulence genes, including *stx1, stx2, eae, hlyA*, and *fliC* (Singh and Jiang, 2015).

The ability of STEC to survive in manure-amended soil has also been assessed. Variation in length of survival in different types of soil as well as among strains has been observed (Franz *et al.*, 2008; Semenov *et al.*, 2008). Long-term survival (> 200 days) was observed for some STEC O157 strains in manure-amended soil, and *rpoS* expression was 2.7-fold higher for those strains in soil when compared to cells grown in broth (van Hoek *et al.*, 2013). In addition, the metabolic profiles for these strains were different from those strains surviving fewer than 200 days. Significantly higher oxidization of α-ketobutyric acid, α-hydroxybutyric acid, propionic acid, β-methyl-D-glucoside, and L-arabinose was observed for the strains with longer survival in soil (Franz *et al.*, 2011). These organic acids are often found in ruminants, and the ability of STEC O157 to utilize these compounds *in vivo* might help their survival *in vitro*, such as in manure-amended soils. In soils, STEC and other microbes are expected to encounter a number of environmental stresses, such as low moisture, low nutrients (non-manure amended), and UV exposure. The effect of repeated transfer of STEC from inoculated soil to uninoculated soil on expression of key stress resistance genes was selection for populations with mutations in *rpoS*, some leading to increased *rpoS* levels, others reducing *rpoS* levels (Ravva *et al.*, 2014). Taken together, these data suggest that strategies utilized by STEC O157:H7 to survive in environments encountered after shedding from cattle may vary by strain, and include stress responses, as well as metabolic functions.

STEC associated with fresh produce

There has been considerable interest in understanding how STEC are able to interact with fresh produce, specifically leafy greens, following a number of high profile outbreaks linked to STEC contaminated spinach (Grant *et al.*, 2008) and lettuce (Ackers *et al.*, 1998; Hilborn *et al.*, 1999; Slayton *et al.*, 2013). Researchers have shown that STEC can associate with plant roots (Hou *et al.*, 2013) and leaves (Feng, 2014), and have used omics data to begin to characterize the STEC responses that occur in these plant-associated environments. STEC O157:H7 EDL 933 associated with the roots of romaine lettuce grown in a hydroponic system significantly induced expression of genes involved in biofilm formation and colanic acid production (*ybiM* and *ycfR*), as well as the general stress response (*rpoS*) and the cell envelope stress response regulated by RpoE (Hou *et al.*, 2013). Exposure of STEC O157:H7 Sakai to spinach root exudates led to significant up-regulation of genes encoding curli, known to play a role in attachment to plant surfaces (Crozier *et al.*, 2016). Together, these data suggest that the plant root environment stimulates STEC O157:H7 to produce various factors involved in bacterial attachment, though the specific signals that initiate this response remain unknown.

Changes in gene expression for STEC O157:H7 Sakai associated with butter lettuce plants under simulated pre-harvest conditions (actively growing plants in growth chambers) were quantified over time, and significant up-regulation of stress response genes (*cspA, asr, htpX, lpxP*), multidrug resistance genes (*marRAB*), and genes involved in the degradation of inhibitory phenolic compounds (*azoR, acpD*) was observed (Van der Linden *et al.*, 2016). In addition to activation of *azoR* and *acpD* on pre-harvest butter lettuce, these genes were also induced by STEC O157:H7 EDL 933 when associated with radish sprouts (Landstorfer *et al.*, 2014). Ability to degrade secondary plant metabolites, such as salicylic acid and jasmonic acid, that are produced by plant defence systems, is likely to contribute to STEC survival on fresh produce. Activation of stress responses were also seen in STEC O157:H7 EDL 933 inoculated onto the surface of post-harvest romaine lettuce leaves. Over a 3-day period, significant up-regulation of genes encoding proteins involved in biofilm formation (*ybiM, bssS*), attachment (*csgAB*), general stress response (*rpoS, uspA, hdeABD, gadBC, gadEWA*), and cell envelope stress response (*rpoE*) occurred (Fink *et al.*, 2012).

Damage to produce surfaces can facilitate survival, and even growth of STEC under certain conditions. Lettuce infected with *Bremia lactucae*, the causative agent of downy mildew disease, allows significantly higher growth of STEC O157:H7 compared to non-infected healthy lettuce (Simko *et al.*, 2015). Lesions on lettuce leaves due to mechanical or physical damage, as well as due to colonization by bacteria that lead to soft rot, can facilitate growth of STEC O157:H7 on post-harvest lettuce (Brandl, 2008). Researchers have utilized omics technologies to characterize how STEC respond to lysates of lettuce and spinach, to understand more about how the pathogen can adapt and grow when lesions are present on produce. The short-term (30 minutes exposure) response to lettuce leaf lysates included significant induction of the OxyR and SoxS regulons, indicating the pathogen is experiencing oxidative stress (Kyle *et al.*, 2010). The lettuce lysate was found to contain 25 μM H_2O_2, likely resulting from the oxidative burst that is a response of plant cells to mechanical damage. A number of osmotic stress response genes (*betAB, caiAT*) and antimicrobial resistance genes of the MarR regulon were also induced. In extract of lettuce leaf cell wall polysaccharides after 1 hour exposure, STEC O157:H7 Sakai significantly up-regulated genes involved in the TCA cycle, purine synthesis, and translation, indicating that the longer-term response to such extracts involves preparation for growth (Crozier *et al.*, 2016). Exposure of STEC O157:H7 EDL 933 to spinach leaf lysates for 5 hours at 37°C led to growth of the pathogen, though the overall transcriptional profile was most similar to that of the same strain grown in minimal medium, indicating the low nutrient composition of the spinach lysate (Landstorfer *et al.*, 2014).

Across these different studies, some common responses can be identified. One is the significant activation of *ycfR* expression. YcfR, also known as BhsA, is a putative membrane protein and has promoter binding sites for SoxS and CRP (Zhang *et al.*, 2007). YcfR is strongly induced under multiple stress conditions, and also contributes to biofilm formation. YcfR was induced in STEC O157:H7 associated with lettuce roots (Hou *et al.*, 2013), post-harvest lettuce leaves (Fink *et al.*, 2012), and exposed to lettuce leaf lysate (Kyle *et al.*, 2010). Subsequent phenotypic assessment of *ycfR* deletion mutants showed significantly lower association with lettuce roots and significantly lower colonization of lettuce leaves (Fink *et al.*, 2012; Hou *et al.*, 2013). Small molecule screens have been used to effectively identify biofilm inhibitors for *Salmonella* (Koopman *et al.*, 2015), and a similar approach could be used to identify potential means for YcfR inhibition that could lead to improved ability to reduce STEC in fresh produce.

STEC responses to stresses in foods

STEC on ground beef

STEC O157:H7 became a well-known pathogen following nationwide outbreaks linked to contaminated undercooked ground beef (Riley *et al.*, 1983; CDC, 1993). These outbreaks led to development of intervention strategies applied during slaughter to minimize cross-contamination between hides and carcasses, and between different carcasses (Elder *et al.*, 2000). The number of outbreaks and recalls of ground beef due to STEC contamination have decreased over time, though still occur. Data from outbreaks of STEC O157 show that beef is a major commodity contributing to outbreaks. In the years 1982–2002, 47% of STEC O157 outbreaks were attributed to beef (Rangel *et al.*, 2005). Similarly, in the years 1998 to 2008, 58% of STEC O157 outbreaks were attributed to beef (Gould *et al.*, 2013), while in 2009–2010, 28.3% of STEC O157 outbreaks were linked to beef (CDC, 2013). Understanding how STEC adapts to potential stresses when present in ground beef could lead to development of additional interventions that further reduce the presence of this pathogen in ground beef.

A comparison of global transcript differences between STEC O157:H7 Sakai in laboratory medium and in ground beef extract showed up-regulation of genes coding proteins involved in protein biosynthesis, transporters, membrane bound acid response proteins, polysaccharide biosynthesis, and fimbriae (Fratamico *et al.*, 2011). Genes coding stress response proteins were down-regulated, including *sodC*, part of the oxidative stress response, *osmB*, part of the osmotic stress response, and *gadB*, part of the acid stress response, indicating that cells in ground beef extract are experiencing a minimal amount of stress and that the main response is initiation of growth in this environment (Fratamico *et al.*, 2011). Gene expression of STEC O157:H7 EDL 933 in ground beef at 37°C was compared to that in LB at the same temperature, and *stx1*, *gadE*, *groEL*, and *lfpA* were significantly up-regulated due to the association with ground beef (Slanec and Schmidt, 2011). When the ground beef was heated to 48°C for 10 minutes, the relative expression of stress associated genes *hcpA*, *groEL* and, *ipfA*, as well as virulence associated genes *eae* and *toxB* increased 5-fold compared to ground beef at 37°C (Slanec and Schmidt, 2011). These studies show that STEC experienced minimal stress in ground beef, and that virulence genes were up-regulated under the conditions examined. Further studies would enhance our understanding of how adaptation to this food environment impacts subsequent survival in the food supply.

STEC in fruit juice

E. coli, including STEC, is known to be able to survive low pH for lengthy periods of time (Lin *et al.*, 1996). STEC has been transmitted to people via acidic foods such as apple juice and apple cider (Besser *et al.*, 1993; Cody *et al.*, 1999; Hilborn *et al.*, 2000), as well as fermented sausages (CDC, 1995; CDC, 2011). STEC O157:H7 Sakai exposed to a model apple juice at pH 3.5 induced expression of acid stress response genes, osmotic stress response genes, and the CpxRA cell envelope stress response regulon (Bergholz *et al.*, 2009). STEC strains of serogroups O26, O103, O104, O111, and O157 exposed to pineapple juice at pH 3.8 were found to significantly increase expression of two key acid-resistance genes, *gadA* and *adiA* (Kim *et al.*, 2016). These genes encode a glutamate decarboxylase and arginine decarboxylase, respectively, and both play a significant role in reducing acidification of the cytoplasm (De Biase and Lund, 2015). Exposure of STEC to fruit juices, such as apple juice and pineapple juice, have been shown to lead to increased resistance to subsequent exposure to low pH conditions simulating those present in the stomach (Bergholz and Whittam, 2007; Kim *et al.*, 2016).

STEC responses to stresses encountered during food processing

STEC and high pressure processing

High-pressure processing (HPP) can be an effective method to inactivate STEC in a number of foods, including ground beef (Sheen *et al.*, 2015), strawberry puree (Hsu *et al.*, 2014), and salami (Porto-Fett *et al.*, 2010). At sublethal pressures of 100 MPa, STEC O157:H7 was found to activate the general stress response controlled by RpoS and the cell envelope stress response controlled by RpoE (Malone *et al.*, 2006). Loss of function of RpoS or RpoE, as well as RpoS-regulated Dps and OtsA, led to significantly lower survival of STEC O157:H7 at lethal levels of pressure (Robey *et al.*, 2001; Malone *et al.*, 2006). Dps is a DNA-binding protein that is produced in response to a number of stress conditions, and plays a role in protecting bacteria from reactive oxygen species (Almiron *et al.*, 1992). HPP is known to trigger the oxidative stress response in *E. coli*, likely due to cells releasing an oxidative burst (Aertsen *et al.*, 2005). STEC O157:H7 virulence factor genes did not change upon exposure to HPP (Malone *et al.*, 2006), and the presence or absence of specific sets of virulence genes such as *stx1*, *stx2*, *eae*, or *ehxA*, however, had no effect on lethality across different STEC serotypes when treated with HPP at 350MPa at 4°C (Sheen *et al.*, 2015).

STEC response to antimicrobial ingredients

There has been considerable interest in recent years in examining plant- and microbial-derived compounds for antimicrobial activity and use in foods. A number of compounds have been demonstrated to have antimicrobial activity (Holley and Patel, 2005), and it is important to understand how foodborne pathogens, such as STEC, attempt to adapt to these stresses. The response of STEC O157:H7 over time to cinnamaldehyde exposure was assessed using transcriptomics, and the initial response was activation of the oxidative stress response and inhibition of cell division, as well as activation of alcohol dehydrogenase and aldehyde reductase (Visvalingam *et al.*, 2013). Cinnamaldehyde is hydrophobic, and it is able to diffuse across the cell membrane. The increased expression of alcohol dehydrogenase and aldehyde reductase convert cinnamaldehyde into the less toxic cinnamic alcohol, and as cinnamaldehyde levels decrease, cells are able to resume growth (Visvalingam *et al.*, 2013). Understanding the mechanisms activated by STEC during exposure to this antimicrobial demonstrates how the organism is able to overcome the initial stress posed by the compound. Even though initial exposure may lead to reduction in cell numbers, the remaining cells will be able to grow once the compound is degraded. Another consideration when assessing antimicrobial compounds for use in foods is if exposure to the compound influences virulence gene expression. Exposure of STEC O157:H7 to oregano oil or carvacrol led to significant decreases in expression of *ler*, the regulator of LEE expression, *stx2B*, *fliC*, and *luxS*, involved in quorum sensing (Mith *et al.*, 2015). These data support the potential use of this antimicrobial compound to inactivate STEC as well as suppress virulence gene expression in cells that may survive the initial killing effects of the compound.

The use of specific probiotic bacteria for their production of antimicrobial compounds and ability to outcompete pathogens on foods is also an active area of research (Jordan *et al.*, 2014). STEC O157:H7 EDL 933 grown in the presence of *Lactobacillus*, *Pediococcus* and *Bifidobacterium* had significantly lower expression of *stx2A* compared to when grown individually (Carey *et al.*, 2008). Similarly, STEC O157:H7 exposed to supernatants from *L. acidophilus* had significantly lower expression of the LEE and reduced production of the quorum sensing molecule AI-2 (Medellin-Pena *et al.*, 2007). Promising studies demonstrating reductions in STEC carriage in ruminants treated with probiotics (Stanford *et al.*, 2014; Rigobelo *et al.*, 2015) highlight the need to further investigate the effects these strains may have on modulating STEC gene expression *in vivo*.

Using microbiome data to characterize interactions between STEC and other bacteria

In the past decade, the microbiomes of many different environments have been extensively studied. Recent research has focused on utilizing microbiome data to understand how native microbiome composition and diversity may be impacted by the presence of STEC. Contamination of cattle hides with STEC is a significant source of STEC entry into the beef supply. Chopyk and colleagues (2016) examined if the microbial composition of cattle hides was associated with STEC prevalence. They found that a low microbial community diversity on cattle hides was significantly associated with the presence of the 'big six' STEC serotypes (O26, O45, O103, O111, O121, and O145), as well as O157. This finding supports further investigation into the relationship between microbiome diversity and STEC prevalence, which could lead to data that can be used in either risk assessments or development of new interventions.

Another pre-harvest environment that is a concern for STEC contamination and has a distinct microbiome is that of leafy greens. Due to the pathogenic properties of STEC, research on STEC and leafy greens is typically conducted in a greenhouse or other controlled environment. Williams and Marco (2014) examined the microbiomes of greenhouse-grown and field-grown romaine lettuce and found that the greenhouse-grown plants had less microbial diversity compared to those plants grown in the field. They were able to successfully transfer the microbiomes of field-grown lettuce to greenhouse lettuce plants, and studied the impact of inoculation with STEC O157:H7 on the phyllosphere microbiome. Phyllosphere bacterial communities were disrupted by the introduction of STEC, with relative amounts of *Microbacterium* increasing (Williams and Marco, 2014). It is interesting to note that *Microbacterium* has been shown to have antagonistic activity against STEC O157:H7 (Lopez-Velasco *et al.*, 2012), and further work may lead to development of this organism as a pre-harvest biocontrol agent. Alterations in the microbiome of post-harvest spinach during refrigerated storage have also been examined. The overall microbiome changed during storage, with increased amounts of psychrotrophs over time. Bacterial diversity was found to decrease when spinach was stored at 10°C, which was correlated with increased growth of STEC O157:H7 (Lopez-Velasco *et al.*, 2010). These types of data on the interactions between STEC and the native microbiota of a given food-related environment will help provide a more complete picture of each of these complex environments, and may provide new ideas for biological-based control strategies for this pathogen.

Using omics data for intervention strategies

Using omics data in predictive modelling

Microbial growth, survival, and death can be quantitatively predicted in foods and food-related environments after exposure to a single or to multiple stresses, and these data can be used for various risk assessment purposes, known as Quantitative Microbiological Risk Assessment (QMRA). Phenotypic characteristics, such as cell growth/death have been assessed under various stress conditions, such as those described in the above sections, including survival on fresh produce during storage (Delbeke *et al.*, 2015), in ground beef processing facilities (Dourou *et al.*, 2011), and in fruit juices (Yuk *et al.*, 2008). QMRA is available for STEC O157 in different food matrices, soil, and manure, and entails modelling phenotypic characteristics, such as cell growth/death under variables, including temperature, a_w, and pH (Ongeng *et al.*, 2014). In addition to these types of variables, omics data can be included to fill in gaps in knowledge related to variation in pathogen physiology under different conditions (Membre and Guillou, 2016). Franz and colleagues (2008) modelled the transfer of STEC O157:H7 from manure-amended soil to lettuce, and their model included the density of *E. coli* O157 in manure before and after storage, as well as the time interval between application to soil, and predicted a 2.8% likelihood that one or more lettuce plant would be contaminated per hectare. Gene expression studies have shown that STEC surviving in manure-amended soil had ~2.5 fold higher *rpoS* expression (van Hoek *et al.*, 2013). In addition, the metabolic profiles for these strains were distinct from those STEC that were poor survivors in soil (Franz et al., 2011). These findings could be implemented in a QRMA analysis to include our understanding of STEC physiological responses in this environment, and how it may impact transfer to fresh produce in the field.

QMRA has also been conducted for post-harvest lettuce involving STEC contamination during a model commercial scale washing, which aimed to provide a better understanding of improved sanitation strategies (Munther et al., 2015). In this study, the variables considered were time that STEC-contaminated lettuce was exposed to chlorine, amount of free chlorine, number of washes, and levels of STEC contamination. Another study modelled the growth of STEC on packaged lettuce under simulated transport and storage conditions (Zeng et al., 2014). The variables included here were temperature fluctuations over time. As described above, we know the stress responses that STEC O157:H7 induces both on pre-harvest and on post-harvest lettuce (Fink et al., 2012; Van der Linden et al., 2016). These data demonstrate that activation of these stress responses contribute to survival of STEC under both conditions, and these data can be modelled as a potential variable when developing risk assessment models.

Another way to capture and include omics data into predictive modelling is to use omics data, typically gene expression data, and link expression patterns for a subset of genes as indicators of pathogen robustness in a given environment (Abee et al., 2011). In this case, the gene expression indicates activation of a set of stress responses in response to that environment, and can also lead to greater survival of the pathogen in subsequent stresses. Quantitative correlation of biomarkers, the set of indicator genes, and the adaptive phenotypes has been demonstrated in laboratory medium-based stress environments for *Bacillus cereus* (den Besten et al., 2010). This type of data can then be used in predictive modelling to include variables that account for the ability of a pathogen to adapt to different environments, which can contribute significantly to survival through the stresses encountered in the food supply.

Using omics data to identify novel inhibitors and control strategies

Modulation of gene and protein expression in response to stress indicates activation or repression of a specific physiological response, and this can be used to determine the physiological state of STEC under different conditions (Bergholz et al., 2014). Activation of these stress responses when STEC is on a food or in a specific food-related environment has the potential to impact resistance or sensitivity of the pathogen to subsequent processing treatments and could be used to modify control strategies. For example, the significant up-regulation of genes involved in STEC O157:H7 cell envelope stress response on lettuce could lead to increased resistance of the pathogen to decontamination treatments that damage the cell envelope. In this case, decontamination treatments could be modified in a way that would pose a completely different stress to the pathogen, one that it may not be adapted for.

Omics data could lead to development of new intervention strategies by describing stress response regulators of importance across different food-related environments. These regulators could then serve as targets that could be evaluated with small molecule screens to identify potential molecules that would serve as inhibitors of the regulator. This type of regulator inhibition was first described for *Vibrio cholerae*, where a small molecule inhibitor of ToxT, the regulator of cholera toxin was demonstrated to eliminate expression of the toxin (Hung et al., 2005). This concept has been applied to identifying inhibitors of stress responses as well, with a small molecule that inhibits SigB, controlling the general stress response, in *Listeria monocytogenes* (Palmer et al., 2011). Specifically targeting and disrupting stress responses could lead to reduced survival of STEC in different environments and lead to a reduction in transfer to other stages of the food supply.

Future trends

Data needed for other important STEC serotypes

As is clear from the data highlighted in this chapter, there is a significant lack of omics data from non-O157 STEC serotypes. STEC O157 strains are a genetically distinct clade of *E. coli* (Reid et al., 2000; Tenaillon et al., 2010), and variation in stress response phenotypes have been described among some of the major STEC serotypes of interest, such as O26, O111, O121, O103, O145, and O45 (Bergholz and Whittam, 2007; Corliss *et al.*, 2015; Hsu *et al.*, 2015). Characterizing the differences in phenotypic responses at the omics level among these different serotypes will provide data that can lead to refinement of interventions and risk assessments that will ensure control measures are effective against the potential range of responses.

Issues related to STEC in low-moisture foods

STEC in dry foods has been linked to a few outbreaks, such as O157 in hazelnuts (Miller *et al.*, 2012) and raw cookie dough, with raw flour suspected as the source of contamination (Neil *et al.*, 2012). The most recent STEC outbreak in a low-moisture food has been STEC O121 and STEC O26 in raw flour (CDC, 2016). However, phenotypic studies on STEC behaviour in such low-moisture foods are minimal, and omics data for STEC in such foods are non-existent. While the osmotic stress response for *E. coli* is relatively well characterized, the majority of data are from laboratory media-based environments. Phenotypic studies conducted on survival of STEC have shown that STEC O157 can survive for more than a year at various storage temperatures on artificially inoculated almonds and pistachios (Kimber *et al.*, 2012). Understanding how STEC survive on these types of foods will provide insights into how this emerging concern might be addressed.

Characterizing population heterogeneity in complex environments

Heterogeneity in stress responses within a population has been described for a number of food-borne pathogens, including STEC (Wood *et al.*, 2013). This heterogeneity is typically exhibited as a tailing response when a population is exposed to a lethal stress, but some portion of the population persists (Abee *et al.*, 2016). Formation of these persister cells, as well as induction of portions of the population into the viable-but-non-culturable (VBNC) state, is an area that is just beginning to be explored for STEC. STEC have been shown to enter the VBNC state in the produce phyllosphere (Dinu and Bach, 2011), and this is currently an understudied factor influencing spread of the pathogen in the food supply.

Typically, low levels of pathogens are present in a given food or food-related environment, and that low level negatively impacts the ability to effectively model pathogen growth using growth models designed for large starting inocula. Characterizing the heterogeneity in growth at the level of single cells has been another rapidly growing area of research (Ding *et al.*, 2016). Additional information on the physiological mechanisms a cell is using to grow under different stresses, such as data from single-cell transcriptomics or proteomics experiments, has the potential to significantly improve understanding of pathogen behaviour at this level, which represents a level that more practically reflects the presence of pathogens such as STEC in the food supply.

Web resources

EcoCyc: www.ecocyc.org A database of annotated genomes, transcriptional units, regulatory networks and metabolic pathways for *E. coli* and many other bacteria (Keseler *et al.*, 2013).

PATRIC: www.patricbrc.org A database of genomic and transcriptomic data from pathogenic *E. coli* as well as other pathogenic bacteria. Provides tools for analysis of user generated genomes and transcriptomes, as well as comparison with publicly available datasets (Wattam *et al.*, 2014).

References

Abban, S., Jakobsen, M., and Jespersen, L. (2012). Attachment behaviour of *Escherichia coli* K12 and *Salmonella* Typhimurium P6 on food contact surfaces for food transportation. Food Microbiol. *31*, 139–147. https://doi.org/10.1016/j.fm.2012.04.003

Abee, T., Koomen, J., Metselaar, K.I., Zwietering, M.H., and den Besten, H.M. (2016). Impact of pathogen population heterogeneity and stress-resistant variants on food safety. Annu Rev Food Sci Technol 7, 439–456. https://doi.org/10.1146/annurev-food-041715-033128

Abee, T., Wels, M., de Been, M., and den Besten, H. (2011). From transcriptional landscapes to the identification of biomarkers for robustness. Microb. Cell Fact. 10 (Suppl. 1), S9. https://doi.org/10.1186/1475-2859-10-S1-S9

Ackers, M.L., Mahon, B.E., Leahy, E., Goode, B., Damrow, T., Hayes, P.S., Bibb, W.F., Rice, D.H., Barrett, T.J., Hutwagner, L., *et al.* (1998). An outbreak of *Escherichia coli* O157:H7 infections associated with leaf lettuce consumption. J. Infect. Dis. *177*, 1588–1593.

Aertsen, A., De Spiegeleer, P., Vanoirbeek, K., Lavilla, M., and Michiels, C.W. (2005). Induction of oxidative stress by high hydrostatic pressure in *Escherichia coli*. Appl. Environ. Microbiol. *71*, 2226–2231. https://doi.org/10.1128/AEM.71.5.2226-2231.2005

Allen, K.J., and Griffiths, M.W. (2012). Impact of hydroxyl- and superoxide anion-based oxidative stress on logarithmic and stationary phase *Escherichia coli* O157:H7 stress and virulence gene expression. Food Microbiol. *29*, 141–147. https://doi.org/10.1016/j.fm.2011.09.014

Almirón, M., Link, A.J., Furlong, D., and Kolter, R. (1992). A novel DNA-binding protein with regulatory and protective roles in starved *Escherichia coli*. Genes Dev. *6*, 2646–2654.

Bergholz, T.M., Moreno Switt, A.I., and Wiedmann, M. (2014). Omics approaches in food safety: fulfilling the promise? Trends Microbiol. *22*, 275–281. https://doi.org/10.1016/j.tim.2014.01.006

Bergholz, T.M., Vanaja, S.K., and Whittam, T.S. (2009). Gene expression induced in *Escherichia coli* O157:H7 upon exposure to model apple juice. Appl. Environ. Microbiol. *75*, 3542–3553. https://doi.org/10.1128/AEM.02841-08

Bergholz, T.M., and Whittam, T.S. (2007). Variation in acid resistance among enterohaemorrhagic Escherichia coli in a simulated gastric environment. J. Appl. Microbiol. 102, 352-362. https://doi.org/10.1111/j.1365-2672.2006.03099.x

Bertin, Y., Chaucheyras-Durand, F., Robbe-Masselot, C., Durand, A., de la Foye, A., Harel, J., Cohen, P.S., Conway, T., Forano, E., and Martin, C. (2013). Carbohydrate utilization by enterohaemorrhagic *Escherichia coli* O157:H7 in bovine intestinal content. Environ. Microbiol. *15*, 610-622. https://doi.org/10.1111/1462-2920.12019

Besser, R.E., Lett, S.M., Weber, J.T., Doyle, M.P., Barrett, T.J., Wells, J.G., and Griffin, P.M. (1993). An outbreak of diarrhea and hemolytic uremic syndrome from *Escherichia coli* O157:H7 in fresh-pressed apple cider. JAMA *269*, 2217–2220.

Besser, T.E., Shaikh, N., Holt, N.J., Tarr, P.I., Konkel, M.E., Malik-Kale, P., Walsh, C.W., Whittam, T.S., and Bono, J.L. (2007). Greater diversity of Shiga toxin-encoding bacteriophage insertion sites among *Escherichia coli* O157:H7 isolates from cattle than in those from humans. Appl. Environ. Microbiol. *73*, 671–679. https://doi.org/10.1128/AEM.01035-06

Bielaszewska, M., Mellmann, A., Zhang, W., Kock, R., Fruth, A., Bauwens, A., Peters, G., and Karch, H. (2011). Characterisation of the *Escherichia coli* strain associated with an outbreak of haemolytic uraemic syndrome in Germany, 2011: a microbiological study. Lancet Infect. Dis. *11*, 671–676. https://doi.org/10.1016/S1473-3099(11)70165-7

Blanco, M., Blanco, J.E., Mora, A., Dahbi, G., Alonso, M.P., Gonzalez, E.A., Bernardez, M.I., and Blanco, J. (2004). Serotypes, virulence genes, and intimin types of shiga toxin (verotoxin)-producing *Escherichia coli* isolates from cattle in Spain and identification of a new intimin variant gene (eae-xi). J. Clin. Microbiol. *42*, 645–651.

Brandl, M.T. (2008). Plant lesions promote the rapid multiplication of *Escherichia coli* O157:H7 on postharvest lettuce. Appl. Environ. Microbiol. *74*, 5285–5289. https://doi.org/10.1128/AEM.01073-08

Brooks, J.T., Sowers, E.G., Wells, J.G., Greene, K.D., Griffin, P.M., Hoekstra, R.M., and Strockbine, N.A. (2005). Non-O157 Shiga toxin-producing *Escherichia coli* infections in the United States, 1983-2002. J. Infect. Dis. *192*, 1422–1429. https://doi.org/10.1086/466536

Buvens, G., Posse, B., De Schrijver, K., De Zutter, L., Lauwers, S., and Pierard, D. (2011). Virulence profiling and quantification of verocytotoxin-producing *Escherichia coli* O145:H28 and O26:H11 isolated during an ice cream-related hemolytic uremic syndrome outbreak. Food-borne Pathog. Dis. *8*, 421-426. https://doi.org/10.1089/fpd.2010.0693

Carey, C.M., Kostrzynska, M., Ojha, S., and Thompson, S. (2008). The effect of probiotics and organic acids on Shiga-toxin 2 gene expression in enterohemorrhagic *Escherichia coli* O157:H7. J. Microbiol. Methods *73*, 125–132. https://doi.org/10.1016/j.mimet.2008.01.014

Carruthers, M.D., and Minion, C. (2009). Transcriptome analysis of *Escherichia coli* O157:H7 EDL933 during heat shock. FEMS Microbiol. Lett. *295*, 96–102. https://doi.org/10.1111/j.1574-6968.2009.01587.x

CDC. (1993). Update: multistate outbreak of *Escherichia coli* O157:H7 infections from hamburgers – western United States, 1992-1993. MMWR Morb. Mortal. Wkly. Rep. *42*, 258-263.

CDC. (1995). *Escherichia coli* O157:H7 outbreak linked to commercially distributed dry-cured salami –Washington and California, 1994. MMWR Morb. Mortal. Wkly. Rep. *44*, 157-160.

CDC. (2011). 'Multistate outbreak of *E. coli* O157:H7 infections associated with Lebanon bologna (Final Update).' from http://www.cdc.gov/ecoli/2011/lebanon-bologna-3-23-11.html.

CDC. (2013). Surveillance for food-borne disease outbreaks – United States, 2009-2010. MMWR Morb. Mortal. Wkly. Rep. *62*, 41-47.

CDC. (2015). 'Multistate outbreak of shiga toxin-producing *Escherichia coli* O157:H7 infections linked to Costco rotisserie shicken salad (Final Update).' from http://www.cdc.gov/ecoli/2015/o157h7-11-15/index.html.

CDC. (2016). 'Multistate outbreak of shiga toxin-producing *Escherichia coli* infections linked to flour.' from https://www.cdc.gov/ecoli/2016/o121-06-16/index.html.

Chiang, S.M., and Schellhorn, H.E. (2012). Regulators of oxidative stress response genes in *Escherichia coli* and their functional conservation in bacteria. Arch. Biochem. Biophys. *525*, 161–169. https://doi.org/10.1016/j.abb.2012.02.007

Chopyk, J., Moore, R.M., DiSpirito, Z., Stromberg, Z.R., Lewis, G.L., Rneter, D.G., Cernicchiaro, N., Moxley, R.A., and Wommack, K.E. (2016). Presence of pathogenic *Escherichia coli* is correlated with bacterial community diversity and composition on pre-harvest cattle hides. Microbiome. 4, 9. doi: 10.1186/s40168-016-0155-4.

Cody, S.H., Glynn, M.K., Farrar, J.A., Cairns, K.L., Griffin, P.M., Kobayashi, J., Fyfe, M., Hoffman, R., King, A.S., Lewis, J.H., *et al.* (1999). An outbreak of *Escherichia coli* O157:H7 infection from unpasteurized commercial apple juice. Ann. Intern. Med. *130*, 202–209.

Cogan, T.A., Bloomfield, S.F., and Humphrey, T.J. (1999). The effectiveness of hygiene procedures for prevention of cross-contamination from chicken carcases in the domestic kitchen. Lett. Appl. Microbiol. *29*, 354–358.

Cooley, M.B., Jay-Russell, M., Atwill, E.R., Carychao, D., Nguyen, K., Quinones, B., Patel, R., Walker, S., Swimley, M., Pierre-Jerome, E., Gordus, A.G., and Mandrell, R.E. (2013). Development of a robust method for isolation of shiga toxin-positive *Escherichia coli* (STEC) from fecal, plant, soil and water samples from a leafy greens production region in California. PLOS ONE. *8*, e65716. https://doi.org/10.1371/journal.pone.0065716

Corliss, B., Brooks, J.C., Martin, J.N., Echeverry, A., Parks, A.R., Pokharel, S., and Brashears, M.M. (2015). The influence of beef quality characteristics on the internalization and thermal susceptibility of shiga toxin-producing *Escherichia coli* (STEC) in blade-tenderized beef steaks. Meat Sci. *110*, 85–92. https://doi.org/10.1016/j.meatsci.2015.06.014

Cote, R., Katani, R., Moreau, M.R., Kudva, I.T., Arthur, T.M., DebRoy, C., Mwangi, M.M., Albert, I., Raygoza Garay, J.A., Li, L., Brandl, M.T., Carter, M.Q., and Kapur, V. (2015). Comparative analysis of super-shedder strains of *Escherichia coli* O157:H7 reveals distinctive genomic features and a strongly aggregative adherent phenotype on bovine rectoanal junction squamous epithelial cells. PLOS ONE. *10*, e0116743. https://doi.org/10.1371/journal.pone.0116743

Crowe, S.J., Mahon, B.E., Vieira, A.R., and Gould, L.H. (2015). Vital signs: multistate food-borne outbreaks – United States, 2010-2014. MMWR Morb. Mortal. Wkly. Rep. *64*, 1221–1225. https://doi.org/10.15585/mmwr.mm6443a4

Crozier, L., Hedley, P.E., Morris, J., Wagstaff, C., Andrews, S.C., Toth, I., Jackson, R.W., and Holden, N.J. (2016). Whole-transcriptome analysis of verocytotoxigenic *Escherichia coli* O157:H7 (Sakai) suggests plant-species-specific metabolic responses on exposure to spinach and lettuce extracts. Front Microbiol. *7*, 1088. https://doi.org/10.3389/fmicb.2016.01088

Davis, M.A., Hancock, D.D., Rice, D.H., Call, D.R., DiGiacomo, R., Samadpour, M., and Besser, T.E. (2003). Feedstuffs as a vehicle of cattle exposure to *Escherichia coli* O157:H7 and *Salmonella enterica*. Vet. Microbiol. *95*, 199–210.

De Biase, D., and Lund, P.A. (2015). The *Escherichia coli* acid stress response and its significance for pathogenesis. Adv. Appl. Microbiol. *92*, 49–88. https://doi.org/10.1016/bs.aambs.2015.03.002

Delbeke, S., Ceuppens, S., Jacxsens, L., and Uyttendaele, M. (2015). Survival of *Salmonella* and *Escherichia coli* O157:H7 on strawberries, basil, and other leafy greens during storage. J. Food Prot. *78*, 652–660. https://doi.org/10.4315/0362-028X.JFP-14-354

den Besten, H.M., Arvind, A., Gaballo, H.M., Moezelaar, R., Zwietering, M.H., and Abee, T. (2010). Short- and long-term biomarkers for bacterial robustness: a framework for quantifying correlations between cellular indicators and adaptive behavior. PLOS ONE. *5*, e13746. https://doi.org/10.1371/journal.pone.0013746

Ding, T., Liao, X.Y., Dong, Q.L., Xuan, X.T., Chen, S.G., Ye, X.Q., and Liu, D.H. (2016). Predictive modeling of microbial single cells: A review. Crit. Rev. Food Sci. Nutr. [Epub ahead of print]. https://doi.org/10.1080/10408398.2016.1217193

Dinu, L.D., and Bach, S. (2011). Induction of viable but nonculturable *Escherichia coli* O157:H7 in the phyllosphere of lettuce: a food safety risk factor. Appl. Environ. Microbiol. *77*, 8295–8302. https://doi.org/10.1128/AEM.05020-11

Dong, T., and Schellhorn, H.E. (2009a). Control of RpoS in global gene expression of *Escherichia coli* in minimal media. Mol. Genet. Genomics *281*, 19–33. https://doi.org/10.1007/s00438-008-0389-3

Dong, T., and Schellhorn, H.E. (2009b). Global effect of RpoS on gene expression in pathogenic *Escherichia coli* O157:H7 strain EDL933. BMC Genomics *10*, 349. https://doi.org/10.1186/1471-2164-10-349

Dourou, D., Beauchamp, C.S., Yoon, Y., Geornaras, I., Belk, K.E., Smith, G.C., Nychas, G.J., and Sofos, J.N. (2011). Attachment and biofilm formation by *Escherichia coli* O157:H7 at different temperatures, on various food-contact surfaces encountered in beef processing. Int. J. Food Microbiol. *149*, 262–268. https://doi.org/10.1016/j.ijfoodmicro.2011.07.004

Elder, R.O., Keen, J.E., Siragusa, G.R., Barkocy-Gallagher, G.A., Koohmaraie, M., and Laegreid, W.W. (2000). Correlation of enterohemorrhagic *Escherichia coli* O157 prevalence in feces, hides, and carcasses of beef cattle during processing. Proc. Natl. Acad. Sci. U.S.A. *97*, 2999-3003.

Erickson, M.C., Liao, J., Cannon, J.L., and Ortega, Y.R. (2015). Contamination of knives and graters by bacterial food-borne pathogens during slicing and grating of produce. Food Microbiol. *52*, 138–145. https://doi.org/10.1016/j.fm.2015.07.008

Ethelberg, S., Olsen, K.E., Scheutz, F., Jensen, C., Schiellerup, P., Enberg, J., Petersen, A.M., Olesen, B., Gerner-Smidt, P., and Molbak, K. (2004). Virulence factors for hemolytic uremic syndrome, Denmark. Emerg. Infect. Dis. *10*, 842-847. https://doi.org/10.3201/eid1005.030576

FDA. (2016). 'FDA investigates multistate outbreak of *E. coli* O26 infections linked to Chipotle Mexican Grill restaurants.' from http://www.fda.gov/Food/RecallsOutbreaksEmergencies/Outbreaks/ucm470410.htm.

Feng, P. (2014). Shiga toxin-producing *Escherichia coli* (STEC) in fresh produce – a food safety dilemma. Microbiol. Spectr. *2*, EHEC–0010–2013. https://doi.org/10.1128/microbiolspec.EHEC-0010-2013

Ferens, W.A., and Hovde, C.J. (2011). *Escherichia coli* O157:H7: animal reservoir and sources of human infection. Food-borne Pathog. Dis. *8*, 465–487. https://doi.org/10.1089/fpd.2010.0673

Fink, R.C., Black, E.P., Hou, Z., Sugawara, M., Sadowsky, M.J., and Diez-Gonzalez, F. (2012). Transcriptional responses of *Escherichia coli* K-12 and O157:H7 associated with lettuce leaves. Appl. Environ. Microbiol. *78*, 1752–1764. https://doi.org/10.1128/AEM.07454-11

Franz, E., Semenov, A.V., Termorshuizen, A.J., de Vos, O.J., Bokhorst, J.G., and van Bruggen, A.H. (2008). Manure-amended soil characteristics affecting the survival of *E. coli* O157:H7 in 36 Dutch soils. Environ. Microbiol. *10*, 313–327. https://doi.org/10.1111/j.1462-2920.2007.01453.x

Franz, E., van Hoek, A.H., Bouw, E., and Aarts, H.J. (2011). Variability of *Escherichia coli* O157 strain survival in manure-amended soil in relation to strain origin, virulence profile, and carbon nutrition profile. Appl. Environ. Microbiol. *77*, 8088-8096. https://doi.org/10.1128/AEM.00745-11

Fratamico, P.M., Wang, S., Yan, X., Zhang, W., and Li, Y. (2011). Differential gene expression of *E. coli* O157:H7 in ground beef extract compared to tryptic soy broth. J. Food Sci. *76*, M79–87. https://doi.org/10.1111/j.1750-3841.2010.01952.x

Garcia, L., Henderson, J., Fabri, M., and Oke, M. (2006). Potential sources of microbial contamination in unpasteurized apple cider. J. Food Prot. *69*, 137–144.

Goldwater, P.N., and Bettelheim, K.A. (1998). New perspectives on the role of *Escherichia coli* O157:H7 and other enterohaemorrhagic *E. coli* serotypes in human disease. J. Med. Microbiol. *47*, 1039–1045. https://doi.org/10.1099/00222615-47-12-1039

Gould, L.H., Walsh, K.A., Vieira, A.R., Herman, K., Williams, I.T., Hall, A.J., and Cole, D. (2013). Surveillance for food-borne disease outbreaks – United States, 1998–2008. MMWR Surveill. Summ. *62*, 1–34.

Grad, Y.H., Lipsitch, M., Feldgarden, M., Arachchi, H.M., Cerqueira, G.C., Fitzgerald, M., Godfrey, P., Haas, B.J., Murphy, C.I., Russ, C., *et al.* (2012). Genomic epidemiology of the *Escherichia coli* O104:H4 outbreaks in Europe, 2011. Proc. Natl. Acad. Sci. U.S.A. *109*, 3065–3070. https://doi.org/10.1073/pnas.1121491109

Grant, J., Wendelboe, A.M., Wendel, A., Jepson, B., Torres, P., Smelser, C., and Rolfs, R.T. (2008). Spinach-associated *Escherichia coli* O157:H7 outbreak, Utah and New Mexico, 2006. Emerging Infect. Dis. *14*, 1633–1636. https://doi.org/10.3201/eid1410.071341

Griffin, P.M., and Tauxe, R.V. (1991). The epidemiology of infections caused by *Escherichia coli* O157:H7, other enterohemorrhagic *E. coli*, and the associated hemolytic uremic syndrome. Epidemiol. Rev. *13*, 60-98.

Gunther, N.W., Liu, Y., Nunez, A., Paul, M., and Uhlich, G.A. (2014). Nonlabeled quantitative proteomic comparison identifies differences in acid resistance between *Escherichia coli* O157:H7 curli production variants. Food-borne Pathog. Dis. *11*, 30–37. https://doi: 10.1089/fpd.2013.1570.Hengge-Aronis, R., Klein, W., Lange, R., Rimmele, M., and Boos, W. (1991). Trehalose synthesis genes are controlled by the putative sigma factor encoded by *rpoS* and are involved in stationary-phase thermotolerance in *Escherichia coli*. J. Bacteriol. *173*, 7918-7924.

Hengge-Aronis, R., Lange, R., Henneberg, N., and Fischer, D. (1993). Osmotic regulation of rpoS-dependent genes in *Escherichia coli*. J. Bacteriol. *175*, 259-265.

Herbert, L.J., Vali, L., Hoyle, D.V., Innocent, G., McKendrick, I.J., Pearce, M.C., Mellor, D., Porphyre, T., Locking, M., Allison, L., Hanson, M., Matthews, L., Gunn, G.J., Woolhouse, M.E., and Chase-Topping, M.E. (2014). *E. coli* O157 on Scottish cattle farms: evidence of local spread and persistence using repeat cross-sectional data. BMC Vet. Res. *10*, 95. https://doi: 10.1186/1746-6148-10-95.Herdt, J., and Feng, H. (2009). Aqueous antimicrobial treatments to improve fresh and fresh-cut produce safety. In Microbial Safety of Fresh Produce, X. Fan, B.A. Niemira, C.J. Doona, F.E. Feeherry, and R.B. Gravani. Ames, Iowa, ed. (Wiley-Blackwell), pp. 169-190.

Hilborn, E.D., Mermin, J.H., Mshar, P.A., Hadler, J.L., Voetsch, A., Wojtkunski, C., Swartz, M., Mshar, R., Lambert-Fair, M.A., Farrar, J.A., *et al.* (1999). A multistate outbreak of *Escherichia coli* O157:H7 infections associated with consumption of mesclun lettuce. Arch. Intern. Med. *159*, 1758–1764.

Hilborn, E.D., Mshar, P.A., Fiorentino, T.R., Dembek, Z.F., Barrett, T.J., Howard, R.T., and Cartter, M.L. (2000). An outbreak of *Escherichia coli* O157:H7 infections and haemolytic uraemic syndrome associated with consumption of unpasteurized apple cider. Epidemiol. Infect. *124*, 31-36.

Holley, R.A., and Patel, D. (2005). Improvement in shelf-life and safety of perishable foods by plant essential oils and smoke antimicrobials. Food Microbiol. *22*, 273–292. doi: 10.1016/j.fm.2004.08.006.

Hou, Z., Fink, R.C., Sugawara, M., Diez-Gonzalez, F., and Sadowsky, M.J. (2013). Transcriptional and functional responses of *Escherichia coli* O157:H7 growing in the lettuce rhizoplane. Food Microbiol. *35*, 136–142. https://doi.org/10.1016/j.fm.2013.03.002

Hsu, H., Sheen, S., Sites, J., Cassidy, J., Scullen, B., and Sommers, C. (2015). Effect of high pressure processing on the survival of Shiga toxin-producing *Escherichia coli* (Big Six vs. O157:H7) in ground beef. Food Microbiol. *48*, 1–7. https://doi.org/10.1016/j.fm.2014.12.002

Hsu, H., Sheen, S., Sites, J., Huang, L., and Wu, J.S. (2014). Effect of high pressure treatment on the survival of Shiga toxin-producing *Escherichia coli* in strawberry puree. Food Microbiol. *40*, 25–30. https://doi.org/10.1016/j.fm.2013.11.019

Hung, D.T., Shakhnovich, E.A., Pierson, E., and Mekalanos, J.J. (2005). Small-molecule inhibitor of *Vibrio cholerae* virulence and intestinal colonization. Science *310*, 670–674. doi: 10.1126/science.1116739.

Hussein, H.S. (2007). Prevalence and pathogenicity of Shiga toxin-producing *Escherichia coli* in beef cattle and their products. J. Anim. Sci. 85 (Suppl. 13), E63–72.

Islam, M., Doyle, M.P., Phatak, S.C., Millner, P., and Jiang, X. (2004). Persistence of enterohemorrhagic *Escherichia coli* O157:H7 in soil and on leaf lettuce and parsley grown in fields treated with contaminated manure composts or irrigation water. J. Food Prot. *67*, 1365–1370. doi: 10.4315/0362-028X-67.7.1365.

Jay, M.T., Cooley, M., Carychao, D., Wiscomb, G.W., Sweitzer, R.A., Crawford-Miksza, L., Farrar, J.A., Lau, D.K., O'Connell, J., Millington, A., *et al.* (2007). *Escherichia coli* O157:H7 in feral swine near spinach fields and cattle, central California coast. Emerg. Infect. Dis. *13*, 1908-1911. doi: 10.3201/eid1312.070763.

Johnson, J.Y., Thomas, J.E., Graham, T.A., Townshend, I., Byrne, J., Selinger, L.B., and Gannon, V.P. (2003). Prevalence of *Escherichia coli* O157:H7 and *Salmonella* spp. in surface waters of southern Alberta and its relation to manure sources. Can. J. Microbiol. *49*, 326–335. https://doi.org/10.1139/w03-046

Jordan, K., Dalmasso, M., Zentek, J., Mader, A., Bruggeman, G., Wallace, J., De Medici, D., Fiore, A., Prukner-Radovcic, E., Lukac, M., *et al.* (2014). Microbes versus microbes: control of pathogens in the food chain. J. Sci. Food Agric. *94*, 3079–3089. https://doi.org/10.1002/jsfa.6735

Karmali, M.A., Petric, M., Lim, C., Fleming, P.C., Arbus, G.S., and Lior, H. (1985). The association between idiopathic hemolytic uremic syndrome and infection by verotoxin-producing *Escherichia coli*. J. Infect. Dis. *151*, 775–782.

Keseler, I.M., Mackie, A., Peralta-Gil, M., Santos-Zavaleta, A., Gama-Castro, S., Bonavides-Martinez, C., Fulcher, C., Huerta, A.M., Kothari, A., Krummenacker, M*et al.* (2013). EcoCyc: fusing model organism databases with systems biology. Nucleic Acids Res. *41*, 605-612. https://doi: 10.1093/nar/gks1027.

Kim, G.H., Breidt, F., Fratamico, P., and Oh, D.H. (2015). Acid resistance and molecular characterization of *Escherichia coli* O157:H7 and different non-O157 shiga toxin-producing *E. coli* serogroups. J. Food Sci. *80*, M2257–2264. https://doi.org/10.1111/1750-3841.12996

Kim, G.H., Fratamico, P., Breidt, F., and Oh, D.H. (2016). Survival and expression of acid resistance genes in shiga toxin-producing *Escherichia coli* acid adapted in pineapple juice and exposed to synthetic gastric fluid. J. Appl. Microbiol. *121*, 1416–1426. htpps://doi: 10.1111/jam.13223.

Kim, J., Nietfeldt, J., and Benson, A.K. (1999). Octamer-based genome scanning distinguishes a unique subpopulation of *Escherichia coli* O157:H7 strains in cattle. Proc. Natl. Acad. Sci. U.S.A. *96*, 13288–13293.

Kimber, M.A., Kaur, H., Wang, L., Danyluk, M.D., and Harris, L.J. (2012). Survival of *Salmonella*, *Escherichia coli* O157:H7, and *Listeria monocytogenes* on inoculated almonds and pistachios stored at −19, 4, and 24° C. J. Food Prot. 75, 1394–1403. https://doi.org/10.4315/0362-028X.JFP-12-023

King, T., Kocharunchitt, C., Gobius, K., Bowman, J.P., and Ross, T. (2014). Global genome response of *Escherichia coli* O157:H7 Sakai during dynamic changes in growth kinetics induced by an abrupt temperature downshift. PLOS ONE. *9*, 99627. https://doi: 10.1371/journal.pone.0099627.

King, T., Lucchini, S., Hinton, J.C., and Gobius, K. (2010). Transcriptomic analysis of *Escherichia coli* O157:H7 and K-12 cultures exposed to inorganic and organic acids in stationary phase reveals acidulant- and strain-specific acid tolerance responses. Appl. Environ. Microbiol. *76*, 6514–6528. htpps://doi: 10.1128/AEM.02392-09.

Kocharunchitt, C., King, T., Gobius, K., Bowman, J.P., and Ross, T. (2012). Integrated transcriptomic and proteomic analysis of the physiological response of *Escherichia coli* O157:H7 Sakai to steady-state conditions of cold and water activity stress. Mol. Cell. Proteomics, *11*, 111.009019.

htpps://doi: 10.1074/mcp.M111.009019.

Kocharunchitt, C., King, T., Gobius, K., Bowman, J.P., and Ross, T. (2014). Global genome response of *Escherichia coli* O157:H7 Sakai during dynamic changes in growth kinetics induced by an abrupt downshift in water activity. PLOS ONE. *9*, 90422. https://doi.org/10.1371/journal.pone.0090422

Koopman, J.A., Marshall, J.M., Bhatiya, A., Eguale, T., Kwiek, J.J., and Gunn, J.S. (2015). Inhibition of *Salmonella enterica* biofilm formation using small-molecule adenosine mimetics. Antimicrob. Agents Chemother. *59*, 76–84. https://doi: 10.1128/AAC.03407-14.

Kudva, I.T., Krastins, B., Torres, A.G., Griffin, R.W., Sheng, H., Sarracino, D.A., Hovde, C.J., Calderwood, S.B., and John, M. (2015). The *Escherichia coli* O157:H7 cattle immunoproteome includes outer membrane protein A (OmpA), a modulator of adherence to bovine rectoanal junction squamous epithelial (RSE) cells. Proteomics. *15*, 1829–1842. https://doi: 10.1002/pmic.201400432.

Kudva, I.T., Stanton, T.B., and Lippolis, J.D. (2014). The *Escherichia coli* O157:H7 bovine rumen fluid proteome reflects adaptive bacterial responses. BMC Microbiol. *14*, 48. https://doi: 10.1186/1471-2180-14-48.

Kyle, J.L., Parker, C.T., Goudeau, D., and Brandl, M.T. (2010). Transcriptome analysis of *Escherichia coli* O157:H7 exposed to lysates of lettuce leaves. Appl. Environ. Microbiol. *76*, 1375–1387. https://doi.org/10.1128/AEM.02461-09

Laidler, M.R., Tourdjman, M., Buser, G.L., Hostetler, T., Repp, K.K., Leman, R., Samadpour, M., and Keene, W.E. (2013). *Escherichia coli* O157:H7 infections associated with consumption of locally grown strawberries contaminated by deer. Clin. Infect. Dis. *57*, 1129–1134. https://doi.org/10.1093/cid/cit468

Landstorfer, R., Simon, S., Schober, S., Keim, D., Scherer, S., and Neuhaus, K. (2014). Comparison of strand-specific transcriptomes of enterohemorrhagic *Escherichia coli* O157:H7 EDL933 (EHEC) under eleven different environmental conditions including radish sprouts and cattle feces. BMC Genomics. *15*, 353. https://doi: 10.1186/1471-2164-15-353.

Lange, R., and Hengge-Aronis, R. (1991). Identification of a central regulator of stationary-phase gene expression in *Escherichia coli*. Mol. Microbiol. *5*, 49–59. https://doi.org/10.1111/j.1365-2958.1991.tb01825.x

Large, T.M., Walk, S.T., and Whittam, T.S. (2005). Variation in acid resistance among shiga toxin-producing clones of pathogenic *Escherichia coli*. Appl. Environ. Microbiol. *71*, 2493–2500. https://doi.org/10.1128/AEM.71.5.2493-2500.2005

Law, D. (2000). Virulence factors of *Escherichia coli* O157 and other Shiga toxin-producing *E. coli*. J. Appl. Microbiol. *88*, 729–745.

Lim, J.Y., Li, J., Sheng, H., Besser, T.E., Potter, K., and Hovde, C.J. (2007). *Escherichia coli* O157:H7 colonization at the rectoanal junction of long-duration culture-positive cattle. Appl. Environ. Microbiol. *73*, 1380–1382. https://doi.org/10.1128/AEM.02242-06

Lim, J.Y., Yoon, J., and Hovde, C.J. (2010). A brief overview of *Escherichia coli* O157:H7 and its plasmid O157. J. Microbiol. Biotechnol. *20*, 5–14.

Lin, J., Smith, M.P., Chapin, K.C., Baik, H.S., Bennett, G.N., and Foster, J.W. (1996). Mechanisms of acid resistance in enterohemorrhagic *Escherichia coli*. Appl. Environ. Microbiol. *62*, 3094–3100.

Lopez-Velasco, G., Davis, M., Boyer, R.R., Williams, R.C., and Ponder, M.A. (2010). Alterations of the phylloepiphytic bacterial community associated with interactions of *Escherichia coli* O157:H7 during storage of packaged spinach at refrigeration temperatures. Food Microbiol. *27*, 476-486. https://doi.org/10.1016/j.fm.2009.12.010

Lopez-Velasco, G., Tydings, H.A., Boyer, R.R., Falkinham, J.O., and Ponder, M.A. (2012). Characterization of interactions between *Escherichia coli* O157:H7 with epiphytic bacteria *in vitro* and on spinach leaf surfaces. Int. J. Food Microbiol. *153*, 351–357. https://doi.org/10.1016/j.ijfoodmicro.2011.11.026

Luo, Y.G. (2007). Fresh-cut produce wash water reuse affects water quality and packaged product quality and microbial growth in Romaine lettuce. Hortscience. *42*, 1413-1419.

Magnuson, R.D. (2007). Hypothetical functions of toxin-antitoxin systems. J. Bacteriol. *189*, 6089–6092.

Malone, A.S., Chung, Y.K., and Yousef, A.E. (2006). Genes of *Escherichia coli* O157:H7 that are involved in high-pressure resistance. Appl. Environ. Microbiol. *72*, 2661–2671. https://doi.org/10.1128/AEM.72.4.2661-2671.2006

Manning, S.D., Motiwala, A.S., Springman, A.C., Qi, W., Lacher, D.W., Ouellette, L.M., Mladonicky, J.M., Somsel, P., Rudrik, J.T., Dietrich, S.E., *et al.* (2008). Variation in virulence among clades of *Escherichia coli* O157:H7 associated with disease outbreaks. Proc. Natl. Acad. Sci. U.S.A. *105*, 4868–4873. https://doi.org/10.1073/pnas.0710834105

Mathusa, E.C., Chen, Y., Enache, E., and Hontz, L. (2010). Non-O157 Shiga toxin-producing *Escherichia coli* in foods. J. Food Prot. 73, 1721–1736.

Matthews, L., Low, J.C., Gally, D.L., Pearce, M.C., Mellor, D.J., Heesterbeek, J.A., Chase-Topping, M., Naylor, S.W., Shaw, D.J., Reid, S.W., *et al.* (2006). Heterogeneous shedding of *Escherichia coli* O157 in cattle and its implications for control. Proc. Natl. Acad. Sci. U.S.A. *103*, 547–552. https://doi.org/10.1073/pnas.0503776103

Medellin-Peña, M.J., Wang, H., Johnson, R., Anand, S., and Griffiths, M.W. (2007). Probiotics affect virulence-related gene expression in *Escherichia coli* O157:H7. Appl. Environ. Microbiol. 73, 4259–4267. https://doi.org/10.1128/AEM.00159-07

Mei, G.Y., Tang, J., Carey, C., Bach, S., and Kostrzynska, M. (2015). The effect of oxidative stress on gene expression of Shiga toxin-producing *Escherichia coli* (STEC) O157:H7 and non-O157 serotypes. Int. J. Food Microbiol. *215*, 7–15. https://doi.org/10.1016/j.ijfoodmicro.2015.07.029

Membre, J.M., and Guillou, S. (2016). Lastest developments in food-borne pathogen risk assessment. Curr. Opin. Food Sci. *8*, 120-126.

Michino, H., Araki, K., Minami, S., Takaya, S., Sakai, N., Miyazaki, M., Ono, A., and Yanagawa, H. (1999). Massive outbreak of *Escherichia coli* O157:H7 infection in schoolchildren in Sakai City, Japan, associated with consumption of white radish sprouts. Am. J. Epidemiol. *150*, 787–796.

Miller, B.D., Rigdon, C.E., Ball, J., Rounds, J.M., Klos, R.F., Brennan, B.M., Arends, K.D., Kennelly, P., Hedberg, C., and Smith, K.E. (2012). Use of traceback methods to confirm the source of a multistate *Escherichia coli* O157:H7 outbreak due to in-shell hazelnuts. J. Food Prot. *75*, 320–327. https://doi.org/10.4315/0362-028X.JFP-11-309

Mith, H., Clinquart, A., Zhiri, A., Daube, G., and Delcenserie, V. (2015). The impact of oregano (*Origanum heracleoticum*) essential oil and carvacrol on virulence gene transcription by *Escherichia coli* O157:H7. FEMS Microbiol. Lett. *362*, 1–7. https://doi.org/10.1093/femsle/fnu021

Monaghan, J.M., and Hutchison, M.L. (2016). Ineffective hand washing and the contamination of carrots after using a field latrine. Lett. Appl. Microbiol. *62*, 299–303. https://doi.org/10.1111/lam.12549

Munther, D., Luo, Y., Wu, J., Magpantay, F.M., and Srinivasan, P. (2015). A mathematical model for pathogen cross-contamination dynamics during produce wash. Food Microbiol. *51*, 101-107. https://doi.org/10.1016/j.fm.2015.05.010

Nadya, S., Delaquis, P., Chen, J., Allen, K., Johnson, R.P., Ziebell, K., Laing, C., Gannon, V., Bach, S., and Topp, E. (2016). Phenotypic and genotypic characteristics of shiga toxin-producing *Escherichia coli* isolated from surface waters and sediments in a Canadian urban-agricultural landscape. Front. Cell. Infect. Microbiol. *6*, 36. https://doi.org/10.3389/fcimb.2016.00036

Naylor, S.W., Low, J.C., Besser, T.E., Mahajan, A., Gunn, G.J., Pearce, M.C., McKendrick, I.J., Smith, D.G., and Gally, D.L. (2003). Lymphoid follicle-dense mucosa at the terminal rectum is the principal site of colonization of enterohemorrhagic *Escherichia coli* O157:H7 in the bovine host. Infect. Immun. *71*, 1505-1512.

Neil, K.P., Biggerstaff, G., MacDonald, J.K., Trees, E., Medus, C., Musser, K.A., Stroika, S.G., Zink, D., and Sotir, M.J. (2012). A novel vehicle for transmission of *Escherichia coli* O157:H7 to humans: multistate outbreak of *E. coli* O157:H7 infections associated with consumption of ready-to-bake commercial prepackaged cookie dough – United States, 2009. Clin. Infect. Dis. *54*, 511–518. https://doi.org/10.1093/cid/cir831

Nonaka, G., Blankschien, M., Herman, C., Gross, C.A., and Rhodius, V.A. (2006). Regulon and promoter analysis of the *E. coli* heat-shock factor, sigma32, reveals a multifaceted cellular response to heat stress. Genes Dev. *20*, 1776–1789. https://doi.org/10.1101/gad.1428206

Omisakin, F., MacRae, M., Ogden, I.D., and Strachan, N.J. (2003). Concentration and prevalence of *Escherichia coli* O157 in cattle feces at slaughter. Appl. Environ. Microbiol. *69*, 2444–2447.

Ongeng, D., Haberbeck, L.U., Mauriello, G., Ryckeboer, J., Springael, D., and Geeraerd, A.H. (2014). Modeling the fate of *Escherichia coli* O157:H7 and *Salmonella enterica* in the agricultural environment: current perspective. J. Food Sci. *79*, R421–7. https://doi.org/10.1111/1750-3841.12392

Palmer, M.E., Chaturongakul, S., Wiedmann, M., and Boor, K.J. (2011). The *Listeria monocytogenes* sigmaB regulon and its virulence-associated functions are inhibited by a small molecule. MBio. *2*, 6. https://doi.org/10.1128/mBio.00241-11

Paton, A.W., Ratcliff, R.M., Doyle, R.M., Seymour-Murray, J., Davos, D., Lanser, J.A., and Paton, J.C. (1996). Molecular microbiological investigation of an outbreak of hemolytic-uremic syndrome caused by dry fermented sausage contaminated with shiga-like toxin-producing *Escherichia coli*. J. Clin. Microbiol. *34*, 1622-1627.

Porto-Fett, A.C., Call, J.E., Shoyer, B.E., Hill, D.E., Pshebniski, C., Cocoma, G.J., and Luchansky, J.B. (2010). Evaluation of fermentation, drying, and/or high pressure processing on viability of *Listeria monocytogenes*, *Escherichia coli* O157:H7, *Salmonella* spp., and *Trichinella spiralis* in raw pork and Genoa salami. Int. J. Food Microbiol. *140*, 61-75. https://doi.org/10.1016/j.ijfoodmicro.2010.02.008

Rangel, J.M., Sparling, P.H., Crowe, C., Griffin, P.M., and Swerdlow, D.L. (2005). Epidemiology of *Escherichia coli* O157:H7 outbreaks, United States, 1982-2002.

Emerging Infect. Dis. *11*, 603–609. https://doi.org/10.3201/eid1104.040739

Ravva, S.V., Cooley, M.B., Sarreal, C.Z., and Mandrell, R.E. (2014). Fitness of outbreak and environmental strains of *Escherichia coli* O157:H7 in aerosolizable soil and association of clonal variation in stress gene regulation. Pathogens *3*, 528–548. https://doi.org/10.3390/pathogens3030528

Reid, S.D., Herbelin, C.J., Bumbaugh, A.C., Selander, R.K., and Whittam, T.S. (2000). Parallel evolution of virulence in pathogenic *Escherichia coli*. Nature *406*, 64–67. https://doi.org/10.1038/35017546

Rhodius, V.A., Suh, W.C., Nonaka, G., West, J., and Gross, C.A. (2006). Conserved and variable functions of the sigmaE stress response in related genomes. PLOS Biol. *4*, e2. https://doi.org/10.1371/journal.pbio.0040002

Rigobelo, E.E., Karapetkov, N., Maestá, S.A., Avila, F.A., and McIntosh, D. (2015). Use of probiotics to reduce faecal shedding of Shiga toxin-producing *Escherichia coli* in sheep. Benef. Microbes *6*, 53–60. https://doi.org/10.3920/BM2013.0094

Riley, L.W., Remis, R.S., Helgerson, S.D., McGee, H.B., Wells, J.G., Davis, B.R., Hebert, R.J., Olcott, E.S., Johnson, L.M., Hargrett, N.T., *et al.* (1983). Hemorrhagic colitis associated with a rare *Escherichia coli* serotype. N. Engl. J. Med. *308*, 681–685. https://doi.org/10.1056/NEJM198303243081203

Robey, M., Benito, A., Hutson, R.H., Pascual, C., Park, S.F., and Mackey, B.M. (2001). Variation in resistance to high hydrostatic pressure and *rpoS* heterogeneity in natural isolates of *Escherichia coli* O157:H7. Appl. Environ. Microbiol. *67*, 4901–4907.

Russell, J.B., Diez-Gonzalez, F., and Jarvis, G.N. (2000). Potential effect of cattle diets on the transmission of pathogenic *Escherichia coli* to humans. Microbes Infect. *2*, 45–53.

Sammartano, L.J., Tuveson, R.W., and Davenport, R. (1986). Control of sensitivity to inactivation by H2O2 and broad-spectrum near-UV radiation by the *Escherichia coli katF* locus. J. Bacteriol. *168*, 13–21.

Scallan, E., Hoekstra, R.M., Angulo, F.J., Tauxe, R.V., Widdowson, M.A., Roy, S.L., Jones, J.L., and Griffin, P.M. (2011). Food-borne illness acquired in the United States – major pathogens. Emerging Infect. Dis. *17*, 7–15. https://doi.org/10.3201/eid1701.P11101

Schumann, W. (2016). Regulation of bacterial heat shock stimulons. Cell Stress Chaperones *21*, 959–968. https://doi.org/10.1007/s12192-016-0727-z

Semenov, A.V., Franz, E., van Overbeek, L., Termorshuizen, A.J., and van Bruggen, A.H. (2008). Estimating the stability of *Escherichia coli* O157:H7 survival in manure-amended soils with different management histories. Environ. Microbiol. *10*, 1450–1459. https://doi.org/10.1111/j.1462-2920.2007.01558.x

Sheen, S., Cassidy, J., Scullen, B., and Sommers, C. (2015). Inactivation of a diverse set of shiga toxin-producing *Escherichia coli* in ground beef by high pressure processing. Food Microbiol. *52*, 84–87. https://doi.org/10.1016/j.fm.2015.07.001

Shere, J.A., Bartlett, K.J., and Kaspar, C.W. (1998). Longitudinal study of *Escherichia coli* O157:H7 dissemination on four dairy farms in Wisconsin. Appl. Environ. Microbiol. *64*, 1390–1399.

Shere, J.A., Kaspar, C.W., Bartlett, K.J., Linden, S.E., Norell, B., Francey, S., and Schaefer, D.M. (2002). Shedding of *Escherichia coli* O157:H7 in dairy cattle housed in a confined environment following waterborne inoculation. Appl. Environ. Microbiol. *68*, 1947–1954.

Simko, I., Zhou, Y., and Brandl, M.T. (2015). Downy mildew disease promotes the colonization of romaine lettuce by *Escherichia coli* O157:H7 and *Salmonella enterica*. BMC Microbiol. *15*, 19. https://doi.org/10.1186/s12866-015-0360-5

Singh, R., and Jiang, X. (2015). Expression of stress and virulence genes in *Escherichia coli* O157:H7 heat shocked in fresh dairy compost. J. Food Prot. *78*, 31–41. https://doi.org/10.4315/0362-028X.JFP-13-529

Slanec, T., and Schmidt, H. (2011). Specific expression of adherence-related genes in *Escherichia coli* O157:H7 strain EDL933 after heat treatment in ground beef. J. Food Prot. *74*, 1434–1440. https://doi.org/10.4315/0362-028X.JFP-11-018

Slayton, R.B., Turabelidze, G., Bennett, S.D., Schwensohn, C.A., Yaffee, A.Q., Khan, F., Butler, C., Trees, E., Ayers, T.L., Davis, M.L., *et al.* (2013). Outbreak of shiga toxin-producing *Escherichia coli* (STEC) O157:H7 associated with romaine lettuce consumption, 2011. PLOS ONE. *8*, e55300. https://doi.org/10.1371/journal.pone.0055300

Small, P., Blankenhorn, D., Welty, D., Zinser, E., and Slonczewski, J.L. (1994). Acid and base resistance in *Escherichia coli* and *Shigella flexneri*: role of *rpoS* and growth pH. J. Bacteriol. *176*, 1729–1737.

Smith, J.L., Fratamico, P.M., and Gunther, N.W. (2014). Shiga toxin-producing *Escherichia coli*. Adv. Appl. Microbiol. *86*, 145–197. https://doi.org/10.1016/B978-0-12-800262-9.00003-2

Stanford, K., Bach, S., Baah, J., and McAllister, T. (2014). A mixture of *Lactobacillus casei*, *Lactobacillus lactis*, and *Paenibacillus polymyxa* reduces *Escherichia coli* O157:H7 in finishing feedlot cattle. J. Food Prot. *77*, 738-744. https://doi.org/10.4315/0362-028X.JFP-13-433

Strachan, N.J., Fenlon, D.R., and Ogden, I.D. (2001). Modelling the vector pathway and infection of humans in an environmental outbreak of *Escherichia coli* O157. FEMS Microbiol. Lett. *203*, 69–73.

Taormina, P.J., Beuchat, L.R., Erickson, M.C., Ma, L., Zhang, G., and Doyle, M.P. (2009). Transfer of *Escherichia coli* O157:H7 to iceberg lettuce via simulated field coring. J. Food Prot. 72, 465–472.

Tenaillon, O., Skurnik, D., Picard, B., and Denamur, E. (2010). The population genetics of commensal *Escherichia coli*. Nat. Rev. Microbiol. *8*, 207–217. https://doi.org/10.1038/nrmicro2298

Tozzoli, R., and Scheutz, F. (2014). Diarrhoeagenic *Escherichia coli* infections in humans. In Pathogenic *E. coli*: Molecular and Cellular Microbiology. S. Morabito, ed. (Caister Academic Press, Norfolk, UK), pp. 1–18.

Tuttle, J., Gomez, T., Doyle, M.P., Wells, J.G., Zhao, T., Tauxe, R.V., and Griffin, P.M. (1999). Lessons from a large outbreak of *Escherichia coli* O157:H7 infections: insights into the infectious dose and method of widespread contamination of hamburger patties. Epidemiol. Infect. *122*, 185-192.

Uhlich, G.A., Keen, J.E., and Elder, R.O. (2001). Mutations in the csgD promoter associated with variations in curli expression in certain strains of *Escherichia coli* O157:H7. Appl. Environ Microbiol. *67*, 2367–2370. https://doi.org/10.1128/AEM.67.5.2367-2370.2001

Van der Linden, I., Cottyn, B., Uyttendaele, M., Vlaemynck, G., Heyndrickx, M., Maes, M., and Holden, N. (2016). Microarray-based screening of differentially expressed genes of *E. coli* O157:H7 Sakai during preharvest survival on butterhead lettuce. Agriculture-Basel. *6*, 1. https://doi.org/ARTN 610.3390/agriculture6010006

van Hoek, A.H., Aarts, H.J., Bouw, E., van Overbeek, W.M., and Franz, E. (2013). The role of *rpoS* in *Escherichia coli* O157 manure-amended soil survival and distribution of allelic variations among bovine, food and clinical isolates. FEMS Microbiol. Lett. 338, 18–23. https://doi.org/10.1111/1574-6968.12024

Vanaja, S.K., Springman, A.C., Besser, T.E., Whittam, T.S., and Manning, S.D. (2010). Differential expression of virulence and stress fitness genes between *Escherichia coli* O157:H7 strains with clinical or bovine-biased genotypes. Appl. Environ. Microbiol. *76*, 60-68. https://doi.org/10.1128/AEM.01666-09

Veses-Garcia, M., Liu, X., Rigden, D.J., Kenny, J.G., McCarthy, A.J., and Allison, H.E. (2015). Transcriptomic analysis of Shiga-toxigenic bacteriophage carriage reveals a profound regulatory effect on acid resistance in *Escherichia coli*. Appl. Environ. Microbiol. *81*, 8118–8125. https://doi.org/10.1128/AEM.02034-15

Vidovic, S., Mangalappalli-Illathu, A.K., and Korber, D.R. (2011) Prolonged cold stress response of *Escherichia coli* O157 and the role of *rpoS*. Int. J. Food Microbiol. *146*, 163–169. https://doi.org/10.1016/j.ijfoodmicro.2011.02.018

Visvalingam, J., Hernandez-Doria, J.D., and Holley, R.A. (2013). Examination of the genome-wide transcriptional response of *Escherichia coli* O157:H7 to cinnamaldehyde exposure. Appl. Environ. Microbiol. 79, 942–950. https://doi.org/10.1128/AEM.02767-12

Wachtel, M.R., Whitehand, L.C., and Mandrell, R.E. (2002). Association of *Escherichia coli* O157:H7 with preharvest leaf lettuce upon exposure to contaminated irrigation water. J. Food Prot. *65*, 18–25.

Wang, S., Deng, K., Zaremba, S., Deng, X., Lin, C., Wang, Q., Tortorello, M.L., and Zhang, W. (2009). Transcriptomic response of *Escherichia coli* O157:H7 to oxidative stress. Appl. Environ. Microbiol. *75*, 6110–6123. https://doi.org/10.1128/AEM.00914-09

Wang, X., Kim, Y., Hong, S.H., Ma, Q., Brown, B.L., Pu, M., Tarone, A.M., Benedik, M.J., Peti, W., Page, R., *et al.* (2011). Antitoxin MqsA helps mediate the bacterial general stress response. Nat. Chem. Biol. *7*, 359–366. https://doi.org/10.1038/nchembio.560

Wattam, A.R., Abraham, D., Dalay, O., Disz, T.L., Driscoll, T., Gabbard, J.L., Gillespie, J.J., Gough, R., Hix, D., Kenyon, R., *et al.* (2014). PATRIC, the bacterial bioinformatics database and analysis resource. Nucleic Acids Res. *42*, D581-591. https://doi.org/10.1093/nar/gkt1099

Weber, H., Polen, T., Heuveling, J., Wendisch, V.F., and Hengge, R. (2005). Genome-Wide analysis of the general stress response network in *Escherichia coli*: {sigma}s-dependent genes, promoters, and sigma factor selectivity. J. Bacteriol. *187*, 1591–1603.

White-Ziegler, C.A., Um, S., Perez, N.M., Berns, A.L., Malhowski, A.J., and Young, S. (2008). Low temperature (23 degrees C) increases expression of biofilm-, cold-shock- and RpoS-dependent genes in *Escherichia coli* K-12. Microbiology. *154*, 148–166. https://doi.org/10.1099/mic.0.2007/012021-0

Wick, L.M., and Egli, T. (2004). Molecular components of physiological stress responses in *Escherichia coli*. Adv. Biochem. Eng. Biotechnol. *89*, 1–45.

Williams, T.R., and Marco, M.L. (2014). Phyllosphere microbiota composition and microbial community transplantation on lettuce plants grown indoors. mBio. *5*, e01564-14. doi:10.1128/mBio.01564-14.

Wood, T.K., Knabel, S.J., and Kwan, B.W. (2013). Bacterial persister cell formation and dormancy. Appl. Environ. Microbiol. *79*, 7116–7121. https://doi.org/10.1128/AEM.02636-13

Yuk, H.G., Jo, S.C., Seo, H.K., Park, S.M., and Lee, S.C. (2008). Effect of storage in juice with or without pulp and/or calcium lactate on the subsequent survival of *Escherichia coli* O157:H7 in simulated gastric fluid. Int. J. Food Microbiol. *123*, 198–203. https://doi.org/10.1016/j.ijfoodmicro.2008.01.013

Zeng, W., Vorst, K., Brown, W., Marks, B.P., Jeong, S., Perez-Rodriguez, F., and Ryser, E.T. (2014). Growth of *Escherichia coli* O157:H7 and *Listeria monocytogenes* in packaged fresh-cut romaine mix at fluctuating temperatures during commercial transport, retail storage, and display. J. Food Prot. 77, 197–206. https://doi.org/10.4315/0362-028X.JFP-13-117

Zhang, X.S., Garcia-Contreras, R., and Wood, T.K. (2007). YcfR (BhsA) influences *Escherichia coli* biofilm formation through stress response and surface hydrophobicity. J. Bacteriol. *189*, 3051–3062. https://doi.org/10.1128/JB.01832-06

Application of Omics Technologies for *in Silico* Modelling to Understand Stress-triggered Physiology of *Escherichia coli* and to Develop Novel Therapeutics

12

Zuyi Huang*[1], Qian Jia[2] and Thomas K. Wood[3,4]

[1]Department of Chemical Engineering, Villanova University, PA, USA.
[2]Department of Health and Exercise Science, Rowan University, Glassboro, NJ, USA.
[3]Department of Chemical Engineering, Pennsylvania State University, University Park, PA, USA.
[4]Department of Biochemistry and Molecular Biology, Pennsylvania State University, University Park, PA, USA.

*Correspondence: zuyi.huang@villanova.edu

https://doi.org/10.21775/9781910190777.12

Abstract

Food-borne pathogens survive under oxidative stress conditions by changing their physiology and forming persister cells; persister cells are dormant cells that can awaken to cause chronic infections. While extensive experimental research has been conducted to investigate the genes and metabolic pathways associated with persister cell formation, no systems-level modelling approach has been developed to integrate these genes and other intracellular components to accelerate the identification of novel therapeutic targets for eliminating persister pathogens. In this work, we present the first genome-scale metabolic modelling approach to evaluate the ability of *Escherichia coli* to form persister cells under oxidative conditions and then identify gene targets that may be implemented to suppress persister cell formation. In particular, genomics data was integrated with a genome-scale metabolic model of *E. coli* to identify the reactions associated with persister cell formation. The change of activity levels of these reactions was then compared with the change of expression levels of genes obtained in persister cells to quantify the ability of the pathogen to form persister cells. Since *rpoS*-controlled genes and reactions have been reported for their important role in regulating persister cell formation, these genes were knocked out in the developed *in-silico* platform to evaluate the capability of these mutants to form persister cells. The results showed that the fluxes of *rpoS*-controlled reactions increase under oxidative stress and that the *rpoS*-deletion mutant showed an increased persister cell formation. In addition, 52 single mutants were found to have reduced ability to form persister cells, and these genes are mainly involved in cell membrane ion transport and in energy and carbon and lipid metabolism. They can be used as therapeutic targets for eliminating persister pathogens. Finally, the challenges and limitations of genome-scale modelling are also discussed.

Introduction

E. coli O157:H7 is one of the most common food-borne pathogens causing human disease in the USA (Ferens and Hovde, 2011). Specifically, it is reported that *E. coli* O157:H7 causes approximately 1.15 infections per 100,000 people in the USA (Johnson *et al.*, 2014). It is thus important to develop effective approaches to eliminate *E. coli* in the food process chain. *E. coli* survives in hostile environments mainly by forming biofilms. In particular, most pathogens become dormant around the bottom of biofilms. The dormant cells, so-called persisters, halt their metabolic activity and thus escape from the treatment of antibiotics that typically target active pathogens (Lewis, 2012). It is the persister cells that can wake up and cause recalcitrant chronic infections. Oxidative stress, which is one of the major mechanisms implemented by antimicrobial compounds to eliminate pathogens may not always have the ability to inactivate them. Instead, a low percentage of pathogens stimulated by oxidative stress may induce Lon-mediated degradation of antitoxins (Maisonneuve *et al.*, 2011; Wang *et al.*, 2011). Since the degradation of antitoxins releases toxins that degrade most mRNA transcripts (Wang and Wood, 2011), including those from stress responsive genes, the degradation of antitoxins makes the cell lack protein production and then become dormant. Since pathogens may stay asleep or awake, it is crucial for us to identify genes or metabolic pathways for effective manipulation of persister cell formation. Measuring the change of gene expression levels in persister cells is one of the most extensively used approaches to investigate the highly interactive regulation pathways leading to the dormant cell formation. For instance, transcriptomics or genomics data have been obtained for *E. coli* under stress conditions like oxidative conditions (Wang *et al.*, 2011) and bile acid stress (Kwan *et al.*, 2015). These data show that persister cell formation may be regulated by toxin/antitoxin (TA) systems that are ubiquitous within bacterial genomes (Kim and Wood, 2010; Lewis, 2008; Maisonneuve *et al.*, 2011; Wang and Wood, 2011). TA systems typically consist of a stable protein toxin paired with a labile RNA or protein antitoxin. The presence of a sufficient amount of antitoxin generally prevents the activity of the toxins that can disrupt essential cellular processes (e.g. translation via specific mRNA degradation) (Schuster and Bertram, 2013). TA systems regulate persister cell formation via the RpoS-based pathway (Hong *et al.*, 2012; Kwan *et al.*, 2015; Wang *et al.*, 2011). In particular, whole-transcriptome analyses of persister cells generated by two different methods (wild-type cells pretreated with hydrogen peroxide and *rpoS* deletion) indicate the importance of suppressing *rpoS* in persister cell formation. In other words, the *rpoS*-deletion *E. coli* showed significant increase in the number of persister cells when the pathogens were exposed to oxidative stress (Hong *et al.*, 2012). The decrease in the ability of the cell to respond to antibiotic stress through its RpoS-based regulation may be a potential reason for this phenomenon.

While extensive omics data have been published for persister cells, little work has been done to integrate these data and evaluate them on the entire metabolic network. Although these genes and reactions are distributed among various metabolic pathways, considering all of them together can provide a better version of how intracellular metabolism changes for the formation of persister cells. This motivated us to integrate all these molecular components together to perform a systems-level investigation of *E. coli*'s physiological change under stress conditions. Genome-scale models are generally the platform used for integrating genomics data with metabolic networks, as these models explicitly contain the metabolic reactions along with the corresponding genes. On the basis of the microorganism genomes, genome-scale metabolic models have been developed for various microorganisms, such as *E. coli* (Orth *et al.*, 2011), *Staphylococcus aureus* (Becker and Palsson, 2005), *Pseudomonas aeruginosa* (Oberhardt *et al.*, 2008), and *Salmonella* Typhimurium (Thiele *et al.*, 2011). These models have been widely implemented to study the biomass growth rates under different nutrient conditions (Monk *et al.*, 2013), to predict the phenotype of gene-knockout mutants (Price *et al.*, 2004), and to study the adaption of pathogens treated by mRNA-targeting antibiotics (Xu *et al.*, 2016) and by olive-extracted oleuropein (Li *et al.*, 2016). While most of this modeling work focuses on planktonic pathogens that can move freely, our group presented the first modelling platform to quantify the biofilm formation of *P. aeruginosa* single mutants (Xu *et al.*, 2013) under various nutrient conditions (Xu *et al.*, 2015). Here, we further extended our platform to quantify the trend of *E. coli*'s physiological change under stress conditions, with a special focus on identifying gene targets that can be inhibited to reduce persister formation. We mainly focused on oxidative stress in this work, as it is one of the major stresses imposed by most traditional antibiotics. It was found that bile acid induced elevated oxidative stress in the absence of farnesoid X receptor (Nomoto *et al.*, 2009) and that bile acid induced apoptosis as a result of oxidative stress with increased ROS generation (Barrasa *et al.*, 2011). The genes that were regulated by oxidative stress were identified from our experimental data for *E. coli* treated by H_2O_2 (Wang *et al.*, 2011) and bile acids (Kwan *et al.*, 2015). The expression levels of these genes were then used to constrain the fluxes of the corresponding metabolic reactions to represent influence from the oxidative stress on *E. coli* metabolism. Genomics data obtained in *E. coli* persister cells (Hong *et al.*, 2012) were implemented to identify the persister-associated reactions. The change of fluxes of these persister-associated reactions was then compared to the change of expression levels of persister-associated genes to indicate the trend of *E. coli* mutants to form persister cells.

Background Information

Toxin–antitoxin (TA) and persister cell formation

TA systems are diverse and abundant in bacterial chromosomes, and they allow the cell to respond to a specific stress or group of stresses in a highly regulated, elegant fashion (Wang and Wood, 2011). For example, *E. coli* has at least 37 TA systems (Tan *et al.*, 2011). Among these, *E. coli* TA systems, the MqsR/MqsA TA pair was the first to be shown to have a definitive role in persistence; deletion of the *mqsRA* locus or deletion of *mqsR* alone decreases persister cell formation and production of MqsR increases the number of persisters (Kim and Wood, 2010). The details of this and other TA systems have been described by Van Melderen and Saavedra De Bast (2009). Most of these TA systems are heavily involved in the stress response via the RpoS-controlled pathways that encode the master regulator of stress. For instance, *rpoS* can be repressed by antitoxin MqsA, which in turn reduces the concentration of c-di-GMP, leading to biofilm formation (Wang *et al.*, 2011). Under oxidative stress, antitoxin MqsA is degraded by Lon protease, leading to the up-regulation of *rpoS* and c-di-GMP and the formation of biofilms to survive under the stress condition (Fig. 12.1) (Wang *et al.*, 2011). The general stress response master regulator RpoS is thus important for TA systems and microbial stress response.

Persisters comprise a subpopulation of bacteria in the biofilm that are dormant and highly tolerant

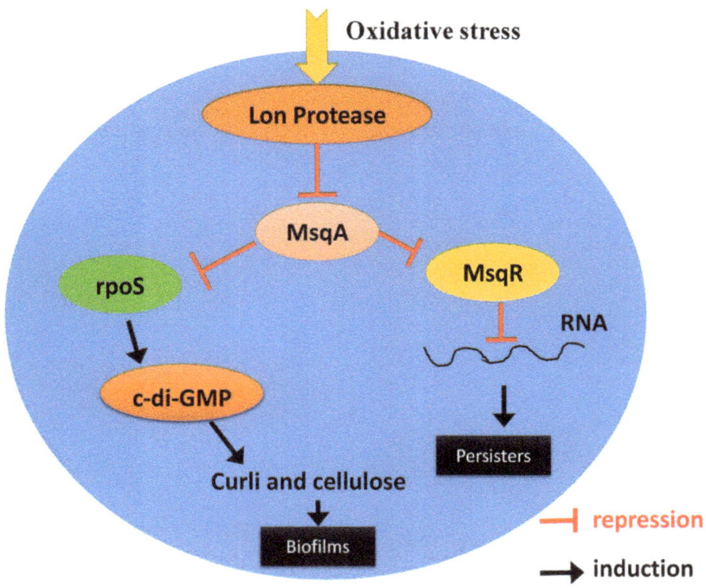

Figure 12.1 The pathway triggered by the oxidative stress (Wang *et al.*, 2011). Oxidative stress activates Lon protease, which, in turn, degrades MqsA. Since MsqA represses both *rpoS* and *msqR*, the degradation of MsqA can up-regulate both rpoS and toxin MsqR. The rpoS then induces c-di-GMP that leads to the formation of curli and cellulose, two important components of biofilms. As the MsqR concentration increases, the toxicity of MsqR represses mRNA and induces persister formation.

to antibiotics (Lewis, 2010). As mentioned above, external stress up-regulates *rpoS* in most pathogens, which, in turn, increases biofilms and keeps the cells relatively active in terms of metabolism. On the other hand, a small portion of pathogens may induce Lon-mediated degradation of antitoxins so that excessive toxins degrade most mRNA transcripts (including stress responsive genes). This results in the formation of persister cells (Fig. 12.1). Persisters are thought to be less sensitive to antibiotics that mainly attack pathogens undergoing cellular activities. RpoS-controlled genes were found to play an important role in determining the final growth mode of *E. coli*. In particular, our previous work shows that *rpoS*-deletion mutants have a much stronger capability to form persister cells than the wild-type cells (Hong *et al.*, 2012). In this work, we further extend our research and develop a systematic and rational investigation of the genes or pathways that can be manipulated to reduce the persister population.

Genome-scale models and flux balance analysis

Genome-scale models are typically derived from the genome sequence of microorganisms via the following procedure, shown in Fig. 12.2. In Step 1, the genome sequence of the target microorganism typically can be found in genome sequence databases. For example, the NCBI genome database (https://www.ncbi.nlm.nih.gov/genome/browse/) provides open-access sequences for eukaryotes, prokaryotes, viruses, plasmids, and organelles. If the target microorganism cannot be found in the existing sequence databases, its genome sequence can be measured by high-throughput genome sequencing techniques (Loman *et al.*, 2012). In Step 2, the genome sequence is annotated so that individual genes and the enzymes encoded by these genes are identified. Extensive research has been conducted for the gene/enzyme annotation so that the biological functions of genes/enzymes are determined. The Kyoto Encyclopedia of Genes and Genomes (KEGG) is the most widely used database that contains the detailed information of genes, enzymes, metabolic reactions, and pathways for sequenced microorganisms (Kanehisa and Goto, 2000). The enzyme encoded by the gene, along with its biological function, can be searched in KEGG. The existing tools for genome annotation include Artemis (Rutherford *et al.*, 2000), Blast2GO (Conesa *et al.*, 2005), Prokka (Seemann, 2014), DAVID (Dennis *et al.*, 2003), and IMG ER (Markowitz *et al.*, 2009). Other methods have also been described (Stein, 2001). In Step 3, metabolic reactions catalysed by the enzymes annotated in Step 2 are further identified and connected in order to get the entire metabolic network for the target microorganism. The KEGG database typically is the first resource to search for the metabolic reaction for a target enzyme. However, KEGG itself cannot put all the reactions together in a model automatically. Moreover, the reactions obtained from KEGG may

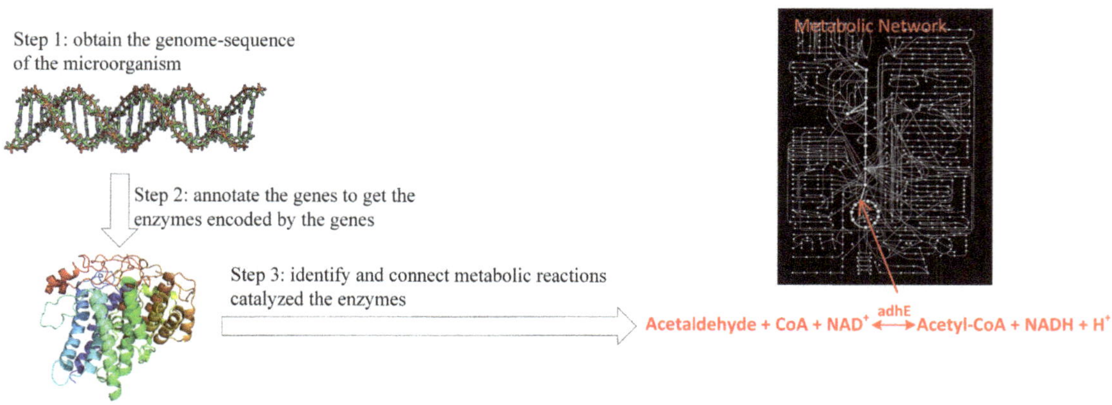

Figure 12.2 The general procedure to derive genome-scale models from microorganism sequence. The reaction of acetyl-CoA synthesis is used as an example of the metabolic reaction contained in the genome-scale model. The enzyme (i.e. AdhE) that catalyses the reaction is explicitly listed in the model. A genome-scale metabolic model typically consists of hundreds of highly interacted reactions.

not be complete. To avoid the significant manual intervention, the open platform KBase (https://kbase.us/) can be used instead, as it can export the genome-scale model in the Systems Biology Markup Language (SBML) format (Hucka *et al.*, 2003). In addition to KEGG and KBase, there are other tools that can be used to develop genome-scale models from genome sequence as the input. Schellenberger *et al.* (2010) gives a detailed review on these existing tools.

The value of genome-scale models lies in their ability to make predictions of phenotypes from genotypes. In other words, genome-scale models should be able to predict phenotypes (e.g. the biomass growth rate, production rates of target metabolites, biofilm formation, and persister cell formation) upon the given medium nutrient information and expression levels of metabolic genes (i.e. genomics data). Flux balance analysis (FBA) is one of the most commonly used approaches to simulate the genome-scale model to make predictions of microbial phenotypes (Orth *et al.*, 2010). In FBA, the genome-scale model is represented by a stoichiometric matrix **S** shown in Equation 12.1. Since the metabolic reactions generally complete in the time scale of millisecond, it is assumed that the metabolic network reaches steady-state in FBA. Therefore, the sum of the fluxes of each metabolite in all reactions it is involved in, which is actually the accumulation rate of the metabolite, is set to zero in Equation 12.1

$$S \cdot V = 0 \tag{12.1}$$

where **V** is equal to $[v_1, v_2, v_3, \ldots, v_n]$, which is a vector containing the fluxes of individual metabolic reactions (i.e. v_i, $i = 1, 2, 3, \ldots, n$). **S** is the stoichiometric matrix with m rows and n columns. m and n are the number of metabolites and metabolic reactions, respectively. The number of metabolic reactions (n) is generally larger than the number of metabolites (m). Therefore, there is no unique solution in Equation 12.1. It has been proven by Dr Palsson's group at UCSD that microorganisms tend to grow as much as possible from the available nutrients in their growing environment (Edwards *et al.*, 2001). The biomass growth rate is thus typically set as the objective function to maximize. Since microbial biomass consists of macromolecular components like protein, RNA and lipid, there is no biomass growth reaction that can be developed from the microorganism's genome sequence. Therefore, the biomass growth reaction in the form of Equation 12.2 needs to be developed first before performing FBA.

$$c_1M_1 + c_2M_2 + \ldots + c_kM_k \rightarrow \text{biomass} \tag{12.2}$$

where c_i are the stoichiometric coefficients of the biomass precursors M_i, which include, but are not limited to, amino acids, DNA precursors (i.e. dATP, dCTP, dGTP, dTTP), RNA precursors (i.e. ATP, CTP, GTP, UTP), lipid precursors (e.g. phosphatidylglycerol, cardiolipin, lipoteichoic acid), cell wall components (i.e. crosslinked peptidoglycan and teichoic acid), and pool solutes (e.g. K^+, Na^+, Mg^{2+}, Orthophosphate, Acetyl-CoA, Succinyl-CoA, CoA, FAD, NAD^+, NADH, $NADP^+$, and NADPH) (Heinemann *et al.*, 2005). Coefficients c_i depend on the biomass composition of microorganisms that can be measured via experiment. When no data are available for the biomass composition of the target microorganism, the values of c_i for other strains can be borrowed, assuming similarity of biomass composition between the target microorganism and other strains in the same species. These pre-assumed c_i can be further refined from the biomass growth data. In addition to the biomass objective reaction, the exchange reactions of extracellular metabolites should also be included before performing FBA. It is these exchange reactions that provide the nutrient resource for the microorganism's growth and offer the avenue to export the waste metabolites. Equation 12.3 shows an example of an exchange reaction in which the external metabolite Ex-M_i is either provided by or secreted to the surrounding environment. If Ex-M_i is up-taken by the microorganisms from the environment, the flux flows from the right (i.e. the environment) to the left (i.e. the external metabolite that will be transferred into the cell in the reaction shown in Equation 12.4). In this case, the flux is negative if the microorganism up-take Ex-M_i from the environment. On the other hand, if the metabolite is secreted from the cell to the environment, the flux flows from the left to the right in Equation 12.3, with a positive value. The flux of Equation 12.3 can be constrained to limit the uptake of the external metabolite, which can mimic the availability of extracellular nutrients in the surrounding environment. The flux of Equation 12.3

can also be used to determine the consumption or accumulation of the external metabolite Ex-M_i.

$$\text{Ex-}M_i \leftrightarrow \tag{12.3}$$

$$M_i \leftrightarrow \text{Ex-}M_i \tag{12.4}$$

The framework of FBA is outlined in Fig. 12.3. In addition to the exchange and transport reactions shown in Equations 12.3 and 12.4, there are hundreds of intracellular metabolic reactions that bridge the extracellular metabolites to the biomass precursors shown in Equation 12.2. The lower bounds of irreversible reactions are set to zeros so that the fluxes of these reactions can only flow from the left to the right. FBA is performed via the following steps: 1) the lower and upper bounds of exchange reactions for extracellular nutrient compounds are set on the basis of their availability; and 2) the maximal biomass growth rate is determined by running the linear optimization programming routine in the linear solution space constrained by the metabolic network stoichiometry (i.e. Equation 12.1) and the inequality constraints (i.e. the lower and upper bounds of metabolic reaction fluxes). The lower and upper bounds of nutrients for minimal and rich media for certain microorganisms have been published. For example, the bounds of exchange reaction rates for *Salmonella* Typhimurium, *E. coli*, and *S. aureus* can be found in Becker and Palsson (2005), Orth *et al.* (2011), and Thiele *et al.* (2011), respectively. The linear optimization programming can be solved by the MATLAB-based COBRA toolbox (Schellenberger *et al.*, 2011) developed by Dr Palsson's group at UCSD. It is the most popular free platform for performing FBA. In addition to determining the biomass growth rate, the FBA platform can be used to perform gene-knockout simulation. In order to knockout a single gene, the lower and upper bounds of the reactions associated with that gene are set to zero in FBA. FBA has been widely used to determine the essential genes, the knockout of which leads to zero biomass growth rate. Moreover, genome-scale models can be used as a platform to identify the genes/pathways for engineering microorganisms to produce the target chemicals. While COBRA toolbox provides routines for single gene knockout for maximizing the single objective (i.e. the biomass growth rate), the OptKnock program (Burgard *et al.*, 2003) can knockout multiple genes and optimize two objective functions, one for the biomass growth rate and the other for the target chemical produced by the microorganism. DFBAlab presented by (Gomez *et al.*, 2014) further extends FBA to optimize more than two objective functions and to address the issue that FBA may not provide consistent flux solutions in replicated simulations.

Methods used

The latest *E. coli* genome-scale model that consists of 1367 genes, 2583 metabolic reactions, and 1805 metabolites, was used in this work (Orth *et al.*,

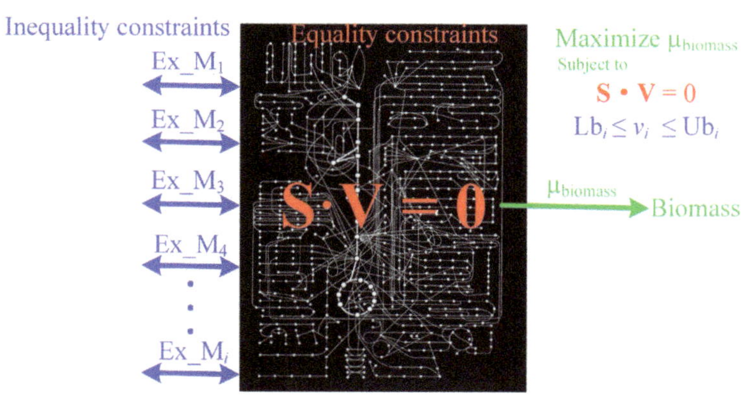

Figure 12.3 The framework of flux balance analysis. The lower and upper bounds of extracellular metabolites Ex_M_i are set according to the availability of nutrients in the surrounding environment. The biomass growth rate is set as the objective function to maximize in FBA, on the basis of the inequality and equality constraints from the nutrient up-take rates and the stoichiometric mass balance.

2011). We developed a systems-level platform to integrate this model with our existing transcriptomics and genomics data under stress conditions from both wild-type and persister *E. coli*, for the first time, to quantify the trend of persister formation under oxidative stress conditions and to identify potential therapeutic targets for reducing persister cell formation. The outline of the proposed platform is shown in Fig. 12.4. Transcriptomics and genomics data from Hong *et al.* (2012), Kwan *et al.* (2015), and Wang *et al.* (2011) were first used to identify the reactions that are associated with persister cell formation and stress response. The reactions representing the influence from oxidative stress on microbial metabolism were constrained by the expression levels of the genes associated with these reactions. The flux change in the persister-associated reactions was then quantified to predict the capability of *E. coli* to form persister cells under the stress conditions. In particular, the correlation of the flux change in those persister-associated reactions, with the change in the expression levels of corresponding genes, was used as the indicator of persister cell formation. A high correlation means *E. coli* adjusts its intracellular metabolism in a similar way as shown in the genomics data obtained from persister cells (Hong *et al.*, 2012). Therefore, a high correlation means that *E. coli* has a high chance of forming persister cells. Using the flux change of persister-associated reactions as the indicator of persister cell formation, we further studied the capability of gene-knockout mutants of *E. coli* to identify potential gene targets to regulate persister cell formation. The details of the steps shown in Fig. 12.4 are given in the following subsections.

Integrate gene expression data with metabolic models to constrain stress response in *E. coli*

The transcriptomics data for *E. coli* treated by H_2O_2 (i.e. oxidative stress) (Wang *et al.*, 2011) and the genomics data of *E. coli* during bile acid stress (Kwan *et al.*, 2015) were overlaid to the genome-scale metabolic network so that the reactions representing the influence from the stress on *E. coli* metabolism, called stress-associated reactions in the following text, were identified. Similarly, our genomics data obtained from *E. coli* in persister cells (Hong *et al.*,

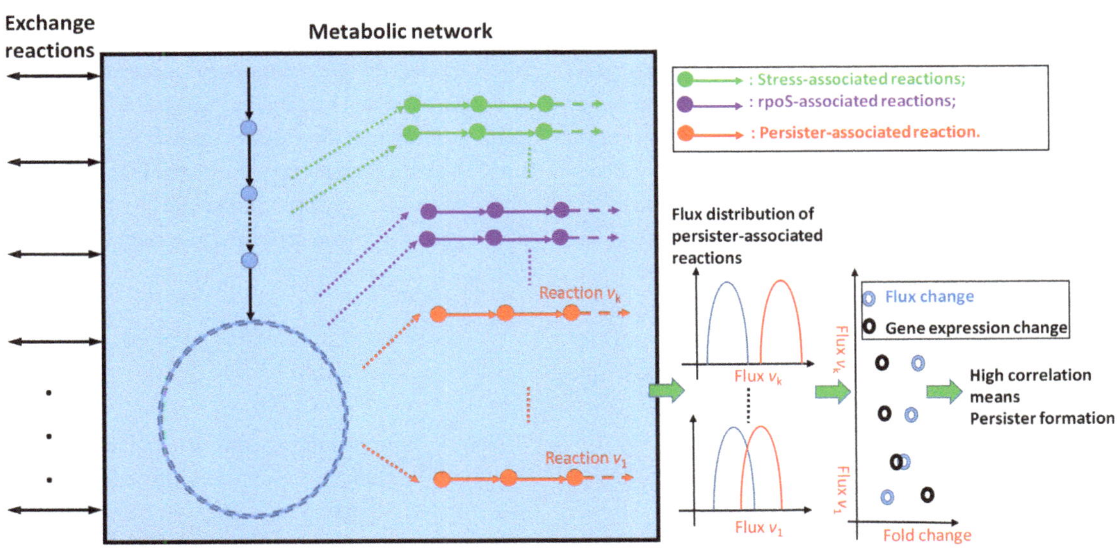

Figure 12.4 Outline of the developed framework to integrate omics data with a genome scale model to quantify the persister formation of *E. coli* under oxidative stress. The transcriptomics and genomics data from references (Hong *et al.*, 2012; Kwan *et al.*, 2015; Wang *et al.*, 2011) were first overlaid with an *E. coli* genome-scale model to identify the stress-associated, RpoS-associated, and persister-associated reactions. The stress-associated reactions were then constrained and the change of fluxes of persister-associated reactions is quantified. The obtained flux change was compared with the change of gene expression levels in persister cells. If the flux change was not correlated with the gene expression change (as shown in this figure), this indicated low persister formation, and vice versa. The RpoS-associated reactions were then blocked to study how *rpoS*-deletion influences persister formation.

2012) were overlaid onto the metabolic network to identify the persister-associated reactions and the RpoS-associated reactions. Extensive research has been conducted to constrain metabolic fluxes using genomics data. Specifically, an up-regulated gene indicates increased fluxes in the metabolic reactions that are catalysed by the enzyme encoded by that gene, and vice versa. A detailed comparison of the existing approaches for integrating gene expression data with metabolic models has been reported (Machado and Herrgård, 2014). The MATLAB codes were also developed for these approaches in this reference. Since the number of metabolic genes that have expression data is relatively low in this work, a simpler approach shown in Equations 12.5 to 12.10 was used in this work to constrain the metabolic fluxes according to the gene expression. Specifically, if the ratio of gene expression levels between the changed condition and the control condition, i.e. r_i, is larger than one, the lower bound of the associated reaction Lb_i is constrained by Equation 12.5 if the flux for the control condition (i.e. $v_{c,i}$) is positive. The rationale behind this is that a higher regulated-gene (i.e. $r_i > 1$) indicates a higher activity (i.e. flux) in the same direction as that for the control direction. Therefore, the upper bound Ub_i is constrained by Equation 12.6 if the control flux $v_{c,i}$ is negative. On the other hand, if the gene is down-regulated, Ub_i is constrained by Equation 12.7 if $v_{c,i}$ is positive so that the reaction activity is limited. If the control flux $v_{c,i}$ is negative, the lowest flux value Lb_i is limited by Equation 12.8. If r_i is negative, the reaction flux direction is reversed by Equations 12.9 and 12.10, so that Ub_i for positive $v_{c,i}$ is negative and Lb_i for negative $v_{c,i}$ is positive. Other than the rules shown in Equations 12.5 to 12.10, Lb_i is set to zero for irreversible reactions, and Ub_i is always larger or equal to Lb_i.

$$Lb_i \geq r_i \times v_{c,i}, \text{ if } r_i > 1 \text{ and } v_{c,i} > 0 \quad (12.5)$$

$$Ub_i \leq r_i \times v_{c,i}, \text{ if } r_i > 1 \text{ and } v_{c,i} < 0 \quad (12.6)$$

$$Ub_i \leq r_i \times v_{c,i}, \text{ if } 0 < r_i < 1 \text{ and } v_{c,i} > 0 \quad (12.7)$$

$$Lb_i \geq r_i \times v_{c,i}, \text{ if } 0 < r_i < 1 \text{ and } v_{c,i} < 0 \quad (12.8)$$

$$Ub_i \leq r_i \times v_{c,i}, \text{ if } r_i < 0, \text{ and } v_{c,i} > 0 \quad (12.9)$$

$$Lb_i \geq r_i \times v_{c,i}, \text{ if } r_i < 0, \text{ and } v_{c,i} < 0 \quad (12.10)$$

Investigate the change in flux distribution to identify potential drug targets for reducing persister cell formation

Equations 12.5 to 12.10 were used to impose constraints from the gene expression data on metabolic fluxes which can be then determined from the flux balance analysis shown in Fig. 12.3. Since the metabolic network is generally redundant, i.e. parallel reactions existing to connect two metabolites in the metabolic network, there are multiple flux solutions that can return the same biomass growth rate. This issue is rooted in the singular stoichiometric matrix in which the number of columns (i.e. the reaction number) is larger than the number of rows (i.e. the metabolite number). A common approach to address this issue is implementing sampling approaches to obtain the distribution of fluxes of individual metabolic reactions and then use the statistical parameters of the distribution (e.g. mean, standard deviation, and the histogram profile) to estimate the most likely value of individual metabolic fluxes. In this work, the artificial centring hit-and-run (ACHR) routine from COBRA toolbox was implemented to sample the potential solutions of metabolic fluxes in the *E. coli* model constrained by Equations 12.5 to 12.10. In particular, the gene expression data under stress condition (Kwan *et al.*, 2015; Wang *et al.*, 2011) was used to constrain the related metabolic reactions, and the fluxes of persister-associated reactions were sampled to calculate the change of the mean values of the fluxes in these reactions, according to Equation 12.11.

$$\Delta\bar{f}_i = \frac{\left(\bar{f}_{i,changed} - \bar{f}_{i,control}\right)}{\bar{f}_{i,control}} \quad (12.11)$$

where $\bar{f}_{i,changed}$ and $\bar{f}_{i,control}$ are the mean values of the fluxes of Reaction *i* for the changed condition (e.g. oxidative stress) and the control condition (e.g. no stress and gene knockout), respectively. $\Delta\bar{f}_i$ is the relative change of mean flux values of Reaction *i*. A similar approach shown in Equation 12.12 was used to calculate $\Delta\bar{g}_i$, the relative change of expression levels of the genes associated with Reaction *i*.

$$\Delta\bar{g}_i = \frac{\left(\bar{g}_{i,changed} - \bar{g}_{i,control}\right)}{\bar{g}_{i,control}} \quad (12.12)$$

where $\bar{g}_{i,changed}$ and $\bar{g}_{i,control}$ are the mean values of expression levels of genes associated with Reaction i for the changed condition and the control condition, respectively. Since those differentially expressed genes in persister cells imply the trend of persister formation, a similar change in fluxes of the related reactions should be seen in persister cells. Therefore, the correlation factor *Cor* between the gene expression and flux changes is quantified by the following equation:

$$Cor = \frac{\mathbf{\Delta F}^{T} \times \mathbf{\Delta G}}{||\mathbf{\Delta F}||\,|\mathbf{\Delta G}||} \qquad (12.3)$$

where $\mathbf{\Delta F}$ is a vector containing the change of fluxes in individual reactions, i.e. $[\Delta\bar{f}_1, \Delta\bar{f}_2, \ldots, \Delta\bar{f}_k]$. Similarly, $\mathbf{\Delta G}$ is a vector of $[\Delta\bar{g}_1, \Delta\bar{g}_2, \ldots, \Delta\bar{g}_k]$. A high correlation *Cor* between the flux change and the gene expression change indicates the changed condition (represented by the model-predicted fluxes) has a high chance to result in the persister formation (represented by the gene expression data).

On the basis of the correlation factor defined by Equation 12.13, the persister formation of *E. coli* under oxidative stress was predicted for both control and *rpoS*-deletion conditions. The RpoS-associated reactions were also used to indicate the change in RpoS-controlled pathway that is triggered by the oxidative stress. 100,000 samples were obtained for each case scenario using the ACHR routine from COBRA toolbox to ensure the sampling convergence. In addition, *in-silico* single-gene knockout simulations were implemented to identify gene targets that can be used to reduce the persister cell formation. Specifically, the flux changes of persister-associated reactions were calculated by Equation 12.11 for each gene mutant. The flux change was then compared to the gene expression change using Equation 12.12. A large correlation *Cor* indicates that the mutant may become a persister cell. The essential genes, whose mutants return zero growth rates, and the genes associated with oxidative stress, RpoS-controlled reactions, and persister formation were excluded in this gene-mutant investigation. It generally takes approximately 20 minutes for ACHR routine to obtain 100,000 sample solutions for one gene mutant, and these solutions take approximately 100 MB space in the computer. On the other hand, there were approximately 1000 genes for the knockout simulation. In order to address these issues, the following simple approach was first used to screen gene targets for further ACHR sampling: the fluxes $f_{i,changed}$ obtained from FBA substituted $\bar{f}_{i,changed}$ in Equation 12.11. In this way, the fluxes from one solution, instead of the mean values of the sampled fluxes, were used for the screening. The 300 genes whose mutants returned the lowest correlation *Cor* with the gene expression change in persister cells in Equation 12.13 were further selected for ACHR sampling. The sampled solutions were then analysed by Equations 12.11 and 12.13 to identify the gene mutants that may not induce persister cell formation.

Results

The genes triggered by stress (Kwan *et al.*, 2015; Wang *et al.*, 2011) were overlaid onto the *E. coli* genome-scale model. Since some genes are involved in signaling pathways (e.g. membrane sensing pathway), not all genes are associated with metabolic reactions. The genes that were found in the *E. coli* metabolic model and their expression levels are shown in Table 12.1. The enzymes and biological functions of these ten genes are also listed. The fluxes of these genes would be constrained according to their express levels in the next step to impose the influence from the oxidative stress on *E. coli* metabolism. Similarly, the genes identified from persister cells (Hong *et al.*, 2012) were listed in Table 12.2. The fluxes of these persister-associated reactions would be used to indicate the trend of *E. coli* mutants to form persister cells. Moreover, the RpoS-controlled genes were identified in the metabolic network (Table 12.3). The fluxes of the reactions associated with these RpoS-associated genes would be used to indicate the strength of RpoS regulation. In addition, the persister formation capability of *rpoS*-deletion would be quantified later.

The fluxes of reactions associated with the stress genes listed in Table 12.1 were constrained via Equations 12.5 to 12.10 according to the expression levels of the genes. The fold change of RpoS-associated reactions was quantified via Equation 12.11 to investigate how oxidative stress regulates the RpoS-controlled pathway at the wild-type stationary phase. Since the RpoS-controlled genes shown in

Table 12.1 Genes and their expression change in *E. coli* due to oxidative or acid stresses

b number	Enzyme	Functions of enzymes	Fold change	Reference
b1732	KatE	Catalase HPII, Heme d-containing	6.0	Wang *et al.* (2011)
b0344	LacZ	Beta-D-galactosidase	3.0	Wang *et al.* (2011)
b2156	LysP	Lysine transporter	2.60	Kwan *et al.* (2015)
b4024	LysC	Lysine-sensitive aspartokinase 3	2.60	Kwan *et al.* (2015)
b0031	DapB	Dihydrodipicolinate reductase	2.00	Kwan *et al.* (2015)
b0613	CitG	Triphosphoribosyl-dephospho-CoA synthase	2.00	Kwan *et al.* (2015)
b1858	ZnuC	Zinc transport system ATP-binding protein	–2.00	Kwan *et al.* (2015)
b4077	GltP	Glutamate/aspartate:proton symporter	1.9	Kwan *et al.* (2015)
b0693	SpeF	Ornithine decarboxylase isozyme, inducible	1.9	Kwan *et al.* (2015)
b1857	ZnuA	Zinc transport system substrate-binding protein	–1.9	Kwan *et al.* (2015)

Table 12.2 Differentially expressed genes in persistent *E. coli* BW25113 with *rpoS* knocked out relative to BW25113 wild-type in the stationary phase. Only genes that have been included in the *E. coli* metabolic model are listed

b number	Enzyme	Functions of enzymes	Fold change	Reference
b1661	Cfa	Cyclopropane-fatty-acyl-phospholipid synthase	–13.96	Hong *et al.* (2012)
b1732	KatE	Catalase HPII, Heme d-containing	–17.98	Hong *et al.* (2012)
b1646	SodC	Superoxide dismutase, Cu, Zn, periplasmic	–13.93	Hong *et al.* (2012)
b2579	GrcA	Autonomous glycyl radical cofactor	13.47	Hong *et al.* (2012)
b1276	AcnA	Aconitate hydratase 1	–15.61	Hong *et al.* (2012)
b3012	DkgA	2,5-diketo-D-gluconate reductase A	–17.78	Hong *et al.* (2012)
b2661	GabD	Succinate-semialdehyde dehydrogenase I, NADP-dependent	–15.11	Hong *et al.* (2012)
b2663	GabP	Gamma-aminobutyrate transporter	–9.84	Hong *et al.* (2012)
b2662	GabT	4-aminobutyrate aminotransferase, PLP-dependent	–12.88	Hong *et al.* (2012)
b0871	PoxB	Pyruvate dehydrogenase	–11.27	Hong *et al.* (2012)
b1444	PatD	Gamma-aminobutyraldehyde dehydrogenase	–8.28	Hong *et al.* (2012)
b1440	YdcS	Polyhydroxybutyrate (PHB) synthase, ABC transporter periplasmic binding protein homolog	–8.94	Hong *et al.* (2012)
b1094	AcpP	Acyl carrier protein (ACP)	16.24	Hong *et al.* (2012)
b1927	AmyA	Cytoplasmic alpha-amylase	–9.39	Hong *et al.* (2012)
b0062	AraA	L-arabinose isomerase	10.5	Hong *et al.* (2012)
b4139	AspA	Aspartate ammonia-lyase	16.51	Hong *et al.* (2012)
b3496	DtpB	Dipeptide and tripeptide permease B	14.14	Hong *et al.* (2012)
b2779	Eno	Enolase	10.13	Hong *et al.* (2012)
b0583	EntD	Enterobactin synthetase component D	12.2	Hong *et al.* (2012)
b3429	GlgA	glycogen synthase	–7.18	Hong *et al.* (2012)
b0783	MoaC	Molybdopterin biosynthesis, protein C	15.96	Hong *et al.* (2012)
b2429	MurP	N-acetylmuramic acid permease, EIIBC component, PTS system	13.45	Hong *et al.* (2012)
b1897	OtsB	trehalose-6-phosphate phosphatase, biosynthetic	–10.65	Hong *et al.* (2012)
b1684	SufA	Fe–S cluster assembly protein	–9.56	Hong *et al.* (2012)
b3616	Tdh	L-threonine 3-dehydrogenase, NAD(P)-binding	10.16	Hong *et al.* (2012)
b2465	TktB	Transketolase 2, thiamine triphosphate-binding	–15.08	Hong *et al.* (2012)

Table 12.2 Continued

b number	Enzyme	Functions of enzymes	Fold change	Reference
b1197	TreA	Periplasmic trehalase	−7.61	Hong *et al.* (2012)
b1710	BtuE	Glutathione peroxidase	−7.26	Hong *et al.* (2012)
b2215	OmpC	Outer membrane porin protein C	42.93	Hong *et al.* (2012)
b3708	TnaA	Tryptophanase/L-cysteine desulfhydrase, PLP-dependent	−4.53	Hong *et al.* (2012)
b1478	AdhP	Alcohol dehydrogenase, propanol-preferring	−9.69	Hong *et al.* (2012)

Table 12.3 List of RpoS-controlled genes that have been included in the *E. coli* metabolic model

b number	Enzyme	Functions of enzymes	Reference
b1478	AdhP	Ethanol-active dehydrogenase/acetaldehyde-active reductase	Hong *et al.* (2012)
b1927	AmyA	Cytoplasmic alpha-amylase	Hong *et al.* (2012)
b3012	DkgA	2,5-diketo-D-gluconate reductase A	Hong *et al.* (2012)
b2661	GabD	Succinate-semialdehyde dehydrogenase I, NADP-dependent	Hong *et al.* (2012)
b2663	GabP	Gamma-aminobutyrate transporter	Hong *et al.* (2012)
b1732	KatE	Catalase HPII, heme d-containing	Hong *et al.* (2012)
b1897	OtsB	Trehalose-6-phosphate phosphatase, Biosynthetic	Hong *et al.* (2012)
b0871	PoxB	Pyruvate dehydrogenase (quinone)	Hong *et al.* (2012)
b1646	SodC	Superoxide dismutase, Cu, Zn, periplasmic	Hong *et al.* (2012)
b1684	SufA	Fe–S cluster assembly protein	Hong *et al.* (2012)
b2465	TktB	Transketolase 2, thiamine triphosphate-binding	Hong *et al.* (2012)
b1197	TreA	Alpha,alpha-trehalase	Hong *et al.* (2012)
b1440	YdcS	Polyhydroxybutyrate (PHB) synthase, ABC transporter periplasmic binding protein homolog	Hong *et al.* (2012)

Table 12.3 were up-regulated under oxidative stress conditions, the related metabolic fluxes should be up-regulated. Fig. 12.5 shows the fold change of the fluxes of the RpoS-controlled reactions. The red line represents no change in metabolic fluxes. While there are four reactions with reduced fluxes, the increased magnitudes are generally larger than the decreased ones, resulting in 2.76 as the summation of all these flux fold changes. This indicates that the fluxes of RpoS-controlled reactions were generally up-regulated under the oxidative stress condition, which is consistent with the up-regulation of RpoS-controlled genes under the same condition shown in our data (Hong *et al.*, 2012).

The RpoS-controlled genes were further inhibited to study the influence of RpoS-controlled reactions on the persiter cell formation. Fig. 12.6 shows the flux change in persister-associated reactions was more significant when RpoS-controlled genes were deleted in the simulation. In particular, the regions high-lighted by the five rectangles in Fig. 12.6 show that the flux change (red color) is similar to the change shown in gene express data (black color) for persister cells. It is thus not surprising that the correlation factor *Cor* of the *rpoS*-deletion mutant, i.e. 0.22, is higher than the one for the wild-type (blue color), i.e. −0.09. The small negative *Cor* for the wild-type means that the oxidative stress itself may not induce persister cell formation. On the other hand, the *rpoS*-deletion mutant has a higher chance to form persister cells. This is also consistent with the experimental data shown by Hong *et al.* (2012). Here are some potential reasons for the low correlation factor (0.22) of the *rpoS*-deletion mutant. First, the gene expression levels are not necessarily highly correlated with the enzyme levels, as the genomics data cannot reflect the post-translational change from the gene expression to the synthesis of enzymes. On the contrary, the metabolic fluxes are highly correlated with the

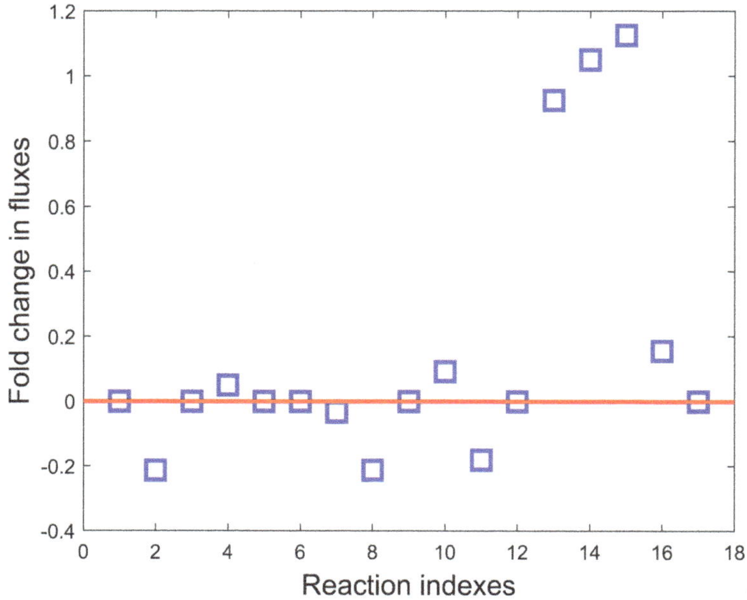

Figure 12.5 Fold changes in fluxes of the RpoS-controlled metabolic reactions in *E. coli* treated by 20 mM H_2O_2 at the wild-type stationary phase. The sum of all flux fold change is 2.76, indicating an overall flux increase in RpoS-controlled reactions.

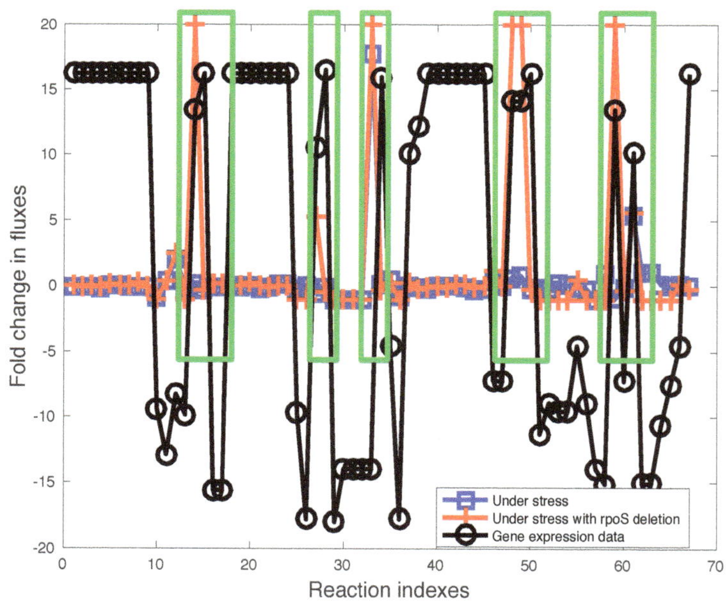

Figure 12.6 Comparison of the fold changes in gene expression levels (black) to the flux fold changes of persister-associated reactions under oxidative stress (blue), and oxidative stress and *rpoS* deletion (red). The correlation is calculated via Equation 12.13. The deletion of *rpoS* increases the correlation between fluxes and gene-expression levels in the persister cells from −0.09 to 0.22. The change in fluxes and gene expression levels have high correlation in the rectangle regions.

enzyme levels. In addition, there may be some noise in the genomics data that also contributes to the low correlation factor. Second, while the *E. coli* model used in this work is the most comprehensive genome-scale model for this microorganism, it only contains the metabolic genes and pathways. The signal transduction pathways that are implemented by the microorganisms to sense the oxidative stress, convey the stress information, and trigger stress response (e.g. Lon protease) are not included in this model. The only constraints imposed from the oxidative stress were implemented by limiting the fluxes of metabolic reactions associated with the ten oxidative stress genes shown in Table 12.1. With the high redundancy in the *E. coli* metabolic model, the constraints from these ten genes, among the total 1367 genes, may not be enough to accurately convey the stress information to the microorganism. Similarly, only 13 RpoS-controlled genes (Table 12.2) were found in the *E. coli* metabolic model. Knockout of these 13 genes may not be enough to impose the influence from the *rpoS*-deletion on the microorganism metabolism.

It is important to identify gene targets to manipulate the formation of persister cells for treating chronic human diseases. Figs. 12.5 and 12.6 show that the developed metabolic modeling platform can make a reasonable prediction of the oxidative stress response and the persister cell formation of *rpoS*-deletion mutants. We further implemented the developed modeling platform to identify genes whose mutants had limited capability to form persister cells. Fig. 12.7 shows the flux change of persister-associated reactions in a gene-knockout mutant. Compared to the large change in the expression levels of persister-associated genes, little change is observed in the fluxes of persister-associated reactions. This implies that the mutant has little chance to form persister cells. On the basis of Equation 12.13, the gene mutants were ranked according to their correlation factor *Cor*. Due to the large number of genes (i.e. approximately 1000 genes) in the ranking, the gene mutants with the lowest *Cor* (ranging from −0.6 to −0.2) were analysed. The information of other gene mutants can be provided upon request. These genes are mainly distributed in the following metabolic pathways (Fig. 12.8): 26 genes in energy metabolism, eight genes in carbon metabolism, seven genes for iron component transports, six genes in lipid metabolism, three reactions in nucleotide metabolism, one reaction for cofactor synthesis, and one reaction for other metabolism (i.e. chaperones and folding catalysts). Energy metabolism has been reported as

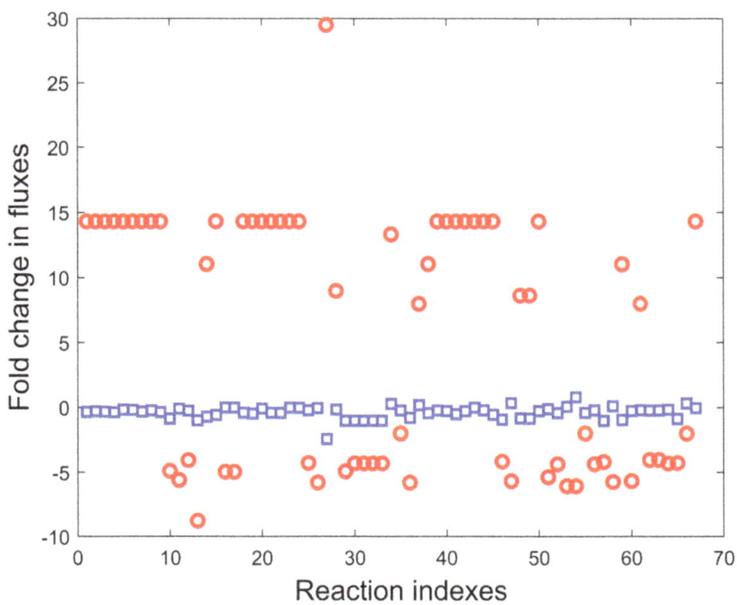

Figure 12.7 Fold changes in fluxes of persister-associated reactions in a single mutant (e.g. *aas* knockout mutant). A low similarity between the fluxes and the gene-expression levels in persister cells indicate that the single-mutant has a different metabolism from persister cells, that is, the mutant may not form persister cells.

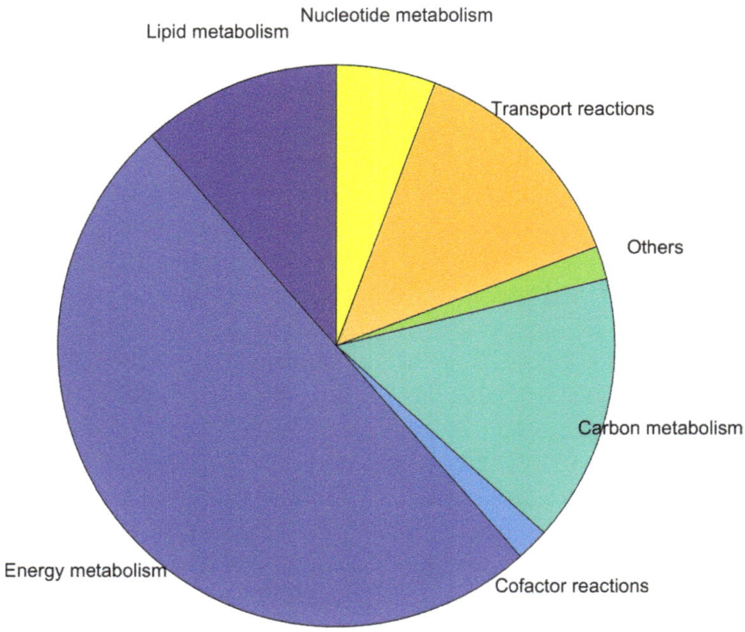

Figure 12.8 The distribution of genes whose knockout mutant may not form persister cells.

one of the major metabolic systems that have a large impact on persister cell formation (Amato *et al.*, 2014). Carbon and lipid metabolism are important to microbial survival. This may explain their importance in persister cell formation. The iron exchange reactions are mainly related to pores ion channels that have been found in the planar lipid bilayers that are produced by persister-promoting bacterial toxin (Gurnev *et al.*, 2012). The genes and their biological functions in cell metabolism are shown in Fig. 12.8 and are listed in Table 12.4. These 52 genes are mainly related to oxidative phosphorylation (energy metabolism); glycerophospholipid, glycerolipid, and fatty acid degradation and biosynthesis (lipid metabolism); glycolysis, pyruvate, citrate cycle, starch and sucrose metabolism (carbon metabolism); pores ion channels (outer membrane transport); pyrimidine metabolism (nucleotide metabolism), and chaperones and folding catalysts. While these genes may be used as the targets for further experimental investigation, some of them have been reported for their important role in regulating the persister cell formation. For example, the *atp* mutations would reduce ATP levels, and persisters are dormant and found to have low ATP levels (Cheng *et al.*, 2014). The genes *atpA* through *atpI* were found important for persister cell formation in Table 12.4. The degradation of the ATP synthase subunit gene *atpE* and *ompA* by GhoS masks GhoT toxicity and increases persistence of *E. coli* (Wang *et al.*, 2012). The *ppk1* mutant was found to generate less persister cell population as compared to the wild-type *P. aeruginosa* (Sarabhai *et al.*, 2015). In addition, Ppk is required to make polyphosphate and is supposed to be induced by ppGpp. The polyphosphate in turn is supposed to activate Lon protease, which then activates toxins by degrading antitoxins (Maisonneuve *et al.*, 2013). In *ppk* knocked-out mutants, Lon protease cannot up-regulate toxin levels so that persister cells are less likely to form. Gene *plsB* is a persister maintenance gene whose mutation reduces the level of persistence of *E. coli* (Spoering *et al.*, 2006). Enzyme *tesB* is a broad-specificity thioesterase that can synthesize HHQ (2-heptyl-4(1H)-quinolone), a signal molecule that influences virulence factor production and biofilm maturation of *P. aeruginosa* (Drees and Fetzner, 2015). A similar molecule that belongs to the same signaling molecule family as HHQ, named 2-aminoacetophenone (2-AA), was reported to mediate persister cell accumulation (Que *et al.*, 2013). HHQ may have similar effects on regulating the persister cell formation.

Table 12.4 Genes whose knockout mutants are predicted to not form persister cells

b number	Enzymes	Functions of enzymes	Pathways (Metabolisms)	Correlation factors
b2836	Aas	Acyl-acyl carrier protein synthetase	Fatty acid degradation (Lipid)	−0.6004
b3732	AtpD	F1 sector of membrane-bound ATP synthase, beta subunit	Oxidative phosphorylation (Energy)	−0.5571
b3734	AtpA	F1 sector of membrane-bound ATP synthase, alpha subunit	Oxidative phosphorylation (Energy)	−0.5513
b3738	AtpB	F0 sector of membrane-bound ATP synthase, subunit a	Oxidative phosphorylation (Energy)	−0.5297
b3736	AtpF	F0 sector of membrane-bound ATP synthase, subunit b	Oxidative phosphorylation (Energy)	−0.5171
b3733	AtpG	F1 sector of membrane-bound ATP synthase, gamma subunit	Oxidative phosphorylation (Energy)	−0.5156
b3737	AtpE	F0 sector of membrane-bound ATP synthase, subunit c	Oxidative phosphorylation (Energy)	−0.5032
b3731	AtpC	F1 sector of membrane-bound ATP synthase, epsilon subunit	Oxidative phosphorylation (Energy)	−0.5007
b3739	AtpI	ATP synthase, membrane-bound accessory factor	Oxidative phosphorylation (Energy)	−0.4836
b2501	Ppk	Polyphosphate kinase, component of RNA degradosome	Oxidative phosphorylation (Energy)	−0.4734
b3735	AtpH	F1 sector of membrane-bound ATP synthase, delta subunit	Oxidative phosphorylation (Energy)	−0.4675
b0957	OmpA	Outer membrane protein A	Pores ion channels	−0.4525
b1319	OmpG	Outer membrane porin G	Pores ion channels	−0.4521
b0875	AqpZ	Aquaporin Z	Aquaporins and small neutral solute transporters	−0.4486
b2278	NuoL	NADH:quinone oxidoreductase, prokaryotes	Oxidative phosphorylation (Energy)	−0.4195
b3875	OmpL	Outer membrane porin L	Pores ion channels	−0.4150
b0452	TesB	Acyl-CoA thioesterase 2	Biosynthesis of unsaturated fatty acids (Lipid)	−0.3582
b2287	NuoB	NADH:ubiquinone oxidoreductase, chain B	Oxidative phosphorylation (Energy)	−0.3500
b2285	NuoE	NADH:ubiquinone oxidoreductase, chain E	Oxidative phosphorylation (Energy)	−0.3479
b2282	NuoH	NADH:ubiquinone oxidoreductase, membrane subunit H	Oxidative phosphorylation (Energy)	−0.3362
b2276	NuoN	NADH:ubiquinone oxidoreductase, membrane subunit N	Oxidative phosphorylation (Energy)	−0.3320
b0431	CyoB	Cytochrome o ubiquinol oxidase subunit I	Oxidative phosphorylation (Energy)	−0.3278
b0429	CyoD	Cytochrome o ubiquinol oxidase subunit IV	Oxidative phosphorylation (Energy)	−0.3192
b2280	NuoJ	NADH:ubiquinone oxidoreductase, membrane subunit J	Oxidative phosphorylation (Energy)	−0.3185
b0430	CyoC	Cytochrome o ubiquinol oxidase subunit III	Oxidative phosphorylation (Energy)	−0.3102
b2286	NuoC	NADH:ubiquinone oxidoreductase, fused CD subunit	Oxidative phosphorylation (Energy)	−0.3035
b0432	CyoA	Cytochrome o ubiquinol oxidase subunit II	Oxidative phosphorylation (Energy)	−0.2999

Table 12.4 Continued

b number	Enzymes	Functions of enzymes	Pathways (Metabolisms)	Correlation factors
b2279	NuoK	NADH:ubiquinone oxidoreductase, membrane subunit K	Oxidative phosphorylation (Energy)	−0.2939
b3825	PldB	Lysophospholipase L2	Glycerophospholipid metabolism (lipid)	−0.2903
b2284	NuoF	NADH:ubiquinone oxidoreductase, chain F	Oxidative phosphorylation (Energy)	−0.2892
b1101	PtsG	Fused glucose-specific PTS enzymes	Glycolysis (Carbohydrate)	−0.2730
b2277	NuoM	NADH:ubiquinone oxidoreductase, membrane subunit M;	Oxidative phosphorylation (Energy)	−0.2688
b2835	LplT	Lysophospholipid transporter	Lipid transporters	−0.2673
b2281	NuoI	NADH:ubiquinone oxidoreductase, chain I	Oxidative phosphorylation (Energy)	−0.2580
b0033	CarB	Carbamoyl-phosphate synthase large subunit	Pyrimidine metabolism (Nucleotide)	−0.2522
b2500	PurN	Phosphoribosylglycinamide formyltransferase 1	Purine metabolism (Nucleotide)	−0.2490
b0789	ClsB	Cardiolipin synthase 2	Glycerophospholipid (lipid)	−0.2479
b3409	FeoB	Ferrous iron transport protein B	Transporters	−0.2354
b3916	PfkA	6-phosphofructokinase I	Glycolysis (Carbon)	−0.2351
b2265	MenF	Isochorismate synthase 2	Ubiquinone and other terpenoid-quinone biosynthesis (cofactors and vitamins)	−0.2286
b0351	MhpF	Acetaldehyde-CoA dehydrogenase II, NAD-binding	Pyruvate metabolism (Carbohydrate)	−0.2222
b2132	BglX	Beta-D-glucoside glucohydrolase, periplasmic	Starch and sucrose metabolism (Carbohydrate)	−0.2220
b1241	AdhE	Acetaldehyde dehydrogenase	Glycolysis (Carbohydrate)	−0.2151
b2342	FadI	Beta-ketoacyl-CoA thiolase, anaerobic, subunit	Fatty acid degradation (lipid)	−0.2131
b3781	TrxA	Thioredoxin 1	Chaperones and folding catalysts	−0.2118
b0115	AceF	Pyruvate dehydrogenase	Glycolysis (Carbohydrate)	−0.2092
b0888	TrxB	Thioredoxin reductase, FAD/NAD(P)-binding	Pyrimidine metabolism (Nucleotide)	−0.2081
b4041	PlsB	Glycerol-3-phosphate O-acyltransferase	Glycerolipid metabolism (lipid)	−0.2072
b2925	FbaA	Fructose-bisphosphate aldolase, class II	Glycolysis (Carbohydrate)	−0.2064
b0813	RhtA	Threonine and homoserine efflux system	electrochemical potential-driven transporters	−0.2031
b3236	Mdh	Malate dehydrogenase, NAD(P)-binding	Citrate cycle (Carbohydrate))	−0.2014
b2288	NuoA	NADH:ubiquinone oxidoreductase, membrane subunit A	Oxidative phosphorylation (Energy)	−0.2006

Conclusions and discussion

Pathogens adjust their metabolism and physiology under stress conditions such as oxidative stress that is created by antibiotics. One major strategy used by pathogens to survive in hostile environments is through the formation of persister cells. While extensive experimental research has been conducted on the stress response and persister cell formation, we developed the first genome-scale modeling approach to integrate the metabolic network and omics data to quantify the capability of *E. coli* to form persister cells and to identify potential therapeutic targets for manipulating persister cell formation. The model correctly predicted the enhanced activities in RpoS-controlled reactions under oxidative stress. The *rpoS*-deletion mutant was predicted to have a higher chance to form persister cells than the wild-type *E. coli*. This prediction is consistent with existing experimental data. On the basis of gene-knockout simulations, 52 gene targets were identified to reduce persister cell formation. Some of these genes have been reported for their important role in regulating microbial persistence, while others can be used as the targets for further experimental investigation.

As mentioned in the Results section, the developed genome-scale modeling approach may be limited by the lack of correlation between gene expression levels and metabolic fluxes. In order to improve the prediction of the genome-scale modeling platform, proteomics data can be used to constrain metabolic fluxes. The biomass growth rate was regarded as the objective function to maximize in flux balance analysis, while microorganisms under stress conditions generally have slow metabolism. We addressed this issue by constraining the metabolic fluxes with omics data under stress conditions so that the adjustment of microbial metabolism, instead of microbial growth, was the first priority in FBA. An objective function other than the microbial growth rate needs to be investigated to further improve the prediction ability of genome-scale modeling approach for stress conditions. In addition, signaling pathways for stress response should be integrated with the genome-scale metabolic model so that metabolic fluxes can be regulated by signaling pathways. The dynamic flux balance analysis (DFBA) platform (Mahadevan *et al.*, 2002) can be used for this purpose. In particular, the signaling protein and extracellular metabolite concentrations are used as inputs to constrain the related metabolic fluxes for FBA, which returns the exchange fluxes of those extracellular metabolites that can be used to update the extracellular metabolite concentrations. Therefore, the kinetic model for signaling proteins is discretized into multiple time steps in which FBA is performed and integrated with the signaling model. While this approach can provide valuable time profiles of metabolites and signaling proteins, the major challenge is that extra experimental data is needed to estimate the kinetic parameters in the signaling model. In addition, FBA may return fluxes with sharp changes for two adjacent time steps, as FBA does not return a unique solution for each time step. This may make the simulated time profiles spike. The DFBAlab (Gomez *et al.*, 2014) may be used to resolve this issue.

References

Amato, S.M., Fazen, C.H., Henry, T.C., Mok, W.W.K., Orman, M.A., Sandvik, E.L., Volzing, K.G., and Brynildsen, M.P. (2014). The role of metabolism in bacterial persistence. Front Microbiol. 5.

Barrasa, J.I., Olmo, N., Perez–Ramos, P., Santiago–Gomez, A., Lecona, E., Turnay, J., and Lizarbe, M.A. (2011). Deoxycholic and chenodeoxycholic bile acids induce apoptosis via oxidative stress in human colon adenocarcinoma cells. Apoptosis *16*, 1054–1067.

Becker, S.A., and Palsson, B.O. (2005). Genome–scale reconstruction of the metabolic network in Staphylococcus aureus N315: an initial draft to the two–dimensional annotation. BMC Microbiol. 5.

Burgard, A.P., Pharkya, P., and Maranas, C.D. (2003). Optknock: a bilevel programming framework for identifying gene knockout strategies for microbial strain optimization. Biotechnol. Bioeng. *84*, 647–657.

Cheng, H.Y., Soo, V.W., Islam, S., McAnulty, M.J., Benedik, M.J., and Wood, T.K. (2014). Toxin GhoT of the GhoT/GhoS toxin/antitoxin system damages the cell membrane to reduce adenosine triphosphate and to reduce growth under stress. Environ. Microbiol. *16*, 1741–1754.

Conesa, A., Gotz, S., Garcia–Gomez, J.M., Terol, J., Talon, M., and Robles, M. (2005). Blast2GO: a universal tool for annotation, visualization and analysis in functional genomics research. Bioinformatics *21*, 3674–3676.

Dennis, G., Sherman, B.T., Hosack, D.A., Yang, J., Gao, W., Lane, H.C., and Lempicki, R.A. (2003). DAVID: Database for Annotation, Visualization, and Integrated Discovery. Genome Biol. *4*, P3.

Drees, S.L., and Fetzner, S. (2015). PqsE of Pseudomonas aeruginosa Acts as Pathway–Specific Thioesterase in the Biosynthesis of Alkylquinolone Signaling Molecules. Chem. Biol. *22*, 611–618.

Edwards, J.S., Ibarra, R.U., and Palsson, B.O. (2001). In silico predictions of *Escherichia coli* metabolic capabilities are consistent with experimental data. Nat. Biotechnol. *19*, 125–130. https://doi.org/10.1038/84379

Ferens, W.A., and Hovde, C.J. (2011). *Escherichia coli* O157:H7: animal reservoir and sources of human infection. Foodborne Pathog. Dis. *8*, 465–487. https://doi.org/10.1089/fpd.2010.0673

Gomez, J.A., Höffner, K., and Barton, P.I. (2014). DFBAlab: a fast and reliable MATLAB code for dynamic flux balance analysis. BMC Bioinformatics *15*, 409. https://doi.org/10.1186/s12859-014-0409-8

Gurnev, P.A., Ortenberg, R., Dorr, T., Lewis, K., and Bezrukov, S.M. (2012). Persister–promoting bacterial toxin TisB produces anion-selective pores in planar lipid bilayers. Febs. Lett. *586*, 2529–2534.

Heinemann, M., Kummel, A., Ruinatscha, R., and Panke, S. (2005). In silico genome-scale reconstruction and validation of the Staphylococcus aureus metabolic network. Biotechnol. Bioeng. *92*, 850–864.

Hong, S.H., Wang, X.X., O'Connor, H.F., Benedik, M.J., and Wood, T.K. (2012). Bacterial persistence increases as environmental fitness decreases. Microb. Biotechnol. *5*, 509–522.

Hucka, M., Finney, A., Sauro, H.M., Bolouri, H., Doyle, J.C., Kitano, H., Arkin, A.P., Bornstein, B.J., Bray, D., Cornish-Bowden, A., et al. (2003). The systems biology markup language (SBML): a medium for representation and exchange of biochemical network models. Bioinformatics *19*, 524–531.

Johnson, N.B., Hayes, L.D., Brown, K., Hoo, E.C., Ethier, K.A., Centers for Disease, C., and Prevention (2014). CDC National Health Report: leading causes of morbidity and mortality and associated behavioral risk and protective factors--United States, 2005-2013. MMWR Suppl *63*, 3–27.

Kanehisa, M., and Goto, S. (2000). KEGG: kyoto encyclopedia of genes and genomes. Nucleic Acids Res. *28*, 27–30.

Kim, Y., and Wood, T.K. (2010). Toxins Hha and CspD and small RNA regulator Hfq are involved in persister cell formation through MqsR in *Escherichia coli*. Biochem. Biophys. Res. Commun. *391*, 209–213. https://doi.org/10.1016/j.bbrc.2009.11.033

Kwan, B.W., Lord, D.M., Peti, W., Page, R., Benedik, M.J., and Wood, T.K. (2015). The MqsR/MqsA toxin/antitoxin system protects *Escherichia coli* during bile acid stress. Environ. Microbiol. *17*, 3168–3181. https://doi.org/10.1111/1462-2920.12749

Lewis, K. (2008). Multidrug tolerance of biofilms and persister cells. Curr. Top. Microbiol. Immunol. *322*, 107–131.

Lewis, K. (2010). Persister cells. Annu. Rev. Microbiol. *64*, 357–372. https://doi.org/10.1146/annurev.micro.112408.134306

Lewis, K. (2012). Persister cells: molecular mechanisms related to antibiotic tolerance. Handb. Exp. Pharmacol. 121–133.

Li, X., Liu, Y., Jia, Q., LaMacchia, V., O'Donoghue, K., and Huang, Z. (2016). A systems biology approach to investigate the antimicrobial activity of oleuropein. J. Ind. Microbiol. Biotechnol. *43*, 1705–1717. https://doi.org/10.1007/s10295-016-1841-8

Loman, N.J., Constantinidou, C., Chan, J.Z., Halachev, M., Sergeant, M., Penn, C.W., Robinson, E.R., and Pallen, M.J. (2012). High-throughput bacterial genome sequencing: an embarrassment of choice, a world of opportunity. Nat. Rev. Microbiol. *10*, 599–606. https://doi.org/10.1038/nrmicro2850

Machado, D., and Herrgård, M. (2014). Systematic evaluation of methods for integration of transcriptomic data into constraint-based models of metabolism. PLOS Comput. Biol. *10*, e1003580. https://doi.org/10.1371/journal.pcbi.1003580

Mahadevan, R., Edwards, J.S., and Doyle, F.J. (2002). Dynamic flux balance analysis of diauxic growth in *Escherichia coli*. Biophys. J. *83*, 1331–1340.

Maisonneuve, E., Castro-Camargo, M., and Gerdes, K. (2013). (p)ppGpp controls bacterial persistence by stochastic induction of toxin-antitoxin activity. Cell *154*, 1140–1150.

Maisonneuve, E., Shakespeare, L.J., Jørgensen, M.G., and Gerdes, K. (2011). Bacterial persistence by RNA endonucleases. Proc. Natl. Acad. Sci. U.S.A. *108*, 13206–13211. https://doi.org/10.1073/pnas.1100186108

Markowitz, V.M., Mavromatis, K., Ivanova, N.N., Chen, I.M., Chu, K., and Kyrpides, N.C. (2009). IMG ER: a system for microbial genome annotation expert review and curation. Bioinformatics *25*, 2271–2278. https://doi.org/10.1093/bioinformatics/btp393

Monk, J.M., Charusanti, P., Aziz, R.K., Lerman, J.A., Premyodhin, N., Orth, J.D., Feist, A.M., and Palsson, B.O. (2013). Genome-scale metabolic reconstructions of multiple Escherichia coli strains highlight strain-specific adaptations to nutritional environments. Proc. Natl. Acad. Sci. U. S. A. *110*, 20338–20343.

Nomoto, M., Miyata, M., Yin, S., Kurata, Y., Shimada, M., Yoshinari, K., Gonzalez, F.J., Suzuki, K., Shibasaki, S., Kurosawa, T., *et al.* (2009). Bile acid-induced elevated oxidative stress in the absence of farnesoid X receptor. Biol. Pharm. Bull. *32*, 172–178.

Oberhardt, M.A., Puchalka, J., Fryer, K.E., dos Santos, V.A.P.M., and Papin, J.A. (2008). Genome-scale metabolic network analysis of the opportunistic pathogen Pseudomonas aeruginosa PAO1. J. Bacteriol. *190*, 2790–2803.

Orth, J.D., Conrad, T.M., Na, J., Lerman, J.A., Nam, H., Feist, A.M., and Palsson, B.Ø. (2011). A comprehensive genome-scale reconstruction of *Escherichia coli* metabolism - 2011. Mol. Syst. Biol. *7*, 535. https://doi.org/10.1038/msb.2011.65

Orth, J.D., Thiele, I., and Palsson, B.Ø. (2010). What is flux balance analysis? Nat. Biotechnol. *28*, 245–248. https://doi.org/10.1038/nbt.1614

Price, N.D., Reed, J.L., and Palsson, B.Ø. (2004). Genome-scale models of microbial cells: evaluating the consequences of constraints. Nat. Rev. Microbiol. *2*, 886–897.

Que, Y.A., Hazan, R., Strobel, B., Maura, D., He, J.X., Kesarwani, M., Panopoulos, P., Tsurumi, A., Giddey, M., Wilhelmy, J., et al. (2013). A Quorum Sensing Small Volatile Molecule Promotes Antibiotic Tolerance in Bacteria. PLOS One *8*.

Rutherford, K., Parkhill, J., Crook, J., Horsnell, T., Rice, P., Rajandream, M.A., and Barrell, B. (2000). Artemis: sequence visualization and annotation. Bioinformatics *16*, 944–945.

Sarabhai, S., Harjai, K., Sharma, P., and Capalash, N. (2015). Ellagic acid derivatives from Terminalia chebula Retz. increase the susceptibility of Pseudomonas aeruginosa to stress by inhibiting polyphosphate kinase. J Appl. Microbiol. *118*, 817–825.

Schellenberger, J., Park, J.O., Conrad, T.M., and Palsson, B.O. (2010). BiGG: a Biochemical Genetic and Genomic knowledgebase of large scale metabolic reconstructions. BMC Bioinformatics *11*, 213.

Schellenberger, J., Que, R., Fleming, R.M., Thiele, I., Orth, J.D., Feist, A.M., Zielinski, D.C., Bordbar, A., Lewis, N.E., Rahmanian, S., et al. (2011). Quantitative prediction of cellular metabolism with constraint-based models: the COBRA Toolbox v2.0. Nat. Protoc. *6*, 1290–1307.

Schuster, C.F., and Bertram, R. (2013). Toxin–antitoxin systems are ubiquitous and versatile modulators of prokaryotic cell fate. FEMS Microbiol. Lett. *340*, 73–85. https://doi.org/10.1111/1574-6968.12074

Seemann, T. (2014). Prokka: rapid prokaryotic genome annotation. Bioinformatics *30*, 2068–2069. https://doi.org/10.1093/bioinformatics/btu153

Spoering, A.L., Vulic, M., and Lewis, K. (2006). GlpD and PlsB participate in persister cell formation in *Escherichia coli*. J. Bacteriol. *188*, 5136–5144.

Stein, L. (2001). Genome annotation: from sequence to biology. Nat. Rev. Genet. *2*, 493–503. https://doi.org/10.1038/35080529

Tan, Q., Awano, N., and Inouye, M. (2011). YeeV is an *Escherichia coli* toxin that inhibits cell division by targeting the cytoskeleton proteins, FtsZ and MreB. Mol. Microbiol. *79*, 109–118. https://doi.org/10.1111/j.1365-2958.2010.07433.x

Thiele, I., Hyduke, D.R., Steeb, B., Fankam, G., Allen, D.K., Bazzani, S., Charusanti, P., Chen, F.C., Fleming, R.M.T., Hsiung, C.A., et al. (2011). A community effort towards a knowledge-base and mathematical model of the human pathogen Salmonella Typhimurium LT2. BMC Syst. Biol. *5*.

Van Melderen, L., and Saavedra De Bast, M. (2009). Bacterial toxin–antitoxin systems: more than selfish entities? PLOS Genet. *5*, e1000437. https://doi.org/10.1371/journal.pgen.1000437

Wang, X.X., Kim, Y., Hong, S.H., Ma, Q., Brown, B.L., Pu, M.M., Tarone, A.M., Benedik, M.J., Peti, W., Page, R., et al. (2011). Antitoxin MqsA helps mediate the bacterial general stress response. Nat. Chem. Biol. *7*, 359–366.

Wang, X., Lord, D.M., Cheng, H.Y., Osbourne, D.O., Hong, S.H., Sanchez-Torres, V., Quiroga, C., Zheng, K., Herrmann, T., Peti, W., *et al.* (2012). A new type V toxin–antitoxin system where mRNA for toxin GhoT is cleaved by antitoxin GhoS. Nat. Chem. Biol. *8*, 855–861.

Wang, X.X., and Wood, T.K. (2011). Toxin–Antitoxin Systems Influence Biofilm and Persister Cell Formation and the General Stress Response. Appl. Environ. Microb. *77*, 5577–5583.

Xu, Z., Fang, X., Wood, T.K., and Huang, Z.J. (2013). A systems-level approach for investigating Pseudomonas aeruginosa biofilm formation. PLOS One *8*, e57050.

Xu, Z., Islam, S., Wood, T.K., and Huang, Z. (2015). An Integrated Modeling and Experimental Approach to Study the Influence of Environmental Nutrients on Biofilm Formation of Pseudomonas aeruginosa. Biomed. Res. Int. *2015*, 506782.

Xu, Z., Ribaudo, N., Li, X., Wood, T.K., and Huang, Z. (2016). A Genome-Scale Modeling Approach to Investigate the Antibiotics-Triggered Perturbation in the Metabolism of Pseudomonas aeruginosa. IEEE Life Science Letters *2*, 39–42.

Current Books of Interest

Index

A

B

C

F

G

O

P

Q

R

S

www.ingramcontent.com/pod-product-compliance
Ingram Content Group UK Ltd.
Pitfield, Milton Keynes, MK11 3LW, UK
UKHW062000041026
15591UKWH00016B/1094